H. W. Cake
Oct. 1st 1878.

WELLS'S

PRINCIPLES AND APPLICATIONS

OF

CHEMISTRY;

FOR THE

USE OF ACADEMIES, HIGH-SCHOOLS, AND COLLEGES:

INTRODUCING

THE LATEST RESULTS OF SCIENTIFIC DISCOVERY AND RESEARCH, AND ARRANGED WITH SPECIAL REFERENCE TO THE PRACTICAL APPLICATION OF CHEMISTRY TO THE ARTS AND EMPLOYMENTS OF COMMON LIFE.

WITH TWO HUNDRED AND FORTY ILLUSTRATIONS.

BY

DAVID A. WELLS, A.M.,

AUTHOR OF "WELLS'S NATURAL PHILOSOPHY," "SCIENCE OF COMMON THINGS," EDITOR OF THE "ANNUAL OF SCIENTIFIC DISCOVERY," ETC.

IVISON, BLAKEMAN, TAYLOR & CO.,

NEW YORK AND CHICAGO.

1872.

ELECTROTYPED BY
T. B SMITH & SON,
82 & 84 Beekman-street, N. Y.

PREFACE.

THIS work has been prepared with special reference to the wants of students in Academies, Seminaries, and Colleges, aiming to furnish just that information which will prove most useful and practical in their future employments and relations of life.

The great general principles of Chemistry, and the more important of the elements and their compounds, have been accordingly very fully discussed; while, on the other hand, the custom adopted in many text-books of enumerating and describing compounds which have no practical value and little scientific interest, has been disregarded.

To enable the student to understand more clearly the relations which Chemistry sustains to the industrial operations of the age, and to the past and present progress of civilization, greater attention has been given to the history of the science than has heretofore been customary in elementary text-books.

Special care has also been taken to present the very latest results of scientific discovery and research, in this country and Europe, and to take advantage of the most approved methods of experimentation and instruction.

An unusually large number of illustrations has been introduced, with the double purpose of rendering the study of the science more intelligible and attractive to the pupil, and of facilitating the instructions of teachers, especially of those not enjoying the advantage of large apparatus.

In respect to originality the author makes little pretension beyond the arrangement and classification of subjects, and the selection of illustrations. Among the authorities to which he is especially indebted he would mention FARADAY, Prof. MILLER, of King's College, London, GRAHAM, REGNAULT, and HAYES.

NEW YORK, May, 1858.

CONTENTS.

INORGANIC CHEMISTRY.

CHAPTER V.

CHAPTER VI.

CHAPTER VII.

CHAPTER VIII.

CHAPTER IX.

CHAPTER X.

CHAPTER XI.

CHAPTER XII.

CHAPTER XIII.

CHAPTER XIV.

CHAPTER XV.

ORGANIC CHEMISTRY.

CHAPTER XVI.

CHAPTER XVII.

CHAPTER XVIII.

CHAPTER XIX.

CHAPTER XX.

CHAPTER XXI.

CHAPTER XXII.

CHAPTER XXIII.

CHAPTER XXIV.

CHAPTER XXV.

PRINCIPLES OF CHEMISTRY.

INTRODUCTION.

1. **Matter** is the general name which has been given to that substance which, under an infinite variety of forms, affects our senses. We apply the term matter to every thing that occupies space, or that has length, breadth, and thickness.

The forms and combinations of matter seen in the animal, vegetable, and mineral kingdoms of nature, are numberless, yet they are all composed of a very few simple substances or elements.

2. **Simple Substances.**—By a simple substance, or element, we mean one which has never been derived from, or separated into, any other kind of matter.

Sulphur, gold, silver, iron, oxygen, and hydrogen, are examples of simple substances or elements; and are so considered because we are unable to decompose them, convert them into, or create them from, other bodies.

No known force has yet extracted any thing from sulphur but sulphur, or from gold but gold; but if by any method these substances could be broken up into two or more factors, or component parts, they would cease to be regarded as elementary.

The number of the elements, or simple substances, with which we are at present acquainted, is sixty-two.

These substances are not all equally distributed over the surface of the earth: many of them are exceedingly rare, and known only to chemists. Of the whole number, from ten to fifteen only are concerned in the formation of

QUESTIONS.—What is matter? What is a simple substance, or element? What is the number of the elements? How are the elements distributed?

the great bulk of all the familiar objects we see around us.

The atmosphere is made up of two—oxygen and nitrogen—with, comparatively speaking, mere traces of carbon and hydrogen: two of these, again—oxygen and hydrogen—give rise to water, a substance covering three fourths of the surface of our planet; while the great rock masses of the earth, are mainly compounds of eight simple substances, viz., oxygen, silicon, aluminium, calcium, potassium, sodium, chlorine, and iron. In the composition also of animal or vegetable structures, the same, or a still greater simplicity is observed.

3. **Compound Bodies.**—A Compound Body is one that can be separated into two or more elements, or simple substances.

4. **Atoms.**—All matter is supposed to be composed of exceedingly minute particles, which can not be subdivided, or separated into parts. Such ultimate particles are termed ATOMS.

No one has ever seen an atom; no one has ever been able to recognize through the agency of the senses a portion of matter so small that it could not in some way be made smaller; yet the evidence on this subject, derived mainly from modern investigations in chemistry, is of such a character that there can be no reasonable doubt that all matter is ultimately composed of indivisible parts, or atoms. The nature of this evidence will be mentioned hereafter.

Simple, or elementary bodies, have simple atoms, and compound bodies compound atoms. The atoms of each substance undoubtedly differ in weight, and may possibly differ in size and form.

Mol′e-cules.—We use the term MOLECULE, or PARTICLE of matter, to designate very small quantities of a substance, not meaning, however, the ultimate atoms. A molecule, or particle of matter, may be supposed to be formed of several atoms united together.

The extent to which matter can be divided, and perceived by the senses, is most wonderful. A grain of musk will fill the air of a room for years with fragrant particles, without suffering any considerable loss of weight. In the manufacture of gilt wire, used for embroidery, the amount of gold employed to cover a foot of wire does not exceed the 720,000th part of an ounce.

QUESTIONS.—What is a Compound Body? What is supposed to be the ultimate constitution of matter? What are atoms? What is a molecule? Illustrate the divisibility of matter.

The manufacturers know this to be a fact, and regulate the price of their wire accordingly. But if the gold which covers one foot is the 720,000th part of an ounce, the gold on an inch of the same wire will be only the 8,640,000th part of an ounce. We may divide this inch into one hundred pieces, and yet see each piece distinctly without the aid of a microscope: in other words, we see the 864,000,000th part of an ounce. If we now use a microscope, magnifying five hundred times, we may clearly distinguish the 432,000,000,000th part of an ounce of gold, each of which parts will be found to have all the characters and qualities which are found in the largest masses of gold.

Some years since, a distinguished English chemist made a series of experiments to determine how small a quantity of matter could be rendered visible to the eye, and by selecting a peculiar chemical compound, small portions of which were easily discernible, he came to the conclusion that he could distinctly see the billionth part of a grain. This quantity may be represented in figures thus, 1,000,000,000,000, but the mind can form no rational conception of it.

5. **Porosity.**—No two atoms of matter are supposed to touch, or be in actual contact with each other, and the openings or spaces which exist between them are called PORES. This property of bodies, according to which their atoms are thus separated by vacant places, is called POROSITY.

The reasons for believing that the atoms or particles of matter do not actually touch each other, are, that every form of matter, so far as we are acquainted with it, can, by pressure, be made to occupy a smaller space than it originally filled. Therefore, as no two particles of matter can occupy the same space at the same time, the space, by which the size or volume of a body may be diminished by pressure, must, before such diminution took place, have been filled with openings, or pores. Again, all bodies expand or contract under the influence of heat and cold. Now, if the atoms were in absolute contact with each other, no such movements could take place.

The porosity of liquids may be proved by mixing together equal measures of strong alcohol and water; when the resulting compound will be found to occupy considerably less space than its two constituents did separately:—in other words, a gallon of each liquid mixed will not form two gallons of compound.

6. **Inertia.**—Matter of itself has no power to change its

QUESTIONS.—What are pores? Are the particles of matter in absolute contact? How may the porosity of liquids be shown? Has matter any power in itself to change its condition?

state, or form. If a body is at rest, it can not of itself commence moving; and if a body be in motion, it can not of itself stop, or come to rest. Motion, or cessation of motion in a body, or any other change of its condition, requires a power to exist independent of itself.

As the cause of all the changes observed to take place in the material world, we admit the existence of certain forces, or agents, which govern and control all matter.

7. **Force** is whatever produces or opposes motion or change in matter.

Causes of Change.—All the changes which we observe to take place in matter may be referred to the following causes, or forces:—The ATTRACTION OF GRAVITATION, MOLECULAR FORCES—or forces acting *only* between molecules, or particles of matter at insensible distances—FORCES developed through the agencies of LIGHT and HEAT, the ATTRACTIVE and REPULSIVE FORCES of ELECTRICITY and MAGNETISM—and finally, a force or power which exists only in living animals and plants, which is called VITAL FORCE.

Concerning the real nature of these forces, we are entirely ignorant. We suppose, or say, they exist, because we see their effects upon matter. In the present state of science, it is impossible to know whether they are merely properties of matter, or whether they are forms of matter itself, existing in an exceedingly minute, subtile condition, without weight, and diffused throughout the whole universe. The general opinion, however, among scientific men, at the present day, is, that these forces, or agents, are not matter, but properties, or qualities, of matter.

8. **Gravitation.**—The Force of Gravitation, or the Attraction of Gravitation, is the name applied to that force by which all the bodies in the universe at sensible distances attract and tend to approach each other. Gravitation differs from all other forces in the fact that its influence is universal; that it acts at all times, upon all matter, and at all distances.

The force of gravitation belongs equally to the smallest atom and to the largest world, producing those attractions which bind masses of matter to-

---

QUESTIONS.—What occasions change in matter? What is force? Enumerate the great forces of nature. What do we know concerning these forces? What is gravitation? Is the force of gravitation universal?

gether, and restrict the motions of the planets to regular orbits. It is the force which draws a small body, free to move, toward a larger.

Terrestrial Gravitation is that force by which all bodies upon the earth are attracted toward its center.

The measure of this force, or the strength with which a body upon the earth is attracted toward its center, is called Weight.

The attractive force which the earth exerts upon a body is proportioned to its mass, or to the quantity of matter contained in it, and as weight is merely the measure of such attraction, it follows that a body of a large mass will be attracted strongly, and possess great weight, while, on the contrary, a body made up of a small quantity of matter, will be attracted in a less degree, and possess less weight. We recognize this difference of attraction by calling the one body heavy and the other light.

9. **Varieties of Force.** — Mo-lec'ular, or, as they are sometimes called, Internal Forces, are distinguished from all the other Forces which act upon matter, in this respect —that they act upon particles or molecules of matter at immeasurably small distances only.

The forces developed through the agencies of heat, light, electricity, and magnetism, are diverse in their nature, and affect different forms of matter differently. They differ from the force of gravitation inasmuch as their influence does not appear to be universal or constant, and is apparently limited by distance. They differ, especially, from molecular forces, inasmuch as their influence upon matter is exerted at sensible distances.

It is not at all certain that the forces which act upon matter, as above enumerated, are all separate and independent. Their connection with each other is most intimate, and there is reason for believing that some of them are only different manifestations of the same agent, or principle.

10. **Molecular Forces.**—Under the designation of molec'-ular forces are especially included four different manifestations of force, or, as they are usually called, varieties of

---

QUESTIONS.—What is Terrestrial Gravitation? What is Weight? What are the peculiarities of molecular forces? What are the peculiarities of the forces developed by the agencies of light, heat, electricity, and magnetism? Is it certain that the forces enumerated are all independent principles? What are included under the head of molecular forces?

attraction. These are COHESION, ADHESION, CAP'ILLARY ATTRACTION, and AFFINITY.

Although essentially differing from each other, these forces all agree in one remarkable particular—and that is, their influence upon matter is exerted only at distances which are immeasurable, or insensible. If the particles of a body are separated from each other to the slightest appreciable degree, the influence of these attractive forces is instantly neutralized or destroyed.

11. **Cohesion**, or COHESIVE ATTRACTION, is that force which binds together atoms of the same kind of matter to form one uniform mass.

The force which holds together the atoms of a mass of iron, wood, or stone, is cohesion, and the atoms are said to cohere to each other.

The effort required to break a substance is a measure of the intensity or strength of the cohesive force exerted by its particles.

When the Attraction of Cohesion between the particles of a substance is once destroyed, it is generally impossible to restore it. Having once reduced a mass of wood or stone to powder, we can not make the minute particles cohere again by merely pushing them into their former position.

In some instances, however, this may be accomplished by resorting to various expedients. Iron may be made to cohere to iron by heating the metal to a high degree, and hammering the two pieces together. The particles are thus driven into such intimate contact, that they cohere and form one uniform mass. This property is called WELDING, and belongs only to two metals, iron and platinum.

12. **Adhesion**, or Adhesive Attraction, is that force which causes *unlike* particles of matter to adhere, or remain attached to each other when united.

Dust floating in the air sticks to the wall or ceiling, through the force of adhesion. When we write on a wall with a piece of chalk, or charcoal, the particles, worn off from the material, stick to the wall and leave a mark, through the force of adhesion. Two pieces of wood may be fastened together by means of glue, in consequence of the adhesive attraction between the particles of the wood and the particles of glue.

13. **Cap'illary Attraction** is that variety of molecular force which manifests itself between the surfaces of solids and liquids.

QUESTIONS.—What is cohesion? What is a measure of the force of cohesion? What is welding? What is adhesion? Give examples of adhesion. What is capillary attraction?

The ordinary definition of Cap'illary Attraction is, that form of attraction which causes liquids to ascend above their level in capillary tubes. This, however, is not strictly correct, as this force not only causes an elevation, but also a depression of liquids in tubes, and is at work wherever fluids are in contact with solid bodies.

FIG. 1.

The name "Capillary Attraction" originated from the circumstance that this class of phenomena was first observed in small glass tubes, the bore of which was not thicker than a hair, and which were hence called *Capillary Tubes*, from the Latin word *capillus*, which signifies a hair.

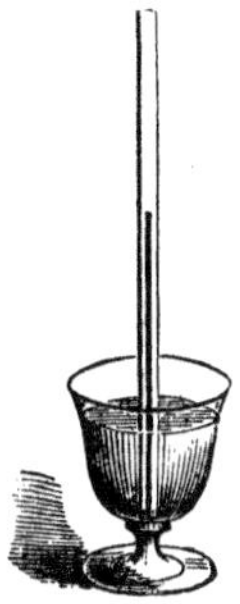

The simplest method of exhibiting capillary attraction is to immerse the end of a piece of thermometer tube in water (see Fig. 1) which has been tinted with ink. The liquid will be seen to ascend, and will remain elevated in the tube at a considerable height above the surface of the liquid in the vessel.

The height to which water will rise in capillary tubes is in proportion to the smallness of their diameters.

FIG. 2.

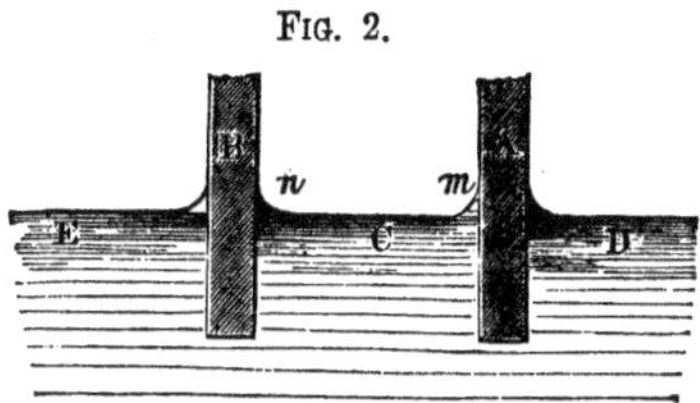

This is clearly shown by the following simple experiment. If two plates of glass, A and B, Fig. 2, be plunged into water at their lower extremities, with their faces vertical and parallel, and at a certain distance asunder, the water will rise at the points *m* and *n*, where it is in contact with the glass; but at all intermediate points, beyond a small distance from the plates, the general level of the surfaces E, C, and D, will correspond.

FIG. 3.

If the two plates, A and B, are brought near to each other, as in Fig. 3, the two curves, *m* and *n*, will unite, so as to form a concave surface, and the water at the same time between them will be raised above the general level, E and D, of the water in the vessel. If the plates be brought still nearer together, as in Fig. 4, the water between them will rise still higher, the force which sustains the column being increased as the distance between the plates is diminished.

Illustrations of capillary attraction are most familiar in the experience of every-day life. The wick of a lamp, or candle, lifts the oil, or melted grease

QUESTIONS.—How may the phenomena of cap'illary attraction be illustrated? To what height will water rise in capillary tubes? What are familiar examples of capillary attraction?

Fig. 4.

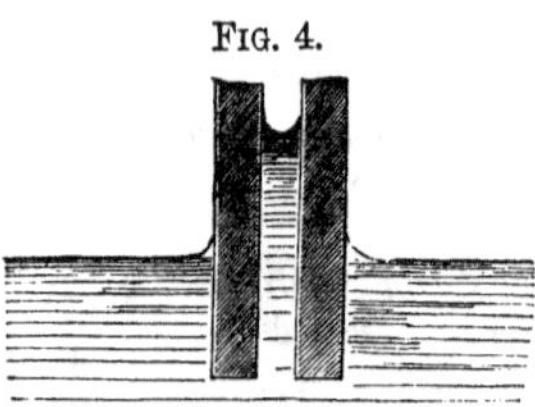

which supplies the flame from a surface often two or three inches below the point of combustion.

When one end of a sponge, or a lump of sugar is brought into contact with water, the liquid, by capillary attraction, will rise, or soak up above its level, into the interior of the sponge, or sugar, until all its pores are filled.

14. **Affinity** is that variety of molec′ular force or attraction which unites atoms of unlike substances into compounds possessing new and distinct properties.

Oxygen, for example, unites with iron, and forms oxyd of iron, or iron-rust, a substance possessing different and distinct properties from either iron or oxgyen. In like manner, oxygen and hydrogen, two gases not to be distinguished in appearance from common air, unite to form water, a liquid.

When the particles of different substances are united together by the force of affinity, the compound formed possesses properties entirely different from that of its constituents, and in no respect resembles a mixture, which is merely a mechanical union of bodies—as when salt is mixed with sand. The forces of adhesion or capillary attraction may closely unite unlike particles of matter together, but they do not effect any change in the nature or properties of the particles acted upon. Affinity, on the contrary, entirely changes the properties of the unlike particles which it unites, and by so doing produces combinations which possess entirely different qualities.

The action of gravity and of the several molecular forces may be illustrated by referring to a particular form of matter, as, for example, water. The force of affinity binds together the atoms of the elements, oxygen and hydrogen, to constitute an atom, or molecule of water; cohesion unites the particles of water thus formed into drops, or larger masses; adhesion causes the union of water with the surfaces of different substances, thereby producing the phenomenon which we call "wetting;" capillary attraction causes water to rise above its level, or "soak up" as it is termed, in a sponge, or other porous substance; while the force of gravity causes coherent quantities of water to fall as rain, or to move down inclined surfaces in the form of rivers, brooks, etc.

15. **Repulsion.**—In opposition to the several attractive forces which act upon matter, a repulsive force also exists, the tendency of which is to separate the particles of matter from one another.

Questions.—What is affinity? What are illustrations of affinity? How do the compounds of matter formed through the force of affinity differ from a mixture? How do the forces of adhesion and capillary attraction differ from affinity? How may the action of gravity and the molecular forces be illustrated? What force acts in opposition to the attractive forces?

The resistance experienced in attempting to compress a substance is the result of the opposition of the repulsive force which pervades its particles and the effort required to effect a compression is a measure of the intensity of the repulsive force.

A dew-drop resting upon a leaf is not in actual contact with its surface, but is sustained at a little distance above it by the force of repulsion. Ingenious experimentation has proved that when two glasses, one slightly convex and the other flat, are placed upon each other, and pressed together with a force of 1,000 pounds to the square inch, they still remain at a distance from each other of the thickness of the top of a soap-bubble before it bursts, or at least 1-4450th of an inch. If we compress a certain quantity of gas, as common air, and then allow it to dilate, by removing all restraint, it will expand without limit, and fill every really empty space which is open to it. This takes place through the agency of an internal repulsive force, which tends to drive the particles from one another.

It is not definitely known whether the repulsive force, which appears to influence, under certain circumstances, the particles of all matter, is a separate and independent principle, or whether it is the result of the action of heat or of electricity, or of both these forces combined. Heat, in its influence upon matter, always acts as a repulsive force, and is always opposed to the influence of cohesion.

16. **Elasticity.**—That property of bodies known as Elasticity is the result of the joint action of the repulsive and attractive forces; and substances are said to be more or less elastic, according to the facility with which they regain their original form and dimension after the force which has compressed or extended them is removed.

17. **Three Forms of Matter.**—According as the attractive or repulsive forces prevail, all bodies will assume one of three forms or conditions—the SOLID, the LIQUID, or the A′ER-I-FORM,* or GASEOUS condition.

18. **Solids.**—A solid body is one in which the particles are so strongly held together by the attractive force of cohesion, that the body maintains its form or figure under all ordinary circumstances.

If the force of cohesion acted exclusively upon matter, every substance would possess insuperable solidity, hardness, and tenacity.

* A′er-i-form, having the form, or resemblance, of air.

QUESTIONS.—What evidence is there of the existence of a repulsive force? What is elasticity? What are illustrations of the influence of a repulsive force? Under what three forms or conditions does matter exist? What is a solid body?

19. **Liquids.**—A liquid body is one in which the particles of matter are held together so slightly by the force of cohesion that they move upon each other with the greatest facility.

Hence a liquid can never be made to assume any particular form except that of the vessel in which it is inclosed.

20. **Gaseous Bodies.**—An a'er-i-form or gaseous body is one in which the particles of matter are not held together by any force of cohesive attraction, and but for the restraining influence of the force of gravity would entirely separate and move off from one another.

A gaseous body is generally invisible, and, like the air surrounding us, affords to the sense of touch no evidence of its existence when in a state of complete repose. Gaseous bodies may be confined in vessels, whence they exclude liquids, or other bodies, thus demonstrating their existence, though invisible, and also their impenetrability.

21. **Change of Condition.**—Most substances can be made to assume successively the form of a solid, a liquid, or a gas. In solids, the attractive force is the strongest; the particles keep their places, and the solid retains its form. But if we heat a solid body, as for example a piece of ice, or sulphur, we weaken the force of cohesion which binds the particles together, and allow the repulsive force to prevail; the particles of the solid thereby become movable upon themselves, and we say the body melts, or becomes liquid. In liquids the attractive and repulsive forces are nearly balanced, but if we supply an additional quantity of heat, we destroy the attractive force altogether, and increase the repulsive force to such an extent that the liquid assumes the form of a gas, or vapor, in which the separate particles tend to fly off from each other. By reversing the process, or, in other words, by withdrawing the heat, we can diminish or destroy the repulsive force, and cause the attractive force again to predominate—the body returning to its former conditions, first of a liquid, then of a solid.

FIG. 5.

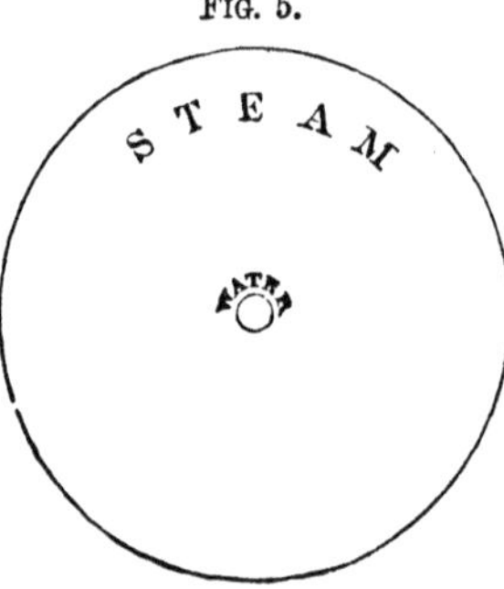

The power of the repulsive force generated by heat is strikingly illustrated in the conversion of water into steam. In a cubic inch of water converted into steam, the particles will repel each other to such an extent, that the space

---

QUESTIONS.—What is a liquid? What is an a'er-i-form, or gaseous body? Under what circumstances will a body assume the form of a solid, a liquid, or a gas? What experiment illustrates the repulsive power of heat.

occupied by the steam will be 1700 times greater than that occupied by the water. Fig. 5 illustrates the comparative difference between the bulk of steam and the bulk of water.

22. **Ethereal Condition.**—Recent investigations in science have rendered it probable, that matter, in addition to the three separate states or consistencies in which it is ordinarily presented to us—solid, liquid, and gaseous—exists also in a fourth state, which is called the ethereal. It is supposed that all space—that existing between the planets and other heavenly bodies, equally with that existing between the atoms, or molecules of every substance, even the most dense—is pervaded by an extremely rare, imponderable, and highly elastic medium, or fluid form of matter, termed ETHER. This substance, like air, is believed to be capable of motion, and of receiving and transmitting vibrations, which vibrations by their action on the ordinary forms of matter, are supposed to produce the phenomena of heat, light, electricity, etc., in the same manner as the vibrations of air occasioned by a sounding body produce the phenomena of sound.

23. **Matter Indestructible.**—All the researches and investigations of science teach us that it is impossible by natural operations, to either create or destroy a single particle of matter. The power to create and destroy matter belongs to the DEITY alone. The quantity of matter which exists, in and upon the earth has never been diminished by the annihilation of a single atom.

When a body is consumed by fire, there is no destruction of matter: it has only changed its form and position. When an animal or vegetable dies and decays, the original form vanishes, but the particles of matter, of which it was once composed, have merely passed off to form new bodies and enter into new combinations.

24. **Force Indestructible.**—Recent investigations in science seem to prove that force is equally as indestructible as matter; or, in other words, that there is no such thing as a destruction of force; consequently the amount of force in operation in the earth, and possibly throughout the universe, never varies in quantity, but remains always the same.

Some of the reasons which have led to a belief in the indestructibility of force may be stated as follows:—

The only mode in which we can judge of the existence of a force is from the effects it produces, and of these effects, that which is the most evident to

QUESTIONS.—What is the supposed ethereal condition of matter, or what are the peculiarities of matter in this condition? Is matter indestructible? Is force indestructible? What reasons induce us to believe that a force can not be destroyed?

our senses is the power either of producing motion, of arresting it, or of altering its direction: whatever is capable of effecting these results is considered as a form of force. Motion, therefore, may be considered as the indicator of force, and wherever we perceive motion, we may be certain that some force is operating. Now it will be found, that in all cases in which work is performed —or, to state it in other words, in all cases in which force is exerted and apparently made to disappear—that it has expended itself either in setting into action some other force, or else it has produced a definite and certain amount of motion. This motion when used will again give rise to an equal amount of the force which originally produced it. For example, we burn coal in the air; the force of affinity causes the particles of coal to unite with the oxygen of the air; the coal changes its form, and a quantity of heat remains, which heat represents the chemical force expended. The heat thus developed is now ready to do work: it may be employed in converting water into steam, and the steam so obtained can, through the medium of machinery, be applied to the production of motion. Motion may again be made to produce heat—as through friction, for example—and recent experiments seem to show that the amount of heat so developed by motion, would, if collected and measured, prove to be equal in amount to that which produced the motion. The heat produced by motion is generally dissipated and lost for practical purposes, but it is not absolutely lost. It has been absorbed, or diffused through space, or converted into some other form of force, which in turn takes part in some of the great operations of nature, or again ministers to the wants and necessities of man. Numerous other facts in support of the view that force, like matter, changes but is never destroyed, might be adduced. The subject is one of great interest, and has a practical bearing on many of the operations of chemistry.

25. **Classification of Forces.**—All the changes which take place in matter through the agency of the several forces which act upon it are considered under three general divisions, or departments of science, viz., Physics, or Natural Philosophy, Animal and Vegetable Physiology, and Chemistry.

26. **Natural Philosophy.**—Physics, or Natural Philosophy, is that department of science which considers generally, all those changes and phenomena which are observed to take place in matter through the agency of the forces of gravitation, cohesion, adhesion, capillary attraction, molecular repulsion, light, heat, electricity, and magnetism, and these several forces have been termed the Physical Forces.

27. **Physiology.**—Animal and Vegetable Physiology is

QUESTIONS.—Under what three general divisions are the forces which act upon matter considered? What forces are considered under the department of Natural Philosophy? What forces under the department of Animal and Vegetable Physiology?

department of science which treats of the changes and phenomena observed to take place in matter through the agency of the vital force.

28. **Chemistry.**—Chemistry is that department of science which relates exclusively to all those changes and phenomena which take place in matter through the agency or influence of the force of affinity.

29. **Chemical Action.**—Chemical Action is the term used to designate all those operations—the result of the force of affinity—by which the form, solidity, color, taste, smell, and action of substances become changed ; so that new bodies, with quite different properties, are formed from the old.

30. **Properties of Matter.**—The properties which characterize material objects in general, may be classed under two heads, viz., physical and chemical properties.

Physical Properties.—The Physical Properties of an object are those by which it is most readily defined, or distinguished from some other object. The form of a body ; its condition as a solid, a liquid, or a gas ; its color, hardness, tenacity, and its relations to heat and electricity, are examples of its physical properties. Physical properties are independent of the action which the body exerts upon other bodies.

Chemical Properties.—The Chemical Properties of a body are those which relate essentially to its action upon other bodies, and to the changes which the body either experiences itself, or causes to take place in other bodies by contact with them.

The physical properties of such a substance as sulphur, are, its peculiar odor, its yellow color, its brittleness, its crystalline structure, its specific gravity, the facility with which it exhibits electrical attraction when rubbed, and the like similar qualities, all of which are independent in a great degree of each other, and are so distinctive in their character that our senses inform us at once that the substance in question is sulphur, and not some other form of matter.

If we would now enumerate the chemical properties of sulphur, it would

---

QUESTIONS.—What is chemistry? Define chemical action. Under what two heads may the properties of material objects be classed? What are the physical properties of a body? What are chemical properties? Illustrate the distinction between the physical and chemical properties of sulphur.

be necessary to refer to those operations by which the body becomes changed and loses its distinctive character—such as the ease with which it takes fire, its insolubility in water, and its solubility in oil of turpentine, and the rapidity with which it unites with iron, silver, copper, and many other of the metals.

Had there been but one kind of matter in the universe, it could have possessed only physical properties, and the laws of Natural Philosophy would have explained all the phenomena and changes which could possibly have taken place in it. As the character or composition of this one form of matter, moreover, could not, under the circumstances, have been changed by the action of any different substance upon it, it could not have possessed any chemical properties, and no idea could have been formed by an intelligent being of any such department of knowledge as chemistry.

The connection, however, between Chemistry and Natural Philosophy is most intimate; and all chemical changes are influenced to such an extent by the action of the physical forces, that a knowledge of the principles of Natural Philosophy is requisite for a proper understanding of the nature of chemical phenomena. Especially is this the case as respects the forces manifested through the agency of Heat, Light, Electricity, and Magnetism; and a brief review of these subjects is generally regarded as a necessary introduction to the study of the science of Chemistry. The first part of this work is therefore devoted to a consideration of the nature and action of the physical forces so far as they are concerned in producing chemical changes, or in characterizing chemical phenomena.*

CHAPTER I.

ON THE CONNECTION OF GRAVITY, COHESION, ADHESION, AND CAPILLARY ATTRACTION WITH CHEMICAL ACTION.

SECTION I.

GRAVITY.

31. Connection of Gravity with Chemical Phenomena.—The influence of the force of gravity on matter is never

* It has been assumed, in the preparation of this work, that the student is conversant with the general principles of Natural Philosophy, and no attempt has therefore been made to treat the subjects of Heat, Light, Electricity, and Magnetism in any other than a general manner, and with special reference to their connection with chemical phenomena.

QUESTIONS.—What connection is there between Natural Philosophy and Chemistry? Is gravity influenced by changes in the condition of matter?

affected by any change which may take place in the form or condition of the matter itself.

A pound of water is attracted by the influence of gravity toward the center of the earth with a certain degree of force, and as weight is the measure of gravity, we express the exact amount of this attractive force, by saying that the water weighs a pound. If we deprive this particular quantity of water of heat, sufficient to freeze and convert it into ice—a solid—it will still weigh a pound; if we convert the same quantity of water into steam by the addition of heat, it will occupy a space seventeen hundred times greater than before—yet the steam produced will be attracted by the force of gravity equally with the water from which it is derived, and will continue to weigh a pound.

As the action of gravity, therefore, is never suspended, and as the smallest particle of matter can not be annihilated by any operation, we are enabled to test the accuracy of every chemical process, and ascertain the true composition of bodies by proving the weight of the compound to be equal to the weight of the substances which produce it.

32. **Use of the Balance.**—The balance is to the chemist what the compass is to the mariner, and before its introduction as a means of verifying experiments, the whole science of Chemistry was a collection of disconnected and separate facts and theories.

Until within a comparatively recent period it was supposed that common air, or gases, did not possess weight; and this error, which was necessarily accompanied with most absurd notions respecting the constitution of air and gases, prevailed until the experiment of weighing them was tried, when they were found to be attracted by gravity equally with all other kinds of matter.

Less than a hundred years ago it was generally taught and believed that when a body was burned, a portion of its substance was lost. Lavoisier, an eminent French philosopher, proved the contrary by carefully burning a body, and then weighing all that was left unconsumed by the fire, and all the invisible products that escaped. He found, that instead of there being a loss of matter, there was a gain, and thus by a simple experiment overthrew at once ideas respecting the nature of fire and combustion that had prevailed for centuries previous.

The great distinction, according to Professor Liebig, between Chemistry and Natural Philosophy, is that the one weighs and the other measures.

33. **Two Great Systems of Weights.**—Two great systems of weights are recognized throughout the civilized world in

---

QUESTIONS.—Give an illustration. What relation does the balance sustain to the operations of the chemist? What facts illustrate the use of the balance in effecting chemical discoveries? What two systems of weights are recognized?

Chemistry and in all other operations. These are known as the English and French Systems.

34. The English System of Weights.—The smallest denomination of weight made use of in the English System (the one generally used in the United States) is a grain. The Parliament of England passed a law in 1286, that 32 grains of wheat, well dried, should weigh a pennyweight. Hence the name grain applied to this measure of weight.

It was afterward ordered that a pennyweight should be divided into only 24 grains. Grain weights for practical purposes are made by weighing a thin plate of metal of uniform thickness, and cutting out, by measurement, such a proportion of the whole as will weigh one grain. In a like manner, weights may be obtained for chemical purposes which weigh only the 1,000th part of a grain.

Seven thousand grains constitute a pound avoirdupois, and from this pound all measures of capacity have been derived by Act of the English Parliament.

Thus a standard gallon is by law as much distilled water as will weigh ten pounds, or 70,000 grains; and a measure holding exactly this quantity of water is a gallon measure. By subdividing the gallon we obtain smaller measures, quarts, pints, etc.

35. French System of Weights.—The French System of Weights is constructed on a different plan, and is distinguished for its great simplicity—all its divisions being made by ten. It is, therefore, sometimes called the decimal system.

On the continent of Europe this system of weights is almost universally adopted for all scientific operations, and is gradually being introduced into England and the United States. It is, therefore, highly important that the principles upon which it is based should be understood.

The basis of the French System is an invariable dimension of the globe, viz., a fourth part of the earth's meridian, or a fourth part of a circle passing round the earth (lengthwise), and intersecting at the poles.

QUESTIONS.—What is the smallest weight recognized in the English System? What is a pound avoirdupois? How are measures of capacity derived from measures of weight? What is an English gallon? What is the distinguishing peculiarity of the French system of weights? Where is the French system used? What is its basis?

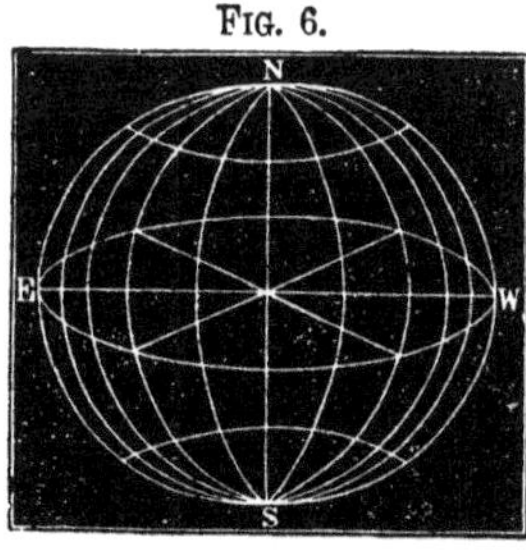

FIG. 6.

The circle N E S W, Fig. 6, represents a meridian of the earth; and a fourth part of this circle, or the distance N E, constitutes the dimension on which the French System is founded.

This distance, which was accurately measured, is divided into ten million equal parts; and a single ten millionth part adopted as a measure of length, and called a *metre.*

A metre is about three feet and a quarter in length, or about thirty-nine English inches. By multiplying or dividing the metre by ten, all the larger and smaller measures of length are obtained. For indicating measures smaller than a metre, Latin terms are used; for indicating measures larger than a metre, Greek terms. Thus—

Smaller Measures.	Larger Measures.
Metre.	Metre.
Decimetre = 1-10th metre.	Decametre = 10 metres.
Centimetre = 1-100th metre.	Hectometre = 100 metres.
Millimetre = 1-1000th metre.	Kilometre = 1,000 metres.
	Myriametre = 10,000 metres.

The system of weights was formed from measures of length in the following manner. A box, in the form of a cube, was taken, measuring one centimetre in every direction. This, filled with distilled water at its greatest density (at a temperature of 39° Fahrenheit's thermometer), was taken as the unit of the decimal weights, and called a *gramme**—a quantity equal to about fifteen English grains. The gramme, multiplied and divided by ten, gives all the other larger or smaller weights. Thus—

Smaller Weights.	Larger Weights.
Gramme.	Gramme.
Decigramme = 1-10th gramme.	Decagramme = 10 grammes.
Centigramme = 1-100th gramme.	Hectogramme = 100 grammes.
Milligramme = 1-1,000th gramme.	Kilogramme = 1,000 grammes.
	Myriagramme = 10,000 grammes.

The kilogramme corresponds in its use in all commercial transactions with the English pound. Its weight is equal to about 2¼ pounds avoirdupois.

* Pronounced *Gram.*

QUESTIONS.—What is the Metre of the French system? How are the larger and smaller measures of length derived from the Metre? How is the system of weights derived from the measures of length? What is a Kilogramme?

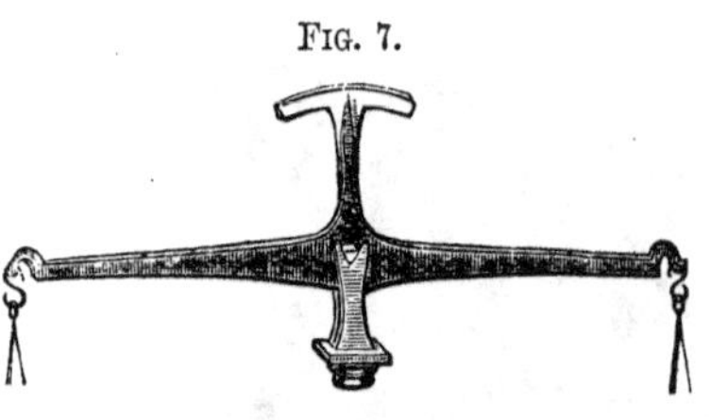

FIG. 7.

36. **Construction of the Balance.**—The balance used for all delicate chemical experiments is constructed in the most perfect manner. The point of support of the beam (see Fig. 7) is a wedge of hardened steel with a sharp, knife-like edge, which rests upon a flat plate of polished agate. The points of support of the two scale-pans are often constructed in a similar manner.

In all nice experiments the balance must be screened from currents of air, and the bodies weighed must have nearly the same temperature as that of the surrounding atmosphere—otherwise currents of air, ascending and descending, will be produced, which will impair the accuracy of the weight.

Balances are at the present time constructed for chemical operations, so delicate and exact, that they are able to indicate the weight of a thousandth part of a grain.

For the experiments described in this book, a common apothecaries' balance is all that is requisite.

37. **Weight Compared with Bulk.**—If equal bulks of matter of different kinds be compared together, they will be found to differ greatly in weight. Platinum, the heaviest body known, is upward of 200,000 times as dense, bulk for bulk, as hydrogen.

Specific Gravity.—The specific gravity, or specific weight of a body, is its weight as compared with the weight of an equal bulk of some other substance, assumed as the standard of comparison.

Absolute Weight.—The absolute weight of a body is the weight of its entire mass, considered without any reference to its bulk, or volume.

The weight of a body, as determined by the ordinary process of weighing, is its absolute weight.

Pure water, at a temperature of 60° Fahrenheit, has been selected as the standard for comparing the weights of equal bulks of different solids and liquids; and common air, dry, and at a temperature of 60° Fahrenheit, and

---

QUESTIONS.—What are the peculiarities of the balance as used for chemical investigations? What precautions are to be observed in nice experiments? How do equal bulks of different substances compare? What is specific gravity, or specific weight? What is absolute weight?

30 inches pressure of the barometer, as the standard for comparing the weights of equal volumes of different gases and vapors.

Attention is given to temperature, and to the pressure of the atmosphere, because the bulk of all substances sensibly varies with changes in these conditions.

Water having been selected as the standard of comparison, the question to be settled in the determination of the specific gravity of a body is simply this —how much heavier or lighter is a given bulk of a substance, than an equal bulk of water? The solution of the problem may be found by the following general rule:

38. Weigh first the body in air, and afterward weigh it when suspended in water. It will be observed to weigh less in water than in air. Subtract the weight in water from the weight in air, and divide the weight in air by the difference; the quotient will be the specific gravity required.

Fig. 8.

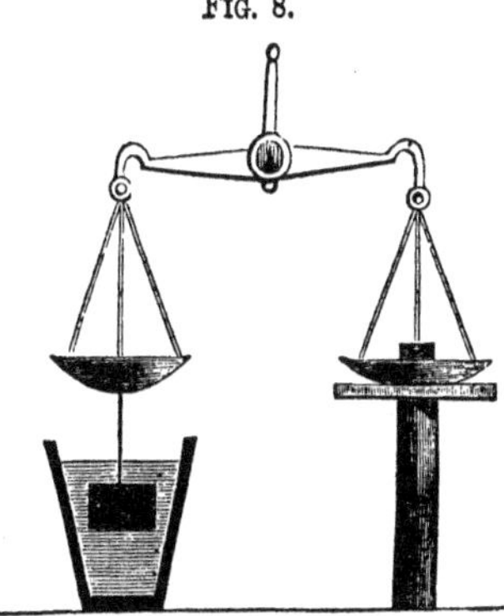

This rule is based upon the fact, that a solid when weighed in water loses weight equal to the water it displaces; and the bulk of the water displaced is exactly equal to its own.

Suppose a piece of gold weighs in the air 19 grains, and in water 18 grains; the loss of weight in water will be 1; $19 \div 1 = 19$, the specific gravity of gold.

Fig. 8 represents the arrangement of the balance for taking specific gravities, and the manner of suspending the body in water from the scale-pan, or beam, by means of a fine thread, or hair.

39. The specific gravity of liquids is easily determined in the following manner. A bottle capable of holding exactly 1,000 grains of distilled water is obtained, filled with water, and balanced upon the scales. The water is then removed, and its place supplied with the liquid whose specific gravity we wish to determine, and the bottle and contents again weighed. The weight of the fluid, divided by the weight of the water, gives the specific gravity required.

QUESTIONS.—What are the standards of specific gravity? How may the specific gravity of solids be determined? Upon what principle is this rule founded? How may the specific gravity of liquids be determined?

Thus, a bottle holding 1,000 grains of distilled water, will hold 1,845 grains of sulphuric acid; 1,845÷1,000=1.845, the specific gravity of sulphuric acid; or this liquid is 1.845 times heavier than an equal bulk of water.

40. The specific gravity of liquids may also be obtained without the aid of a balance, by means of an instrument called the HYDROMETER.

FIG. 9.

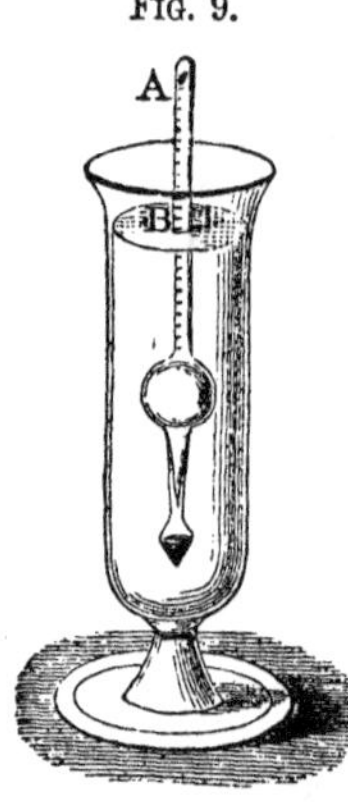

The Hydro'meter.—This consists of a hollow glass tube, on the lower part of which a spherical bulb is blown, the latter being filled with a suitable quantity of small shot, or quicksilver, in order to cause it to float in a vertical position. The upper part of the tube contains a scale graduated into suitable divisions. (See Fig. 9.)

It is obvious that the hydrometer will sink to a greater or less depth in different liquids; deeper in the lighter ones, or those of small specific gravity, and not so deep in those which are denser, or which have great specific gravity. The specific gravity of a liquid may, therefore, be estimated by the number of divisions on the scale which remain above the surface of the liquid. Tables are constructed, so that, by their aid, when the number on the scale at which the hydrometer floats in a given liquid is determined by experiment, the specific gravity is expressed by figures in a column directly opposite that number in the table.

The liquid whose specific gravity is to be determined, is usually, for convenience, placed in a narrow vessel or jar (see Fig. 9), and the zero point on the scale of the hydrometer is always placed at that point where the instrument will float in pure water. The numbers on the scale read either up or down, according as the liquid to be tested is either heavier or lighter than water.

For the testing of alcohol and spirituous liquors, a particular form of hydrometer is used, called the "alcoholo'meter." This is so graduated as to indicate the number of parts of pure alcohol in a hundred of liquid;—perfectly pure, or, as it is called, "absolute" alcohol, being 100, and pure water 1.

In the arts, a French hydro'meter, known as *Beaumé's*, and an English instrument known as Twaddell's, so called from their makers, are much used. Dealers and manufacturers of spirituous liquors, syrups, oils, leys for soap-making, etc., in buying, selling, or compounding, are accustomed to indicate the strength or quality of their products, by saying that they stand at so many degrees Beaumé, or Twaddell.

QUESTIONS.—What is a hydrometer? Upon what principle may the specific gravity of a liquid be determined by the hydrometer? How is the hydrometer graduated? What is an alcoholometer? What are the instruments known as Beaumé and Twaddell?

The practical value of the hydrometer in the arts as a labor-saving invention, is very great. The soap-maker, by dipping the instrument into his ley, and noticing the point at which it floats, knows at once by experience whether it is of sufficient strength to convert his grease into soap; the salt-boiler, by a like observation, is enabled to judge how long his brine must be boiled before salt will deposit at the bottom of his kettles; and the bleacher has in a like manner a sure check against the use of bleaching liquors of strength sufficient to damage his fabrics. So in very many other industrial processes also the hydrometer is equally useful.

41. **Specific Gravity of Gases.**—In principle, the method of determining the specific gravity of gases is the same as that used in the case of solids. A flask, or globe, is first weighed empty, then when filled with air, and a third time, when the gas whose specific gravity is sought for, has been substituted for air. The difference between these respective weights furnishes the data for calculating the specific gravity required.

42. The specific gravity of a body constitutes one of its most important and distinguishing physical characteristics. Thus, for instance, the mineral known as iron pyri'tes resembles gold in color so closely, that it is often mistaken by the unskilled for that metal. It may, however, be at once distinguished from gold by the difference in specific gravity, an equal bulk of gold being nearly four times as heavy.

SECTION II.

COHESION.

43. **Cohesion and Chemical Action.**—The force with which like particles of matter are held together by the influence of cohesion, or what is termed the "strength of materials," although of great importance in all the operations of the mechanic, the engineer, and the architect, has comparatively little to do with Chemistry. Variations, however, in the cohesion and aggregation of the particles of a particular substance, modify, to a considerable extent, the nature and rapidity of chemical action upon it.

Thus gunpowder, for example, when in the form of a hard cake, or as fine dust, burns comparatively slowly, as in what is termed a slow match, or fuse;

QUESTIONS.—Explain the practical value of the hydrometer as a labor-saving expedient. How may the specific gravity of a gas be determined? Is the specific gravity of a substance one of its important characteristics? What is the relation of the force of cohesion to chemical action?

but in the form of fine grains, each portion quickly ignites, and an almost instantaneous explosion occurs.

As a general rule, the cohesion of a body diminishes as its temperature increases. A heated liquid forms smaller drops than a cold one. Sulphur, of all bodies, is an exception to this rule, its consistency increasing, after melting, as its temperature rises.

In liquids, notwithstanding the freedom with which their particles glide over each other, there still exists an appreciable amount of cohesion. This is shown by the fact that every detached drop of a liquid, as a dew-drop upon a leaf, always assumes a rounded form—a globe or sphere being the figure which will contain the greatest amount of matter within a given surface.

This influence of cohesion is beautifully shown in the case of two liquids which do not mix with each other, but which have precisely the same specific gravity, as is the case with oil and alcohol of a certain degree of dilution. If a little oil be poured into weak alcohol, it remains suspended within it in the form of a perfect spherical mass.*

44. **Limpid and Vis'cous Liquids.**—Liquids, according to the difference of cohesive force which exists among their particles, have received the distinctive names of *limpid* and *vis'cous.*

Limpid Liquids are those which, like ether, alcohol, etc., display great mobility of their particles. Bubbles produced in such liquids by agitation, quickly rise to the surface, break, and disappear.

Vis'cous Liquids are those in which the particles are held together so strongly, by the force of cohesion, that they move sluggishly upon one another. Oil, syrup, gum-water, etc., are examples of viscous liquids.

* This experiment may be successfully and easily performed by the teacher in the following manner:—Oil will float upon the surface of water, but will sink to the bottom of strong alcohol; if we, therefore, pour a portion of alcohol into a glass, and put in a globule of oil (olive oil is preferable), the spirit will float above it, and the oil will have the form of a flattened spheroid. If we now add a little water, and mix it carefully with the spirit without breaking the floating mass of the oil, it will be seen to swim higher up in the spirituous medium and present less flatness, and by continuing to carefully add water, we may at last bring the oil to the very center of the fluid, where it will assume the form of a perfect sphere.

QUESTIONS.—What relation exists between cohesion and temperature? Does the cohesive force influence the particles of liquids? Why is a dew-drop spherical in shape? What experiment illustrates the cohesion of liquids? Into what two classes may liquids be divided? What is a limpid liquid? What is a viscous liquid?

45. Variations of Cohesion in Solids.—Those properties of solid bodies which we denominate hardness, softness, brittleness, malleability, and ductility, are occasioned by variations of the cohesive force. The cause of these variations, or the reason why one metal should be malleable and another ductile, or why the same substances should possess, under different circumstances, different degrees of hardness, is not fully understood.

The most trifling variations in the external circumstances to which a body is subjected, will often produce the most extraordinary differences in its hardness, brittleness, ductility, and malleability. A piece of steel slowly cooled from a red heat is soft, and may be easily cut with a file, or stamped with a die; but the same piece of steel, if heated to redness and suddenly cooled, becomes extremely hard, and as brittle as glass. Gold is one of the most ductile of metals, but if a mass of melted gold be exposed to the mere fumes of antimony, it loses its ductility altogether.*

46. Hardness.—The hardness of a body is measured by its power of scratching other substances.

The variations in the degree of hardness presented by different bodies, often furnish the mineralogist and chemist with a valuable physical test, by which they are enabled to distinguish one mineral from another. For the purpose of facilitating such comparisons a table has been constructed, by taking ten well-known minerals, and arranging them in such a way that each is scratched by the one that follows it. Such a table is known as the Scale of Hardness; and by comparing any unknown mineral with this scale, its comparative degree of hardness may be at once determined.

For example, suppose a body neither to scratch nor to be scratched by pure quartz, or rock crystal, which is No. 7 of the table, its hardness is said to be 7; if, however, it should scratch quartz, and not scratch the topaz, which is No. 8 of the table, its hardness would be said to be between 7 and 8. Very many different minerals have the same external appearance, and by the sight alone can not be distinguished from each other; but by the employ-

* See Crystallization.

QUESTIONS.—What physical properties of bodies are due to variations of the cohesive force? How is the hardness of a body measured? What is the scale of hardness? How is the scale of hardness used, and what are its advantages in determining the character of minerals?

ment of this test, a difference in their physical or chemical composition may be at once recognized.*

SECTION III.

ADHESION AND CAPILLARY ATTRACTION.

47. **Adhesion and Chemical Action.**—The force of adhesion is exerted between substances in every form or condition. When it occurs between solids, it is the principal cause of that resistance to motion which is termed *friction.*

As a general rule, friction is greater between surfaces of the same substances than between those of unlike substances. Thus an iron axle moving in an iron box or socket, experiences a greater amount of friction than if revolving in a brass socket.

We reduce the amount of friction between two surfaces by interposing some substances, like grease, oil, black-lead, etc., the particles of which have very little cohesion.

The valuable properties of cements and mortars depend entirely upon the operation of the force of adhesion. The fact, also, that different kinds of cement are required for joining together different materials, proves that adhesion acts with varying degrees of force between different kinds of matter.

Thus, glue or gum may be used for joining pieces of paper or wood, but

---

* The following is the scale of hardness generally adopted:—

1. Talc.	6. Feldspar.
2. Compact gypsum.	7. Limpid quartz.
3. Calcareous spar.	8. Topaz.
4. Fluor spar.	9. Sapphire, or Corundum.
5. Apatite (phosphate of lime).	10. Diamond.

Each of these minerals is harder than those which precede it, and is softer than any which follow it.

Teachers and pupils can, with the exception of No. 10, the diamond, easily obtain the materials necessary to construct the scale of hardness as above given. It may also be obtained, put up in a neat box, of most philosophical instruments dealers, at a trifling expense.

QUESTIONS.—In what manner is the force of adhesion exerted? What is friction? Under what circumstances is friction the greatest? How may friction be diminished? To what are the valuable properties of cements and mortars due? What facts prove the varying force of adhesion?

they will not answer for cementing glass or china; while for the union of marble, brick, or stone, a cement containing lime is required.

Generally, the force of adhesion is inferior in strength to the force of cohesion: but in some instances the opposite is true.

Thus, in detaching glue from the surface of wood, it not unfrequently happens that portions of the wood are torn off by the glue, on account of the force of adhesion between the two bodies proving stronger than the force of cohesion between the particles of the wood.

The property of water to adhere to solid surfaces and wet them, and the rapid diffusion of a drop of oil over the surface of water, are illustrations of the force of adhesion between solids and liquids, and between different liquids.

Some experiments seem to show that the force of adhesion may even overcome the force of affinity under some circumstances. Thus, when vinegar is filtered through pure quartz sand, the first portion that runs through is deprived of nearly all its acid, and the vinegar will not pass through unchanged until the sand has become charged with acid.

48. **Surface Action.**—As adhesion takes place solely between the surfaces of bodies, it is evident that whatever circumstances affect surface must essentially influence the exertion of the force of adhesion. It has accordingly been found that by minutely subdividing a body, and thus increasing its extent of surface, we generally increase the effect of adhesion.

A cubic inch of matter cut into little cubes, each 1-2400th of an inch on the edge, will exhibit a surface of exactly 100 square feet.

All pulverized bodies, by reason of their great extent of surface, attract moisture, or the vapor of water, and also air, so that by exposure to the atmosphere they increase in weight to a considerable extent.

A most striking illustration of the fact that extent of surface facilitates the action of adhesion is found in the case of charcoal. When wood is heated apart from the air, certain portions of matter which compose its structure are driven off by the action of the heat, and the charcoal, which remains behind, is left full of little pores, or openings. In this way an enormous extent of surface is acquired, so much so, that a cubic inch of good charcoal is estimated to have a surface of at least a hundred square feet. By reason of such an extended surface, the effect of the force of adhesion existing between char-

---

QUESTIONS.—Why is not glue suitable for cementing glass or china? Does the force of adhesion ever prove superior to the force of cohesion? What are illustrations of adhesion between solids and liquids? What influence has surface upon adhesion? Why do most pulverized substances attract moisture? How does charcoal illustrate the influence of surface upon the force of adhesion?

coal and various liquids and gases is greatly increased. Thus, it has been found that freshly-burned charcoal is capable, through the force of adhesion alone, of absorbing or condensing upon its surface from 80 to 90 times its own bulk of certain gases; and that it absorbs, when exposed to moist air, so much water, as to increase in weight by nearly one fifth.

All coloring matters of vegetable or animal origin, and many other substances, have likewise the property of adhering to charcoal—a circumstance which has been turned to great practical advantage in the arts.

Other substances beside charcoal, exert, by reason of their peculiar extension of surface, a similar influence on the force of adhesion. Metallic platinum, finely divided, is even more remarkable in its effects than charcoal, and is capable of absorbing eight hundred times its bulk of oxygen gas. This oxygen must be contained within it in a state of condensation very like that of a liquid. In a like manner, every porous body attracts, through the force of adhesion, air and moisture to a greater or less degree, the action of the force being proportioned to the extent of the porosity, or the surface exposed. A field whose soil is finely divided and kept porous by a high state of cultivation, suffers less from drought than one similarly situated which is partially or wholly uncultivated. It is not improbable, also, that plants are assisted in obtaining nutriment from the air, through the influence of an adhesive force acting between the surfaces of their leaves and the constituents of the atmosphere.

49. **Capillary Attraction.**—The phenomena produced by the agency of the force of capillary attraction are similar in character to those produced by the force of adhesion. Indeed, according to some authorities, capillary attraction is merely a variety of adhesion.* The fact, however, that capillary attraction both elevates and depresses the surfaces of liquids, seems to prove that there are essential differences between these two forces.

The two distinguishing manifestations of capillary attraction may be clearly illustrated by the following experiments:—

If a liquid be poured into a vessel, as water in glass, whose sides are of such a nature as to be wetted by it, the liquid will be elevated above the general

* According to the latest and best sustained hypothesis on this subject, the phenomena of capillary attraction are due not only to an adhesive attraction between the liquid and the solid, but also to a contractile force existing on the free surface of every liquid, and which is increased or diminished in a given direction by the convexity or concavity of this surface.

QUESTIONS.—What other facts illustrate the influence of surface action? Do the phenomena of capillary attraction resemble those of adhesion? How may the two distinguishing manifestations of capillary force be exhibited?

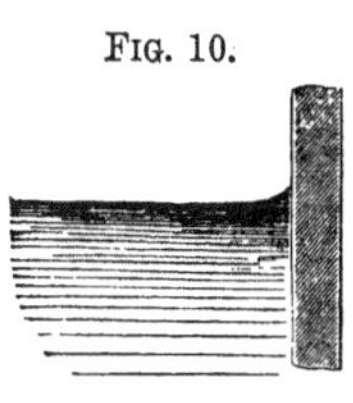
FIG. 10.

level of its surface at the points where it touches the sides of the vessel. This is shown in Fig. 10.

If, however, the liquid is poured into a vessel whose sides are of such a nature that they are not wetted by it, as in the case of quicksilver in glass vessel, then the liquid will be depressed below the general level of its surface at the points where it comes in contact with the sides of the vessel. This is shown in Fig. 11.

FIG. 11.

In like manner, if we plunge a small tube of glass into water, the liquid will rise in it above the general level; but if we plunge it into mercury, the liquid will be depressed below the general level, or will not enter the tube at all.

It has been proved by experiment, that water, through the force of capillary attraction, can be made to pass through a crevice the width of which is less than one half of the millionth of an inch.

Notwithstanding the force which capillary attraction exerts to cause liquids to rise, or pass into tubes of small diameter, it can not of itself establish a flowage, or continuous current. If, however, a part of the liquid be removed from the end of the capillary tube by evaporation, or other agency, an additional portion will be pushed forward by capillary force to supply its place, and in this way a current may be established.

An illustration of this is seen in the case of an oil-lamp, the wick of which may be regarded as a bundle of capillary tubes. So long as the lamp remains unlighted, the wick, although full of oil never overflows; but when the lamp is lighted, and the oil burned off from the top, a current is at once created.

Different liquids do not appear to be equally susceptible to the action of the capillary force. Thus, if we represent the height to which water will ascend in a capillary tube by 100, the height to which alcohol will ascend in the same tube will be only 40, and a solution of common salt in water, 88.

50. **Filtration.**—The process of filtration, or the separation of impurities from liquids by straining, or filtering them through some porous substance, is the result of the

---

QUESTIONS.—Can capillary force produce a current of liquid through the pores of a substance? Are all liquids elevated to the same height in capillary tubes? Upon what does the process of filtration depend?

action of capillary force. The pores, or interstices which exist between the particles of the substance used as a filter, are really little capillary tubes through which the liquid passes, leaving the solid impurities contained in it behind.

When a drop of ink, or chocolate falls upon cloth, or blotting-paper, it produces a dark central spot surrounded by a circle of a paler colored liquid. This is due to the fact that the particles of the liquid *only* are enabled to diffuse themselves, or "spread," as it is termed, through the pores of the material. That appearance of the skin which accompanies a contusion, and is termed "black and blue," is a similar phenomenon—the result of a separation of the coloring and denser matters of the blood from the watery portions, by a process of filtration through the pores of the tissues.

In chemical operations, coarse sand, or cloth, is sometimes used to form filters, but most generally a variety of porous, or unsized paper (blotting-paper), is employed. Writing-paper can not be used for filtration, as its pores are filled up with glue, or starch. For a like reason, ink does not "spread" on this kind of paper.

A paper filter is prepared by folding a circular piece of unsized paper into the form of a quadrant, which is then opened to form a cone. It is generally fitted into a funnel, which is supported upon a stand. (See Fig. 12.)

FIG. 12.

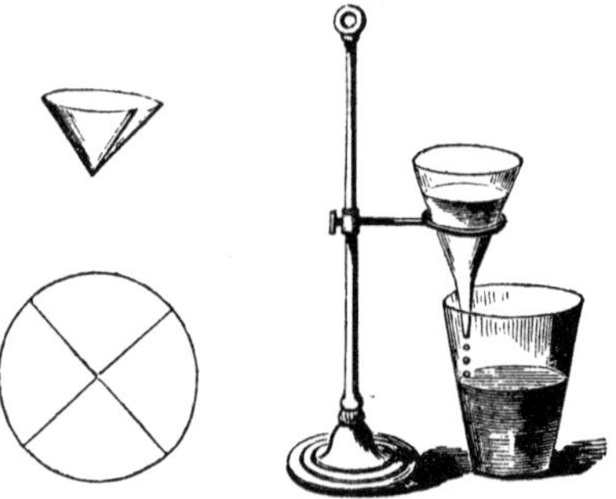

A filtered liquid is termed a *filtrate*.

51. **En'dosmosis.**—When two liquids which are capable of mixing with each other, as alcohol and water, are sep-

QUESTIONS.—Why can not "sized," or writing-paper, be used for filtration? How is a paper filter prepared? What is a filtrate? What is endosmosis?

arated by a substance, or partition which is porous, each will pass through the partition in opposite directions, in order to mix with the other. The exchange, however, always takes place in unequal proportions, so that the volume of one liquid increases while that of the other diminishes. This phenomenon is known by the name of *Endosmosis.*

The name *Endosmose*, derived from the Greek, and signifying "*to go in*," is applied to designate the stronger current, because it penetrates into the opposite liquid; while the name *Exosmose*, which signifies "*to go out*," is applied to the weaker current.

The phenomena of endosmosis may be illustrated by the following experiments:—If some alcohol be placed in a bladder, the neck of which is tightly tied, and the bladder be sunk in a vessel of water, the water will pass into the bladder to such an extent as to distend it, even to bursting.

FIG. 13.

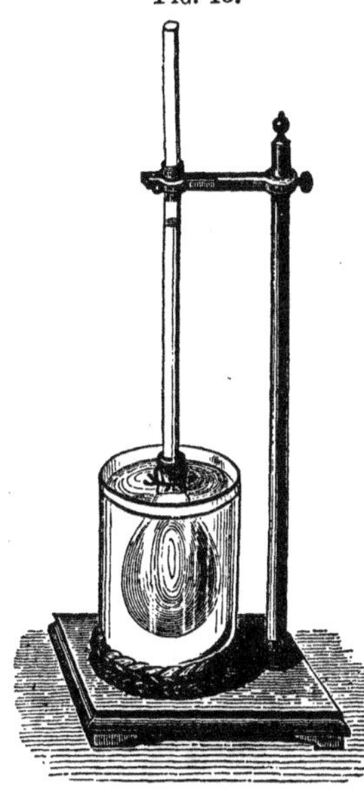

The same result may be also shown more effectively by means of an instrument called the endosmometer. This consists (see Fig. 13) of a bladder filled with alcohol, which is tightly fastened to one end of a tube and inserted in a vessel of water—the tube being sustained in a vertical position. As the water introduces itself through the pores of the bladder the liquid rises in the glass tube, and, if the action be continued sufficiently long, it will rise to the top and overflow. Such an instrument as this may be kept in operation a long time, the liquid flowing continually over the top of the tube. At the same time that the water is passing from without into the bladder to reach the alcohol, a very small quantity of alcohol is passing through the bladder in a contrary direction to reach the water.

The explanation of this phenomenon of endosmosis is as follows:—The pores of the bladder, or any other like substance, are merely short capillary tubes through which the water passes by the force of capillary attraction. If the bladder be distended with air and sunk under water, the water will fill the tubes, but will not discharge itself in the interior, since capillary force

---

QUESTIONS.—What is the origin and derivation of the name? What experiments illustrate the phenomena of endosmosis? How is endosmosis explained?

alone can not establish a continuous movement. But when the bladder is filled with alcohol, the case is different; since the alcohol dissolves away the water as fast as it reaches the interior, and thus produces a constant and rapid current.

The reason that the water passes *in* more rapidly than the alcohol passes *out*, is due to the fact that the water adheres more strongly to the walls of the bladder than the alcohol does—and of any two liquids, that which most freely wets the porous dividing partition will always flow in the stronger current.

Any two liquids may be used to exhibit the action of endosmosis, provided that they have different degrees of attraction for the bladder, and a strong tendency to mix with each other. Thus, in the above experiment a solution of gum, of salt, or of sugar in water, might have been substituted in place of the alcohol.

Very thin plates of slate-stone, or of baked clay, may be also used in place of a bladder, or membrane.

The force with which a liquid will pass through a pore to mingle with another liquid beyond is very great—occurring in some instances in opposition to a pressure of from forty to seventy pounds upon a square inch.

An India-rubber bottle, filled with sulphuric ether, and carefully closed, will gradually empty itself if placed in either alcohol or water. If filled with alcohol, it distends itself in ether, but empties itself in water; if filled with water, it distends when placed in either alcohol or ether.

If a bladder containing equal parts of alcohol and water be hung up in the air, the water will gradually escape through the membrane, leaving the strong spirit behind. In the same manner, if strong alcohol be placed in a wine-glass covered with porous paper, the water contained in it escapes, and the spirit increases in strength.

Endosmotic action exercises an important influence in many of the operations of chemistry, and of animal and vegetable life.

The power which plants possess of absorbing nutritive matter from the soil, through the delicate fibers of their roots, is supposed to be due in part to the action of endosmosis.

All nutriment taken up by the organs of the body, reaches the interior of the system by passing *through* animal membranes in the fluid state. The food we eat passes from the mouth through the throat to the stomach. The structure of the membranes which line the throat is such, that fluids can not pass through them, but the walls of the stomach and of the intestines are

---

Questions.—What determines the rapidity of the two currents in endosmotic action? Under what circumstances will different liquids exert this action? Does endosmosis exert an influence upon chemical and physiological operations? What are illustrations of this fact?

differently constituted, and at these points endosmotic action is continually and energetically going on within us.

Endosmotic action takes place between different gases much more powerfully than between different liquids. No matter what the thickness, or thinness of the porous substance separating two gases may be, currents are established through it, until the media on both sides have the same chemical composition.

The following simple experiment shows this action:—If we tie over the mouth of a glass jar filled with carbonic acid gas, a thin sheet of India rubber, and expose the whole to the air, the carbonic acid will pass out so fast that the cover will be depressed by the external pressure of the atmosphere almost to the bottom of the jar. (See *a*, Fig. 14.) If, on the contrary, we fill the jar with air, and place it in an atmosphere of carbonic acid, the movement takes place in an opposite direction—a little air flows out of the bottle into the carbonic acid, but so large a quantity of the gas passes the opposite way, that the India rubber swells out, and caps the bottle like a dome. (See *b*, Fig. 14.)

FIG. 14.

52. **Diffusion of Gases.**—Connected with this subject is another interesting class of phenomena, known as the diffusion of gases.

When two liquids which are wanting in any attraction for each other, as oil and water, are mixed together, they separate after standing at rest, and arrange themselves according to their specific gravities, the heaviest at the bottom and the lightest at the top. If, however, a light and heavy gas are once mixed together, no separation takes place, but the two remain permanently intermingled.

It has also been found that every gas, or gaseous mixture, possesses the power of diffusing itself equally

---

QUESTIONS.—Does endosmotic action take place between different gases? What are illustrations of it? What is meant by the diffusion of gases? How is each gas affected as regards the presence of another gas?

through every other gas with which it is brought in contact, and this, too, in opposition to the action of their weight, or gravity.

FIG. 15.

Thus, carbonic acid gas is twenty-two times heavier than hydrogen gas, but if a jar filled with hydrogen be placed with its mouth downward over the mouth of a jar filled with carbonic acid, as shown in Fig. 15, the two will diffuse themselves so completely that in a few moments each jar will contain equal quantities of both gases.

Each gas appears to act as void, or empty space for another, or, in other words, it spreads, or expands into the space occupied by another gas, as if it were a vacuum. The same law applies also to vapors.

Thus, as much steam can be forced into a space filled with dry air, as into a space absolutely devoid of air, or any other substance.

This force, or law, regulating the diffusion of gases, is one of great practical importance in the operations of nature, and is often referred to as a most remarkable evidence of design on the part of the Creator. Thus, carbonic acid, which is a deadly poison when inhaled, is one and a half times heavier than common air. The atmosphere contains about one part in two thousand of this gas, uniformly diffused through it—the same quantity being present in air collected on the tops of the highest mountains and on the level surface of the earth. If the law which produces such a complete diffusion were suspended, this heavy gas would accumulate under the influence of gravitation as a bed or layer in the lower part of the atmosphere, and render the immediate surface of the earth uninhabitable.

By reason of this same law of diffusion, the carbonic acid gas which is abundantly formed in every process of combustion and in respiration, and the noxious gases discharged from sewers, and from all decaying matter, are silently and speedily dispersed, and prevented from accumulating.

The equable diffusion of vapor of water through the atmosphere, in accordance with the same law, is no less important than the diffusion of gases. But for such diffusion, the whole surface of the earth would have assumed the condition of an arid desert. Water is 800 times more dense than air, yet the particles of water in the form of vapor ascend into the atmosphere, and diffusing themselves everywhere throughout its substance, give rise to the phenomena of dew and rain.

It is through the operation of this principle, also, that we are enabled to

---

QUESTIONS.—What practical bearing has the law of diffusion upon the constitution of the atmosphere? What upon the condition of the earth's surface? How is it that we are enabled to perceive the odor of volatile substances at a distance?

perceive and enjoy at a distance the fragrant odors which arise from volatile substances; and were its action suspended, the sense of smell would be nearly unknown to us.

53. **Diffusion of Liquids.**—Liquids of different densities, which are susceptible of mixing, will, when brought in contact, gradually become intermingled, by a law somewhat resembling that which governs the diffusion of gases.

Thus, if pure water be carefully poured upon a strong solution of salt or of sugar, the lighter fluid will at first float upon the surface of the heavier; but after a time the two will mingle together more or less uniformly. In like manner, a drop of ink, or other similar coloring matter, will diffuse itself through a large quantity of water.

54. **Solution.**—When the adhesion between the particles of a solid and those of a liquid is more powerful than the force of cohesion which binds together the particles of the solid, the power of cohesion will be entirely overcome, or suspended, and the substance is said to dissolve, or undergo solution in the liquid. In this way sugar or salt dissolves in water, rosin or camphor in alcohol, and lead or silver in mercury.

A body is said to be insoluble when the adhesive force exerted by a liquid upon its particles, is not strong enough to overcome the cohesive force which binds them together.

Any thing which weakens the force of cohesion in a solid favors solution. Thus, if a substance be reduced to a powder, it dissolves more quickly, both from the larger extent of surface which it exposes to the action of the liquid, and from the partial destruction of cohesion between its particles. In the same way heat, by diminishing the force of cohesion, generally promotes the process of solution. Some substances, however, as lime, for example, dissolve more freely in cold than in warm water.

55. **Saturation.**—When a liquid has dissolved as much of a solid as it is capable of doing, it is said to be saturated. When this occurs, the force of adhesion, between the liquid and the solid becomes reduced to an equality with the force of cohesion between the particles of the solid, and the act of solution ceases.

QUESTIONS.—What is understood by the diffusion of liquids? What are illustrations of liquid diffusion? What is solution? When is a body said to be insoluble? What circumstances favor the solution of a solid? What is saturation?

56. **Precipitation.**—When a solid body dissolves in a liquid, the property of cohesion is not destroyed, but merely overcome, or suspended by the superior force of adhesion. If this latter force is in turn weakened, or overcome, the force of cohesion acquires an ascendancy, and the particles in solution unite again to form a solid. A solid thus reproduced and separated from a liquid, is called a Precipitate.

Thus, the common solution of camphor is formed by dissolving the camphor gum in alcohol. If water be added to this solution, the alcohol at once mixes with the water, and abandons the camphor, which immediately resumes its solid form, and falls to the bottom of the vessel—it is precipitated.

The precipitation of a solid from its solution may also be effected by several other methods:—

Especially may this be accomplished by changing the character of the substance held in solution, by bringing in contact with it another body with which it is able to unite chemically, and form an insoluble compound. Thus, lime is somewhat soluble in water, but if we bring carbonic acid gas in contact with it while in solution, the two substances unite together by the action of the chemical force of affinity, and overcome the adhesion which the water previously had for the lime. The compound of carbonic acid and lime thus produced, being solid and insoluble, is immediately precipitated.

The above case illustrates a general law in chemistry, which may be stated as follows:—

Two substances which, when united, form an insoluble compound, generally combine and produce the same compound when they meet in solution.

This law is practically taken advantage of in chemical operations for separating the different constituents of a compound from each other, or for detecting the presence of a body when in solution with other substances. Thus, if it is desirable to know whether a perfectly clear spring-water contains lime, carbonic acid gas is introduced into it. This uniting immediately with the lime, forms an insoluble compound, which is precipitated. On the other hand, by reversing the process and introducing a solution of lime, we may be able to detect the presence of carbonic acid under the same circumstances.

The depression of the temperature of a solution will sometimes cause the cohesion of the particles of the solid dissolved to acquire an ascendancy over the force of adhesion. Thus, alum dissolved in hot water will resume in part

---

QUESTIONS.—What is a precipitate? Give an illustration. How may precipitation be effected by changing the character of a substance? What general law governs the precipitation of substances from their solutions? How is this law practically applied in chemical operations? How may precipitation be effected through a depression of the temperature of a solution?

its solid form as the solution is cooled; and when brandy is exposed to intense cold, many degrees below that necessary to freeze water, the spirituous portion retains its liquid form, and separates from the aqueous part, which solidifies as ice.

A remarkable illustration of this action is to be found in the fact that ice formed by the freezing of sea-water is, under all ordinary circumstances, fresh, and entirely destitute of salt. The great ice-fields which cover the ocean in the Arctic and Antarctic regions, are always composed of fresh-water ice. Indeed, water in the act of freezing separates completely from every thing which it previously held in solution. Even the air contained in water is expelled in the act of freezing, and becoming entangled in the thickening fluid, gives rise to the minute bubbles generally observed in blocks of ice. For a like reason, the ice formed by the congelation of a solution of indigo is colorless.

Elevation of temperature will also effect the separation of bodies in solution. When, for instance, a solution of common salt in water is exposed to the action of heat, the repulsive power of this agent overcomes not only the cohesion of the water, but also its adhesion to the salt; the water assumes the æriform state, and passes off as steam, while the salt, deprived of its solvent, resumes the solid state.

57. **Solution and Chemical Combination.**—A clear distinction exists between a solution and a chemical combination, which latter, in ordinary language, is often termed a solution.

A simple solution is occasioned by the action of the force of adhesion exerted between the particles of the solid and the liquid with which it is brought in contact. In all cases of simple solution, the properties of both the solid and the liquid are retained.

Thus sugar, whether in a mass in the hand, or dissolved in water, is the same substance; so also when camphor is dissolved in alcohol, the solution partakes of the properties of both, having the smell and taste of both camphor and spirit.

When a solid disappears in a liquid through the influence of a chemical force exerted between the particles of the two substances, the compound is not a true solution, but a chemical combination, in which the properties of both the solid and liquid are essentially changed.

QUESTIONS.—What are illustrations of this principle? Why is ice, formed by the freezing of sea-water, fresh? What is the occasion of the numerous bubbles observed in blocks of ice? How may precipitation be effected by an elevation of temperature? State and illustrate the difference between solution and chemical combination.

Thus, iron placed in diluted acid disappears in it, but the resulting liquid does not contain finely divided iron, but a finely divided compound of iron and the acid, which possesses entirely different properties from either of its constituents.

Solution differs also from chemical combination in the varying proportions in which it occurs, according to temperature, etc. Thus, a given quantity of water at the boiling temperature will dissolve nearly four hundred times more saltpetre than it can at a temperature of 60°; but in chemical combinations the proportions in which bodies unite are fixed and invariable.

SECTION IV.

CRYSTALLIZATION.

58. **Crystals.**—The particles of most substances, in passing from a liquid to a solid condition, have a tendency to arrange themselves into regular and symmetrical forms, each different substance assuming always a peculiar shape, from which it never essentially varies. Such regular geometrical solids are termed Crystals.

The number of known crystalline forms is much smaller than the number of substances which are capable of crystallizing, and it therefore follows that crystals of various kinds of matter may possess the same form. No substance, however, has ever been found to be capable of assuming indifferently any form, but most substances are restricted to one form of crystal and its modifications. This circumstance enables us, very often, to identify a substance, or determine its composition, simply by the shape of its crystals. For example, common salt always crystallizes in cubes, alum in octohedrons, saltpeter in six-sided prisms, Epsom salts in four-sided prisms, and so on.

59. **Amor'phous Bodies.**—A solid whose particles are arranged irregularly, and which possesses no definite external form, is said to be amorphous (*i. e.*, without form).

Every solid body is either amorphous, or crystalline, and many bodies exist in both of these conditions. Thus, carbon, in the form of charcoal and lampblack, is amorphous, but in the form of the diamond it is crystalline.

60. **Formation of Crystals.**—The usual method of obtaining crystals is to form a strong solution of the substance in hot water, as most bodies dissolve more freely in water when it is at an elevated temperature than when

---

QUESTIONS.—What are crystals? Can a substance in crystallizing assume indifferently any form? What are amorphous bodies? Can a substance be both crystalline and amorphous? What is the usual method of obtaining crystals?

cold. As the liquid cools, and the force of cohesion gradually begins to resume the ascendancy, the separated particles of the solid have time to select, as it were, the arrangement they will assume, and crystals are formed.

When a solid is melted, or made to assume a liquid form by heating, and allowed to cool quietly, its particles also, in most instances, assume a crystalline arrangement.

Illustrations of this may be seen in the crystalline fracture of zinc and antimony. Sulphur, also, crystallizes beautifully on cooling after fusion.

Water, in freezing, or assuming the solid condition, often shoots into beautiful crystals, as may be seen by examining the snow-flakes which fall during a period of intense cold, beneath a microscope. These crystals may also, under favorable circumstances, be seen with the naked eye, by placing the flake upon a dark body cooled below 32° F. Fig. 16 represents some of the varied and beautiful forms of snow crystals.

FIG. 16.

The same crystals which appear in snow, exist also in ice, but they are so blended together that their symmetry is lost in the compact mass. When water freezes, its particles all arrange themselves in ranks and lines which cross each other at angles of 60 and 120 degrees. This may be seen by examining the surface of water in a saucer while freezing.

If we fracture thin ice, by allowing a pole, or weight to fall upon it, the fracture will have more, or less of regularity, being generally in the form of a star, with six equidistant radii, or angles of 60°.

Another beautiful illustration of the crystallization of water in freezing, is seen in the frost-work upon windows in winter, caused by the congelation

QUESTIONS.—What peculiarities of crystallization does water present in freezing? Has ice a crystalline structure? What occasions the symmetrical arrangement of frost-work upon windows, etc., in winter?

of the vapor of the room when it comes in contact with the cold surface of the glass. All these frost-work figures are limited by the laws of crystallization, and the lines which bound them, form among themselves no angles but those of 30°, 60°, and 120°.

When a substance has been converted, through the action of heat, into a vapor or gas, and then by cooling is caused to change back again at once into the solid state, its particles arrange themselves so as to form crystals.

Thus, camphor, or sulphur, if heated in a glass tube, will be first converted into vapor, and then deposited in a ring of crystals higher up, at the first point where the temperature is sufficiently low.

In general, it is important to the process of crystallization that the liquid from which the solid body is separating should not be shaken or disturbed, but when the forces of cohesion and adhesion are nearly balanced, as in a saturated solution, it seems necessary that some slight motion should be given to the liquid in order to initiate the process, which does not commence at all in a state of absolute rest.

Thus, a saturated hot solution of Glauber's salt, if allowed to cool in perfect stillness, will remain liquid as long as the stillness is preserved, but the slightest movement or tremor—even a wave of the hand through the air in its vicinity—will instantly transform the solution into a solid mass, some of the water entering into the composition of the crystals, and some being retained by interstices in their structure. In the same manner, water may be cooled eight, or ten degrees below the freezing point and yet remain liquid; but the slightest disturbance, even a vibration of the vessel, will cause it to freeze (crystallize) instantaneously.

The more slowly a liquefied body is brought back to a solid state, and the more the liquid is kept at rest after the process of crystallization has commenced, the smaller will be the number and the larger the size of the crystals produced; but when the solution is caused to solidify very quickly, the crystals are numerous, but small and imperfect.

In the first case the particles of the solidifying body have time to arrange themselves regularly upon each other; but in the latter instance the solidifi-

---

QUESTIONS.—Under what other circumstances may crystallization take place? What are important requisites in the process of crystallization? What facts illustrate these conditions? Under what circumstances will crystals be perfect, and when imperfect?

cation takes place so rapidly that the particles attach themselves irregularly, and interlace with each other in every direction. In this consists the difference between "sugar," or "rock candy" and loaf, or granulated sugar; between the fine grained statuary marble and crystallized "spar."

Crystals have always a tendency to fasten upon any foreign substance that occupies a prominent position in the liquid which affords them, a circumstance which is applied to many useful purposes in the arts.

Illustrations of this are seen in the formation of the somewhat familiar ornament known as the "alum basket," and in the strings which are stretched across the vessels in which pure solutions of sugar crystallize in the manufacture of "rock candy." When only two or three very minute crystals can be deposited, it is usual to place a piece of thread or some other suitable substance in the liquor; and upon this support the crystals, if anywhere, will be found. In this way the chemist is enabled to draw together and collect readily the smallest quantities that can be thrown down from a solution.

Nothing can be more beautiful than to watch the progress of crystallization as it takes place when we suspend a series, or network of threads in a hot saturated solution of alum, and then allow the liquor to cool slowly. The minute invisible atoms are gradually drawn together toward the foundation thus afforded, and presently little glittering specks may be discerned entangled among the fibers, or studding the network of the threads. If the process be well managed, these specks increase steadily in size, by the regular addition of fresh atoms to every part; but if the temperature be not attended to, or the solution be improperly disturbed, they increase chiefly in numbers, and the larger crystals are apt to be disfigured by adhering to small ones.*

61. Purification by Crystallization.—A substance in crystallizing has a tendency to purify, or separate itself from any foreign substances which may have been mingled with it. Crystalline form is, therefore, to some extent, a

* "The beautiful crystalline masses that are sometimes seen in druggists' windows, can not be produced without the greatest care and attention, each crystal being separated from the liquor when it has attained a sufficient size, and being placed alone in a shallow pan, perfectly glazed, at a temperature carefully regulated, and under a solution of a specified strength. It is then turned over from day to day, as otherwise the face in contact with the pan would be prevented from increasing, and a deformed crystal would result. It is also carefully supplied with fresh solution from time to time: because, if that around it were exhausted, its most prominent angles would be re-dissolved. By neglecting these precautions, deformed or monstrous crystals are obtained, and are exhibited, perhaps, as often as the perfect ones. Crystalline masses of the blue sulphate of copper, the red chromate of potash, of alum, and some other chemical compounds, may be produced of almost any magnitude that is desired."

QUESTIONS.—How does interrupted crystallization affect the physical character of a body? What curious tendency do crystals exhibit in separating from a solution? What practical application of this is made in the arts?

guaranty of purity, or at least of the absence of adulteration; and hence, in medicine, and in the arts, many substances are subjected to tedious and expensive processes for no other purpose than to cause them to assume this form.

Sea-water, in addition to salt, contains a variety of other substances, but by the process of evaporating the salt water and crystallizing the salt, most of these impurities are separated. A single crystallization gives the salt sufficiently pure for commercial purposes, but to render it perfectly pure, it is necessary to re-dissolve the first crystals in pure water and repeat the process of crystallization several times.

This principle may be demonstrated by a simple experiment. If we dissolve a small quantity of common salt and saltpetre in warm water, and allow the solution to evaporate slowly, the two substances, which are intimately united in the solution, will separate completely from each other in crystallizing—the saltpetre assuming the form of long needles or prisms, and the common salt the form of cubes. It is in this way that saltpeter is purified preparatory to being used in the manufacture of gunpowder.

If two bodies, however, which crystallize in the same form, be mingled in solution. they can not be separated from each other by crystallization.

The difference in the crystallizing properties of silver and lead has been taken advantage of in a recent invention for separating a small quantity of silver which exists in almost all the ores of lead. The two metals are melted and allowed to cool slowly; the silver, forming into crystals more easily than the lead, solidifies first, and the lead remaining is poured off.

62. **Change in Bulk.**—Many substances in crystallizing, or in passing from a liquid to a solid state, experience a change in bulk.

Water, at the moment of congelation, increases in bulk, and expands with an almost irresistible force. As an illustration, the following experiment may be quoted:—Cast-iron bomb-shells, thirteen inches in diameter and two inches thick, were filled with water, and their apertures or fuse-holes firmly plugged with iron bolts. Thus prepared, they were exposed to the severe cold of a Canadian winter, at a temperature of about 19° below zero. At the moment the water froze, the iron plugs were violently thrust out, and the ice protruded, and in some instances the shells burst asunder, thus demonstrating the enormous interior pressure to which they were subjected by water assuming a solid state.

QUESTIONS.—Can two substances in solution be separated from each other by the act of crystallization? What are practical illustrations of this principle? Under what circumstances will crystallization fail to effect separation? What physical change frequently accompanies crystallization? Illustrate this action in the case of water.

One thousand parts of water at 32° become dilated by freezing to 1063 parts.

Iron, in passing from a melted to a solid state, expands in the same manner as water, a fact which renders this metal most suitable for castings.

Other substances, however, present equally remarkable instances of contraction in passing from a liquid to a solid state, of which gold and lead are illustrations; hence it is impossible to obtain with either of these metals a fine casting from a mould.

63. **Mother Liquor.**—When a substance separates itself in part from a liquid by crystallization, the solution remaining behind is termed the Mother Liquor.

64. **Water of Crystallization.**—Some substances are not capable of assuming a crystalline form until they have chemically combined with a certain definite amount of water, termed the WATER OF CRYSTALLIZATION. This water is not essential to the chemical composition of the substance, but merely to its existence in the form of crystals.

Thus, a crystal of alum contains nearly one half its weight of water chemically combined with it. Without this water, alum could not assume the crystalline form, although it would retain all its chemical properties unchanged. The existence of the water of crystallization in alum may be experimentally shown by placing a small crystal of this substance upon a hot surface, when it will be observed to foam and melt, and finally settle down into a white porous mass. The foaming is occasioned by the evaporation of the water of crystallization.

65. **Ef-flo-res'cence.**—Some substances containing water of crystallization, part with it on exposure to the atmosphere, and crumble down to a fine powder. This action is termed Efflorescence.

If we place half an ounce of crystalline Glauber's salts in a warm place, it will soon lose its transparency, and finally crumble into a white powder, weighing hardly a quarter of an ounce. This loss of weight is entirely owing to the evaporation of the chemically combined water which imparted to the salt its transparency and crystalline form. Common salt, and saltpetre, on the contrary, if treated in a similar way, undergo no change in either appearance, or weight, because they contain no water of crystallization.

66. **Del-i-ques'cence.**—When a crystalline substance,

QUESTIONS.—To what extent will water expand in freezing? Why is iron eminently suitable for fine castings, and gold and lead unsuitable? What is a mother liquor? What is water of crystallization? What is efflorescence? What is deliquescence?

on exposure to air, absorbs water, and becomes converted thereby into a liquid, or semi-liquid mass, it is said to deliquesce, and the phenomenon is termed Deliquescence.

67. **De-crep-i-ta'tion.**—Some substances, when crystallized rapidly from a solution, frequently inclose mechanically within their texture small quantities of the mother liquor, the expansion of which, when heated, bursts the crystals with a sort of crackling explosion. This phenomenon is known by the name of Decrepitation.

This result may be exhibited by throwing a small quantity of common salt, which has been crystallized rapidly, upon a heated surface. If the salt, however, has been crystallized by slow evaporation, it will not decrepitate.

68. **Native Crystals.**—The mineral kingdom presents us with the most splendid examples of crystallized bodies, many of which the chemist is able to artificially reproduce in his laboratory. Within the last few years, M. Ebelman, an eminent French chemist, has succeeded in producing some of the most valuable gems—as, for example, the emerald and the ruby—by mixing together in proper proportions the elementary substances which enter into their composition, and then exposing the compound to the long-continued and intense heat of a furnace used for baking porcelain.

Some native crystals, however, seem to be beyond the power of art to imitate. Of these, the diamond is perhaps the most remarkable. This body consists of pure carbon (the same substance with which we are familiar as charcoal and as black-lead), but which can not be either fused or dissolved, and consequently can not be crystallized by any means at present known. Such means have been eagerly sought for, however, since the discovery of the composition of the diamond, and there seems no reason why they should not at some period be discovered.

The most perfect crystals of gems are met with in nature of only a moderate size. The larger ones are less clear, and wanting in transparency and luster. The emerald, sufficiently pure for jewelry, does not often exceed an inch in length, and seldom so much as this. Transparent sapphires above an inch in length are very rare. Crystals of quartz are sometimes found of very large size. One at Milan measures 3½ feet in length, 5½ in circumference, and weighs 870 pounds.

69. **Formation of Crystals in Solid Bodies.**—A very remarkable change, a variety of crystallization, sometimes takes place in the form and arrangement of the particles of solid bodies, without their undergoing any alteration

---

QUESTIONS.—What is decrepitation? Where are the most splendid examples of crystallized bodies to be met with? Are any native crystals capable of being reproduced by art? What crystallized body can not be imitated? What remarkable change sometimes takes place in the particles of solid bodies?

from the solid to the liquid state. This subject is one of great importance, and its investigation has furnished a partial solution of some phenomena that were once regarded as inexplicable.

The simplest illustration of this action is to be found in the case of sugar. When this substance is melted and allowed to cool, it forms a perfectly transparent, hard mass, without the slightest trace of crystalline arrangement; but after some months it loses its transparency, becomes white, crystalline, and brittle. Similar changes take place also in many other bodies, but in cases of this character the cause which produced the result described is not certainly known, and has been ascribed to the action of several forces.

The following illustrations are of a somewhat different character. If we submit a piece of metal, even the toughest, to long-continued hammering, or jarring, the atoms, or particles of which it is composed, seem to take on a new arrangement, and the metal gradually loses all its tenacity, flexibility, malleability, and ductility, and becomes brittle.

The surface of a fresh fracture, under such circumstances, exhibits a distinctly crystalline structure. The tenacity of a metal thus rendered brittle may be restored again in great measure by heating and slowly cooling—a process known in the arts as "annealing."

A great number of other instances illustrative of the effect of jarring and concussion on the structure of metals, might also be adduced. Coppersmiths, who form vessels of brass and copper by the hammer alone, can work on them only for a short time before they require annealing; otherwise they would crack and fly into pieces.

For similar reasons, a cannon can only be fired a certain number of times before it will burst, and a cannon which has been long in use, although apparently sound, is always condemned and broken up. The tone of a bell, during the two or three first years of use, uniformly increases in strength, owing probably to a change in the arrangement of the particles under the hammering action in ringing.

A more important illustration, and one that more closely affects our interests, is the liability of railroad car-axles and wheels to break from the same cause. A car-axle, after a long lapse of time and use, is almost certain to break.

The explanation of these changes, especially in the case of iron, is as follows:—The particles of cast-iron, as may be seen by the naked eye, are crystals, more or less perfect in form, and aggregated together by the force of cohesion. In the conversion of cast-iron into wrought-iron, each crystal by heating, hammering, and rolling, is gradually elongated into a thread, so that wrought-iron is an aggregation of fibers (fibrous iron, as it is sometimes called), or a series of threads kept together by the force of cohesion. When now a bar of cold,

---

Questions.—Give an illustration. How is the strength of iron and other metals affected by hammering, jarring, etc.? Under what circumstances will cannon burst, and railway axles break? What explanation has been given of these phenomena?

wrought, or fibrous iron is made to vibrate by shocks communicated either by blows of a hammer, or by the rapping of any part of a machine, or by the continued rolling and jarring of a railway car upon the rails, the little fine threads, or fibers snap one by one, and the particles return to their original crystalline, or granular state, and by this change the entire mass is rendered brittle.

70. **Primary Forms of Crystals.**—The apparently innumerable variety of figures which various substances assume in crystallizing, may all be referred to a few regular and fundamental forms.

Each substance has a characteristic form of crystal, which is termed its *Primary Form.*

Variations of this original form, which may take place to any extent so long as a correspondence of angles is preserved, are termed *Secondary Forms.*

The number of primary or fundamental forms to which all other crystalline solids may be referred is six—the cube, the square prism, the right rectangular prism, the oblique rhombic prism, the oblique rhomboidal prism, and the hexagonal prism, or rhombohedron.

The number of secondary forms is almost innumerable, all of which are modifications of the six primary forms.

Thus, carbonate of lime has been found crystallized in more than six hundred different secondary forms, but all of them are related to each other, and are derivable from one original primary figure, the rhombohedron.

The study of the geometrical relations of the different crystalline forms to each other, belongs to the science of crystallography. The investigations of chemistry, however, have contributed much to our knowledge of the laws and forces which govern the production of crystals, and have furnished some explanation of the reason why the several atoms, each invisible on account of its minuteness, should arrange themselves in the same manner, and in the fitting place, so as to build up a cubical or prismatic crystal, rather than an incoherent mass, shapeless and devoid of regularity.

71. **Theory of Crystallization.**—It is supposed that the atoms, or molecules which make up the body of a crystal, are possessed of polarity ; or, in other words, that the two opposite sides of the atoms are like the two opposite poles

---

Questions.—What is understood by primary and secondary forms of crystals? How many primary, or fundamental forms of crystals are recognized? How many secondary forms exist? Give an illustration of the primary and secondary forms of carbonate of lime. Explain the general theory of crystallization.

of a magnet, endowed with opposite forces. The action of these forces compels the atom, in assuming its place in a crystal, to maintain a certain direction as respects the contiguous particles (see Fig. 17), in the same way that the action of the magnetic forces on a bar of steel compels it to maintain a constant direction as regards the poles of the earth.

Fig. 17.

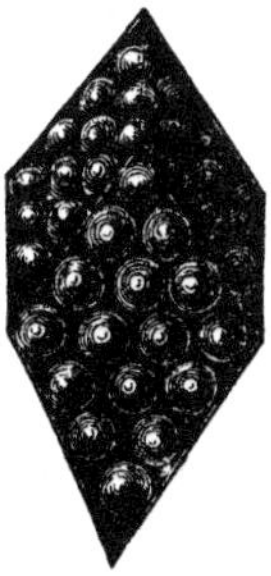

That the strength of the directive force which influences the atoms of matter to assume a symmetrical arrangement is not feeble or insignificant, is clearly shown by the enormous power which crystallizing action exerts. Thus, the expansive force of water in freezing, illustrations of which are most familiar, is due entirely to a re-arrangement of the particles in crystallizing, and a consequent occupation of more space.

The direction in which the supposed polar forces act, or the lines in which the particles arrange themselves in order to build up symmetrical solids, are termed the axes of the crystal.

Variations in the number or arrangement of these lines or axes, necessarily modify the geometrical form of the crystal, and a consideration of the relation which the multitude of crystalline forms sustain to each other through their axes, or symmetrical lines of formation, has enabled us to select six primary forms from which all the others may be derived.

Fig. 18.

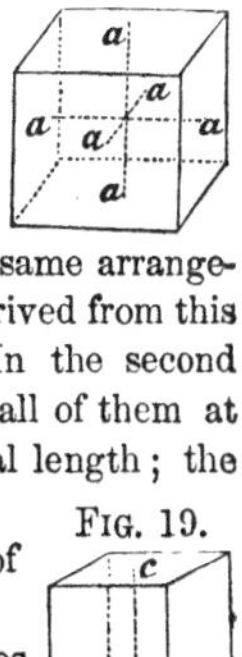

Thus, in the first primary form, Fig. 18, which is the cubical, or regular form, there are three axes, *a a a*, all equal and crossing each other at the center of the crystal at right angles. The same arrangement of axes holds good in all the secondary forms which are derived from this primary form; and in consequence of this are all regular. In the second primary form, the square prism, Fig. 19, there are three axes, all of them at right angles to each other, but only two, *a a*, *a a*, are of equal length; the third, *c c*, being either longer or shorter than the others.

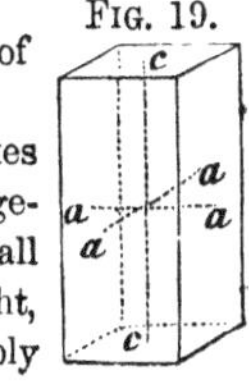

Fig. 19.

Similar variations exist, also, in the number or inclination of the axes of the other primary forms.

Many facts in science seem to prove that the existence of axes in crystals is not imaginary, but real. Thus, when the arrangement of a crystalline body is perfectly symmetrical, as it is in all crystals belonging to the cubical system, the transmission of light, the expansion of heat, the conducting power of heat, and probably

---

Questions.—What are the axes of a crystal? On what ground do we recognize six primary forms of crystalline solids? What facts in science seem to prove that the axes of crystals have a real existence?

the power of transmitting sound, electricity, and magnetism, is uniform in every direction; but when the axes of a crystal are unequal; or, in other words, when the action of the molecular force which has given direction to the atoms and shaped the crystal, is more powerful in one direction than in another, an irregularity in the action of the body on light, and in its expansive and conductive powers for heat, may be immediately traced.

I-so-morph'ism.—The term Isomorphism (equal forms) is applied to those bodies which can be substituted for one another in a chemical compound, without producing any change in the crystalline form of that compound. This property, is restricted to a comparatively few substances.

Thus, an oxyd of zinc may replace or be substituted for oxyd of magnesia, and an oxyd of iron for an oxyd of copper, in a chemical compound, without causing any alteration of crystalline form. As a general rule, however, the change, or substitution of one element of a chemical compound for another of different character, occasions a change in the crystalline form of the compound.

The consideration of isomorphism is of great importance in chemistry, and has added much to our knowledge respecting the nature of the elementary atoms of matter. A study of its principles, among other results, has established the existence of such curious relations between certain of the so-called elementary substances, as to suggest their derivation from some common and unknown form of matter. This subject, under another department, will be again referred to.

72. **Di-morph'ism.**—The rule that all the crystalline figures of any particular substance may be derived from the same ultimate form, is subject to several exceptions. Some substances are capable of assuming two forms of crystals, according to the temperature at which they are produced, which are geometrically incompatible with each other; and this difference of crystalline form is associated with difference of specific gravity, hardness, color, and other properties. Such bodies are termed *Dimorphous* (two-formed).

The crystals of sulphur found in nature, and the crystals obtained by the slow cooling of a melted mass of sulphur, are entirely different. A beautiful instance of this kind is afforded by a compound of iodine and mercury, known as the iodide of mercury. The minute particles of this substance are of a brilliant scarlet color, but by the application of heat their crystalline arrangement is changed, and the change is rendered visible to the eye by the sub-

---

QUESTIONS.—What is isomorphism? Give an illustration. What is dimorphism? What are examples?

stitution of a bright yellow color in place of the scarlet. When the substance has become cool, the application of a slight mechanical force, such as a mere scratch upon a single point, will change the crystalline arrangement back to its original condition, and instantly restore the original color.

Some few substances are even trimorphous; that is, they crystallize in three different forms.

73. **Cleav'age.**—Crystals can not be broken with equal readiness in all directions, but they have a tendency to split or divide according to certain determinate lines. This property is termed the *Cleavage* of the crystal.

Cleavage will often enable us to detect crystalline structure in a body which at first appears as a shapeless mass. Thus, in the case of the very common mineral known as "Iceland spar," which is a variety of carbonate of lime, if we strike gently upon an irregular fragment with a hammer, we shall find that the lines in which fracture occurs are all inclined to each other at angles of 105 degrees, and in consequence of this, the detached particles have all the form of rhombohedrons. In like manner, mica splits only in leaves, and galena, the name applied to the common ore of lead, only in cubes.

This property of crystals has long been known to jewelers, who have profited by it to alter the form of precious stones, in place of resorting to the expensive process of cutting. Thus, the diamond will split with a smooth surface in four directions, and by taking advantage of this, a thin layer on a defective side may be smoothly removed at a single operation.

A property analogous to the cleavage of crystals may be observed in bodies of a different character. Thus, wood splits with greater facility in a direction parallel to its fibers than at right angles to them, or, as it is termed, "across the grain."

QUESTIONS.—What is cleavage? How will cleavage often enable us to detect crystalline structure in an irregular body? What practical application has been made of the cleavage of crystals?

CHAPTER II.

HEAT.

74. **Heat and Chemical Action.**—Almost every form of chemical action is influenced to a greater or less extent by the agency of heat. A general knowledge, therefore, of the principles and applications of heat is essential to a correct understanding of the science of chemistry.

Heat and Caloric.—Heat is a physical agent, known only by its effects upon matter. In ordinary language we use the term heat to express the sensation of warmth. Caloric is the general name given to the physical agent which produces the sensation of warmth, and the various effects of heat observed in matter.

75. **Two Conditions of Heat.**—Heat exists in two very different conditions, as FREE, or SENSIBLE HEAT, and as LATENT HEAT.*

When the heat retained or lost by a body is attended with a sense of increased or diminished warmth, it is called sensible heat.

When the heat retained or lost by a body is not perceptible to our sense, it is called latent heat.†

76. **Measurement of Heat.**—The quantity of heat observed in different substances is measured, and its effects on matter estimated, only by the change in bulk, or appearance, which different bodies assume, according as heat is added or subtracted.

77. **Distinguishing Characteristic of Heat.**—Heat possesses a distinguishing characteristic of passing through and existing in all kinds of matter at all times. So far as

* *Latent*, from the Latin word *lateo*, to be hid.

† The phenomena of latent heat are further considered under the head of Liquefaction and Vaporization.

QUESTIONS.—What relation exists between heat and chemistry? What is heat? Define the meaning of the term caloric. In what two conditions does heat exist? What is free, or sensible heat? What is latent heat? How is heat measured? What is the distinguishing characteristic of heat?

we know, heat is everywhere present, and every body that exists contains it without known limits.

Ice contains heat in large quantities. Sir Humphrey Davy, by friction, extracted heat from two pieces of ice, and quickly melted them, in a room cooled below the freezing-point, by rubbing them against each other.

78. **Temperature.**—The degree of sensible heat a body manifests is called its temperature.

The temperature of a body affords no indication of the real quantity of heat which it contains. A pint of boiling water will raise a thermometer to the same degree as a gallon of the same water; yet it is obvious that the larger quantity of liquid contains the greater amount of heat.

Cold is a relative term expressing only the absence of heat in a degree; not its total absence, for heat exists always in all bodies.

A body may feel hot and cold to the same person at the same time, since the sensation of heat is produced by a body colder than the hand, provided it be less cold than the body touched immediately before; and the sensation of cold is produced under the opposite circumstances, of touching a comparatively warm body, but which is less warm than some other body touched previously. Thus, if a person transfer one hand to common spring water immediately after touching ice, to that hand the water would feel very warm; while the other hand, transferred from warm water to the spring water, would feel a sensation of cold.

It is a very curious fact, that intense cold produces the same sensation as intense heat. Frozen mercury will blister the part to which it is applied in the same manner as red hot iron; and the physiological action of a freezing mixture resembles that of boiling water. Sensations of heat and cold are, therefore, merely degrees of temperature, contrasted by name in reference to the peculiar temperature of the individual speaking of them.

79. **Diffusion of Heat.**—The tendency of heat is to diffuse, or spread itself among all neighboring substances, until all have acquired the same, or a uniform temperature.

A piece of iron thrust into burning coals becomes hot among them, because the heat passes from the coals into the iron, until the metal has acquired an equal temperature.

80. **Heat Imponderable.**—Heat is imponderable, or does not possess any perceptible weight.

If we balance a quantity of ice in a delicate scale, and then leave it to

---

Questions.—What is temperature? Does the temperature of a body indicate the actual quantity of heat it contains? What is cold? May a body feel hot and cold at the same time? What are sensations of heat and cold? In what manner does heat diffuse itself? Does heat possess weight?

melt, the equilibrium will not be in the slightest degree disturbed. If we substitute for the ice boiling water, or red hot iron, and leave this to cool, there will be no difference in the result. Count Rumford, having suspended a bottle containing water, and another containing alcohol, to the arms of a balance, and adjusted them so as to be exactly in equilibrium, found that the balance remained undisturbed when the water was completely frozen, though the heat the water had lost must have been more than sufficient to have made an equal weight of gold red hot.

81. **Theory of Heat.**—The nature, or cause of heat is not clearly understood. Two explanations, or theories, have been proposed to account for the various phenomena of heat, which are known as the mechanical and vibratory theories.

Mechanical Theory.—The mechanical theory supposes heat to be an extremely subtile fluid, or ethereal kind of matter pervading all space, and entering into combination in various proportions and quantities, with all bodies, and producing by this combination all the various effects noticed.

Vibratory Theory.—The vibratory theory, on the contrary, supposes heat to be merely the effect of a species of motion, like a vibration or undulation, produced either in the constituent particles of bodies, or in a subtile, imponderable fluid which pervades them.

When one end of a bar of iron is thrust into the fire and heated, the other end soon becomes hot also. According to the mechanical theory, a subtile fluid coming out of the fire enters into the iron, and passes from particle to particle until it has spread through the whole. When the hand is applied to the bar it passes into it also, and occasions the sensation of warmth. According to the vibratory theory, the heat of the fire communicates to the particles of the iron themselves, or to a subtile fluid pervading them, certain vibratory motions, which motions are gradually transmitted in every direction, and produce the sensation of heat, in the same way that the undulations or vibrations of air, produce the sensation of sound.

The fact that vibrations do occur in masses of metal and other substances during the passage of heat through them, can be demonstrated by experiment. Thus, if an irregularly curved bar of hot brass be laid upon a support of cold lead, the bar will be thrown into a vibratory state, accompanied

---

QUESTIONS.—What two theories have been proposed to account for the origin of heat? What is the mechanical theory? What is the vibratory theory? Illustrate the supposed production of heat in accordance with the two theories.

by a somewhat musical sound and a rocking motion; and this action continues so long as an inequality of temperature exists between the two metals.

There seems to be but little doubt at the present time among scientific men, that the theory which ascribes the phenomena of heat to a series of vibrations, or undulations, either in matter, or a fluid pervading it, is substantially correct. At the same time it is not wholly satisfactory, and neither theory will perfectly explain *all* the facts in relation to heat with which we are acquainted. For the purpose of describing and explaining the phenomena and effects of heat, it is convenient, in many cases, to retain the idea that heat is a substance.

The fact that nature nowhere presents us, neither has art ever succeeded in showing us, heat alone in a separate state, is a strong ground for believing that heat has no separate material existence.* Heat, moreover, can be produced without limit by friction, and intense heat is also produced by the explosion of gunpowder. On the contrary, as arguments in favor of the material existence of heat, we have the fact, that heat becomes instantly sensible on the condensation of any material mass, as if it were squeezed out of it: as when, on reducing the bulk of a piece of iron by hammering, we render it red hot (the greatest amount of heat being emitted with the blows that most change its bulk).

Finally, the laws of the spreading of heat do not resemble those of the spreading of sound, or of any other motion known to us.

82. **Relations of Light and Heat.**—The relation between heat and light is a very intimate one. Heat exists without light, but all the ordinary sources of light are also sources of heat; and by whatever artificial means natural light is condensed, so as to increase its splendor, the heat which it produces is also, at the same time, rendered more intense.

Incandescence.—When a body, naturally incapable of emitting light, is heated to a sufficient extent to become luminous, it is said to be incandescent, or ignited.

Flame.—Flame is an ignited gas issuing from a burning body. Fire is the appearance of heat and light in conjunction, produced by the combustion of inflammable substances.

The ancient philosophers used the term fire as a characteristic of the nature of heat, and regarded it as one of the four elements of nature; air, earth, and water being the other three.

QUESTIONS.—Which theory is generally received? What relations exist between light and heat? Define incandescence. What is flame? What is fire?

SECTION I.

SOURCES OF HEAT.

83. **Sources of Heat.**—The principal sources of heat of which practical advantage may be taken, are the sun, mechanical action, chemical action, and electricity.

84. **The Sun a Source of Heat.**—The greatest natural source of heat is the sun, as it is also the greatest natural source of light.

Although the quantity of heat sent forth from the sun is immense, its rays, falling naturally, are never hot enough, even in the torrid zone, to kindle combustible substances. By means, however, of a burning-glass, the heat of the sun's rays can be concentrated, or bent toward one point, called a focus, in sufficient quantity to set fire to substances submitted to their action.

Fig. 20.

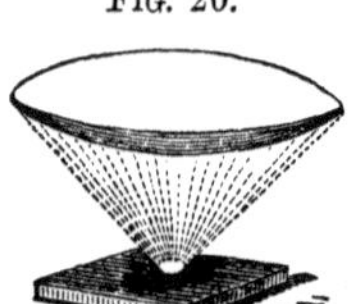

Fig. 20 represents the manner in which a burning-glass concentrates or bends down the rays of heat until they meet in a focus.

The greatest natural temperature ever authentically recorded was at Bagdad, in 1819, when the thermometer (Fahrenheit's) rose to 120° in the shade. On the west coast of Africa the thermometer has been observed as high as 108° F. in the shade. Burckhardt in Egypt, and Humboldt in South America, observed it at 117° F. in the shade.

About 70° below the zero of Fahrenheit's thermometer is the lowest atmospheric temperature ever experienced by the Arctic navigators.

The greatest artificial cold ever measured was 220° F. below zero.

This temperature was obtained some years since by M. Natterer, a German chemist. Professor Faraday has also produced a cold of 166° F. below zero. Neither of these experimenters succeeded in freezing pure alcohol or ether.

The estimated temperature of the space above the earth's atmosphere has been estimated at 58° below zero, Fahrenheit's thermometer.

QUESTIONS.—What are the principal sources of heat? What is the greatest source of heat? What is the greatest natural temperature ever observed? What is the lowest natural temperature observed? What is the greatest artificial cold ever measured?

85. **Mechanical Action**, considered as a source of heat, includes friction and compression, or percussion.

Friction.—The supply of heat which can be derived from friction is apparently unlimited.

Savage nations kindle a fire by the friction of two pieces of dry wood; the axles of wheels revolving rapidly frequently become ignited; and in the boring and turning of metal the chisels often become intensely hot. In all these cases the friction of the surfaces of wood or of metal in contact disturbs the latent heat of these substances, and renders it sensible.

The following interesting experiment was made by Count Rumford, to illustrate the effect of friction in producing heat:—A borer was made to revolve in a cylinder of brass, partially bored, thirty-two times in a minute. The cylinder was inclosed in a box containing 18 pounds of water, the temperature of which was at first 60°, but rose in an hour to 107°; and in two hours and a half the water boiled. The heat thus obtained was calculated to be somewhat greater than that given out during the same period by the burning of nine wax candles, each ¾ths of an inch in diameter.

Recent experiments made by Mr. Joule of England, appear to show that the actual quantity of heat developed by friction is dependent simply upon the amount of force expended, without regard to the nature of the substances rubbed together. He found, as the result of a great number of experiments, that when water was agitated by means of a horizontal brass wheel, which was made to revolve, as the wheels of a clock are, by the descent of a weight, that the temperature of the water was increased by friction against the metal; and that in this way, one pound of water could be raised in temperature one degree by an expenditure of an amount of force sufficient to raise 772 pounds weight to the height of one foot. When cast-iron was rubbed against iron, the force required to produce heat by friction sufficient to elevate the temperature of a pound of water one degree, was found to be equivalent to 775 pounds, and when iron was rubbed against mercury, to 774 pounds.

It thus appears from these experiments, that force expended in producing friction is converted into heat, and that when a pound of water is elevated in temperature one degree, some force equivalent to the raising of a weight of about 772 pounds to the height of one foot is always exerted.*

* This discovery, that heat and mechanical power are mutually convertible, and that the relation between them is definite, 772 foot-pounds of motive power being equivalent to a unit of heat—that is, to the amount of heat requisite to raise a pound of water through one degree of Fahrenheit—is one of the most interesting of modern science, and has led to many important deductions. Thus, force is expended by friction in the ebb and flow of every tide, and must, therefore, reappear as heat. According to the computations of Bessel, the astronomer, 25,000 miles of water flow in every six hours from one quarter of

QUESTIONS.—What does mechanical action, considered as a source of heat, include? What is said of the development of heat by friction? What experiments were made by Count Rumford? What has recently been determined respecting the production of heat by friction?

Compression.—The reduction of matter into a smaller compass by an external or mechanical force, is generally attended with an evolution of heat. To such an act of compression we apply the term condensation.

Heat may be evolved from air by condensation. This may be shown by placing a piece of tinder in a tube, and suddenly compressing the air contained in it by means of a piston. The air being thus condensed, parts with its latent heat in sufficient quantity to set fire to the tinder at the bottom of the tube.

Percussion, which is a combination of friction and compression, is a familiar method of developing heat. An example of this is seen in the use of the common steel and flint, in which the compression occasioned by the violent collision of the two substances elicits heat sufficient to set fire to detached portions of steel. The striking of iron with a hammer, or the subjection of any metal to great and sudden pressure, also develops large quantities of heat.

86. **Chemical Action** is the principal source resorted to for procuring heat artificially. Whenever this occurs with a high degree of intensity, heat is produced, accompanied generally by an evolution of light. A common fire, of wood or coal, is a familiar illustration of the development of heat and light by chemical action.

87. **Electricity.**—The passage of accumulated electricity through various substances, or from one medium to another, generally produces heat. The most intense artificial heat with which we are acquainted, is thus produced by the agency of the electric, or galvanic current. All known substances can be melted or volatilized by it.

Heat so developed has not been employed for practical or economical purposes to any great extent; but for chemical experiments and investigations it has been made quite useful.

88. **Other Sources of Heat.**—In addition to the above-mentioned sources, some heat is derived from the earth

---

the earth to another. The store of mechanical force is thus diminished, and the temperature of our globe augmented by every tide. We do not, however, possess the data which will enable us to calculate the magnitude of these effects.

QUESTIONS.—How may heat be produced by condensation? How by percussion? What is said of chemical action as a source of heat? What of electricity? What other sources of heat are recognized?

itself, and from the stars and planetary bodies. Heat, also, is generated or excited through the organs of a living structure, the result, undoubtedly, of chemical actions which are continually going on in the systems of animals and plants. Heat thus produced is termed vital, or animal heat.

Experimentation has also proved that the simple act of moistening any dry substance is attended with slight, yet constant disengagement of heat. With bodies of mineral origin, when reduced to a fine powder with a view of increasing the extent of surface, the rise of temperature does not exceed from half a degree to two degrees, Fahrenheit's thermometer; but with some animal and vegetable substances, such as cotton, thread, hair, wool, ivory, and well-dried paper, a rise of temperature varying from 2° to even 10° or 14° F. has been observed.

SECTION II.

COMMUNICATION OF HEAT.

89. **Heat** may be communicated in three ways: by CONDUCTION, by CONVECTION, and by RADIATION.

By one or all of these methods, bodies which have been heated, or cooled, gradually return to the temperature of surrounding objects. If the body is hot, heat passes from it to contiguous bodies; if cold, it gains heat at the expense of those substances which possess a higher temperature.

The three methods of communicating heat will be considered in the order above named.

90. **Conduction.**—Heat is said to be communicated by conduction when it is transmitted from particle to particle of a substance, as from the end of an iron bar placed in the fire to that part of the bar most remote from the fire.

Different bodies exhibit a very great degree of difference in the facility or power with which they conduct heat; some substances oppose very little resistance to its passage, while through others it is transmitted slowly, or with great difficulty.

QUESTIONS.—How is heat communicated? In what way is an equilibrium of temperature preserved? What is conduction? Is the power of conduction the same in all bodies?

Fig. 21.

If we place the end of a short rod of glass, and of a rod of iron of equal length, in the flame of a lamp, Fig. 21, we shall soon be sensible that heat reaches the fingers more rapidly through the metal than through the glass; and shall have a clear proof that these two substances differ greatly in their power of conducting heat.

The different conducting power of various solids may be also strikingly shown by taking a series of rods of different materials, but of the same dimensions (see Fig. 22), placing a bit of wax, or phosphorus upon one of their extremities, and applying to the other extremities an equal degree of heat. The wax will melt, or the phosphorus inflame at different times, according to the conducting power of the different solids.

Fig. 22.

91. Conductors and Non-conductors.—All bodies are divided into two classes in respect to their conduction of heat, viz., into conductors and non-conductors. The former are such as allow heat to pass freely through them; the latter comprise those which do not give an easy passage to it.

92. Conduction of Solids.*—Of all known substances, the metals conduct heat with the greatest facility; but they differ considerably when compared with each other. As a general rule, the denser a body is, the better it conducts heat. Light, porous substances, more especially those of a fibrous nature, are the worst conductors of heat. Of all substances, gold is the best conductor of heat, and may

* The following table exhibits the relative conducting power of different substances, the ratio expressing the conducting power of gold being taken at 100:

Substance	Power	Substance	Power
Gold	100·00	Tin	30·38
Platinum	98·10	Lead	17·96
Silver	97·30	Marble	2·34
Copper	89·82	Porcelain	1·22
Iron	37·41	Brick earth	1·13
Zinc	36·37		

Questions.—What experiments illustrate this fact? What are conductors and non-conductors? What are good conductors? What are bad conductors? What substance is the best conductor of heat?

be represented by the number 100; then iron will be 37.4; marble 2.3; and brick clay 1.1.

The conducting power of stones is next to that of the metals, and crystalline stones are better conductors than those which are not crystallized.

93. **Conduction of Liquids.**—Liquids conduct heat in a very limited degree.

This may be satisfactorily proved by a number of simple experiments. If a small quantity of alcohol be poured on the surface of water and inflamed, it will continue to burn for some time. (See Fig. 23.) A thermometer, immersed at a small depth below the common surface of the spirit and the water, will fail to show any increase in temperature.

FIG. 23.

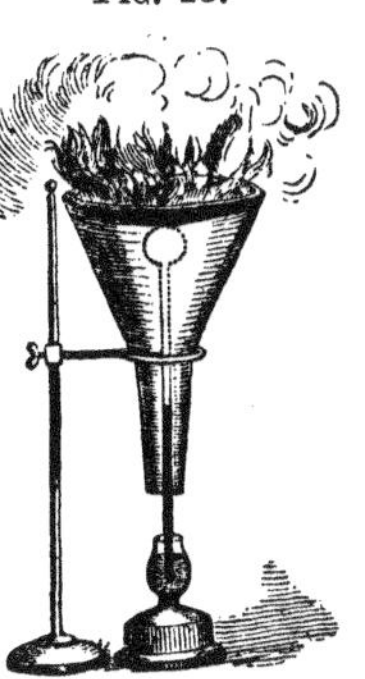

Another and more simple experiment proves the same fact; as when a blacksmith immerses his red-hot iron in a tank of water, the water which surrounds the iron is made boiling hot, while the water not immediately in contact with it remains quite cold.

If a tube nearly filled with water is held over a spirit lamp, as in Fig. 24, in such a manner as to direct the flame against the upper layers of the water, the water at the top of the tube may be kept boiling for a considerable time, without occasioning the slightest inconvenience to the person who holds it.

94. **Conduction of Gases.**—Bodies in the gaseous, or aeriform condition are more imperfect conductors of heat than liquids. Common air, especially, is one of the worst conductors of heat with which we are acquainted.

FIG. 24.

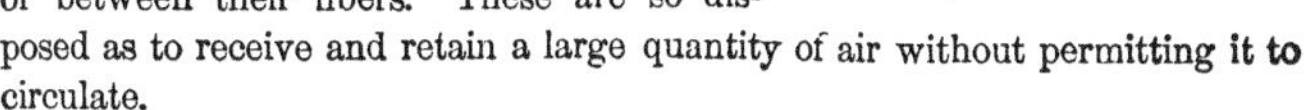

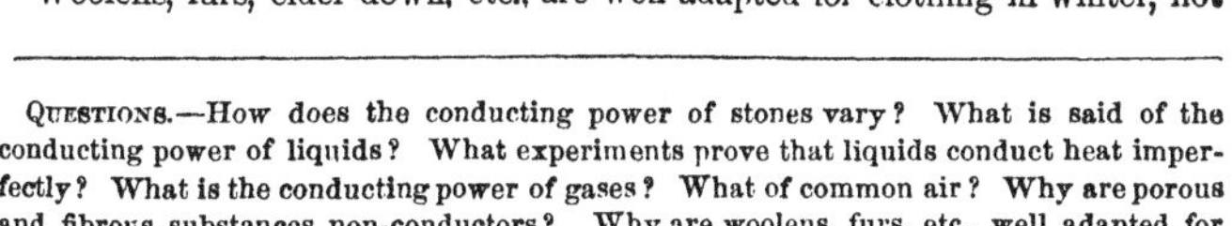

The non-conducting properties of fibrous and porous substances are due almost altogether to the air contained in their interstices, or between their fibers. These are so disposed as to receive and retain a large quantity of air without permitting it to circulate.

Woolens, furs, eider-down, etc., are well adapted for clothing in winter, not

---

QUESTIONS.—How does the conducting power of stones vary? What is said of the conducting power of liquids? What experiments prove that liquids conduct heat imperfectly? What is the conducting power of gases? What of common air? Why are porous and fibrous substances non-conductors? Why are woolens, furs, etc., well adapted for clothing?

because they impart any heat to the body, but on account of the large quantities of air which they contain, imprisoned between their fibers; this renders them non-conductors, and prevents the escape of heat from the body.

Blankets and warm woolen goods are always made with a nap or projection of fibers upon the outside, in order to take advantage of this principle. The nap, or fibers retain air among them, which, from its non-conducting properties, serves to increase the warmth of the material.

The heat generated in the animal system by vital action has constantly a tendency to escape, and be dissipated at the surface of the body, and the rate at which it is dissipated depends on the difference between the temperature of the surface of the body and the temperature of the surrounding medium. By interposing, however, a non-conducting substance between the surface of the body and the external atmosphere, we prevent the loss of heat which would otherwise take place to a greater or less degree.

An apartment is rendered much warmer for being furnished with double doors and windows, because the air confined between the two surfaces opposes by reason of its non-conducting properties, the communication of heat from the interior.

Snow protects the soil in winter from the effects of cold in the same way that fur and wool protect animals, and clothing man. Snow is made up of an infinite number of little crystals, which retain among their interstices a large amount of air, and thus contribute to render it a non-conductor of heat. A covering of snow also prevents the earth from throwing off its heat by radiation. The temperature of the earth, therefore, when covered with snow, rarely descends much below the freezing-point, even when the air is fifteen or twenty degrees colder. Thus roots and fibers of trees and plants are protected from a destructive cold.

As a non-conducting substance prevents the escape of heat from within a body, so it is equally efficacious in preventing the access of heat from without. In an atmosphere hotter than our bodies, the effect of clothing would be to keep the body cool. Flannel is one of the warmest articles of dress, yet we can not preserve ice more effectually in summer than by enveloping it in its folds. Firemen exposed to the intense heat of furnaces and steam-boilers, invariably protect themselves with flannel garments.

Cargoes of ice shipped to the tropics, are generally packed for preservation in sawdust; a casing of sawdust is also one of the most effectual means of preventing the escape of heat from the surfaces of steam-boilers and steam-pipes. Straw, from its fibrous character, is an excellent non-conductor of heat, and is for this reason extensively used by gardeners for incasing plants and trees which are exposed to the extreme cold of winter.

Refrigerators, used for the preservation of animal and vegetable substances in warm weather, are double-walled boxes, with the spaces between the sides filled with powdered charcoal, or some other porous, non-conducting substance.

Questions.—Why are blankets made with a nap? What is the use of clothing? Why do double doors and windows render a room warmer? How does snow protect the soil? Why do persons exposed to intense heat wear flannel? How are refrigerators constructed?

The so-called "fire-proof" safes are also constructed of double or treble walls of iron, with intervening spaces between them filled with gypsum, or "Plaster of Paris." This lining, which is a most perfect non-conductor, prevents the heat from passing from the exterior of the safe to the books and papers within. The idea of applying "Plaster of Paris" in this way for the construction of safes, originated, in the first instance, from a workman attempting to heat water in a tin basin, the bottom and sides of which were thinly coated with this substance. The non-conducting properties of the plaster were so great as to almost entirely intercept the passage of the heat, and the man, to his surprise, found that the water, although directly over the fire, did not get hot.

95. Much of the economy of fuel depends upon a judicious application of the principles which regulate the conduction of heat. An instructive illustration of their importance is exhibited in the manner in which heat may be economized by an appropriate construction of steam-boilers. Thus, one of the most economical forms, which is known as the Cornish, or cylinder boiler, consists of two cylinders, placed one within the other. (See Fig. 25.) Between the two is the space for the water; the inner cylinder contains the furnace, fire-grates, ash-pit, and the flue, or chamber through which the products of combustion pass off. By this arrangement, the heat which would otherwise be conducted away by the fire-bars and the masonry of the ash-pit, is taken up by the surrounding water, and thus economized. The smoke and hot air from the fire also pass through the boiler for its whole length, which is sometimes as much as forty, or even sixty feet, and then return along the outside of the boiler through a chamber of masonry, before they finally escape up the chimney.

FIG. 25.

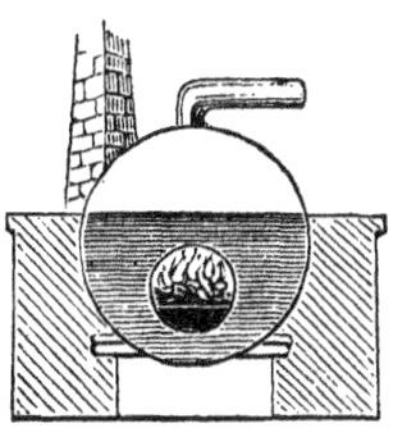

FIG. 26.

In the boiler of a locomotive, Fig. 26, the fire-box is surrounded at the top and two sides by a double casing containing water, and the hot air from the furnace passes through the water in the boiler in numerous small parallel flues, or tubes, which open at one end into the fire-box, and at the other into the smoke-

QUESTIONS.—How are safes rendered fire-proof? Illustrate the application of the principles of conduction in the construction of the cylinder boiler. Also in locomotive boilers.

pipe. By this last arrangement, the heat is, as it were, filtered through the water, and is nearly all communicated to it. Loss of heat from the external surfaces of locomotive-boilers may be also prevented by casing them with wood, or some other non-conducting substance.

96. **Convection.**—Liquids and gases, being non-conductors, can not well be heated like solids, by the communication of heat from particle to particle. Heat, however, is diffused through them with great rapidity by a motion of their particles, which brings them successively in contact with the heated surfaces. This process is termed Convection.

Thus, when heat is applied to the bottom of a vessel containing water, the particles which constitute the lower layers of liquid expand and become lighter, and a double set of currents is immediately established—one of hot particles rising toward the surface, and the other of colder particles descending to the bottom. The portion of liquid which receives heat from below is thus continually diffused through the other parts, and by this motion of the particles the heat is communicated.

Fig. 27.

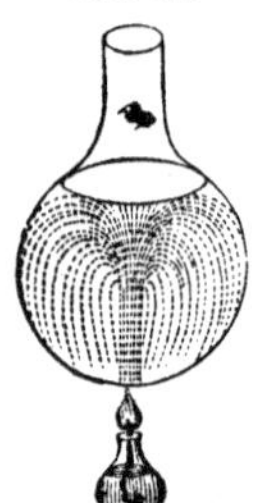

These currents take place so rapidly, that if a thermometer be placed at the bottom and another at the top of a long jar (the fire being applied below), the upper one will begin to rise almost as soon as the lower one. The circulation described may be rendered visible, by adding to a flask of boiling water a small quantity of bran or saw-dust, or a few particles of bituminous coal. (See Fig. 27.)

The process of cooling in a liquid is directly the reverse of that of heating. The particles at the surface, by contact with the air, readily lose their heat, become heavier, and sink, while the warmer particles below in turn rise to the surface.

To heat a liquid, therefore, the heat should be applied at the bottom of the mass; to cool it, the cold should be applied at the top, or surface.

The facility with which a liquid may be heated or cooled, depends in a great degree on the mobility of its particles. Water may be made to retain its heat for a long time by adding to it a small quantity of starch, the particles of which, by their viscidity or tenacity, prevent the free circulation of the heated particles of water. For the same reason soup retains its heat longer than water, and all thick liquids, like oil, molasses, tar, etc., require a considerable time for cooling.

Questions.—What is convection? Illustrate the communication of heat by convection. Explain the process of cooling in liquids. What circumstance greatly influences the heating and cooling of liquids?

97. **Heating of Gases and Vapors.**—Common air, and all gases and vapors, are heated in the same manner as liquids. From every heated substance, an upward current of air is continually rising.

It is in accordance with this principle that we are enabled to readily warm the air of an apartment by means of a stove, or furnace. The air in immediate contact with the hot surface becomes heated and rises, while cooler and heavier air rushes in from all sides to supply its place. This, in turn, becomes heated and ascends, and thus a circulation similar to that which occurs in a flask of boiling water, is established.

98. **Winds and Ocean Currents.**—The processes of circulation produced by heat in liquids and in gases, which have been described, occur upon a vast scale in the atmosphere and in the ocean.

Aërial currents are most powerful at the equator, the warm air of which rises and incessantly flows in the upper regions of the atmosphere toward the poles; while just as constantly, at the earth's surface, currents of cool air, constituting the trade winds, flow from the poles to the equator.

Similar currents are produced by the same cause in the waters of the ocean. Their power may be inferred from the influences which in some cases they exert upon climate. By them the warm water of the Gulf of Mexico is carried to the British Isles, thereby producing a mild, uniform warmth, and a rich moisture; while through similar causes, the floating ice of the North Pole is carried to the coast of Newfoundland, and produces cold.*

99. **Radiation.**—When the hand is placed near a hot body suspended in the air, a sensation of warmth is perceived, even for a considerable distance. If the hand be held beneath the body, the sensation will be as great as upon the sides, although the heat has to shoot down through an opposing current of air approaching it. This effect does not arise from the heat being conveyed by

* Further, by the heat of the sun a portion of the water is converted into vapor, which rises in the atmosphere, is condensed into clouds, or falls as rain or snow upon the earth; collects in the form of springs, brooks, and rivers; and finally reaches the sea again, after having gnawed the rocks, carried away the light earth, and thus performed its part in the geologic changes of the earth; perhaps, beside all this, it has driven our water-mill on its way. If the heat of the sun were withdrawn, there would remain only a single motion of water (provided it remained a liquid), namely, the tides, which are produced by the attraction of the sun and moon.

QUESTIONS.—How are gases and vapors heated? Upon what principle are rooms warmed by stoves and furnaces? What is the occasion of winds and ocean currents? Define radiation.

means of a hot current, since all the heated particles have a uniform tendency to rise; neither can it depend upon the conducting power of the air, because aeriform substances possess that power in a very low degree, while the sensation in the present case is excited almost on the instant. This method of distributing heat, to distinguish it from heat passing by conduction, or convection, is called radiation, and heat thus distributed is termed radiant, or radiated heat.

Heat is communicated through space by radiation in straight lines, and its intensity diminishes as the square of the distance from the center of action increases.

Thus the heating effect of any hot body is nine times less at three feet than at one; sixteen times less at four feet; and twenty-five times less at five.

All bodies radiate heat in some measure, but not all equally well; radiation being generally in proportion to the roughness of the radiating surface. All dull and dark substances are, for the most part, good radiators of heat; but bright and polished substances are generally bad radiators. Color, however, alone, has no effect on the radiation of heat.

A liquid contained in a bright, highly-polished metal pot, will retain its heat much longer than in a dull and blackened one. This is not due to the polish or brightness of the surface, but to the fact that, by polishing, the surface is rendered dense and smooth, and such surfaces do not allow the heat to escape readily. If we cover the polished metal surface with a thin cotton or linen cloth, so as to render the surface less dense, the radiation of heat, and consequent cooling, will proceed rapidly.

Black lead is one of the best known radiators of heat, and on this account is generally employed for the blackening of stoves and hot-air flues. As a high polish is unfavorable to radiation, stoves should not be too highly polished with this substance.

The great supply of heat to the earth from the sun is transmitted by the process of radiation. Some idea of the amount of heat thus received by the earth may be formed from a calculation of Professor Faraday, which indicated that the average amount of heat radiated in a summer's day upon each acre

---

QUESTIONS.—How is heat communicated by radiation? What circumstances influence radiation? What are good and bad radiators? What amount of heat does the earth receive by radiation from the sun?

of land in the latitude of London, is not less than that which would be produced by the combustion of 18,000 pounds of coal.

The radiation of heat goes on at all times, and from all surfaces, whether their temperature be the same as, or different from that of surrounding objects; therefore the temperature of a body falls when it radiates more heat than it absorbs; its temperature is stationary when the quantities emitted and received are equal; and it grows warm when the absorption exceeds the radiation.

If a body, at any temperature, be placed among other bodies, it will affect their condition of temperature, or as we express it, it will affect them *thermally;* just as a candle brought into a room illuminates all bodies in its presence; with this difference, however, that if the candle be extinguished, no more light is diffused by it; but no body can be thermally extinguished. All bodies, however low be their temperature, contain heat, and therefore radiate it.

If a piece of ice be held before a thermometer, it will cause the mercury in its tube to fall, and hence it has been supposed that the ice emitted rays of cold. This supposition is erroneous. The ice and the thermometer both radiate heat, and each absorbs more or less of what the other radiates toward it. But the ice, being at a lower temperature than the thermometer, radiates less than the thermometer, and therefore the thermometer absorbs less than the ice, and consequently falls. If the thermometer placed in the presence of the ice had been at a lower temperature than the ice, it would, for like reasons, have risen. The ice in that case would have warmed the thermometer.

100. **Disposition of Radiant Heat.**—When rays of heat radiated from one body fall upon the surface of another body, they may be disposed of in three ways: 1. They may rebound from its surface, or be reflected; 2. They may be received into its surface, or be absorbed; 3. They may pass directly through the substance of the body, or be transmitted.

101. **Reflection of Heat.**—Polished metallic surfaces constitute the best reflectors of heat; but all bright and light colored surfaces are adapted for this purpose to a greater, or less degree.

Water requires a longer time to become hot in a *bright* tin vessel than in a

QUESTIONS.—Does radiation proceed constantly from all bodies? Why does the mercury of a thermometer sink when brought near ice? When radiant heat falls upon the surface of a body, how may it be disposed of? What surfaces are good reflectors of heat?

dark colored one, because the heat is reflected from the bright surface, and does not enter the vessel.

The power of reflection of heat seems to reside almost exclusively in the surface. A film of gold leaf, not exceeding 1-200,000th of an inch in thickness, answers the purpose of a reflector nearly as well as a mass of solid gold.

102. **Absorption of Heat.**—The power of absorbing heat varies with almost every form of matter. Surfaces are good absorbers of heat in proportion as they are poor reflectors. The best radiators of heat also are the most powerful absorbers, and the most imperfect reflectors.

Dark colors absorb heat from the sun more abundantly than light ones. This may be proved by placing a piece of black and a piece of white cloth upon the snow exposed to the sun; in a few hours the black cloth will have melted the snow beneath it, while the white cloth will have produced little or no effect upon it.

A piece of brown paper submitted to the action of a burning-glass, ignites much more quickly than a piece of white paper. The reason of this is, that the white paper reflects the rays of the sun, and though but slightly heated appears highly luminous; while the brown paper which absorbs the rays, readily becomes heated to ignition. For the same reason a kettle whose bottom and sides are covered with soot, heats water more readily than a kettle whose sides are bright and clean.

Air absorbs heat very slowly, and does not readily part with it. Air is not heated to any extent by the direct rays of the sun. The sun, however, heats the surface of the earth, and the air resting upon it is heated by contact with it, and ascends, its place being supplied by colder portions, which in turn are heated also.

This reluctance of air to part with its heat occasions some very curious differences between its burning temperature and that of other bodies. Metals, which are generally the best conductors, and therefore communicate heat most readily, can not be handled with impunity when raised to a temperature of more than 120° F.; water becomes scalding hot at 150° F.; but air applied to the skin occasions no very painful sensation when its heat is far beyond that of boiling water.

103. **Formation of Dew.**—Dew is the moisture of the air condensed by coming in contact with bodies colder than itself.

As soon as the sun has set in summer, and the earth is no longer receiving new supplies of heat, its surface begins to throw off the heat (which it has

QUESTIONS.—Where does the power of reflecting heat reside in solid bodies? How does the power of absorbing heat vary in different substances? What are good absorbers of heat? What are the peculiarities of air as respects absorption of heat? How is the atmosphere heated? What curious experiments illustrate the retention of heat by the air? What is dew? To what is the formation of dew owing?

accumulated during the day) by radiation; the air, however, does not radiate its heat, and, in consequence, the different objects upon the earth's surface are soon cooled down from 7 to 25 degrees below the temperature of the surrounding atmosphere. The warm vapor of the air, coming in contact with these cool bodies, is condensed and precipitated as dew.

All bodies have not an equal capacity for radiating heat, but some cool much more rapidly and perfectly than others Hence it follows, that with the same exposure, some bodies will be densely covered with dew, while others will remain perfectly dry. Grass, the leaves of trees, wood, etc., radiate heat very freely; but polished metals, smooth stones, and woolen cloth, part with their heat slowly: the former of these substances will therefore be completely drenched with dew, while the latter, in the same situations, will be almost dry.

The surfaces of rocks and barren lands are so compact and hard, that they can neither absorb nor radiate much heat; and (as their temperature varies but slightly) little dew is deposited upon them. Cultivated soils, on the contrary (being loose and porous) very freely radiate by night the heat which they absorb by day; in consequence of which they are much cooled down, and plentifully condense the vapor of the air into dew. Such a condition of things is a remarkable evidence of design on the part of the Creator, since every plant and inch of land which needs the moisture of dew is adapted to collect it; but not a single drop is wasted where its refreshing moisture is not required.

Dew is always formed upon the surface of the material upon which it is found, and does not fall from the atmosphere.

104. **Frost** is frozen dew. When the temperature of the body upon which the dew is deposited sinks below 32° F., the moisture freezes and assumes a solid form, constituting what is called "*frost.*"

105. **Dew-Point.**—The temperature at which the condensation of moisture in the atmosphere commences, or the degree indicated by the thermometer at which dew begins to be deposited, is called the "Dew-Point."

This point is by no means constant or invariable, since dew is only deposited when the air is saturated with vapor, and the amount of moisture required to saturate air of high temperature is much greater than for air of low temperature.

If the saturation be complete, the least diminution of temperature is attended with the formation of dew; but if the air is dry, a body must be

---

QUESTIONS.—Is dew deposited equally upon all substances? Does dew fall? What is frost? What is the dew-point?

several degrees colder before moisture is deposited on its surface; and indeed the drier the atmosphere, the greater will be the difference between the temperature and its dew-point.

Dew may be produced at any time by bringing a vessel of cold water into a warm room. The sides of the vessel cool the surrounding air to such an extent that it can no longer retain all its vapor, or, in other words, the temperature of the air contiguous to the cold surface is reduced below the dew-point; dew therefore forms upon the vessel. A pitcher of water under such circumstances is vulgarly said to "sweat."

106. **Transmission of Heat.**—Heat derived from the sun, like light emanating from the same source, passes through all transparent bodies, without material loss; but heat derived from terrestrial and less intense sources, is in great part arrested by many substances, which allow light to pass freely.

Thus, a plate of glass held between one's face and the sun will not protect it, but held between the face and a fire, it will intercept a large proportion of the heat.

The power of heat to penetrate a dense transparent substance increases in proportion as the temperature of the body from which it is radiated increases.

Rock-salt appears to be the only substance which transmits an equal amount of heat from all sources. It has, hence, been called the "glass of heat," since it permits heat to pass with the same ease that glass does light. Alum, on the contrary, which is nearly transparent, almost entirely intercepts the passage of terrestrial heat. Heat, indeed, will pass more readily through a black glass, so dark that the sun at noonday is scarcely discernible through it, than through a thin plate of clear alum.

Transparent substances of considerable density, such as glass, alum, water, rock-crystal, etc., interfere most with the passage of heat; while transparent substances of little density, as air, the various gases, etc., allow heat to pass with comparatively little interruption.

Those substances which transmit heat most freely, are termed *diathermanous;* and those which intercept the rays of heat more or less completely, *athermanous.*

QUESTIONS.—State the peculiarities which distinguish the transmission of heat derived from different sources? Upon what does the power of heat to penetrate a substance depend? What substance transmits heat most readily? What least so? What terms have been used to indicate the difference in bodies as respects the transmission of heat?

SECTION III.

THE EFFECTS OF HEAT.

107. **Universal Influence of Heat.**—The form of all bodies appears to be materially affected by heat; by its increase solids are converted into liquids, and liquids into vapor; by its diminution vapors are condensed into liquids, and these in turn become solids.

If matter ceased to be influenced by heat, all liquids, vapors, and doubtless even gases, would become permanently solid, and all motion on the surface of the earth would be arrested.

108. **Specific Heat.**—All bodies contain incorporated with them more or less of heat; but equal weights of dissimilar substances require unequal quantities of heat to elevate them to the same temperature.

Thus, if we place a pound of water and a pound of mercury over a fire, it will be found that the mercury will attain to any given temperature much quicker than the water. Or if we perform the converse of this experiment, and take two equal quantities of mercury and water, and having heated them to the same degree of temperature, allow them to cool freely in the air, it will be found that the water will require much more time to cool down to a common temperature than the mercury. The water obviously contains more heat at the elevated temperature than the mercury, and therefore requires a longer time to cool.

Dissimilar substances require, respectively, different quantities of heat to raise their temperature one degree; and the quantity of heat required to raise any substance one degree in temperature, as compared with the quantity required to raise an equal weight of some other substance, selected as a standard of comparison, one degree, is called its specific heat. In like manner, the weight which a body includes under a given volume, is termed its specific weight. Water is adopted as the standard for comparing the different quantities of heat which equal weights of dissimilar substances contain.

QUESTIONS.—What is said respecting the universal influence of heat? Is the same amount of heat contained in all substances? What experiment proves that water contains more heat than mercury? What is specific heat? What standard is adopted for comparing the heat of different substances?

109. **Capacity for Heat.**—A substance is said to have a greater, or less capacity for heat, according as a greater, or less quantity of heat is required to produce a definite change of temperature, or an elevation of temperature of one degree.

In general, the capacity of bodies for heat decreases with their density. Thus mercury has a less capacity for heat than water, because its density is greater. Air that is rarefied, or thin, has a greater capacity for heat than dense air. This circumstance will explain, in part, the reason of the very low temperatures which exist at great elevations in the atmosphere. Persons ascending high mountains, or in balloons, find that the cold increases with the elevation. The reason of this is, that the air in the upper regions of the atmosphere, relieved from superincumbent pressure, is expanded and rarefied; its capacity for heat is, therefore, greatly increased, and it absorbs its own sensible heat.

In all quarters of the globe, the temperature of the air at a certain height is reduced so low by its rarefaction, that water can not exist in a liquid state. This limit, the height of which varies, being the most elevated at the equator, and the most depressed at the poles, is called the line of PERPETUAL SNOW.*

If compressed air be allowed suddenly to expand, by escaping into the atmosphere, the rarefaction produced increases its capacity for heat; it, therefore, absorbs heat most readily, and occasions a sensation of cold. It is on this account that air forcibly expelled from the mouth feels cool.

On the contrary, if we compress a quantity of air, and render it more dense, we diminish its capacity for heat, and it becomes incapable of retaining what was before incorporated into its substance. The proof of this may be found in the fact, that by the sudden compression of a small quantity of air in a suitable vessel we may obtain a sufficient amount of heat to ignite tinder and other inflammable substances.

The capacity for heat increases with the temperature. Thus it requires a greater amount of heat to elevate the temperature of platinum from 212° to 213°, than from 32° to 33°.

A body in a liquid state has a higher specific heat than the same substance when it is in the solid form.

---

* The line of perpetual snow at the equator occurs at a height of about 15,000 feet; at the Straits of Magellan, it occurs at an elevation of only 4,000 feet.

QUESTIONS.—What is understood by capacity for heat? How does the capacity of bodies for heat increase? Why is the temperature of air at high elevations very much reduced? Why does the compression of air produce heat? How does the capacity for heat vary with the temperature?

This is remarkably shown in the case of water, the specific heat of which is double that of ice.

Of all known substances, water has the greatest capacity for heat. This circumstance renders the ocean a great reservoir of heat, and a regulator of temperatures upon the surface of the earth. Thus in hot weather, the water of the ocean, on account of its great capacity for heat, absorbs and retains large quantities from the air; the air, therefore, accumulates heat but slowly. In cold weather, the heat previously absorbed by the ocean is gradually restored to the air, and a sudden reduction of atmospheric temperature is prevented. It is, therefore, mainly on this account that sea-coasts and islands enjoy a more uniform temperature than the interior of continents. In the summer, the proximity of the sea serves to mitigate the heat; in the winter, to diminish the cold. Inland lakes, in like manner, raise the mean temperature. The climate of the shores of Lake Erie is much milder than that of the adjacent inland country, and fruit may be successfully cultivated at Cleveland, upon the southern shore, which fails to ripen in districts further south.

An ocean of mercury would produce very different results, since it is capable of absorbing but a small amount of heat, which it readily parts with at a slight reduction of temperature.

110. **Cal-o-rim'e-try.**—The art of determining the specific heat of various substances is called Calorimetry.

Several different methods may be employed for this purpose. One method consists in inclosing equal weights of different substances, heated to the same temperature, in closed cavities in a block of ice, and measuring the respective quantities of water which they produce by melting the ice.

The same result may also be obtained by what is called the method of mixtures. Thus, if we mix 1 pound of mercury at 66° with 1 pound of water at 32°, the common temperature will be 33°. Here the mercury loses 33° and the water gains 1°; that is to say, the 33° of the mercury only elevates the water 1°, therefore the capacity of water for heat is 33 times that of mercury; or, if we call the capacity or specific heat of water 1, then the capacity, or specific heat of mercury, as compared with water, will be 1-33d, or .303.

In this way the specific heat of a great number of bodies has been determined, and tables constructed in which they are recorded.

111. **Apparent Effects of Heat.**—The three most apparent effects of heat, so far as they relate to the form and dimensions of bodies, are Expansion, Liquefaction, and Vaporization.

112. **Theory of Expansion.**—Heat operates to produce

QUESTIONS.—What substance has the greatest capacity for heat? How do great bodies of water serve to regulate temperature? What is calorimetry? How is the specific heat of bodies determined? What are the three most apparent effects of heat? How does heat produce expansion?

expansion by introducing a repulsive force among the particles of the body it pervades. This repulsive force gives to the particles a tendency to separate, or increase their distance from one another. Hence the general mass of the body is made to occupy a larger space, or expand.

The expansion occasioned by heat is greatest in those bodies which are the least influenced by cohesion. Solids expand less for equal elevations of temperature than either liquids or gases.

The expansion of the same body will continue to increase with the quantity of heat that enters it, so long as the form and chemical constitution of the body is preserved.

113. **Expansion of Solids.**—Solids appear to expand uniformly from the freezing point of water up to 212°, the boiling point of water;—that is to say, the increase of volume which attends each degree of temperature which the body receives is equal. When solids are elevated, however, to temperatures above 212°, they do not dilate uniformly, but expand in an increasing ratio.

Different solids, however, expand very unequally for equal additions of temperature.

Among solids the metals expand the most; but an iron wire increases only 1-282 in bulk when heated from 32° of the thermometer up to 212°. Zinc is the most expansible of the metals, and platinum the most uniform in its rate of expansion at all temperatures. Wood and marble expand but slightly.

Fig. 28.

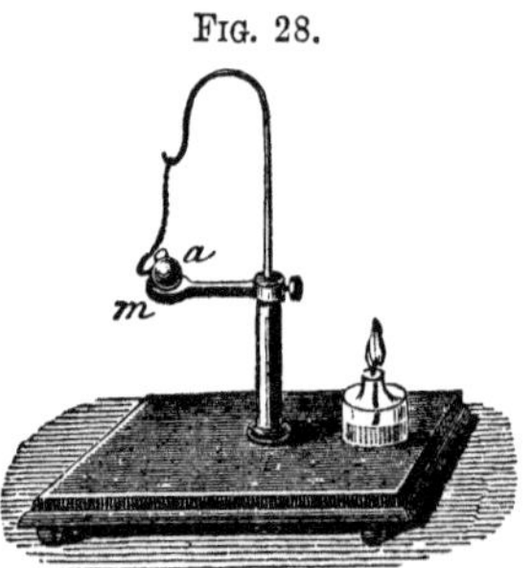

The expansion of solids by heat is clearly shown by the following experiment, Fig. 28. *m* represents a ring of metal, through which, at the ordinary temperature, a small iron or copper ball, *a*, will pass freely, this ball being a little less than the diameter of the ring. If this ball be now heated by the flame of an alcohol lamp, it will expand by heat to such an extent as no longer to pass through the ring.

Questions.—What bodies expand most under the influence of heat? Is the expansion of bodies by heat limited? What is the law of expansion for solids?

Bodies, in general, expanded under the influence of heat, return to their original dimensions in cooling.

Lead, however, is an exception to this rule. From its extreme softness, its particles slide over each other in the act of expansion, and do not return to their original position. "A leaden pipe, used for conveying steam, permanently lengthens some inches in a short time, and the leaden flooring of a sink, which often receives hot water, becomes, in the course of use, thrown up into ridges and puckers."

114. Force of Expansion.—The force with which bodies expand and contract under the influence of the increase or diminution of heat, is apparently irresistible, and is recognized as one of the greatest forces in nature.

The amount of force with which a solid body will expand or contract through the influence of heat, is equal to that which would be required to compress it by mechanical means through a space equal to its expansion, or elongate it through a space equal to its contraction.

A bar of malleable iron, having a section of a square inch, is stretched 1-10,000th of its length by a ton weight; a similar elongation is produced by the influence of about sixteen degrees of heat, Fahrenheit. In this climate, a variation of 80° F. between the cold of winter and the heat of summer not unfrequently takes place. Within these limits, a wrought iron bar ten inches long will vary in length 5-1,000th of an inch; and is capable of exerting a strain of fifty tons upon a square inch.

Experiments made a few years since demonstrated, that Bunker Hill monument is caused to vary each day from a vertical position, by the heat of the sun expanding unequally the granite of which it is constructed.

The expansion of solids by heat is made applicable for many useful purposes in the arts. The tires of wheels, and hoops surrounding water-vats, barrels, etc., are made in the first instance somewhat smaller than the framework they are intended to surround. They are then heated red hot and put on in an expanded condition; on cooling, they contract and bind together the several parts with a greater force than could be conveniently applied by any mechanical means. In like manner, in constructing steam-boilers, the rivets are fastened while hot, in order that they may, by subsequent contraction, bind the plates together more firmly.

In many operations, however, the force of expansion requires to be care-

QUESTIONS.—Is expansion by heat counteracted by cooling? With what force do bodies expand and contract by the increase, or diminution of heat? Mention some instances of expansion in the arts? In what cases is it necessary to guard against the expansion of solids?

fully guarded against. This is especially the case when iron is combined in any structure with less expansible materials.

Iron clamps and bars, built into walls of masonry, frequently weaken, or destroy, by their expansion and contraction, the structure they were intended to support. Iron pipes used for the conveyance of steam or hot water, should not be allowed to abut against a wall, or substance which might be moved, or injured by their expansion.

115. Expansion of Liquids.—Liquids expand through the agency of heat more unequally, and to a much greater extent than solids.

A column of water contained in a cylindrical glass vessel will expand 1-23d in length on being heated from the freezing to the boiling point, while a column of iron, with the same increase of temperature, will expand only 1-846th.

A familiar illustration of the expansion of water by heat is seen in the overflow of full vessels before boiling commences.

Different liquids expand very unequally with an equal increase in temperature.

This may be illustrated by partially filling several glass tubes furnished with bulbs, with different liquids, as ether, alcohol, water, and sulphuric acid, and placing them in a vessel of hot water. Their different rates of expansion will cause them to rise to different heights in the tubes. (See Fig. 29.)

Fig. 29.

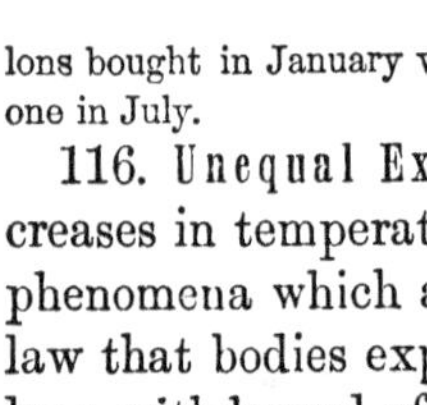

Spirits of wine, on being heated from 32° to 212°, increase in bulk one ninth; the oils expand one twelfth, and water gains one twenty-third. A person buying oil, molasses, and spirits in winter will obtain a greater weight of the same material, in the same measure, than in summer. Spirits, in the height of summer, will measure five per cent. more than in the depth of winter, or twenty gallons bought in January will, under ordinary circumstances, become twenty-one in July.

116. Unequal Expansion of Water.—Water, as it decreases in temperature toward the freezing point, exhibits phenomena which are wholly at variance with the general law that bodies expand by heat and contract by cold, or by a withdrawal of heat.

As the temperature of water is lowered, it continues to contract until it arrives at a temperature of 39° F., when all further contraction ceases. The volume or bulk is observed to remain stationary for a time, but on lowering

---

QUESTIONS.—What is said of the expansion of liquids? What peculiarities of expansion does water exhibit?

the temperature still more, instead of contraction, expansion is produced, and this expansion continues at an increasing rate until the water is congealed.

Water attains its greatest density, or the greatest quantity is contained in a given bulk, at a temperature of 39° F.

As the temperature of water continues to decrease below 39°, the point of its greatest density, its particles, from their expansion, necessarily occupy a larger space than those which possess a temperature somewhat more elevated. The coldest water, therefore, being lighter, rises and floats upon the surface of the warmer water. On the approach of winter this phenomenon actually takes place in our lakes, ponds, and rivers. When the surface water becomes sufficiently chilled to assume the form of ice, it becomes still lighter, and continues to float. By this arrangement, water and ice being almost perfect non-conductors of heat, the great mass of the water is protected from the influence of cold, and prevented from becoming chilled throughout.

A few other liquids beside water expand with a reduction of temperature. Fused iron, antimony, zinc, and bismuth, are examples of such expansion. Mercury is a remarkable instance of the reverse, for when it freezes, it suffers a very great contraction.

The ordinary temperature at which water freezes is 32°, Fahrenheit's thermometer. This rule applies only to fresh water; salt water never freezes until the surface is cooled down to 27°, or five degrees lower than the freezing point of fresh water.

117. **Expansion of Gases.**—Gases are more expansible by heat than either solids, or liquids. All gases and all vapors, except at the point of condensation, are expanded equally by the application of equal additions of heat. The rate of expansion is equal to the 1-490th of the bulk which a gas possesses at 32° F. for every degree of heat which it receives above that point, and for every degree of heat withdrawn from them a contraction to an equal amount takes place.

Thus 490 cubic inches of air at 32° F. becomes 491 cubic inches at 33° F.; at 34° F., 492 cubic inches; at 35°, 493, and so on—the addition of every degree of heat increasing its bulk one cubic inch. In a like manner, by the withdrawal of heat, 490 cubic inches of air would occupy an inch less space at 31° than at 32°; two inches less at 30°, and so on.

By means of this law we can easily calculate the amount of space which a

---

QUESTIONS.—At what temperature does water possess the greatest density? What beneficial results attend the expansion of water in freezing? Do any other liquids expand in cooling beside water? At what temperature does water freeze? In what manner do gases expand? What law governs the expansion of gases? How can we calculate the amount of space a gas occupies at a given temperature?

given volume of gas will occupy, when heated up to any particular temperature; or the contraction which will take place in its volume through a reduction of temperature. A given volume of air possessing the temperature of freezing water, will occupy double the space when heated 490 degrees; and three times the space when heated 980 degrees.

118. **Theory of Heat Measurement.**—As the magnitude of every body changes with the heat to which it is exposed, and as the same body, when subjected to calorific influences under the same circumstances has always the same magnitude, the expansions and contractions which are the constant effects of heat, may be taken as the measure of the cause which produced them.

The instruments for measuring heat are Thermometers and Pyrometers. The former are used for measuring moderate temperatures: the latter for determining the more elevated degrees of heat.

Liquids are better adapted than either solids or gases for measuring the effects of heat by expansion and contraction; since in solids the direct expansion by heat is so small as to be seen and recognized with difficulty, and in air or gases it is too extensive, and too liable to be affected by variations in the atmospheric pressure. From both of these disadvantages liquids are free.

The liquid generally used in the construction of thermometers is mercury, or quicksilver.

Mercury possesses greater advantages for this purpose than any other liquid. It is, in the first place, eminently distinguished for its fluidity at all ordinary temperatures; it is, in addition, the only body in a liquid state whose variations in volume, or magnitude, through a considerable range of temperature are exactly uniform and proportional with every increase and diminution of heat. Mercury, moreover, boils at a higher temperature than any other liquid, except certain oils; and, on the other hand, it freezes at a lower temperature than all other liquids, except some of the most volatile, such as ether and alcohol. Thus a mercurial thermometer will have a wider range than any other liquid thermometer. It is also attended with this convenience, that the extent of temperature included between melting ice and boiling water stands at a considerable distance from the limits of its range, or its freezing and boiling points.

119. **The Mercurial Thermometer** (see Fig. 30) consists

QUESTIONS.—What is the theory of heat measurement? What are the instruments for measuring heat called? Why are liquids best adapted for indicating by expansion and contraction the effects of heat and cold? What liquid is generally employed? What are the advantages of mercury for this purpose? Describe the mercurial thermometer?

essentially of a glass tube with a bulb at one end, partially filled with mercury. The mercury, introduced through an opening in the end of the tube, is afterward boiled, so as to expel all air and moisture, and fill the tube with its own vapor. The open end of the tube is then closed, by fusing the glass, and as the mercury cools it contracts, and collects in the bulb and lower part of the tube, leaving a vacuum above, through which it may again expand and rise on the application of heat. In this condition the thermometer is complete, with the exception of graduation.

Fig. 30.

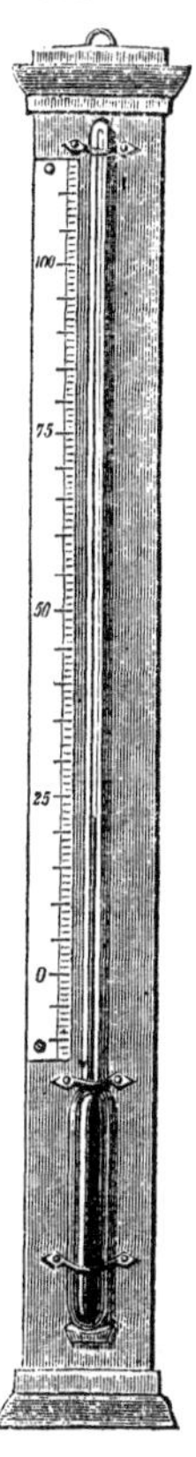

120. **Graduation of Thermometers.**—As thermometers are constructed of different dimensions and capacities, it is necessary to have some fixed rules for graduating them, in order that they may always indicate the same temperature under the same circumstances, as the freezing point, for example. To accomplish this end the following plan has been adopted:—The thermometers are first immersed in melting snow or ice. The mercury will be observed to stop in each thermometer-tube at a certain height; these heights are then marked upon the tubes. Now it has been ascertained that at whatever time and place the instruments may be afterward immersed in melting snow or ice, the mercury contained in them will always fix itself at the point thus marked. This point is called the freezing point of water.

Another fixed point is determined by immersing the instruments in boiling water. It has been found that at whatever time or place the instruments are immersed in pure water, when boiling, provided the barometer stands at the height of thirty inches, the mercury will always rise in each to a certain height. This, therefore, forms another fixed point on the scale, and is called the boiling point.

So far as the determination of the boiling and freezing points of water are concerned, all the varieties of the mercurial thermometer are constructed alike. The interval, however, between these two fixed points is differently divided in different instruments.

Questions.—How are thermometers graduated? In what respect are all thermometers alike?

121. Fahrenheit's Thermometer.—In the thermometer most generally used in the United States and England, and which is known as Fahrenheit's, the interval on the scale between the freezing and boiling points, is divided into 180 equal parts. This division is similarly continued below the freezing point to the place 0, called zero, and each division upward from that is marked with the successive numbers 1, 2, 3, etc. The freezing point will now be the 32d division, and the boiling point will be the 212th division. These divisions are called degrees, and the boiling point will therefore be 212°, and the freezing temperature, 32°.

Thermometers of this character are called Fahrenheit's, from a Dutch philosophical instrument-maker who first introduced this method of graduation in the year 1724.

"The zero of a thermometric scale has no relation to the *real zero* of heat, or the point at which bodies are entirely deprived of heat. Of this point we know nothing, and there is no reason to suppose we have ever approached it. The scale of temperature may be compared to a chain, extended both upward and downward beyond our sight. We fix upon a particular link, and count upward and downward from that link, and not from the beginning of the chain."—GRAHAM.

In indicating thermometrical degrees, the sign — is used to designate those below the zero point, in order to distinguish them from degrees of the same name above the zero point. Thus, 32° means the 32d degree above zero; and —32° the 32d below zero.

FIG. 31.

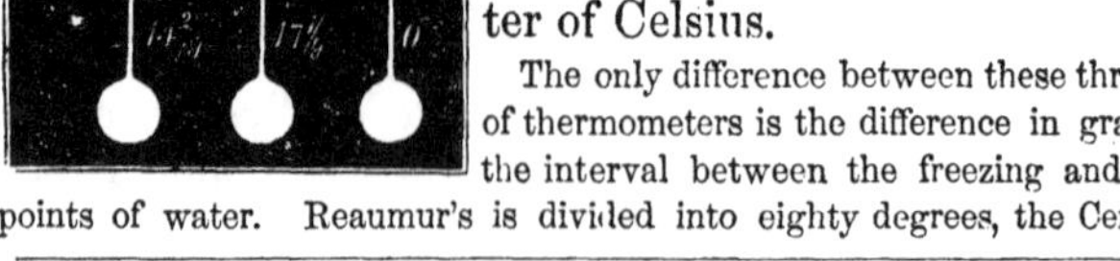

122. Reaumur and Centigrade Thermometers.—In addition to Fahrenheit's thermometer, two others are extensively used, which are known as Reaumur's, and the Centigrade thermometer, or thermometer of Celsius.

The only difference between these three kinds of thermometers is the difference in graduating the interval between the freezing and boiling points of water. Reaumur's is divided into eighty degrees, the Centigrade

QUESTIONS.—Describe the graduation of Fahrenheit's thermometer. What does the zero point indicate? How are degrees below zero distinguished? What other scales are used? Describe the graduation of Reaumur; of Centigrade.

into one hundred, and Fahrenheit's into one hundred and eighty. According to Reaumur, water freezes at 0°, and boils at 80°; according to Centigrade, it freezes at 0°, and boils at 100°; and according to Fahrenheit, it freezes at 32°, and boils at 212°; the last, very singularly, commences counting, not at the freezing point, but 32° below it.*

The difference between these instruments can be easily seen by reference to Fig. 31.

In England, Holland, and the United States, the thermometer most generally used is Fahrenheit's. Reaumur's scale is used in Germany, and the Centigrade in France, Sweden, and some other parts of Europe. The scale of the Centigrade is by far the simplest and most rational method of graduation, and at present it is almost universally adopted for scientific purposes.

The scale employed in the present work is that of Fahrenheit's.

The thermometer was invented about the year 1600; but, like many other inventions, the merit of its discovery is not to be ascribed to one person, but to be distributed among many.

The variety of circumstances under which thermometers are used, have occasioned a considerable variety in their form. The following are some of the most important of these modifications.

123. **The Self-Registering Thermometer** is a form of thermometer contrived for the purpose of ascertaining the extremes of variation which may occur during a particular interval of time, as in the night.

FIG. 32.

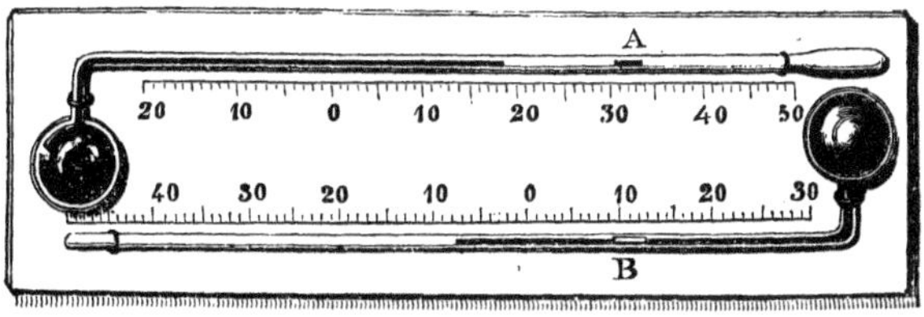

It consists of two horizontal thermometers attached to one frame, as is represented in Fig. 32, the one, A, containing mercury, and the other, B, spirits of wine. On the surface of the mercurial column in the tube is placed a

* The temperatures expressed by one thermometer scale may be easily reduced to that of another, by remembering that 9° of Fahrenheit are equivalent to 5° of Centigrade, or 4° of Reaumur. In converting Fahrenheit to Reaumur, or Centigrade, if the degree be above the freezing point, 32° must be first subtracted, in order to reduce the degrees of the other scales to those of Fahrenheit; but in the conversion of Reaumur or Centigrade to Fahrenheit, 32° must be added.

QUESTIONS.—When was the thermometer invented? What is a self-registering thermometer? Describe its construction?

piece of steel-wire, and on the surface of the spirits of wine, a piece of black enamel, or ivory. As the spirits contract by exposure to cold, the enamel follows it toward the bulb; but when it expands, the enamel remains stationary, and suffers the liquid to pass by it. When the mercury contracts, the steel-wire does not follow; but when the mercury expands, it is forced along. Consequently, it remains at the highest temperature. The position, therefore, of the two indices will indicate the lowest and highest temperatures during any given time. A simpler form of thermometer for indicating maximum temperature, has been constructed by Messrs. Negretti and Zambra, of London, and is known by their names. It is merely an ordinary thermometer, placed horizontally, with a contraction in the tube just above the bulb. When the mercury expands through heat, the expansive force pushes the column past the contraction without difficulty; but when the temperature falls, and the expansive force ceases to act, the contraction in the tube prevents the column from receding. The position of the mercury above the contraction indicates, therefore, the highest temperature attained since the last observation. The mercurial column is restored to its true place, by a slight percussion of the instrument.

124. The Differential Thermometer is a form of thermometer so named because it denotes only differences of temperature between two substances, or two contiguous portions of the same atmosphere.

FIG. 33.

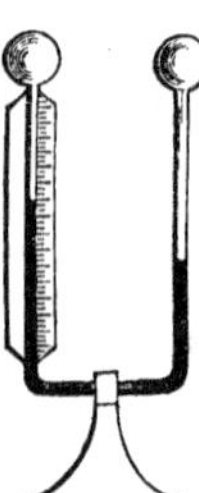

It consists of two glass bulbs on one tube, bent twice at right angles, and supported as represented in Fig. 33. The bulbs contain air, but the tube is nearly filled with sulphuric acid colored red. To one leg of the tube is applied a scale. When the bulbs of this instrument are heated or cooled alike, no change will take place in the columns of liquid, because the air in both bulbs will undergo an equal expansion or contraction; but the instant any inequality of temperature exists between them, as from bringing a heated substance near to one of them, the liquid in the two legs will rise and fall rapidly.

125. Metallic Thermometer.—A very delicate thermometer, known as the metallic, or Breguet's thermometer, is constructed on the principle of the unequal expansion of two metals.

It consists (see Fig. 34) of two equal strips of platinum and silver, firmly soldered together and coiled in the form of a spiral. One end of the spiral is suspended from a fixed point, while the lower end is free and carries an index. Variations of temperature cause the two metals to expand and con-

---

QUESTIONS.—What is the thermometer of Negrettó and Zambra? What is a differential thermometer? Describe its construction. Describe the metallic, or Breguet's thermometer.

tract unequally, and the spiral to twist in opposite directions. These motions imparted to the index, cause it to move over a graduated circle, on which degrees are indicated. So sensitive is this instrument, that when inclosed in a large receiver, which was rapidly exhausted by an air pump, it indicated a reduction of temperature from 66° to 25°=41°, while a mercurial thermometer fell only to 36°.

FIG. 34.

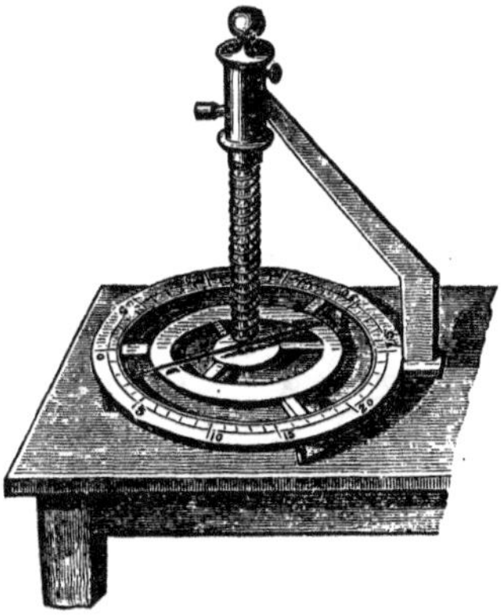

For chemical purposes, thermometers are sometimes constructed in such a way, that the lower part of the scale turns up by a hinge, in order to allow the bulb to be immersed in corrosive liquids. (See Fig. 35.)

126. **Air Thermometers.**—The first thermometer used consisted of a column of air confined in a glass tube over colored water. Heat expands the air and increases the length of the column downward, pushing the water before it: cold produces a contrary effect. The temperature is thus indicated by the height at which the water is elevated in the tube. Fig. 36 represents the principle of the construction of the air thermometer.

FIG. 35.

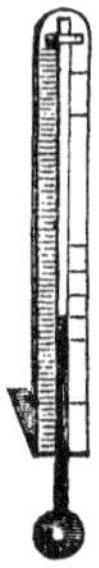

FIG. 36.

Fig. 37 represents an air thermometer filled up with a scale, and termed the thermometer of Sanctorius, from its inventor.

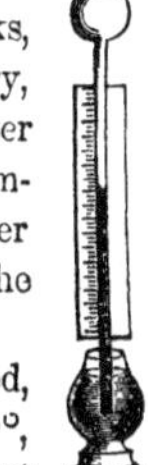
FIG. 37.

127. **Spirit Thermometers.**—As the temperature is lowered, the mercury in Fahrenheit's thermometer gradually sinks, until it reaches a point 39° below zero, where it freezes. Mercury, therefore, can not be made available for measuring cold of a greater intensity. This difficulty is, however, obviated by using a thermometer filled with alcohol colored red, as this fluid, when pure, never freezes, but will continue to sink lower and lower in the tube as the cold increases. Such a thermometer is called a spirit thermometer.

128. **Pyrometers.**—If a Fahrenheit's thermometer be heated, the mercury contained in it will rise in the tube until it reaches 660°, at which temperature it begins to boil. A slight additional heat forms vapor sufficient to burst the tube. Mercury, therefore, can not be used

QUESTIONS.—What was the first thermometer used? How is cold of great intensity indicated? How is heat of great intensity measured? Describe the principle upon which the pyrometer is constructed.

to measure degrees of heat of greater intensity than 660° F. Temperatures greater than this are determined by means of the expansion of solids; and instruments founded upon this principle are commonly called pyrometers.

Fig. 38.

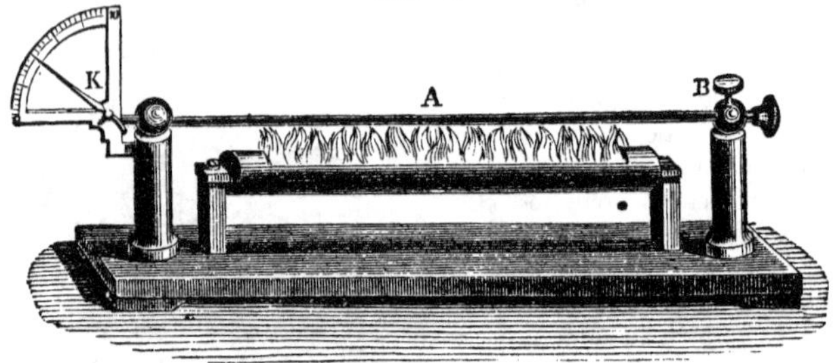

The principle of the construction of the pyrometer is shown in Fig. 38. A represents a metallic bar, fixed at one end, B, but left free at the other, and in contact with the end of a pointer, K, moving freely over a graduated scale. If the bar be heated by the flame of alcohol, the metal expands, and pressing upon the end of the pointer, moves it, in a greater or less degree. In this manner, the effect of heat, applied for a given length of time, to bars of different metals, having the same length and diameter, may be determined.

The pyrometer of practical use is known as Daniel's, and consists of a platinum bar inclosed in a tube of black-lead, closed at the bottom. The whole is then placed in the fire, or in a mass of melted metal, whose temperature it is desirable to ascertain. The platinum expands much more than the case which incloses it, and projecting upward, moves a lever, which drives forward an index over a graduated arc.

A thermometer does not inform us how much heat any substance contains, but it merely points out the difference in the temperature of two or more substances. All we learn by the thermometer is whether the temperature of one body is greater or less than that of another; and if there is a difference, it is expressed numerically—namely, by the degrees of the thermometer. It must be remembered that these degrees are part of an arbitrary scale, selected for convenience, without any reference whatever to the actual quantity of heat present in bodies.

129. Fluidity as an Effect of Heat.—The first effect produced by heat upon solids is expansion. If the heat be augmented, they change their aggregate state, and melt, or become liquid. Many solids become soft before melting, so that they may be kneaded; for instance, wax,

Questions.—Describe Daniel's pyrometer. Does the thermometer inform us how much heat a body contains? After the expansion of bodies by heat, what other effects are next observed?

glass, and iron. In this position, glass can be bent and molded with facility, and iron can be forged or welded.

130. **Liquefaction.**—By liquefaction we understand the conversion of a solid into a liquid by the agency of heat, as solid ice is converted into water by the heat of the sun.

The temperature at which liquefaction takes place is called the melting point, or point of fusion; and that at which liquids solidify, the freezing point, or point of congelation.

The melting point of a given solid is always fixed and constant, but the degree of heat at which different solids melt varies exceedingly.

Thus, platinum is not melted at 3280°; iron melts at about 2800°; lead at 612°; wax, 142°; tallow, 92°; olive oil, 36°; ice, 32°; milk, 30°; oil of turpentine, 14°; mercury, — 39°; liquid ammonia, — 46°; while pure alcohol, having never been solidified, possesses no known melting point.

131. **Vaporization.**—By vaporization is meant the conversion of liquid and solid substances into vapor, through the agency of heat. Thus water, if heated sufficiently, will be converted into steam. It is generally supposed that all solid and liquid substances, under the influence of a sufficient degree of heat, are susceptible of this change.

A gas differs from a vapor in the circumstance that it is not so easily condensed into a liquid, but permanently retains its state under all ordinary conditions of temperature and pressure.

132. **Condensation.**—If a body in a state of vapor lose heat in sufficient quantity, it will pass into a liquid or solid state. Thus, if a certain quantity of heat be abstracted from steam, it will become water. This change is called CONDENSATION.

The change from a state of vapor to a liquid is termed condensation, because, in so doing, the body always undergoes a very considerable diminution of volume, and therefore becomes condensed.

133. **Volatile and Fixed Bodies.**—Substances according to the facility with which they yield vapor, are said to be volatile, or fixed and non-

---

QUESTIONS.—What is liquefaction? What is said respecting the melting point of bodies? What is vaporization? How does a gas differ from a vapor? What is meant by the term condensation as applied to vapors? What is sublimation? What is the distinction between fixed and volatile substances?

volatile. A volatile substance is one which yields vapor readily by the application of heat, and wastes away on simple exposure to the atmosphere. Those substances, on the contrary, are said to be fixed and non-volatile, which have little or no tendency to assume the condition of vapor. Thus, iron is a fixed substance, because it does not suffer a sensible degree of waste when exposed to intense heat. Oils which do not evaporate on simple exposure to the atmosphere are also termed fixed, to distinguish them from those which yield vapor under the same circumstances.

The melting of a solid, or its conversion into a liquid, only occurs when the solid is heated up to a certain fixed point; but the conversion of a liquid into a vapor takes place at all temperatures.

Thus, the vapor of water is continually passing off from the surface of the soil, from the ocean, and from all animal and vegetable productions. The production of vapor also takes place to a very considerable extent from the surface of snow and ice, even when the temperature of the air is far below the freezing point.

This circumstance explains the waste of snow and ice which may be observed during the continuance of severe cold.

134. Vapors Invisible.—The vapor of water, and all other vapors, are invisible and transparent. The water which has become diffused through the air by evaporation only becomes visible when, on returning to its fluid condition, it forms mist, cloud, dew, rain, etc.

Steam, which is the vapor of boiling water, is invisible, but when it comes in contact with air, which is cooler, it becomes condensed into small drops, and is thus rendered visible.

The proof of this may be found in examining the steam as it issues from an orifice, or the spout of a boiling kettle: for a short space next to the opening no steam can be seen, since the air is not able to condense it; but as it spreads and comes in contact with a larger volume of air, the invisible vapor becomes condensed into drops, and is thus rendered visible.

The visible matter popularly called steam, should be, therefore, distinguished from steam proper, or the aeriform state of water. The cloud, or smoke-like matter observed, is really not an air or vapor at all, but a collection of minute bubbles of water, wafted by a current either of true steam, or, more frequently, of mere moist air.

The surface of any watery liquid, whose temperature is about 20° warmer than any superincumbent air, rapidly gives off true steam. It is not necessary, therefore, for the production of steam, that water should be raised to the boiling temperature.

135. Comparative Volume of Vapors.—Liquids in pass-

QUESTIONS.—Do vapors form at all temperatures? Are vapors really visible? Is steam invisible? What is the proof of it? At what temperature is steam produced? What is the comparative volume of vapors?

ing into vapors occupy a much greater space than the substances from which they are produced. Water, in passing from its point of greatest density into steam, expands to nearly 1700 times its volume.

136. **Density of Vapors.**—Vapors are of all degrees of density. The vapor of water may be as thin as air, or almost as dense as water.

The subject of vaporization may be considered under two heads, viz., evaporation and ebullition.

137. **Evaporation.**—When vaporization takes place only from the surface of a body, either because the heat has access to that part, or because the evolution of vapor takes place through the medium of a gas or air present, the action can only be recognized by the diminution of the bulk of the body; this phenomenon is termed Evaporation.

138. **Ebullition.**—When a liquid is heated sufficiently to form steam, the production of vapor takes place principally at that part where the heat enters; and when the heating takes place not from above, but from the bottom and sides, the steam as it is produced rises in bubbles through the liquid, and produces the phenomenon of boiling, or ebullition.

Boiling, therefore, may be defined to be the mechanical agitation of a fluid by its own vapor.

139. **Boiling Point.**—The temperature at which vapor rises with sufficient freedom to cause the phenomenon of ebullition, is called the boiling point.

140. **Conditions of Evaporation.**—Evaporation takes place from the surfaces of bodies only, and is influenced in a great degree by the temperature, dryness, stillness, and density of the atmosphere.*

* It is a common error that the sun's rays are the first source of evaporation, and many persons ignorantly imagine, that because a locality is sunny it is sure to be dry. It can, however, be shown, by a great variety of facts, that the wind has more to do with drying

QUESTIONS.—What is said of the density of vapors? In what two ways may liquids be vaporized? What is evaporation? What is ebullition? Define boiling and the boiling point. What are the conditions of evaporation?

When water is covered by a stratum of dry air, the evaporation is rapid, even when its temperature is low; whereas it goes on very slowly if the atmosphere contains much vapor, even though the air be very warm.

Evaporation is far slower in still air than in a current. The air immediately in contact with the water soon becomes moist, and thus a check is put to evaporation. But if the air be removed by wind from the surface of the water as soon as it has become charged with vapor, and its place supplied with fresh air, then the evaporation continues on without interruption.

Air without vapor (theoretically known as dry air) does not exist in nature, and can not probably be produced by art.

141. Capacity for Absorption.—Air absorbs moisture at all temperatures, and retains it in an invisible state. This power of the air is termed its capacity for absorption.

The capacity of air for moisture increases with the temperature.

---

and evaporation than the sun. In the formation of ice on ponds, for instance, on a windy night in extreme winter, nothing is actually gained, since the ice wastes by evaporation from the surface as fast as it forms beneath. Every housewife knows that wet linen dries more rapidly when flying in the cold wind, than when hanging quietly in the warm sun. The driving blast which accompanies those sudden showers that vex and drench travelers in mountain regions, brings an almost instant remedy when the shower has passed. Air at rest will take up only a limited quantity of moisture, and is speedily saturated. But air in motion is never satisfied, and is constantly abstracting moisture from the soil. It is not the character of the soil, but the constant and unobstructed motion of the air, which reduces open land to barrenness.

"A proper understanding of the influence which trees and forests have upon the fertility of a country, by controling the evaporation of moisture from its surface, is of great practical importance. It is matter of surprise to every one who journeys in Syria or Greece, that the sacred and classic streams should be of such mean dimensions. The circumstance, however, finds an explanation in the fact, that the hills of these countries have been almost entirely deprived of their forests. And the like cause will everywhere produce the like effect. In an open country, the absolute quantity of water which the rivers discharge is not only less than in a wooded country, but the flow is incomparably more irregular and unequal. It has been especially noticed in the Western States, that since the country has been extensively cleared, the alternations of the 'stage' of water in the rivers have been more marked and violent. In New England the effect of an indiscriminate clearing away of forests has been practically illustrated by the constant hinderance of mill-streams from drought and freshets. Many water-privileges which, half a century ago, were valuable and steady, have now become nearly worthless. The dam which was conveniently put up to saw an adjoining forest into profitable plank, now that its excellent work is done, will drive the saw in the summer no longer. Many of the larger New England factories have been compelled to introduce steam-power to supply a deficiency in the volume of water, which a few years ago was not troublesome. The cutting away of forests does not probably diminish the quantity of rain or snow, although some authorities maintain that this is the fact; but it deprives the moisture of its beneficent effect upon the earth, by causing it to be too rapidly abstracted—thus producing pernicious alternations of freshet and drought, which are as fatal to the health of the soil, as to the health of the men who own the soil."

Questions.—Does air exist without vapor? What is understood by the capacity of absorption in air?

A volume of air at 32° can absorb an amount of moisture equal to the hundred and sixtieth part of its own weight, and for every 27 additional degrees of heat, the quantity of water it can absorb at 32° is doubled. Thus a body of air at 32° F. absorbs the 160th part of its own weight; at 59° F., the 80th; at 86° F., the 40th; at 113° F., the 20th part of its own weight in moisture. It follows from this that while the temperature of the air advances in an arithmetical series, its capacity for moisture is accelerated in a geometrical series.

Air is said to be saturated with moisture when it contains as much of the vapor of water as it is capable of holding with a given temperature.

142. **Hy-grom'e-ters.**—Instruments designed for measuring the quantity of moisture contained in the atmosphere, are called HYGROMETERS.*

Many organic bodies have the property of absorbing vapor, and thus increasing their dimensions. Among such may be mentioned hair, wood, whalebone, ivory, etc. Any of these connected with a mechanical arrangement by which the change in volume might be registered, would furnish a hygrometer. The thin, transparent shavings of whalebone, which by bending and rolling up when placed upon the warm hand, constitute the so-called sensitive figures, are illustrations of this principle.

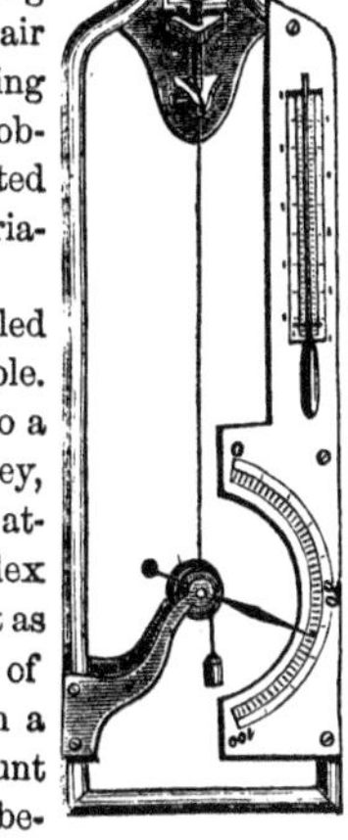

FIG. 39.

If we fix against a wall a long piece of catgut, and hang a weight to the end of it, it will be observed, as the air becomes moist or dry, to alter in length; and by marking a scale the two extremities of which are determined by observation when the air is very dry, and when it is saturated with moisture, it will be found easy to measure the variations.

143. **Hair Hygrometer.**—An instrument called the "Hair Hygrometer," is constructed upon this principle. It consists of a human hair, fastened at one extremity to a screw (see Fig. 39), and at the other passing over a pulley, being strained tight by a silk thread and weight, also attached to the pulley. To the axis of the pulley an index is attached, which passes over a graduated scale, so that as the pulley turns, through the shortening or lengthening of the hair, the index moves. When the instrument is in a damp atmosphere, the hair absorbs a considerable amount of vapor, and is thus made longer, while in dry air it be-

* *Hygrometer*, from the Greek words υγρος (moist) and μετρον (measure).

QUESTIONS.—When is air said to be saturated with moisture? What are hygrometers? Explain the hair hygrometer and its principle of construction?

comes shorter; so that the index is of course turned alternately from one side to the other.

The instrument is graduated by first placing it in air artificially made as dry as possible, and the point on the scale at which the index stops under these circumstances, is the point of greatest dryness, and is marked 0. The hygrometer is then placed in a confined space of air, which is completely saturated with vapor, and under these circumstances the index moves to the other end of the scale: this point, which is that of greatest moisture, is marked 100. The intervening space is then divided into 100 equal parts, which indicate different degrees of moisture.

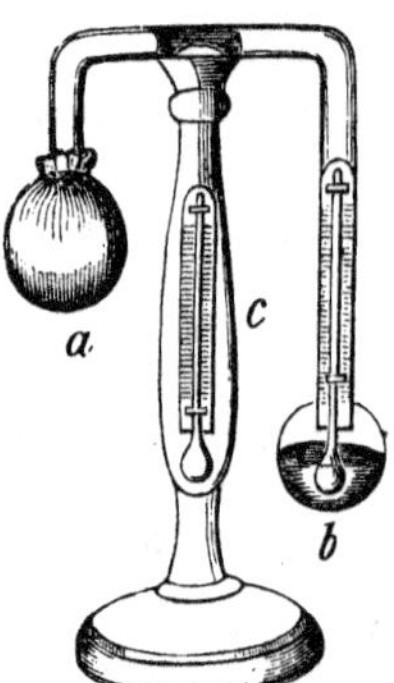

FIG. 40.

144. **Daniel's Hygrometer.**—Another form of hygrometer, known as "Daniel's Hygrometer," determines the moisture in the air by indicating the dew point, or the temperature at which moisture is deposited from the air. It consists of a bent tube of glass, Fig. 40, at the extremities of which two bulbs, *a* and *b*, are blown. The bulb *b* is made of black glass, and contains a little ether, into which dips the ball of a small and delicate thermometer, contained in the cavity of the tube. The whole instrument contains only the vapor of ether, the air having been removed. The bulb *a* is covered over with a piece of muslin. The support of the tube sustains another thermometer, by which we can observe the temperature of the air. When an observation is to be made with this instrument, a little ether is poured on the muslin of the bulb *a;* this evaporates rapidly, and by so doing reduces the temperature of the other bulb, *b*. As soon as this has cooled sufficiently to condense the moisture of the atmosphere, dew will be observed to collect upon it, and the temperature at which the deposition takes place is determined by observing the thermometer included in the tube. If the air is very moist, it is necessary to cool the bulb *b* but very little before dew is deposited upon it; if, however, the air is very dry, the cooling must be carried to a corresponding lower degree. If the air is perfectly saturated, the slightest depression of temperature will cause its moisture to precipitate. Knowing, therefore, the temperature of the dew point, we are enabled by tables calculated for the purpose, to determine the proportional amount of moisture contained in the atmosphere.

145. **Conditions of Ebullition.**—Different liquids boil at different temperatures, but the boiling point of the same liquid is always the same under the same circumstances. The boiling temperature, therefore, constitutes a distinctive characteristic of a liquid, and in practical

QUESTIONS.—Describe Daniel's hygrometer. How does the boiling point of liquids vary?

chemistry often affords a ready method of detecting a difference in the chemical composition of similar liquids.

Thus water, under ordinary circumstances, begins to boil when it is heated up to 212° F.; alcohol at 173°; ether at 96°; syrup at 221°; linseed oil at 640°.

146. **Salinometer.**—Water containing any dissolved matter boils at a higher temperature than when pure—the boiling point on the thermometric scale rising in proportion as the amount of matter dissolved in the water increases. Advantage is taken of this principle in the construction of an instrument known as the "Salinometer," which is especially used by salt-boilers for indicating the quantity of salt held in solution in the water of the boilers. It simply consists of a delicate thermometer arranged in connection with the interior of the boiler, and by means of a properly graduated scale, the percentage of salt held in the water is indicated by the boiling point of the water.

147. **Influence of Atmospheric Pressure on Boiling.**—Liquids, in general, being boiled in open vessels, are subjected to the pressure of the atmosphere. The tendency of this pressure is to prevent and retard the particles of water from expanding to a sufficient extent to form steam. Hence, if the pressure of the atmosphere varies, as it does at different times and places, or if it be increased or diminished by artificial means, the boiling point of a liquid will undergo a corresponding change.

The pressure of the atmosphere at the level of the sea is about fifteen pounds upon each square inch of surface. It varies occasionally at the same place sufficiently to affect the boiling point to the extent of 4½ degrees.

148. **Measurement of Altitudes.**—As we ascend into the atmosphere the pressure is diminished, because there is less of it above us; it therefore follows, that water at different heights in the atmosphere will boil at different temperatures, and it has been found by observation, that an elevation of 550 feet above the level of the sea causes a difference of one degree in its boiling point. Hence the boiling point of water becomes an indication of the height of any station above the sea-level, or in other words, an indication of the atmospheric pressure; and thus by means of a kettle of boiling water and a thermometer, the height of the summit of any mountain may be ascertained with a great degree of accuracy. If the water boils at 211° by the thermometer, the height of the place is 550 feet; if at 210°, the height is 1100 feet, and so on, it being only necessary to multiply 550 by the number of degrees on the thermometer between the actual boiling point and 212°, to ascertain the elevation. In the city of Quito, in South America, water boils at 194° Fahr.; its height above the sea-level, is, therefore, 9,900 feet.

As we descend into mines, the pressure of the atmosphere is increased, there

---

QUESTIONS.—What influence has the pressure of the atmosphere upon the boiling point? How may the height of mountains be determined by the boiling point of water?

being more of it above us than at the surface of the earth. Water, therefore, must be heated to a higher temperature before it will boil, and it has been found that a descent of 550 feet, as before, makes a difference of one degree.

Boiling water is, consequently, not equally hot at all places upon the earth, and, therefore, not every where alike applicable for domestic purposes. Thus at Quito and at the hospital of St. Bernard, in Switzerland, great difficulty is experienced in cooking eggs by boiling.

In a like manner, if by artificial means we increase or diminish the pressure of the atmosphere on the surface of a liquid, we change its boiling point. If water be heated in a vacuum, ebullition will commence at a point 140° lower than in the open air. If a vessel of ether be placed under the receiver of an air-pump, and the atmospheric pressure removed from its surface, the vapor rises so abundantly that ebullition is produced without any increase of temperature.

149. **Pulse-Glass.**—This principle is illustrated by a simple instrument called the pulse-glass, Fig. 41, which consists of a glass tube, *c*, with bulbs, *a* and *b*, blown upon each extremity; the whole is then filled with spirits of wine and its vapor, and hermetically sealed. The pressure of the air being thus removed from the surface of the liquid, the heat of the hand upon either bulb is sufficient to cause a violent ebullition.

Fig. 41.

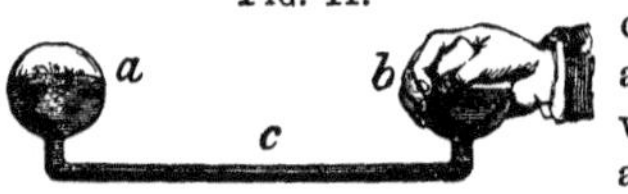

150. **Culinary Paradox.**—The fact that water boils at a reduced temperature under diminished pressure, is illustrated by an experiment known as the culinary paradox. A glass flask, containing boiling water is closed tightly with a cork, and then inverted, as in Fig. 42. The boiling will instantly cease, owing to the pressure of the steam which is formed, upon the surface of the liquid. If we now pour cold water upon the outside of the flask, the steam within is condensed, and a partial vacuum produced, which causes the boiling to recommence with great energy. On the other hand, by pouring hot water upon the outside of the flask, the steam and consequent pressure within is renewed, and the boiling ceases.

Fig. 42.

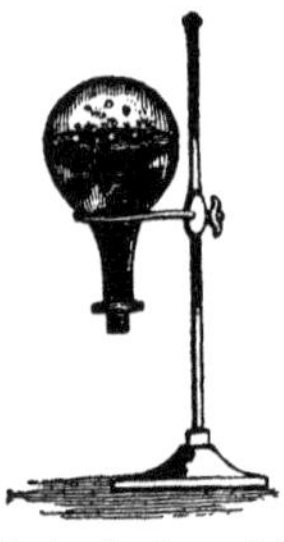

A proof also that steam in escaping from boiling water is obliged to overcome the pressure of the atmosphere, is obtained by repeating the last experiment with a tin canister instead of a globular glass flask. On corking up the canister and pouring cold water over it, the steam within is suddenly condensed, a vacuum is produced, and the canister is instantly crushed in by the pressure of the external air.

QUESTIONS.—How may the boiling point of a liquid be elevated or depressed by artificial means? What is the pulse-glass? What is the culinary paradox? What experiment proves that steam in escaping is obliged to overcome the pressure of the atmosphere?

151. **Sugar Boiling.**—Several beautiful applications in the arts have been made of the principle that liquids boil at a lower temperature when freed from the pressure of the atmosphere than in the open air.

In the refining of sugar, if the syrup is boiled in the open air, the temperature of the boiling point is so high that portions of the sugar become decomposed by the excess of heat, and lost or injured; the syrup is therefore boiled in close vessels from which the air has been previously exhausted, and in this way the water of the syrup may be evaporated at a temperature so low as to prevent all injury from heat.

152. **Influence of Adhesion on the Boiling Point.**—Adhesion of the fluid to the surface of the vessel that contains it, has a marked effect in raising the boiling point. Water boils somewhat more readily in a metallic vessel than in one of glass. If the interior of a vessel be varnished with shell-lac, the boiling will not often occur until a temperature of 221° F. is reached, and then it will take place in bursts, the temperature at each evolution of vapor falling to 212° F. Boiling can be made to take place steadily at 212° in any variety of vessel, by the introduction of a few irregular substances, as little fragments of wire, a few pieces of charcoal, etc. The reason of this is that in a mass of boiling liquid, the formation of vapor takes place principally at the edges of the solid substances with which it may be in contact; and the introduction and presence of irregular surfaces thus facilitate its formation.

153. **Influence of Air on the Boiling Point.**—Recent experiments have shown that the presence of air in solution singularly assists the evolution of vapor. Air dissolved in water acquires, through the agency of heat, a great degree of elasticity, and minute bubbles of it are in consequence thrown off in the interior of a boiling liquid, especially where it is in contact with a rough surface; into these bubbles the steam escapes and rises. Water when boiled for a long time is nearly deprived of air; and in such cases the temperature has been observed to rise even as high as 260°, or 48° above the boiling point, in an open glass vessel, which was then shattered with a loud report by a sudden explosive burst of vapor. In this case, the force of cohesion retains the particles of liquid throughout the mass in contact with each other, in a species of unstable equilibrium; and when this equilibrium is overturned at any one point, the repulsive power of the excess of heat stored up in the mass, suddenly exerts itself, and the explosion is the result of the instantaneous conversion of the liquid into vapor.

The same result takes place when ice, free from air, is melted out of contact with the atmosphere, as under oil. The temperature of the liquid formed gradually rises to about 260° F., when, instead of boiling, it explodes.

If a single drop of water containing air, be allowed to fall into a mass of

---

QUESTIONS.—What practical application of these principles has been made in the arts? What influence does adhesion have upon the boiling point? How may liquids be made to boil steadily? What effect has air dissolved in water upon the evolution of vapor? What curious experiments illustrate this? What takes place when ice free from air is heated out of contact with air?

water free from air, which has been heated to a temperature of 250° or 260° F., the whole volume instantly becomes agitated in a singular manner, and an explosion generally occurs.

154. **Spheroidal State.**—When a drop of water falls upon a surface highly heated, as of metal, it will be observed to roll along the surface without adhering, or immediately passing into vapor. The explanation of this is, that the drop of water does not in reality touch the heated surface, but is buoyed up and supported on a layer of vapor which intervenes between the bottom of the drop and the hot surface. This vapor is produced by the heat which is radiated from the hot substance, before the liquid can come in contact with it, and being constantly renewed, continues to support the drop. The drop generally rolls because the current of air which is always passing over a heated surface drives it forward. The drop evaporates slowly, because the layer of vapor between the hot surface and the liquid prevents the rapid transmission of heat. The liquid resting upon a cushion of steam continually evolved from its lower surface by heat, assumes a rounded, or globular shape, as the result of the gravity of its particles toward its own center.

The designation which has been given to the condition which water and other liquids assume when brought in contact with very hot surfaces, is that of the "spheroidal state."

If the surface upon which the liquid rests is cooled down to such an extent that vapor is not generated rapidly, and in sufficient quantity to support the drop, it will come in contact with the surface, and heat being communnicated by conduction, will transform it instantly into steam.

This is the explanation of the practice adopted by laundresses of touching a flat-iron with moisture to ascertain whether the surface is sufficiently hot. If the temperature of the iron is not elevated sufficiently, the moisture wets the surface, and is evaporated; but at a higher degree of temperature, the moisture is repelled.

The phenomenon of the spheroidal condition of water furnishes an explanation of the feats often performed by jugglers, of plunging the hands with impunity into molten lead, or iron. The hand is moistened, and when passed into the liquid metal the moisture is vaporized, and interposes between the metal and the skin a sheath of vapor. In its conversion into vapor, the moisture absorbs heat, and thus still further protects the skin.

The bulb of a thermometer plunged into liquids while in the spheroidal state, indicates temperatures considerably below the ordinary boiling point. Thus water in a spheroidal state has a temperature of 205°; alcohol, 167°; ether, 93°; sulphurous acid, 13°. When distilled water is allowed to fall drop by drop into sulphurous acid in the spheroidal state, the water is instantly congealed into a spongy mass of ice, even when the containing vessel is red hot.

QUESTIONS.—What takes place when a drop of water falls upon a highly heated surface? What is meant by the spheroidal state? Why can the hand be safely plunged into molten iron? What is the temperature of liquids in the spheroidal state?

155. **Distillation, or Sublimation,** is a process by which one body is separated from another in close vessels, by means of heat, in cases where one of the bodies assumes the form of vapor at a lower temperature than the other; this first rises in the form of vapor, and is received and condensed in a separate vessel. The operation is termed Distillation, when the vapor formed condenses into a liquid, and Sublimation when it condenses into a solid. The product in the first instance is called a *distillate,* and in the second a *sublimate.*

When the product of one distillation is subjected to further distillations, in order to free it to a still greater extent from less volatile substances, the operation is called *rectification.*

By this means very volatile bodies can be easily separated from less volatile ones; as brandy and alcohol from the less volatile water which may be mixed with them. Water of extreme purity can also be obtained by distillation, because the non-volatile and earthy substances contained in all spring waters do not ascend with the vapor, but remain behind in the vessel.

FIG. 43.

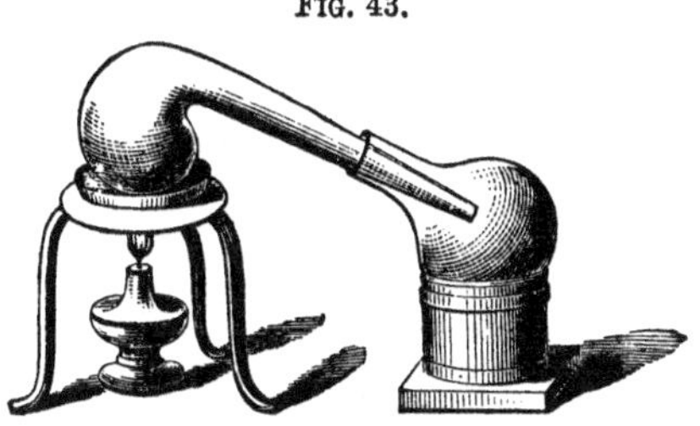

Distillation upon a small scale is effected by means of a peculiar-shaped vessel, called a retort, Fig. 43, which is half filled with a volatile liquid and heated; the steam, as it forms, passes through the neck of the retort into a glass receiver set into a vessel filled with cold water, and is then condensed.

FIG. 44.

When the operation of distillation is conducted on an extensive scale, a large vessel called a "*still*" is used, and, for condensing the vapor, vats are constructed, holding serpentine pipes, called "worms," which present a greater condensing surface than if they had passed directly through the vat. To keep the coil of pipe cool, the vats are kept filled

QUESTIONS.—What is distillation, or sublimation? What is the difference between a distillate and a sublimate? What is rectification? How is distillation effected?

with cold water. In Fig. 44, *a* is a furnace, in which is fixed a copper vessel, or still, to contain the liquid. Heat being applied, the steam rises in the head, *b*, and passes through the worm, *d*, which is placed in a vessel of water, the refrigerator. The vapor thus generated is condensed in its passage, and passes out as a liquid by the external pipe into a receiver.

156. **Drying and Distillation.**—The difference between drying by heat and distillation is, that in one case, the substance vaporized, being of no use, is allowed to escape or become dissipated in the atmosphene; while in the other, being the valuable part, it is caught and condensed into the liquid form. The vapor arising from damp linen, if caught and condensed would be distilled water; the vapor given out by bread while baking, would, if collected, be a spirit like that obtained in the distillation of grain.

157. **Latent Heat.**—When a solid is converted into a liquid, or a liquid into a vapor or gas, heat in large quantity disappears, and ceases for the time to affect the thermometer. It is not, however, absolutely lost, but remains incorporated with the substance of the liquid, or the gas, in an insensible condition. Heat thus disappearing, is termed Latent, or Insensible Heat.

For example, if a thermometer be applied to a mass of snow, or ice just upon the point of melting, it will be found to stand at 32° F. If the ice be placed in a vessel over a fire, and the temperature tested at the moment it has entirely melted, the water produced will have only the temperature of 32°, the same as that of the original ice. Heat, however, during the whole process of melting, has been passing rapidly into the vessel from the fire, and if a quantity of mercury, or a solid of the same size, had been exposed to the same amount of heat, it would have constantly increased in temperature. It is clear, therefore, that the conversion of ice, a solid, into water, a liquid, has been attended with a disappearance of heat.

Again, if a pound of water at 212° F. be mixed with a pound of water at 33° F., we shall obtain two pounds of water at 122°, a temperature exactly intermediate between the temperature of the two. If, however, a pound of ice at 32°, is mixed with a pound of water at 212°, we shall obtain two pounds of water, of which the temperature is only 51°. In this case the water has lost 161°, while the ice has apparently gained but 19°; so that 142° have disappeared, or become latent. Thus, in order to convert a pound of ice at 32° F. into water at 33°, as much heat is required as would be sufficient to raise 142 pounds of water from 32° to 33° F. Water, therefore, may be regarded as ice in combination with a certain quantity of heat.

QUESTIONS.—What is the difference between drying by heat and distillation? What remarkable circumstance characterizes the phenomena of liquefaction and vaporization? Explain what is meant by latent heat? What experiments prove that liquefaction occasions a disappearance of heat?

158. **Heat required to Melt Ice.**—Some idea of the quantity of heat that is required to convert ice into water, without any apparent rise in temperature, may be formed from the fact that the simple conversion of a cube of ice, three feet on the side, into water at 32°, would absorb the whole amount of heat emitted during the combustion of a bushel of coal.

159. **Disappearance of Heat in Vaporization.**—In the conversion of a liquid into gas or vapor, heat disappears to a much greater extent than in the conversion of a solid into a liquid.

The absorption of heat by vaporization, may be easily rendered perceptible to the feelings by pouring a few drops of some liquid which readily evaporates, such as ether, alcohol, etc., upon the hand. A sensation of cold is immediately experienced, because the hand is deprived of heat, which is drawn away to effect the evaporation of the liquid. On the same principle, inflammation and feverish heat in the head may be allayed by bathing the temples with any liquid which evaporates easily, as Cologne water, alcohol, vinegar, etc.

A vessel containing water placed over a source of heat which is tolerably uniform in temperature, receives equal accessions of heat in equal times. The water at first rises steadily in temperature, and at 212° it boils. After this, no matter how much the heat is increased, provided the steam be allowed to escape freely, it becomes no hotter; all the heat which is added serving only to convert the water at 212° into steam or vapor.

This fact is of considerable importance in domestic economy, and attention to it will save much fuel in culinary operations. Soups, etc., made to boil in a gentle way by the application of a moderate heat, are just as hot as when they are made to boil over a strong fire with the greatest violence. When a liquid is once brought to the boiling point, the fire may be reduced, as a comparatively small quantity of heat will be then sufficient to maintain it there.

160. **Latent Heat of Steam.**—If we immerse a thermometer in boiling water, it stands at 212°; if we place it in steam immediately above it, it indicates the same temperature. The question then arises, what becomes of all the heat which is communicated to the water, since it is neither indicated by the water nor by the steam formed from it? The answer is, that it enters into the water and converts it into steam, without raising its temperature. The proof that steam contains more heat than boiling water, is to be found in the fact that if we mix an ounce of water at 212° with five and a half ounces of water at 32°, we obtain six and a half ounces of water at a temperature of about 60°; but if we mix an ounce of steam at 212° with

---

QUESTIONS.—What is the comparative quantity of heat necessary to convert ice into water? To what extent is heat rendered latent by vaporization? What experiments prove that heat disappears in vaporization? Do liquids acquire additional heat after attaining a boiling temperature? What practical application can be made of this principle in domestic economy? What is the sensible heat of steam? What is its latent heat? How may steam at 212° F. be proved to contain more heat than water at the same temperature?

five and a half ounces of water at 32°, we obtain six and a half ounces of water at 212°. The steam, from which the increased heat is all derived, contains as much more heat than the ounce of water at the same temperature, as would be necessary to raise six and a half ounces of water from the temperature of 60° to 212°, or six and a half times as much heat as would be requisite to raise one ounce of water through about 152° of temperature. This quantity of heat will, therefore, be found by multiplying 152° by six and a half, which will give a product of 988°—the excess of heat contained in an ounce of steam at 212° over that contained in an ounce of boiling water at the same temperature.

In round numbers, therefore, one thousand degrees of heat are absorbed in the conversion of water into steam, and this constitutes the latent heat of steam.

The absorption of heat in the process by which liquids are converted into vapor, will explain why a vessel containing a liquid that is constantly exposed to the action of fire, can never receive such a degree of heat as would destroy it. A tin kettle containing water may be exposed to the action of the most fierce furnace, and remain uninjured ; but if it be exposed, without containing water, to the most moderate fire, it will soon be destroyed. The heat which the fire imparts to the kettle containing water is immediately absorbed by the steam into which the water is converted. So long as water is contained in the vessel, this absorption of heat will continue; but if any part of the vessel not containing water be exposed to the fire, the metal will be fused, and the vessel destroyed.

161. **Effects Produced by the Absorption of Heat.**—In the conversion of solids into liquids, and of liquids into gases or vapors, the heat which disappears is the agent by which liquefaction in the one case, and vaporization in the other, are produced; in other words, the absorption of a certain amount of heat is necessary for the production of the change. A liquid, therefore, may be regarded as a compound of a solid and heat, and a vapor as a compound of heat and the liquid from which it was formed.

162. **Freezing Mixtures.**—The absorption of heat consequent on the conversion of solids into liquids, has been taken advantage of in the arts for the production of artificial cold ; and the compounds of different substances which are made for this purpose, are called freezing mixtures.

The most simple freezing mixture is snow and salt. Salt dissolved in water would occasion a reduction of temperature, but when the chemical relations of two solids are such, that both by mixing are rendered liquid, a still

---

QUESTIONS.—Why does a kettle containing water remain uninjured, when exposed to the heat of a fire? What may be considered as the true constitution of liquids and vapors? What are freezing mixtures? Why does a mixture of snow and salt produce a high degree of cold?

greater degree of cold is produced. Such a relation exists between salt and snow, or ice, and therefore the latter substances are used in preference to water. When the two are mixed, the salt causes the snow to melt by reason of its attraction for water, and the water formed dissolves the salt: so that both pass from the solid to the liquid condition. If the operation is so conducted that no heat is supplied from any external source, it follows that the heat absorbed in liquefaction must be obtained from the salt and snow which comprise the mixture, and they must therefore suffer a depression of temperature proportional to the heat which is rendered latent.

In this way a degree of cold equal to 40° below the freezing point of water may be obtained. The application of this experiment to the freezing of ice-creams is familiar to all.

By mixing snow and sulphuric acid together in proper proportions, a temperature of from 70° to 90° below zero can be obtained without difficulty.

A very convenient process for freezing water without the use of ice is to drench finely-powdered sulphate of soda with the undiluted hydrochloric (muriatic) acid of the shops. In this way a very low temperature may be readily obtained. The vessel in which the mixture is made becomes covered with hoar frost, and water in tubes or bottles immersed in the mixture, is speedily frozen.

163. **Greatest Artificial Cold**.—The most intense artificial cold is, however, produced by the rapid evaporation of highly volatile liquids, such as result from the condensation and liquefaction of certain gases. By means of a mixture of liquid nitrous oxyd and sulphuret of carbon, placed under the exhausted receiver of an air-pump, M. Natterer obtained the enormously low temperature of two hundred and twenty degrees below zero.

The cold produced by evaporation is due to the absorption of heat by the newly-formed vapor, and the more rapidly evaporation takes place, the more rapidly is heat abstracted from the evaporating liquid and from surrounding substances.

164. **Freezing by Evaporation**.—Ether may be made to evaporate so rapidly as to freeze water, even in summer. This may be illustrated by filling a small glass tube with water, and surrounding it with cotton, or some other porous substance, soaked in ether. If a current of air be then directed upon the cotton from a common bellows, the ether will evaporate and absorb heat so rapidly, as to convert the water into ice in a few minutes.

FIG. 45.

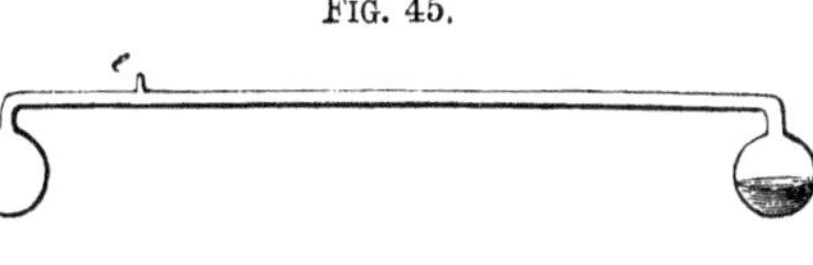

165. **The Cry-oph'o-rus**.—An instrument known as the

QUESTIONS.—By what process may water be frozen in summer without the aid of ice? What is the most intense artificial cold produced? What is the lowest degree of temperature ever observed? To what is the cold produced by evaporation due? How may water be frozen by the evaporation of ether? Explain the action of the cryophorus.

cryophorus, or *frost-bearer*, strikingly illustrates the production of cold by evaporation. It consists of two glass bulbs connected by a tube, and containing a portion of water, as represented in Fig. 45. The air is first expelled from the instrument by boiling the water inclosed, and allowing the steam to escape by a small opening at the extremity of the little projecting tube, *e*. While the instrument is entirely filled with steam, the point *e* is fused by the blow-pipe flame, and the opening hermetically closed. In experimenting with this instrument, the water is all poured into one bulb, and the other, or empty bulb, is placed in a basin containing a mixture of ice and salt. The vapor in the cooled bulb is condensed, but its place is immediately supplied by vapor which rises into the vacuum from the water in the other bulb. A rapid evaporation, therefore, takes place in the water-bulb, and condensation in the empty bulb, until by reason of the condensation and rapid evaporation, the water in the former bulb is cooled so low as to freeze.

Practical Illustrations.—A shower of rain cools the air in summer, because the earth and the air both part with their heat to promote evaporation. In a like manner, the sprinkling of a hot room with water cools it.

The danger arising from wet feet and clothes is owing to the absorption of heat from the body by the evaporation from the surfaces of the wet materials; the temperature of the body is in this way reduced below its natural standard, and the proper circulation of the blood interrupted.

The evaporation which takes place continually from the surface of the skin and the cells of the lungs of animals, is a powerfully cooling agency, and a protection against external heat. When the heat of the body is increased by exercise, or by exposure to high temperatures, perspiration and evaporation take place rapidly. Heat is thereby absorbed and rendered latent in large quantity, and a healthy temperature of the system maintained. It is on this principle that persons are enabled to expose themselves for a time to an atmosphere of very high temperature without serious inconvenience, as in foundries, boiler-rooms of steamers, ovens of manufactories, etc. If, however, the air be moist, or the surface of the skin be varnished, so as to check or prevent perspiration and evaporation, the heat can only be sustained for a few moments.

The air in the spring of the year, when the ice and snow are thawing, is always peculiarly cold and chilly. This is due to the constant absorption of heat from the air by the ice and snow in their transition from a solid to a liquid state.

166. Conversion of Latent into Sensible Heat.—When vapors are condensed into liquids, and liquids are changed

QUESTIONS.—How does a shower of rain cool the air and the earth in summer? How does the drainage of a country promote its warmth? From what does the danger of wet clothes and feet arise? How does perspiration and evaporation from the surface of the skin equalize the temperature of the body? Why is the air in the spring of the year peculiarly cold and chilly? Under what circumstances is latent converted into sensible heat?

into solids, the latent heat contained in them is set free, or made sensible.

If water be taken into an apartment whose temperature is several degrees below the freezing point, and allowed to congeal, it will render the room sensibly warmer. It is, therefore, in accordance with this principle that tubs of water are allowed to freeze in cellars in order to prevent excessive cold.

The large amount of heat latent in water, which it gives forth as it freezes, furnishes a source of heat of the greatest value in mitigating the severity of winter, and in rendering the transitions of atmospheric temperature, from heat to cold and from cold to heat, uniform and gradual.

In the colder regions, every ton of water converted into ice gives out and diffuses in the surrounding region as much heat as would raise a ton of water from 32° to 174°; and, on the other hand, when a rise of temperature takes place, the thawing of the ice absorbs a like quantity of heat: thus, in the one case, supplying heat to the atmosphere when the temperature falls; and, in the other, absorbing heat from it when the temperature rises.

In the winter, the weather generally moderates on the fall of snow; snow is frozen water, and in its formation heat is imparted to the atmosphere, and its temperature increased.

Steam, on account of the latent heat it contains, is well adapted for the warming of buildings, or for cooking. In passing through a line of pipes, or through meat and vegetables, it is condensed, and imparts to the adjoining surfaces nearly 1000° of the latent heat which it contained before condensation.

Steam burns much more severely than boiling water, for the reason that the heat it imparts to any surface upon which it is condensed, is much greater than that of boiling water.

167. **Elastic Force of Vapors.**—All vapors are elastic, like air.

The tendency of vapors to expand is generally considered to be unlimited; that is to say, the smallest quantity of vapor *has a tendency* to diffuse itself through every part of a vacuum, be its size what it may, exercising a greater or less degree of force against any obstacle which may restrain it.

Recent researches of M. Babinet, a French physicist, seem to show, that all gases and vapors entirely lose their elasticity when reduced to a certain degree of tenuity, and that no gas or vapor, formed under the ordinary pressure of the atmosphere, can expand sufficiently to fill an empty space 20,000 times greater than the original volume of the gas or vapor.

QUESTIONS.—How does the freezing of water tend to elevate the temperature of the surrounding atmosphere? Why is steam well adapted for the warming of buildings and for cooking? Why does steam burn more severely than water of the same temperature? What is said of the elasticity of vapors? In what manner do vapors tend to expand?

The force with which a vapor expands is called its elastic force, or tension.

The elasticity or pressure of vapors is best illustrated in the case of steam, which may be considered as the type of all vapors.

168. **Expansive Force of Steam.**—When a quantity of pure steam is confined in a close vessel, its elastic force will exert on every part of the interior of the vessel a certain pressure directed outward, having a tendency to burst the vessel.

When steam is generated in an open vessel its elastic force must be equal to the elastic force or pressure of the atmosphere; otherwise the pressure of the air would prevent it from forming and rising. Steam, therefore, produced from boiling water at 212° F., is capable of exerting a pressure of 15 pounds upon every square inch of surface, or one ton on every square foot, a force equivalent to the pressure of the atmosphere.

If water be boiled under a diminished pressure, and therefore at a lower temperature, the steam which is produced from it will have a pressure which is diminished in an equal degree. If, on the contrary, the pressure under which water boils be increased, the boiling temperature of the water and the pressure of the steam formed will be increased in a like proportion. We have, therefore, the following rule:—

Steam raised from water, boiling under any given pressure, has an elasticity always equal to the pressure under which the water boils.

Steam of a high elastic force can only be made in close vessels, or boilers. The water in a steam-boiler, in the first instance, boils at 212°, but the steam thus generated being prevented from escaping, presses on the surface of the water equally as on the surface of the boiler, and therefore the boiling point of the water becomes higher and higher; or in other words, the water has to grow constantly hotter, in order that the steam may form. The steam thus formed has the same sensible temperature as the water which produces it.

169. **Marcet's Digester.**—The above principles are experimentally proved by means of an apparatus known as Marcet's Digester. This consists of a stout globular vessel of iron, Fig. 46, into which a portion of mercury is poured, and then water sufficient to half fill it. Into the top of the vessel a long glass tube, *b*, is tightly fitted, open at both ends, and dipping into the mercury. This tube is provided with a scale divided into inches. The globular vessel has also two other openings, into one of which a stop-

---

QUESTIONS.—What is the force with which a vapor expands termed? In what manner will steam confined in a close vessel exert a pressure? What is the pressure of steam generated in the open air? What rule governs the elasticity of steam? What arrangements are essential to the production of steam of great elastic force? What relations exist between the temperature of steam formed under pressure and the water which produces it? What is Marcet's digester? What principles may be experimentally proved by this apparatus?

cock, *d*, is screwed, and into the other a thermometer, *c*, having its bulb within the globe. Heat is applied to the vessel, and the water made to boil. So long as free communication with the atmosphere is permitted through the open stop-cock *d*, the temperature of ebullition, as indicated by the thermometer, *c*, continues steady at 212°, and the steam formed exerts a pressure of course equal to one atmosphere, or 15 lbs. to the square inch. On shutting the stop-cock, and continuing the heat, the temperature of the interior rises above 212°. The steam in the upper part of the vessel becomes denser, and as fresh portions continue to rise from the water, the pressure on the surface of the water increases, and this in turn pressing upon the mercury, forces it to ascend in the tube. Now the height of the mercurial column expresses the elastic force or pressure of the steam produced in the boiler at any particular temperature above 212°. Thus the weight of that section of the atmosphere which presses upon the mercury in the open end of the tube is equivalent to the weight of a column of mercury of 30 inches; and this pressure must be overcome by the steam at 212° before it can commence to act upon the mercurial guage at all. For every thirty inches after this that the mercury is forced up into the tube by the steam, it is said to have the pressure, or elastic force of another atmosphere. Thus, when the mercury in the tube stands at 30 inches, the steam is said to be of two atmospheres; at 45 inches, of two and a half; at 60 inches, of three atmospheres, and so on. The boiling point of the water, also, as shown by the thermometer, increases with the pressure of the steam upon its surface. When the mercury stands at 30 inches, or when the pressure on the water is equal to that of an additional atmosphere, the thermometer marks a temperature of 249°; at 60 inches, 273°; at 90 inches, or with a pressure of four atmospheres, 291°, and so on.

FIG. 46.

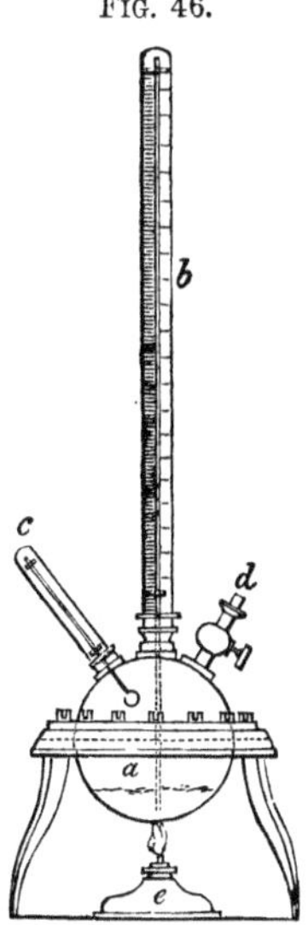

170. **Tables of the Temperature and Pressure of Steam.**—As the relation between the temperature and the pressure of steam, and the varying temperature at which water boils or gives off steam under pressure, are matters of great importance in connection with the steam-engine, the French government many years ago appointed a commission of eminent scientific men to investigate the whole subject. The result of their labors has been embodied in a series of tables, which show at once the pressure of steam formed in contact with water at any given temperature, or conversely, the temperature at any given pressure. It was thus found that the temperature of steam capable of exerting a pressure of twenty five atmos-

---

QUESTION.—**Under what circumstances were the relations between the temperature and pressure of steam investigated?**

pheres, or 375 pounds upon each square inch of boiler surface, was 439°. The temperature of the water producing steam of this pressure, must have been consequently the same.

171. **Determination of Steam-pressure in Boilers.**—The application of these principles affords a ready method of determining the pressure at any moment which steam exerts upon the interior of a boiler, or upon the piston of a steam-engine. Thus, if a thermometer inserted into a steam-boiler indicates a temperature of 212° F., we know that the steam exerts a pressure of one atmosphere, or 15 pounds upon a square inch; if the thermometer stands at 249°, the pressure is 30 pounds · at 273°, 45 pounds; and so on.

172. **Barometer Guage.**—The degree of pressure which steam exerts upon the interior of the boiler is, however, more generally determined by the height to which a column of mercury is elevated and sustained by such pressure. The instrument employed for this purpose is termed a "steam" or "barometer guage." It consists simply of a bent tube, A, C, D, E, Fig. 47, fitted into the boiler at one end, and open to the air at the other. The lower part of the bend of the tube contains mercury, which, when the pressure of steam in the boiler is equal to that of the external atmosphere, will stand at the same level, H, R, in both legs. of the tube. When the pressure of the steam is greater than that of the atmosphere, the mercury is depressed in the leg C D, and elevated in the leg D E. A scale, G, is attached to the long arm of the tube, and by observing the difference of the levels of the mercury in the two tubes, the pressure of the steam may be calculated. Thus, when the mercury is at the same level in both legs, the pressure of the steam balances the pressure of the atmosphere, and is therefore 15 pounds per square inch. If the mercury stands 30 inches higher in the long arm of the tube, then the pressure of the steam is equal to that of two atmospheres, or is 30 pounds to the square inch, and so on.

Fig. 47.

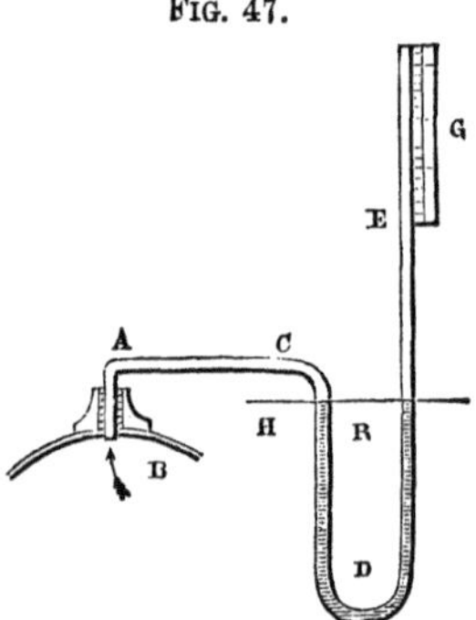

173. **Varying Conditions of Steam-pressure.**—It is to be understood that the relations between the pressure of steam and its temperature which have been pointed out, exist only when the steam is in contact with a body of water from which fresh steam is constantly rising, as in an ordinary steam-boiler. Under such circumstances, the elasticity, or expansive force of the steam, increases rapidly with its increase in temperature, but in

QUESTIONS.—How may the pressure of steam upon the interior of a boiler be determined by means of the thermometer? What is a barometer guage? Under what circumstances do the relations which have been pointed out between the pressure of steam and its temperature exist? In what manner does steam heated apart from water expand?

a greater degree by equal additions of heat at high, than at low temperatures. If, however, the steam is heated apart from water, it follows the law that regulates the expansion of all gaseous bodies, viz., that equal increments of heat expand it equally at all temperatures—this expansion being equal to 1-490th of its volume at 32° F. for every additional degree of heat imparted to it.*

174. **High-pressure Steam.**—Steam generated by water boiling at a very high temperature, is known as high-pressure steam. By this we mean steam condensed, not by the withdrawal of heat, but by pressure, just as high-pressure air is merely condensed air. To obtain double,

* Some very curious experiments which have been made from time to time, seem to show that steam and other vapors, when subjected to extraordinary pressure, do not continue to expand with additions of heat, but actually contract. The first information which was obtained in relation to this subject was from a very dangerous experiment tried many years since in England. A measured quantity of water was placed in a boiler, with all the safety-valves most carefully closed, and every chance for the escape of steam prevented. The fire was now got up, and for some time the steam-guage, as usual, indicated a regularly increasing pressure. At length, however, to the surprise of all, the pressure was seen slowly but gradually to diminish, and although the boiler-plates became nearly red-hot, this remarkable phenomenon continued, and when the boiler had cooled, it was found that no water had escaped.

The experiment was afterward repeated by De la Tour, a French chemist, in a different manner with similar results. He partially filled some very strong glass tubes with water, alcohol, ether, and some other liquids, furnished them with guages, and hermetically sealed them. The tubes were then gradually exposed to heat, until the contained liquids vaporized, and as true steam became transparent, or invisible. Under these circumstances, the law "that the elasticity or expansive force of vapors augments with every additional increase of temperature," was not found to hold good, and the following results were obtained:

All the liquids, by reason of the enormous pressure which the vapor gradually formed from them exerted upon their surfaces, required to be elevated to a high degree of temperature before complete vaporization took place. Ether, which passes into vapor in the open air at a temperature of 96° F., only became vapor at 328°, in a space equal to double its original bulk! At this temperature its vapor should, according to the recognized law of expansion, have exerted a pressure of 200 atmospheres, or more than 3,000 pounds per square inch: it, however, exerted a pressure of only 37 atmospheres, or 555 pounds per square inch. Alcohol, which occupied 2-8ths the capacity of its tube, gradually expanded to double its volume, and then suddenly disappeared in vapor, at a temperature of 404° F.; its calculated pressure was 3,600 pounds per square inch; its real pressure was only 1,700 pounds. Water was found to become vapor in a space equal to about four times its original bulk, at a temperature of about 773°. At this temperature its solvent power was so greatly increased, that it acted most powerfully upon the glass and broke it, and it was found necessary to add carbonate of soda to the water to diminish its action. As the vapors in the tubes cooled, a point was observed at which a sort of cloud filled the tube, and in a few moments after, the liquid suddenly re-appeared.

In explanation of the diminished pressure which vapors of high temperature exert under the above-mentioned conditions, it has been suggested that their particles, by reason of their forced and close contiguity, are partially controlled by a force of cohesion, which in part neutralizes the expansive force imparted by the heat.

QUESTION.—What is high-pressure steam?

triple, or greater pressure of steam, we must have twice, thrice, or more steam under the same volume.

175. **Super-heated Steam.**—Steam which has been heated in a separate state to a high degree of temperature is known as super-heated steam. In this condition it is employed for the production of effects not attainable by the use of ordinary steam; such as the distillation of oils, the carbonization of wood, etc.

In some of the processes recently introduced for the distillations of oils by the use of super-heated steam, the temperature of the steam is elevated to a sufficient degree to melt lead. To effect the carbonization of wood, steam is elevated to a high degree of temperature by passage through red-hot pipes. It is then allowed to enter a vessel containing wood which is intended to be converted into charcoal. The heated steam penetrating into the pores of the wood, drives off the volatile portions, the water, tar, etc., and leaves the pure carbon behind.

In the manufacture of lard on an extensive scale, the carcase of the whole hog is exposed to the action of steam at a very high pressure and temperature. This acting upon the mass of flesh, breaks up and reduces the whole to a fat fluid mass, leaving the bones in the state of powder. Steam of ordinary pressure and temperature, under the same circumstances, would not produce this effect.

176. **Vapor produced by different Liquids.**—Equal bulks of different liquids raised to their respective boiling points, produce very different quantities of vapor.

Water furnishes, bulk for bulk, a much larger amount of vapor than any other liquid; a cubic inch of water at its ordinary boiling point, 212°, expanding to nearly a cubic foot of steam at 212°, or to about 1700 times its volume; a cubic inch of alcohol, on the other hand, at its ordinary boiling temperature, expands only 528 times its volume; ether to 298; and oil of turpentine to 193.

177. **Ratio between Sensible and Latent Heat.**—The sum of the sensible heat of steam, and the quantity of latent heat contained in it, are always the same, since the latent heat of steam diminishes exactly in proportion as its sensible heat rises.

Water may be easily made to boil in a vacuum at the temperature of 100°,

---

QUESTIONS.—What is super-heated steam? For what purposes is it applied? How can wood be carbonized by the use of steam? How is high-pressure steam employed in the manufacture of lard? Is the quantity of vapor produced from equal bulks of liquid the same? What are illustrations of this? What ratio exists between the sensible and latent heat of steam? Is there any economy in evaporating water at a low temperature and under diminished pressure?

but the steam generated is much less dense than that produced at 212° and has a greater latent heat. If water boils at 312°, the amount of heat absorbed (rendered latent) in vaporization, will be less by 100° than if it had boiled at 212°; and, on the contrary, if water be boiled under a diminished pressure, at 112°, the heat absorbed in vaporization will be 100° more than if it had boiled at 212°. Hence there can be no economy of heat in distilling in vacuo.

The sum of the sensible and latent heat of steam being always the same, 1184°, we may very readily ascertain the latent heat of steam at any temperature, by subtracting its sensible heat from this constant number. For example, steam at 280° has a latent heat of 904° (1184−280=904); so also steam at 100° has 1084° of latent heat.

The theory of latent heat, and the principles which govern the formation, expansion, and condensation of vapors, are practically applied in the working of the steam-engine, and in many industrial operations. A further consideration of them is, however, foreign to the object of this work.

178. Liquefaction of Gases.—Gases were formerly considered to be essentially different in their nature from vapors, but comparatively recent experiments have shown that their constitution is similar, and is owing to the latent heat they contain.

Faraday demonstrated the possibility, by the joint action of cold and great pressure, of reducing several of the so-called permanent gases to the liquid and even to the solid state.

The method employed by him was to generate the gas from materials placed in one end of a strong glass tube, bent in the middle, and hermetically sealed, as represented in Fig. 48. The gas, accumulating in a confined space, exerts an enormous pressure in virtue of its expansive force; the effect of which is, that a portion of the gas itself condenses into a liquid in the end of the tube most remote from the materials, which is kept cool by immersion in a freezing mixture This experiment is a somewhat hazardous one, from the liability of the tube to burst under the pressure exerted, and the hands and face of the operator should always be protected by gloves and a mask of wire gauze. In this way chlorine, cyanogen, carbonic acid, and several other gases, may be liquefied.

FIG. 48.

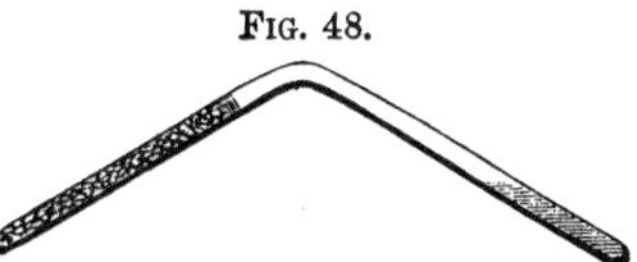

By means of an apparatus of different construction, but involving the same principle, carbonic acid gas can be liquefied and solidified in large quantities. The details of this process will be described under the chemical consideration of this substance.

QUESTIONS.—How may the latent heat of steam be calculated? To what do gases and vapors undoubtedly owe their constitution? Who first liquefied gases? By what means was this accomplished? What gases were thus liquefied?

Some of the gases are liquefiable with much greater facility than others, and a few assume a liquid or solid form by the mere application of cold, as sulphurous acid gas. Others have resisted all attempts to reduce them to a liquid state by subjection to immense pressure aided by the greatest artificial cold. Among these are oxygen, hydrogen, nitrogen, carbonic oxyd, coal gas, etc. Oxygen remained gaseous under a pressure of over 900 pounds to the square inch, and at a temperature of 140° below zero.

179. **Absorption of Gases by Water.**—All gases are absorbed or condensed by water in a greater or less degree, in which case they must certainly assume the liquid form. The quantity absorbed is very different for different gases; and in the same gas the quantity absorbed depends upon the pressure to which the gas is subjected, and the temperature of the water. The colder the water, the greater the quantity of the gas taken up and retained by it.

CHAPTER III.

LIGHT.

180. **Light and its Chemical Relations.**—The general consideration of the laws of light belongs to the science of Optics, a department of Natural Philosophy. Light, however, is an important agent in producing chemical changes, especially in the organized forms of matter; while the physical characters of an object, revealed by the mere mechanical action of light on its structure, are often of the greatest chemical value.

A brief reference to some of the more important laws and physical properties of light, constitutes a proper introduction and preparation for the study of its chemical effects.

SECTION I.

NATURE AND SOURCES OF LIGHT.

181. **Nature of Light.**—Of the real nature of light we know nothing. Two theories or hypothesis, however, have been proposed to account for its phenomena, which are

QUESTIONS.—Are all gases reduced with equal facility? What gases have resisted all attempts to liquefy them? What is said of the absorption of gases by water? What connection is there between light and chemistry? What do we know respecting the real nature of light?

known as the Corpuscular, or Emission theory, and the Undulatory theory.

182. **The Corpuscular Theory** supposes the sensation of light to be occasioned by the transmission of particles of a refined species of matter from the luminous body to the eye.

According to this theory, there is a striking analogy or resemblance between the eye and the organs of smelling. Thus, we recognize the odor of an object in consequence of the material particles which pass from the object to the organs of smelling, and there produce a sensation. In the same manner, a visible object at any distance may be supposed to send forth particles of light, which move to the eye and produce vision, by acting mechanically on its nervous structure, as the odoriferous particles of a rose produce a sensible effect upon the organs of smell.

183. **The Undulatory Theory** supposes that all space, and the interstices of all material objects, are pervaded by an elastic medium, or ether, of inconceivable tenuity. This medium is not light itself, but is susceptible of being thrown into vibrations or undulations by impulses incessantly emanating from all luminous bodies. These, reaching the eye, affect the optic nerve, and produce the sensation which we call light.

According to this theory, there is a striking analogy between the eye and the ear; the vibrations, or undulations of the ethereal medium being supposed to pass along the space intervening between the visible object and the eye, in the same manner as the undulations of the air, produced by a sounding body, are transmitted to the ear.

The corpuscular theory was sustained by Newton, and was for a long time generally believed. Since the commencement of the present century, however, it has been gradually losing ground, and recent experiments instituted by MM. Foucault and Fizeau, of France, conclusively demonstrate its incorrectness. It is now, therefore, entirely discarded by all the leading scientific authorities, and the undulatory theory is received as substantially correct—since it affords the most complete explanation of the facts upon which the science of optics is based. The language, however, which is generally employed in describing optical phenomena is for the most part framed in accordance with the corpuscular theory.

184. **Sources of Light.**—The great natural sources of

QUESTIONS.—Explain the corpuscular theory of light. What analogy does this theory present? Explain the undulatory theory. What analogy, according to this theory, exists between the eye and the ear? Which theory is generally received? What are the sources of light?

light are the sun and the heavenly bodies. All bodies when heated to a sufficient degree become luminous.

All solid bodies begin to emit light in the daytime at the same temperature, viz., 977° of Fahrenheit's thermometer. As the temperature rises, the brilliancy of the light rapidly increases, so that at a temperature of 2600° it is almost forty times as intense as at 1900°. Gases must be heated to a much greater extent before they begin to emit light.

185. **Electric Light.**—The most splendid artificial light known is developed through the agency of electricity.

The electric light, so-called, is produced by fixing pieces of pointed charcoal to the wires connected with opposite poles of a powerful galvanic battery, and bringing them within a short distance of each other. The space between the points is occupied by an arch of flame that nearly equals in dazzling brightness the rays of the sun.

186. **Phosphorescence.**—The term phosphorescence is applied to that property which various bodies possess of emitting a feeble light at ordinary, or low temperatures.

Phosphorescence was formerly supposed to be due to the presence of phosphorus (an elementary substance which emits light in the dark). Hence the origin of the name. The phenomenon is now known to proceed from other agencies.

A great number of bodies possess the property of shining in the dark when they have been previously exposed to the light of the sun. Oyster shells which have been ignited and cooled, especially exhibit phosphorescence. Among other substances which are often luminous in the dark, are white paper (especially when it has been heated nearly to burning), egg-shells, corals, bones, ivory, leather, and the skins of men and animals. The cause of this phenomenon is, probably, that the bodies by being exposed to light, absorb a portion of it unaltered into their substance by adhesion, and subsequently give it out in a dark place.—GMELIN.

The phenomenon of phosphorescence occurs in the most marked degree in living organized bodies. The glow-worms, and several species of flies and beetles, have the power of emitting from their bodies a beautiful pale, bluish white light. The great lantern-fly of South America is especially brilliant—a single insect affording sufficient light to enable a person to read. The appearance of vast luminous tracts in the sea, at night, is a well-known phenomenon. This was formerly ascribed to the motion of the waves, to electricity, or to the formation of gases containing phosphorus, through the putrefaction of marine animals; but it is now generally believed to be due to the presence of an immense number of phosphorescent animalculæ.

QUESTIONS.—At what temperature do solids become luminous? How is the most splendid artificial light produced? What is phosphorescence? Under what circumstances do bodies often become luminous? How is the phenomenon accounted for? What substances exhibit phosphorescence in the most marked degree? What are remarkable instances of phosphorescence in the animal kingdom? To what is the luminous appearance of the sea due?

Sea-fish, in general, soon after death exhibit a luminous appearance, particularly the herring and the mackerel. The light is most intense before putrefaction commences, and gradually disappears as decomposition proceeds. In order to observe the phenomenon more distinctly, the fish should be gutted, and the roes and scales removed. By placing such luminous fish also in weak saline solutions, such as those of Epsom salts or common salt, the solutions even become luminous, and the appearance continues for some days; it is particularly noticeable when the liquids are agitated. The light is quickly extinguished by the addition of pure water, of lime water, and by acids in general.

The decay of wood, when the temperature is moderate and moisture and a small quantity of air are present, is frequently attended with an evolution of light. Wood exhibiting this appearance is familiarly known as "*light wood*," and is of a white appearance. When wood decays in the presence of much moisture and a free access of air, it is reduced to a brown pulverulent mass which is not luminous. The phosphorescence of wood ceases when the temperature falls as low as 42° F., and it is also irrecoverably destroyed by the action of boiling water.

The cause of phosphorescence is not fully understood; it is, however, believed to be the result of a chemical action between the oxygen of the air, or water, and the so-called phosphorescent matter. This matter is capable of separation from the living animal, and is characterized by a remarkable and disagreeable odor.

Light is also developed, under certain circumstances, in the act of crystallization.

If the process of crystallizing certain substances be watched in a darkened room, the separation of each crystal will be observed to be accompanied with a faint flash of light.

SECTION II.

PROPERTIES OF LIGHT.

187. **Propagation of Light.**—Light, from whatever source it may be derived, moves in straight lines, or rays, so long as the medium traversed is uniform in density.

By a medium, we mean the space or substance through which light passes. In taking aim with a gun or arrow, we proceed upon the supposition that light moves in straight lines, and try to make the projectile go to the desired object as nearly as possible by the path along which the light comes from the object to the eye.

QUESTIONS.—What circumstances attend the decomposition of sea-fish? What is said of the luminosity of decayed wood? What is the supposed cause of phosphorescence? Is light ever developed by the act of crystallization? In what manner is light propagated?

Thus, in Fig. 49, the line A B, which represents the line of sight, is also the direction of a line of light passing in a perfectly straight direction from the object aimed at to the eye of the marksman.

FIG. 49.

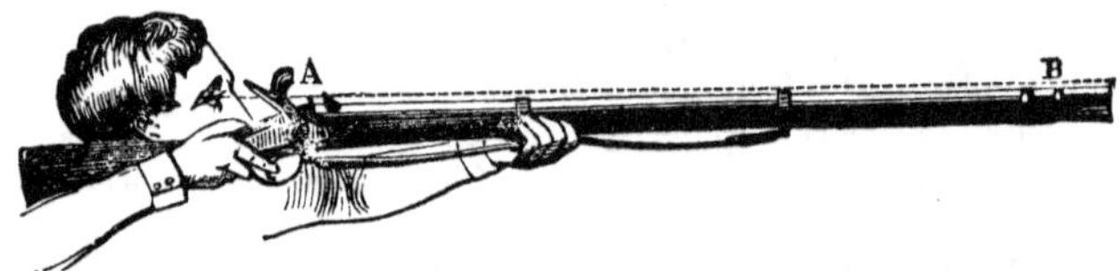

188. **Divergence of Light.**—Rays of light proceeding from a luminous body diverge, or spread out from one another in every direction.

189. **Law of Diminution of Light by Distance.**—When light diverges from a luminous center, its intensity diminishes, not according to the distance, but as the square of the distance.*

Thus, at a distance of two feet, the intensity of light will be one fourth of what it is at one foot; at three feet the intensity will be one ninth of what it is at one foot. In other words, the amount of illumination at the distance of one foot from a single candle would be the same as that from four, or nine candles at a distance of two, or three feet, the numbers four and nine being the square of the distances two, and three, from the center of illumination.

190. **Velocity of Light.**—Light does not pass instantaneously through space, but requires for its passage from one point to another a certain interval of time.

The velocity of light is at the rate of about one hundred and ninety-two thousand miles in a second of time.

191. **Action of Light on Matter.**—When light falls upon any object, it may be disposed of in three ways; 1st, it may be bent back, or reflected; 2d, it may be absorbed into the substance of the body, and disappear; or 3d, it may be transmitted, or pass through the body.

* It is an exceedingly curious fact, that this law of the variation of influence according to the square of the distance, applies to all physical forces which spread or radiate from a center, such as gravitation, heat, light, electricity, magnetism, and sound.

QUESTIONS.—What is meant by the divergence of light? How does the intensity of light diminish by distance? Illustrate this law. What is the velocity of light? How is light falling upon the surface of a body disposed of?

When the portion of light reflected from any surface, or point of a surface, to the eye is considerable, such surface, or point, appears white; when very little is reflected, it appears dark-colored; but when all, or nearly all the rays are absorbed, and none are reflected back to the eye, the surface appears black.

192. **Transparent and Opaque Bodies.**—Bodies which allow the light which falls upon their surfaces to pass through them, are said to be transparent; while those which prevent its passage are said to be opaque.

193. **Luminous Bodies** are those which shine by their own light; such, for example, as the sun, the flame of a candle, metal rendered red hot, etc.

All bodies not in themselves luminous, become visible by reflecting the rays of light.

194. **Law of Reflection of Light.**—The law which governs the reflection of light is exceedingly simple, and is the same as that which governs the motion of an elastic body thrown against a hard, smooth surface. If the light falls perpendicularly upon a flat surface, it is turned back, or reflected perpendicularly, and in the same lines; if it falls obliquely, it is reflected obliquely, the angle of incidence being equal to the angle of reflection.

Thus, in Fig. 50, let A B represent the direction of an incident ray of light falling on a mirror, F C. It will be reflected in the direction B E. If we draw a line, D B, perpendicular to the surface of the mirror, at the point of reflection, B, it will be found that the angle of incidence, A B D, is precisely equal to the angle of reflection, E B D. If the light falls perpendicularly upon the surface, F C, as in the direction D B, it will be reflected in the same line, B D; or in other words, the incident and reflected ray will coincide.

Fig. 50.

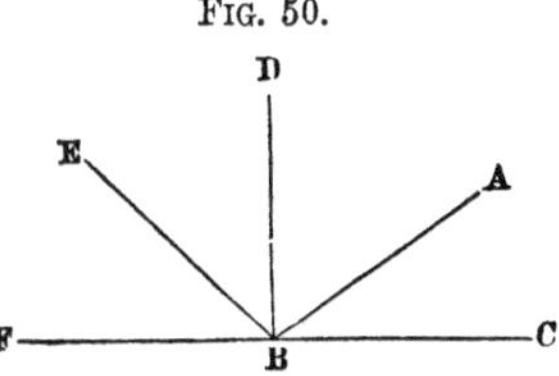

The same law holds good in regard to every form of surface, curved as well as plane, since a curve may be supposed to be formed of an infinite number of little planes.

QUESTIONS.—When is a body light-colored, and when dark? What are transparent and opaque bodies? What are luminous bodies? How are bodies not luminous rendered visible? What is the law of the reflection of light?

195. **Refraction.**—When a ray of light falls *perpendicularly* upon the surface of an uncrystallized transparent substance of uniform density, it continues on its course unchanged; but if it falls upon the surface *obliquely*, its direction is suddenly changed as it enters the transparent object, or medium; it then passes on in its new direction in a straight line, and on quitting the medium, it is again abruptly bent back to its original course, provided the surface of entrance and the surface of exit be parallel to each other. Such a change in the course of a ray of light is termed Refraction.

When the ray of light passes from a rarer to a denser medium (as from air into glass or water), the ray is bent or refracted toward a line perpendicular to that point of the surface on which the light falls; when, on the contrary, the ray passes from a denser to a rarer medium, the ray is bent in the opposite direction, or from the perpendicular.

Fig. 51.

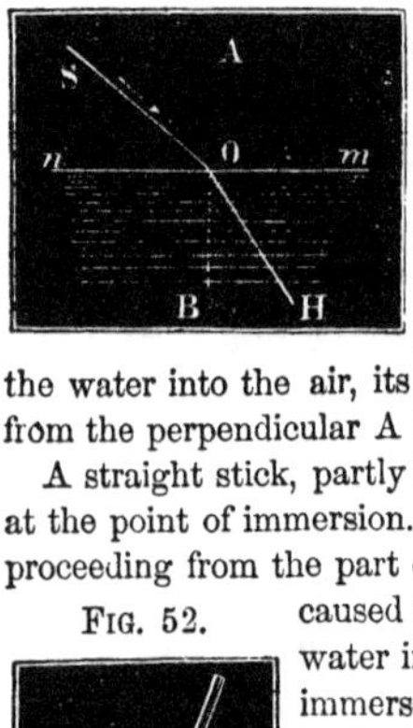

Thus, in Fig. 51, suppose *n m* to represent the surface of water, and S O a ray of light striking upon its surface. When the ray S O enters the water, it will no longer pursue a straight course, but will be refracted, or bent toward the perpendicular line, A B, in the direction O H. The denser the water or other fluid may be, the more the ray S O H will be refracted, or turned toward A B. If, on the contrary, a ray of light, H O, passes from the water into the air, its direction after leaving the water will be further from the perpendicular A B, in the direction O S.

A straight stick, partly immersed in water, appears to be broken or bent at the point of immersion. This is owing to the fact that the rays of light proceeding from the part of the stick contained in the water are refracted, or caused to deviate from a straight line as they pass from the water into the air; consequently that portion of the stick immersed in the water will appear to be lifted up, or to be bent in such a manner as to form an angle with the part out of the water.

Fig. 52.

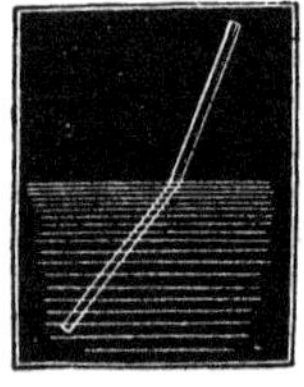

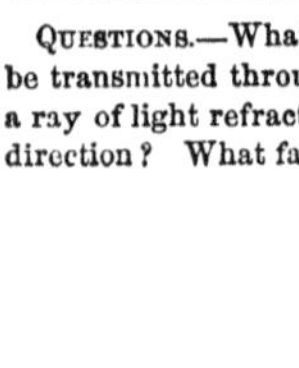

The bent appearance of the stick in water is represented in Fig. 52. For the same reason, a spoon in a glass of water, or an oar partially immersed in water, always appears bent.

Questions.—What is understood by the refraction of light? When will a ray of light be transmitted through a transparent substance without refraction? In what manner is a ray of light refracted in passing from a rarer to a denser medium, and in the reverse direction? What familiar fact illustrates this principle?

196. Variations of Refractive Power.—No law has yet been discovered which will enable us to judge of the refractive power of bodies from their other qualities. As a general rule, dense bodies have a greater refractive power than those which are rare; and the refractive power of any particular substance is increased or diminished in the same ratio as its density is increased or diminished.

Refractive power seems to be the only property, except weight, which is unaltered by chemical combination; so that by knowing the refractive power of the ingredients, we can calculate that of the compound.

All highly inflammable bodies, such as oils, hydrogen, the diamond, phosphorus, sulphur, amber, camphor, etc., have a refractive power from ten to seven times greater than that of incombustible substances of equal density.

Of all transparent bodies the diamond possesses the greatest refractive or light-bending power, although it is exceeded by a few deeply-colored, almost opaque minerals. It is in part to this property that the diamond owes its brilliancy as a jewel.

Many years before the combustibility of the diamond was proved by experiment, Sir Isaac Newton predicted, from the circumstance of its high refractive power, that it would ultimately be found to be inflammable.

The determination of the refracting power of a body is often a valuable guide in estimating its chemical purity. The adulteration of essential oils may in this way be often detected with ease, when it would be otherwise difficult to ascertain it. Thus genuine oil of cloves has a refractive power expressed by the numbers 1,535, while that of an impure and adulterated specimen was not more than 1,498.

197. Double Refraction is a property which certain transparent substances possess, of causing a ray of light in passing through them to undergo two refractions; that is, the single ray of light is divided into two separate rays.

A very common mineral called "Iceland spar," which is a crystallized form of carbonate of lime, is a remarkable example of a body possessing double refracting properties. It is usually transparent and colorless, and its crystals, as shown in Fig. 53, have the geometrical form of a rhomb, or rhomboid;—this term being applied to a solid bounded by parallel faces, inclined to each other at an angle of 105°.

FIG. 53.

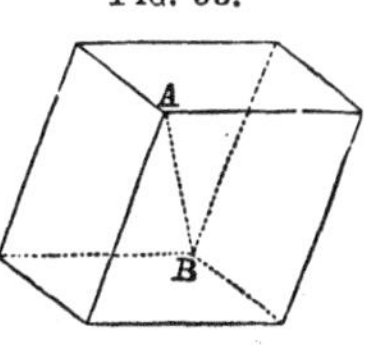

QUESTIONS.—What estimate can we form of the refractive power of a body from its other qualities? What is the refractive property of inflammable substances? What transparent substance possesses the greatest refractive power? How may refraction be used for determining the chemical purity of a substance? What is an illustration of this? What is double refraction? What substance possesses doubly refracting powers in a remarkable degree?

The manner in which a crystal of Iceland spar divides a ray of light into two separate portions is clearly shown in Fig. 54; in which S T represents a ray of light, falling upon a surface of a crystal of Iceland spar, A D E C, in a perpendicular direction. Instead of passing through without any refraction, as it would in case it had fallen perpendicularly upon the surface of glass, the ray is divided into two separate rays, the one, T O, being in the direction of the original ray, and the other, T E, being bent or refracted. The first of these rays, or the one which follows the ordinary law of refraction, is called the "ordinary" ray; the second, which follows a different law, is called the "extraordinary" ray.

FIG. 54.

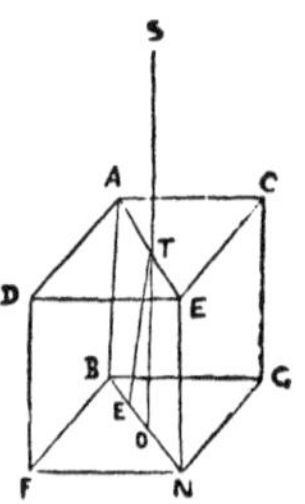

If we look at an object, as a dot, a letter, or a line, through a plate of glass it appears single; but if a double refracting substance, as a plate of Iceland spar, be substituted, a double image will be perceived, as two dots, two letters, two lines, etc. This result of double refraction is represented in Fig. 55.

FIG. 55.

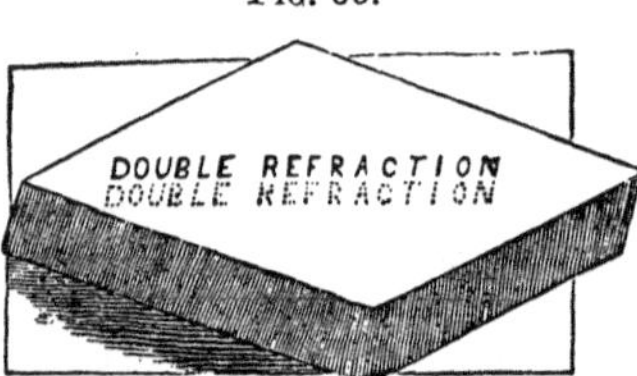

The phenomenon of double refraction is due entirely to the peculiar molecular structure of the medium through which the light passes. This is proved by taking a cube of regularly annealed glass, which produces but one refracted ray, and heating it unequally, or subjecting it to pressure: a change is thereby effected in the arrangement of its parts, and double refraction takes place.

The diamond may be distinguished from all other precious stones, with a single exception (the garnet), by having only a single refraction, the others possessing double refraction, or giving a double image of a taper or small light viewed through their faces. By the same means all precious stones, except diamond and garnet, may be distinguished from artificial ones, by the former having double refraction, and the latter only single refraction.

198. **Polarization.**—Light which has been refracted from certain surfaces, or transmitted through certain substances, under certain special conditions, assumes new properties, and is no longer reflected, refracted, or transmitted as before. This change in the action of light is called Polarization, and a ray thus modified is said to be polarized.

A ray of light which by any method has become polarized, seems to have

QUESTIONS.—To what is this phenomenon due? How may the diamond be distinguished from all other precious stones? What is polarized light? What is the origin and explanation of this term?

acquired a property of possessing sides. If the original ray be supposed to be a cylindrical rod, polished or white all round, which is capable of being reflected from a polished surface whatever part of its circumference may strike that surface, the polarized ray may be compared to a square-shaped rod with four flat sides, two of which (opposite), bright and polished, are capable of reflection, while two, black or dull, are not. Now, the word "poles," in physical science, is often used to denote the ends or sides of any body which have acquired contrary properties, as the opposite ends of a magnet, which are called the positive and negative poles. By analogy, the ray of light whose sides, situated at right angles with each other, were found to be endowed with opposite physical properties, was said to be polarized. The term is unfortunate, but is too firmly engrafted upon science to be changed.

The explanation of the change occasioned by polarization of light may be briefly stated as follows:—According to the undulatory theory, common light is assumed to be produced by vibrations of the ethereal particles in two planes at right angles to the progress of the wave; there are perpendicular vibrations, and there are horizontal vibrations. Polarized light, on the contrary, is light occasioned by vibrations taking place in only one plane—the effect of whatever produces polarization being to suppress all the vibrations which take place in one plane at right angles to the other. Hence the different properties possessed by opposite sides or poles of the ray.

Common light is converted into polarized light, for all practical purposes and for experiment, in three ways—

First,—When it is reflected from glass at an angle of incidence of fifty-six degrees, forty-five minutes from the perpendicular. It is also polarized by reflection from almost any bright non-metallic surface, but the maximum polarizing angle for each different surface is peculiar to itself. When the reflection from glass takes place at the exact angle of 56° 45′, all the light is polarized, but when the angle of reflection deviates from this amount, some of the reflected light will remain unchanged, the quantity unpolarized being in proportion to the deviation.

Secondly,—Light may be polarized by transmission through a bundle consisting of from sixteen to eighteen plates of thin glass or mica.

Thirdly,—Light is polarized by passing through certain transparent crystals, especially those which possess the property of double refraction.

199. **Peculiarities of Polarized Light.**—If a ray of light which has been polarized by reflection from a glass plate is caused to fall upon a second plate, it is not reflected as common light would be. If the plane of the second reflecting surface is so inclined to the first, that the ray falls at an angle of 56°, the ray is not reflected at all, but vanishes; if, on the contrary, the plane of the second reflecting surface is parallel to the first, it is entirely reflected. It is also a peculiar property of polarized light,

---

Questions.—In what three ways may light be polarized? What peculiarities are manifested by light polarized by reflection from glass? How is polarized light affected by certain transparent substances?

that it will not pass through certain substances which are transparent to common light. This is shown in a remarkable manner by a mineral substance called tourmaline, the internal structure of which is such, that a ray of common light which has passed through a thin plate of it, and thereby become polarized, can not pass through a second similar plate, if it is placed at right angles to the first.

For example, in Fig. 56, if a ray of light be caused to pass through a thin plate of tourmaline, as *c d*, in the direction of the line *a b*, and be received upon a second plate, *e f*, placed symmetrically with the first, it passes through both without difficulty; but if the second plate be turned a quarter round, as in the direction *g h*, the light is totally cut off.

FIG. 56.

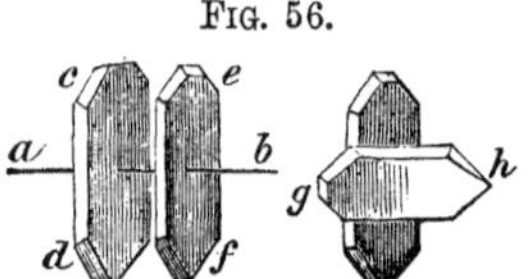

200. **Discovery of Polarized Light.**—The phenomenon of polarized light was discovered in 1808, by Malus, a young engineer officer of Paris. On one occasion, as he was viewing through a double refracting prism of Iceland spar the light of the sun reflected from a glass window in one of the French palaces, he observed some very peculiar effects. The window accidentally stood open like a door on its hinges, at an angle of 56° and Malus noticed that the light reflected at this angle was endowed with properties which distinguish it from ordinary light.

201. **Practical Applications of Polarized Light.**—The principles of polarized light have been applied to the determination of many practical results. Thus, it has been found that all reflected light, come from whence it may, acquires certain properties which enable us to distinguish it from direct light; and the astronomer, in this way, is enabled to determine with infallible precision whether the light he is gazing on (and which may have required hundreds of years to pass from its source to the eye), is inherent in the luminous body itself, or is derived from some other source by reflection.

It has been also ascertained by Arago that light proceeding from incandescent bodies, as red hot iron, glass, and liquids, under a certain angle, in polarized light; but that light proceeding, under the same circumstances, from an inflamed gaseous substance, such as is used in street illumination, is always in a natural state, or unpolarized. Applying these principles to the sun, he discovered that the light-giving substance of this luminary was of the nature of a gas, and not a red hot solid or liquid body.

When we transmit light, whether *common* or *polarized*, through a piece of well annealed glass, it suffers no change, and we see no structure in the glass different from what we would see if we looked through pure water. But if

---

QUESTIONS.—Illustrate this in the case of tourmaline. When and how was polarized light discovered? What are some of the practical applications of polarized light? What is the difference between light emitted from incandescent solids and inflamed gases? What inference has Arago made respecting the constitution of the sun? What information does polarized light impart respecting the structure of bodies?

we make heat pass through the glass, by placing the edge of the plate upon a heated iron, or if we either bend or compress the glass by mechanical force, its structure, or the mechanical condition of its particles, will be changed. If we now transmit common light through the glass thus changed, the change will not be visible; but if we transmit polarized light through it, and allow that light to be reflected from a transparent body at an angle of about 56°, and in a plane at right angles to that in which the common light was reflected and polarized, the observer, looking through the glass, will see the most brilliant colors, indicating the effects of the compressing or dilating forces, or of the contracting or expanding cause—the degree of compression or dilatation, of expansion or contraction, being indicated by the colors displayed at particular parts of the glass. In this way polarized light enables us to discover that certain portions of a body have been subjected to certain mechanical forces, the nature of which must be sought for in the circumstances under which the body has been originally formed, or in which it has been subsequently placed. On this principle, many bodies which are quite transparent to the eye, and which upon examination appear to be perfectly uniform, or homogenous in structure, exhibit, under polarized light, the most exquisite organization.*

---

* "Integumentary substances in particular form a brilliant and interesting class of objects. A section of a horse's hoof has the effect of the richest Brussels' carpet, with a symmetrical pattern that might be copied by the loom.

"The vegetable world has a less brilliant display to make, but is still replete with interest. Cuticles containing flint are often very beautiful; that of the common marestail presents a remarkably neat shawl pattern in stripes. Very curious optical effects are presented by the various starches. The starch called tous-les-mois, having the largest grains, is usually selected for exhibition.

"Crystalline forms, however, afford the most striking exhibitions of the phenomena of polarized light. Salacine, a salt extracted from the bark of the willow, offers, when almost an imperceptible film, the appearance of a pavement consisting not merely of gold, but of lapis lazuli, ruby, emerald, and opal. Chlorate of potash strews the field of view with liberal handfuls of pyramidal jewels. Chromate of potash, which forms a bright yellow solution, presents a remarkable assemblage of club-shaped crystals, which have been compared to vast heaps of constables' staves. Oxalate of potash, like several other combinations of oxalic acid, is a salt of such variety and brilliancy, that its crystals, floating and glowing in a few drops of solution on the slide, look as if their form and color were the result of a Chinese imagination in its happiest moments.

"Fancy yourself living in a region solely illuminated by Aurora borealis—imagine a country where every passing cloud throws a diverse-colored shadow of gorgeous hues across your path; where the air breeds rainbows without the aid of a shower, and where the summer breeze breaks those rainbows into irregular lengths, fragments, and glittering dust, scattering them broadcast over the land, like autumnal leaves swept by a gale from the forest, and you have an approximate, and by no means exaggerated idea of the effects of polarized light on substances capable of being affected by it. For, it is light endowed with extra delicacy, subtlety, and versatility. It renders visible minute details of structure in the most glaring colors; it gauges crystalline films of infinitesimal thinness: it betrays to the student's search, otherwise inappreciable differences of density or elasticity in the various parts of tissues. Indeed, as a detector, polarized light is invaluable, acting the part of a spy under the most unexpected circumstances. It denounces as cotton what you believed to be silk; it demonstrates disease where you supposed health. It adorns objects that are vile and mean, whose destiny is only to be cast out—such as parings of nails, shavings of animals' hoofs, cuticle rubbed or peeled from the stems of

In a similar manner the chemist is able to determine, by the manner in which light is reflected or polarized by a crystallized body, whether it has been adulterated by the addition of foreign substances. Polarized light, also, in certain cases, affords the best means of arriving at a knowledge of the varieties and proportion of sugar in the juices of plants, and in complex saccharine liquids.

202. **Magnetization of Light.**—Recent experiments made by Professor Faraday have proved that magnetism has the power of influencing a ray of light in its passage through transparent bodies. This fact is shown by the following experiment:—A ray of polarized light is passed through a piece of glass, or a crystal, or along the length of a tube filled with some transparent fluid, and the line of its path carefully observed; if, when this is done, the solid or fluid body is brought under powerful magnetic influence, such as may be called into action by the circulation of an electric current around a bar of soft iron, it will be found that the polarized light is disturbed, and that it does not continue to pass through the medium along the same line. "As this effect is most strikingly shown in bodies of the greatest density and diminishes in fluids, the particles of which are easily movable upon each other, and has not as yet been observed in any gaseous medium, the question has arisen, does magnetism act directly upon the ray of light, or only indirectly, by producing a molecular change in the body through which the ray is passing? In the present state of science no satisfactory reply can be given."—ROBERT HUNT.

203. **Decomposition of Light.**—When a beam of light, S A, Fig. 57, from the sun is admitted into a dark room, by a small aperture in the window-shutter, and is intercepted in its passage by a wedge, or solid angle of glass called a prism, it is refracted, or bent from its course as it enters, and again as it issues from the glass. In place of forming a circular spot of white light on the floor of the apartment, as it would have done if allowed to proceed in its original direction, S K, it illuminates with several colors an oblong space, H, on the opposite wall, or on a white screen properly placed to receive it. This oblong colored image is called the prismatic, or solar spectrum.

Newton, who first carefully investigated this remarkable fact, distinguished seven different colors, which gradually shade off one into the other in the following order, commencing at the upper part of the spectrum, viz., violet, indigo, blue, green, yellow, orange, and red.

White light may, therefore, be regarded as the result

plants, offscouring of our kitchens and store-rooms, sugar, acids, and salts—with the most magnificent, the most resplendent tints, such as are seen when the sun streams through the stained glass windows of a Norman cathedral."

QUESTIONS.—Can polarized light be made available in determining the chemical character of a substance? What influence has magnetism on light? What is meant by the decomposition of light? What is the solar spectrum? How are the colors of the spectrum arranged? How may white light be regarded?

of a mixture of rays of different colors, which are unequally acted upon by the prism—each color possessing its own peculiar refrangibility.

Thus the red rays, which are the least refracted, or the least turned from their course by the prism, always occur at the bottom of the spectrum, while the violet, which is the most refracted, occurs at the top; the remaining colors being arranged in the intermediate space in the order of their refrangibility.

FIG. 57.

The seven different rays of light, when once separated and refracted by a prism, are not capable of being analysed by refraction again; but if by means of a convex lens they are collected together and converged to a focus, they will form white light.

204. **Lines in the Solar Spectrum.**—When the solar spectrum is formed in the usual manner upon a white screen, it appears like a continuous band of colored light. By taking certain precautions, however, it may be seen that this luminous band is traversed in the direction of its breadth by numerous dark lines, varying in different parts in width and distinctness; or, in other words, there are interruptions in the spectrum where there is no light of any color. These lines are independent of the refracting medium, and always occur in the same color and at corresponding points of the spectrum.

The position of these dark spaces varies, however, with the source of light. With a few exceptions, each of the fixed stars has a system of lines peculiar to it. The light proceeding from the fixed stars Sirius and Castor agree very nearly in this respect, but differ from the light of the sun. The spectrum, however, which is formed from the light proceeding from the fixed star Pol-

QUESTIONS.—Are the colored rays capable of further decomposition by refraction? What effect results from their union? What lines are seen in the spectrum? What differences have been observed in light emanating from different sources?

lux is the same as that of the sun. Every artificial light, also, shows some peculiarity in this respect.

Recent discoveries have given to these phenomena an entirely chemical character. It has been found that the white light of ordinary flames requires merely to be sent through a certain gaseous medium (such as nitrous acid vapor) to acquire more than a thousand dark lines in its spectrum; and it has hence been inferred, that it is the presence of certain gases in the atmosphere of the sun and of the fixed stars, which occasion the observed deficiencies in the spectra formed from their light. In this way points of resemblance and difference may be traced between the constitution of our sun and the suns of other systems.

In Fig. 58, No. 1 shows the principal dark lines of the pure solar spectrum; No. 2, the alteration occasioned by passing solar light through the vapor of bromine; while No. 3 represents the very different result effected by the peroxyd of nitrogen.

FIG. 58.

205. Calorific and Chemical Elements of Solar Light.—Solar light, in addition to the luminous principle which produces the phenomena of color and is the cause of vision, contains two other principles, viz., heat and actinism, or the chemical principle. These principles are invisible to the eye, and have only been discovered by their effects on other bodies.

The constitution of the solar ray may be compared to a bundle of three sticks, one of which represents heat, another light, and a third the actinic principle.

We know that these three principles exist in every ray of solar light, because we are able to separate them in a great degree from each other. Thus, when we decompose a ray of solar light by means of a prism, and throw the spectrum upon a screen, the luminous, the calorific, and the chemical or actinic radiations, will each be refracted, or bent out of their course in different

QUESTIONS.—What discoveries have given to these lines a chemical character? What three principles are included in solar light? How do we know of the existence of these principles? How are they affected by the prism?

degrees, and will consequently assume different positions upon the screen. In other words, the light of the sun refracted by the prism produces in reality three spectra, one visible and two invisible.

The calorific, or heat radiations, will be refracted least, and their maximum point will be found but slightly thrown out of the right line which the solar ray would have traversed had it not been intercepted by the prism. The heat diminishes with much regularity on each side of this line.

The luminous radiations are subject to a greater degree of refraction; their point of maximum intensity being in the yellow ray, lying considerably above the point of greatest heat. The light diminishes on each side of it, producing orange, red, and crimson colors below the maximum point, and green, blue, and violet above it.

The radiations which produce chemical action are more refrangible than either the calorific or luminous radiations, and the maximum of chemical power is found at that point of the spectrum where light is feeble, and where scarcely any heat can be detected.

The positions in the spectrum of the heat and actinic radiations, which are invisible to the eye, may be found by experiment. Thus, if we place a delicate thermometer in the different rays of the spectrum (§ 203, Fig. 57), it will be found that the indigo and violet rays scarcely affect it at all, while the yellow ray, which is the most luminous, is inferior in heating action to the red ray, which, yielding but little light, possesses the greatest amount of heat. If now the thermometer be carried a little below and just out of the red ray, into the darkened space, it will exhibit the greatest increase in temperature, thus proving the presence of a heating ray in solar light independent of the luminous ray. In a like manner, by substituting a chemically prepared surface, as a piece of photographic paper, for the thermometer, the presence of a chemical ray can be proved in the darkened space at the other end of the spectrum, and near to the blue and violet rays.

206. **Analysis of Heat.**—The heat emanating from the sun or from a bright flame, consists of rays which differ from each other as much as the red, yellow, and blue rays do which constitute white light. Heat radiated from a body having a lower temperature than 800° F., is much less refrangible than red light; but if the temperature of the radiating body be increased, it emits, in addition to the rays previously emitted, others of a higher refrangibility, until at last some few of its rays become as refrangible as the least refrangible rays of light. The body then appears of the same color as the least refrangible rays of light, and is said to be *red* hot. If it be heated more, it emits, in addition to the red, still more refrangible rays, viz., orange; then (at a higher temperature) yellow rays are added, and so on, until when the body is *white* hot, it emits all the colors visible to us; and in some instances (of very intense heat), even the invisible chemical rays, more refrangible than the violet, are emitted, though in less quantity than in the solar rays.

QUESTIONS.—Is heat emanating from various sources uniform in character? How do the rays of heat differ in refrangibility?

Thus light, in one sense, appears to be nothing more than visible heat, and heat invisible light—the constitution of the eye being such that it can perceive one and not the other, in the same way as the ear can appreciate vibrations of sound more rapid than sixteen per second, but not those which are less rapid.

A series of interesting experiments made some years since by Melloni, show very conclusively that heat emanating from different sources differs in its nature, in the same manner as the light of a red body differs from that of a blue. He employed four sources of caloric, two of which were luminous and two non-luminous, or obscure; namely, an oil-lamp without a glass, incandescent platinum, copper heated to 696° F., and a copper vessel filled with water at a temperature of 178° F. Rock-salt transmitted heat in the proportion of 92 rays out of every 100 from each of these sources; but every other substance pervious to radiant heat, whether solid or liquid, transmitted more caloric from sources of high temperature than from such as were low. For instance, a clear and limpid mineral, the fluate of lime, transmitted in the proportion of 78 rays out of 100 from the lamp, 69 from the platinum, 42 from the copper, and 33 from the hot water; while transparent rock crystal transmitted 38 rays in 100 from the lamp, 28 from the platinum, 6 from the copper, and 9 from the hot water. Pure ice transmitted only in the proportion of 6 rays in the 100 from the lamp, and entirely excluded those from other sources.

The discovery of the fact that heat proceeding from the sun or any other luminous body is susceptible of division into rays, differing in nature and in refrangibility, has furnished an explanation of many curious phenomena. Heat from very intense sources is more refrangible and passes more readily through most substances than heat of low intensity. Thus, the heat of the sun passes readily through glass, but the heat of a fire is almost entirely obstructed. Advantage has been taken of this fact by those who have occasion to inspect the progress of operations carried on in furnaces; since they are able, by the use of a glass screen, to protect the face from the scorching rays which the glass absorbs, although it offers no impediment to the transmission of light.

It is a well-known fact that snow which lies near the trunks of trees or other like substances, is melted much more rapidly than that exposed to the action of the direct rays of the sun. The reason of this is, that the heat of the sun, being heat of high intensity and high refrangibility, passes through the snow without experiencing a great degree of absorption; but solar heat, which first falls upon the tree and is then radiated upon the snow, is thereby changed into heat of low refrangibility, and is readily absorbed instead of being transmitted.

207. **Action of the Chemical Rays.**—The chemical principle

QUESTIONS.—Describe the experiments of Melloni. What results have followed the discovery of the analysis of heat? Why will glass transmit heat from the sun, and not from a fire? How does the action of light on snow vary? What is the character of the chemical principle of light?

of light is, without doubt, like the calorific principle, composed of rays of different character, and of different refrangibility. Recent experiments of Professor Stokes of England, seem to show that when the invisible rays which occupy in the spectrum a position beyond the violet, are caused to pass through a solution of quinine, they are changed in refrangibility, and become visible—appearing as a sky-blue light at a point far beyond the usual luminous limit of the spectrum. This phenomenon has been termed the "degradation of light."

The study of the chemical principle contained in the rays of solar light has rendered probable the curious fact, that no substance can be exposed to the sun's rays without undergoing a chemical change; and from numerous examples it would seem that the changes in the molecular condition of bodies which sunlight effects during the daytime, is made up during the hours of night, when the action is no longer influencing them. Thus darkness appears to be essential to the healthy condition of all organized and unorganized forms of matter.

The process of forming Daguerreotype and other photographic pictures, depends solely upon the actinic, or chemical influence of the solar ray.

The term "photography," signifying light drawing, which is the general name given to this art, is unfortunate and ill-chosen, for not only does light not exercise any influence in producing the pictures, but it tends to destroy them.

That the luminous principle is not necessary for the success of the photographic process, may be proved by the experiment of taking a daguerreotype in absolute darkness. This can be accomplished in the following manner:—A large prismatic spectrum is thrown upon a lens fitted into one side of a dark chamber; and as the actinic power resides in great activity at a point beyond the violet ray, where there is no light, the only rays allowed to pass the lens into the chamber are those beyond the limit of coloration, and non-luminous; these are directed upon any object, and from that object radiated upon a highly sensitive photographic surface. In this way a picture may be formed by radiations which produce no effect upon the eye.

It has also been found that the yellow, the orange, and the red rays of light possess the power of retarding by their presence all chemical or photogenic action, in proportion to their predominance; and if unaccompanied by other light, they arrest the effects of the chemical rays altogether. On the contrary, the violet, indigo, and blue rays of light favor chemical action. This is clearly exemplified in the following manner:—If an engraving be covered one half with a yellow glass, and placed in front of a camera for the pur-

QUESTIONS.—What experiments have been made by Mr. Stokes? What curious fact has the study of the chemical principle of light evolved? Upon what does the production of photographic pictures depend? What experiment shows that light is not necessary for the production of a photographic picture? How do the different luminous rays of the solar beam affect the chemical principle? What experiments and facts illustrate their relative action?

pose of representation on a daguerreotype plate, an accurate copy will be shortly obtained of the uncovered portion, while the yellow screen entirely prevents the plate from receiving an impression of the rest. But if the engraving be covered, one half with blue and the other half with yellow glass, while it will be distinctly discernible to the eye through the latter and not at all through the former, the camera will faithfully copy the portion which is invisible, but wholly neglect the other. Again, in a room illuminated solely through red, or orange glass, in which light may fall with dazzling luster, no photographic operations can be conducted; while if blue glass be substituted, the change, while it will dim the effulgence, will enable the photographer to exercise his art with success. In the same way, during certain states of the atmosphere, there may be an abundance of illuminating, but very few photogenic rays.

208. **Influence of Light on Vegetation.**—There are many reasons for supposing that each of the three principles, light, heat, and actinism, included in the solar ray, exercise a distinct and peculiar influence upon vegetation. Thus the luminous principle controls the growth and coloration of plants, the calorific principle their ripening and fructification, and the chemical principle the germination of seeds. Seeds which ordinarily require ten or twelve days for germination, will germinate under a blue glass in two or three. The reason of this is, that the blue glass permits the chemical principle of light to pass freely, but excludes, in a great measure, the heat and the light. On the contrary, it is nearly impossible to make seeds germinate under a yellow glass, because it excludes nearly all the chemical influence of the solar ray.

Further consideration of the chemical effects of light will be postponed until after the chemical properties of the elementary bodies have been described.

CHAPTER IV.

ELECTRICITY.

209. **Electricity** is a subtile agency or force, without weight or form, that appears to be diffused through all nature, existing in all substances without affecting their volume or their temperature, or giving any indication of its presence when in a latent, or ordinary state. When, however, it is liberated from this repose, it is capable of

---

QUESTIONS.—What influence do the three principles of the solar ray exert on vegetation? What is electricity?

producing the most sudden and destructive effects, or of exerting powerful influences by a quiet and long-continued action.

We are unable to say whether electricity is a material substance, a property of matter, or the vibration of an ether. The general opinion at the present day, however, is, that electricity, like light and heat, is the result of some modification, or vibration of that subtile ethereal medium which pervades all space, and which is capable of moving with various degrees of facility through the pores of even the densest substances.

The language which is almost universally adopted in describing electrical phenomena, is based upon the supposition that electricity is a form, or kind of matter, since by the use of this hypothesis, the leading facts of the science may be clearly and simply set forth.

210. **Electricity and Chemical Action.**—The relation which exists between the force of electricity and the operations of chemical affinity is most intimate; and according to some authorities electricity and chemical affinity are merely different manifestations of the same agent.

211. **Excitation of Electricity.**—Electricity may be excited, or called into activity by mechanical action, by chemical action, by heat, and by magnetic influence.

Why the means above enumerated should develop electricity, or excite it from a neutral condition, is a matter at present wholly inexplicable.

212. **Two Conditions of Electricity.**—Electricity in the act of becoming free, as when excited by friction, or when evolved from a galvanic battery, appears to separate into two forces, or, as it is generally termed, into two kinds of electricity. These two forces are identical in their nature and equal in power, but opposite and contrary in their action. When they meet, they do not unite to form a double electrical force, but they mutually neutralize and destroy the power of each other.

The existence and action of these two forces, or kinds of electricity, may be demonstrated by the following simple experiment:—If we take a dry glass rod, rub it well with silk, and present it to a light pith ball, or feather, P,

---

QUESTIONS.—What do we know concerning the real nature of this agent? What is the relation between electricity and chemical action? How may electricity be excited? In what manner does electricity, on being set free, display itself? What is the character of the two forces, or kinds of electricity? How may the existence and action of the two kinds of electricity be demonstrated?

Fig. 59, suspended from a support by a silk thread, the ball or feather will be attracted toward the glass, as seen at G. After it has adhered to it a moment, it will fly off, or be repelled, as P′ from G′. The same thing will also happen if sealing-wax be rubbed with dry flannel, and a like experiment made.

Fig. 59.

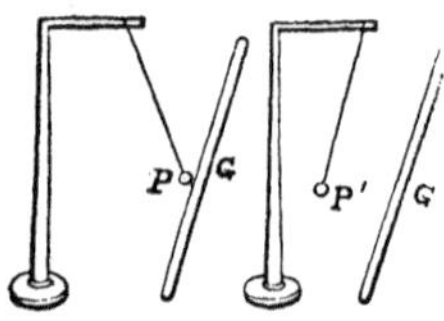

If, however, the action of the glass and the wax be compared together, a remarkable difference between the two will immediately manifest itself, for when the glass repels the ball the sealing-wax will attract it most strongly, and when the wax repels, the glass attracts in like manner; so that if we suspend a light pith ball, or feather, by a silk thread, as in Fig. 60, and present a stick of excited sealing-wax, S, on one side, and a tube of excited glass, G, on the other, the ball will commence vibrating like a pendulum from one to the other, being alternately attracted and repelled by each, the one attracting when the other repels. We therefore conclude that the electricities excited in the glass and wax are different.

Fig. 60.

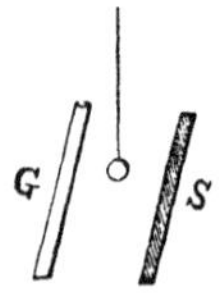

In order to distinguish the two opposite forces or conditions of electricity from each other, that force which is obtained from the glass has been termed vitreous, or positive electricity; and that from the wax, resinous, or negative electricity.

While the terms vitreous and resinous are now rarely used, those of positive and negative are somewhat unfortunate, since they almost unavoidably convey to the learner the impression that the one force is stronger or more potent than the other, whereas the negative electricity has as positive an existence and as substantial power as the opposite electricity.

Electricity may be excited in all bodies. There are no exceptions to this fact, but electricity is developed in some bodies with great ease, and in others with great difficulty. In no case, however, can electricity of one kind be excited without setting free a corresponding amount of electricity of the other kind; hence, when electricity is excited by friction, the rubber always exhibits the one, and the body rubbed, the other.

213. **Fundamental Law of Electricity.**—The fundamental law which governs the relation of the two forces of electricity to each other may be expressed as follows:

Like electricities repel each other, unlike electricities attract each other.

Questions.—By what names do we distinguish the two forces, or kinds of electricity? Why is the use of the terms positive and negative unfortunate? Can one electricity be developed independently of the other? What is the great fundamental law of electricity?

Thus, if two substances are charged with positive electricity, they repel each other; two substances charged with negative electricity also repel each other; but if one is charged with positive and the other with negative electricity, they attract each other.

The attraction which the two opposite electricities have for each other is very great, and their tendency is, therefore, constantly to combine together. From such combination latent, or quiescent electricity results.

214. **Electrified and Non-Electrified Bodies.**—When a body holds its own natural quantity of electricity undisturbed, it is said to be non-electrified.

When an electrified body touches one that is non-electrified, the electricity contained in the former is transferred in part to the latter.

Thus, on touching the end of a suspended silk thread with a piece of excited wax or glass, electricity will pass from the wax or glass into the silk, and render it electrified; and the silk will exhibit the effects of the electricity imparted to it, by moving toward any object that may be placed near it.

215. **Conductors and Non-Conductors.**—Bodies differ greatly in the freedom with which they allow electricity to pass over or through them. Those substances which facilitate its passage are called conductors; those that retard, or almost prevent it, are called non-conductors.

No substance can *entirely* prevent the passage of electricity, nor is there any which does not oppose some resistance to its passage.

Of all bodies, the metals are the most perfect conductors of electricity; charcoal, the earth, water, moist air, most liquids, except oils, and the human body, are also good conductors of electricity.

Gum shellac and gutta percha are the most perfect non-conductors of electricity; sulphur, sealing-wax, resin, and all resinous bodies, glass, silk, feathers, hair, dry wool, dry air, and baked wood, are also non-conductors.

Electricity always passes by preference over the best conductors.

216. **Insulation.**—When a conductor of electricity is surrounded on all sides by non-conducting substances, it is said to be *insulated;* and the non-conducting substances which surround it are called *insulators.*

When a conducting body is insulated, it retains upon its surface the electricity communicated to it, and in this condition it is said to be charged with electricity.

QUESTIONS.—Illustrate it. When is a body said to be electrified, and when non-electrified? What are conductors and non-conductors of electricity? What substances are good conductors? What are bad conductors? When is a conductor said to be insulated? When charged?

217. **Velocity of Electricity.**—The velocity with which the influence of electricity passes through good conductors is so great, that the most rapid motion produced by art appears to be actual rest when compared to it. Some authorities have estimated that frictional electricity will pass through copper wire at the rate of 288,000 miles in a second of time—a velocity greater than that of light. The results obtained, however, by the United States Coast Survey, with galvanic electricity and iron wire, show a velocity of from 15,000 to 20,000 miles per second.

The terms "electric fluid" and "electric current," which are frequently employed in describing electrical phenomena, are calculated to mislead the student into the supposition that electricity is known to be a fluid, and that it flows in a rapid stream along a conductor. Such terms, it should be understood, are founded merely on an assumed analogy between the electric force and a fluid substance. The nature of that force, however, is unknown, and whether its transmission be in the form of a current, or by vibrations, is undetermined.*

218. **Galvanic, or Voltaic Electricity.**—Electricity excited or produced by the chemical action of two or more dissimilar substances upon each other, is termed Galvanic, or Voltaic Electricity, and the department of physical science which treats of this form of electrical disturbance is called Galvanism.

The most simple method of illustrating the production of galvanic electricity is by placing a piece of silver (as a coin) on the tongue, and a piece of zinc underneath. So long as the two metals are kept asunder no effect will be noticed, but when their ends are brought together a distinct thrill will pass through the tongue, a metallic taste will diffuse itself, and, if the eyes are closed, a sensation of light will be evident at the same moment.

This result is owing to a chemical action which is developed the moment

* In a discussion which took place some years since at a meeting of the British Association for the Advancement of Science, respecting the nature of electricity, Professor Faraday expressed his opinion as follows:—"There was a time when I thought I knew something about the matter; but the longer I live, and the more carefully I study the subject the more convinced I am of my total ignorance of the nature of electricity."

"After such an avowal as this," says Mr. Bakewell, "from the most eminent electrician of the age, it is almost useless to say that any terms which seem to designate the form of electricity are merely to be considered as convenient conventional expressions."

QUESTIONS.—What is the velocity of electricity? What is understood by the use of the word current, as applied to electricity? What is galvanic, or voltaic electricity? What is the most simple method of illustrating its production? To what is this result owing?

the two metals touch each other. The saliva of the tongue acts chemically upon, or oxydizes a portion of the zinc, which excites electricity, for no chemical action ever takes place without producing electricity. Upon bringing the ends of the two metals together, a slight current passes from one to the other.

219. **Discovery of Galvanic Electricity.**—The production of electricity by the chemical action of two metals when brought in contact, was first noticed by Galvani, a professor of anatomy at Bologna, Italy, in 1790.

His attention was directed to the subject in the following manner:—Having occasion to dissect several frogs, he hung up their hind legs on some copper hooks, until he might find it necessary to use them for illustration. In this manner he happened to suspend a number of the copper hooks on an iron balcony, when, to his great astonishment, the limbs were thrown into violent convulsions. On investigating the phenomenon, he found that the mere contact of dissimilar metals with the moist surfaces of the muscles and nerves, was all that was necessary to produce the convulsions.

FIG. 61.

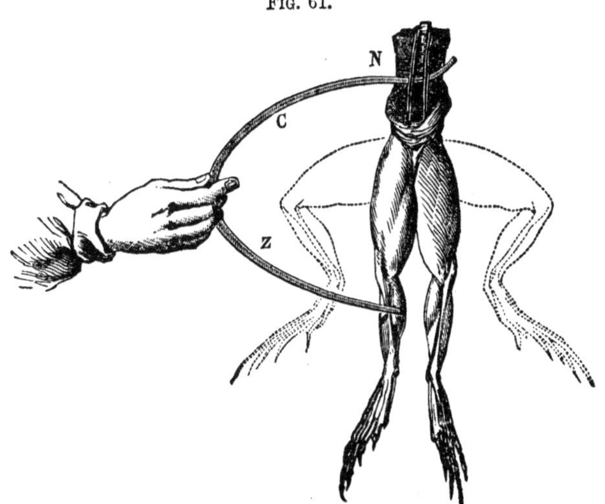

This singular action of electricity, first noticed by Galvani, may be experimentally exhibited without difficulty. Fig. 61 represents the extremities of a frog, with the upper part dissected in such a way as to exhibit the nerves

QUESTIONS.—When and how was galvanic electricity discovered? How may the phenomenon first noticed by Galvani be experimentally repeated?

of the legs, and a portion of the spinal marrow. If we now take two thin pieces of copper and zinc, C Z, and place one under the nerves, and the other in contact with the muscles of the leg, we shall find that so long as the two pieces of metal are separated, so long will the limbs remain motionless; but by making a connection, instantly the whole lower extremities will be thrown into violent convulsions, quivering and stretching themselves in a manner too singular to describe. If the wire is kept closely in contact, these phenomena are of momentary duration, but are renewed every time the contact is made and broken.

Galvani attributed these movements of the muscles to a kind of nervous fluid pervading the animal system, similar to the electric fluid, which passed from the nerves to the muscles, as soon as the two were brought in communication with each other, by means of the metallic connection. He therefore called the supposed fluid animal electricity.

220. **The Voltaic Pile.**—The experiments of Galvani were repeated by Volta, an eminent Italian philosopher, who found that no electrical or nervous excitement took place unless a communication between the muscles and the nerves was made by two different metals, as copper and iron, or copper and zinc. He also observed that all the effects noticed could be produced in a much higher degree by using a number of pieces of different metals and a fluid, or a substance moistened with a fluid. He accordingly arranged a series of copper and zinc plates in a pile with cloths wet in a saline or acid liquid between them, as is represented in Fig. 62. The series commenced with a zinc plate, upon which was placed a copper plate of the same size, and on that a circular piece of cloth previously soaked in water slightly acidulated. On the cloth was laid another plate of zinc, then copper, and again cloth, and so on in succession, until a pile of fifty series of alternate metal plates and moistened cloths was formed, the terminal plate of the series at one end being copper and at the other end zinc. Such an apparatus received the name of a "*Voltaic Pile*," and its effects were soon seen to be of an electrical character.

FIG. 62.

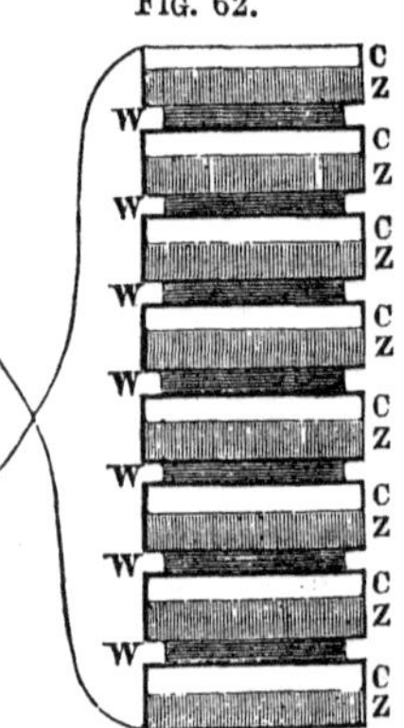

For instance, if the two ends or terminal plates of the pile were touched, one with each hand previously moistened, a sensation similar to that of an electric shock was experienced. If the two ends were connected by means of metallic wires, sparks could be obtained, shocks communicated, and many other electrical effects produced.

QUESTIONS.—To what did Galvani attribute the results by him noticed? What conclusion was arrived at by Volta? What discovery did Volta make? Describe the voltaic pile.

221. **Results of Galvani's and Volta's Discoveries.**—Such is an outline of one of the greatest and most remarkable discoveries of modern times—a discovery which illustrates in a striking manner the importance of cultivating correct habits of observation, and of rightly estimating the relations which exist between a cause and its effect. The attention bestowed by Galvani on the simple circumstance of the twitching of a frog's legs in 1790, led to the discovery of the voltaic pile in 1800, a modification of which constitutes the present galvanic battery. Since the last named period the progress of discovery has been most rapid, embracing the whole science of electro-magnetism, electro-metallurgy, the application of electricity to chemical analysis, to the production of intense heat and light, to the recording of time, to the determination of longitudes, and finally, to the almost instantaneous communication of intelligence by means of the telegraph.

Volta considered that electricity was produced by simple contact of dissimilar metals, positive electricity being evolved from the one, and negative from the other. It is now generally believed that chemical action, taking place between the surfaces in contact, is the sole cause of exciting and continuing the electric currents.

222. **Fundamental Principle of Galvanic Electricity.**—The fundamental principle which forms the basis of the science of galvanic electricity is as follows:

Any two metals, or more generally, any two different bodies which are conductors of electricity, when placed in contact, develop electricity by chemical action—positive electricity flowing from the body which is acted upon most powerfully, and negative electricity from the other.

223. **Electro-positive and Negative Elements.**—In general, that substance which is acted upon most easily is termed the electro-positive element; and the other the electro-negative element.

The electrical force or power generated in this way is called the electro-motive force.

Different bodies placed in contact manifest different electro-motive forces, or develop different quantities of electricity.

Bodies capable of developing electricity by contact may be arranged in a

---

QUESTIONS.—What have been the results of Galvani's and V lta's discoveries? What did Volta suppose to be the origin of the electricity of the pile? What is now believed on this subject? What is the fundamental principle of galvanic electricity? What are electro-positive and electro-negative elements? What is understood by the term electro-motive force? How may bodies capable of exciting electro-motive force be classed?

series in such a manner that any one placed in contact with another holding a lower place in the series, will receive the positive fluid, and the lower one the negative fluid; and the more remote they stand from each other in the order of the series, the more decidedly will the electricity be developed by their contact.

The most common substances used for exciting galvanic electricity may be arranged in such a series as follows:—zinc, lead, tin, antimony, iron, brass, copper, silver, gold, platinum, black lead or graphite, and charcoal.

Thus, zinc and lead, when brought in contact, will produce electricity, but it will be much less active than that produced by the union of zinc and iron, or the same metal and copper, and the last less active than zinc and platinum or zinc and charcoal.

224. **Zamboni's Pile.**—According to the principles above explained, a perfectly dry pile, known from its inventor as Zamboni's pile, may be constructed of sheets of gilded paper and sheet zinc. If several thousand of these be packed together in a glass tube, so that their similar metallic faces shall all look the same way, and be pressed tightly together at each end by metallic plates, it will be found that one extremity of the pile is positive and the other negative. Such a series will last more than twenty years, but it requires as many as 10,000 pairs to afford sparks visible in daylight.

Fig. 63.

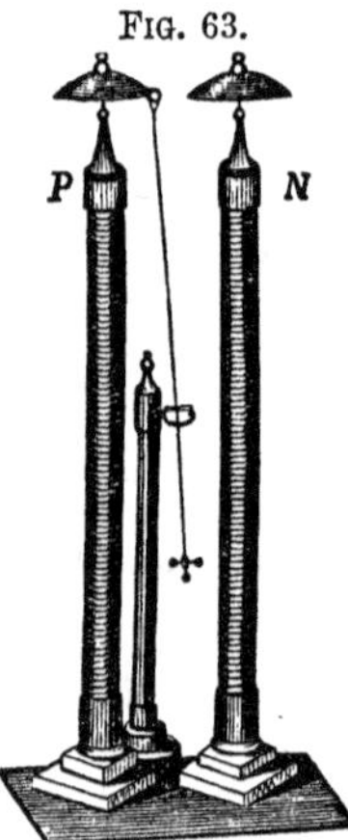

Fig. 63 represents a pair of these piles, so arranged as to produce what has been called a perpetual motion. Two piles, P N, are placed in such a position that the positive extremity of one pile is opposite and near to the negative extremity of the other. Between them a light pendulum is placed, vibrating on an axis and insulated on a glass pillar. This pendulum is alternately attracted to one and then to the other, and thus rings two little bells connected with the positive and negative poles.

In a similar manner, voltaic piles have been constructed entirely of vegetable substances, without resorting to the use of any metal, by placing discs of beet-root and walnut-wood in contact. With such a pile, and a leaf of grass as a conductor, convulsions in the muscles of a dead frog are said to have been produced. Other experimentalists have formed voltaic piles wholly of animal substances.

225. **Practical Production of Galvanic Electricity.**—In the production of galvanic electricity for practical purposes, it is necessary to have a combination of three dif-

QUESTIONS.—Describe the dry, or Zamboni's pile. May a voltaic pile be produced entirely of vegetable or animal substances? What arrangement is necessary for the practical production of galvanic electricity?

ferent conductors, or elements, one of which must be solid and one fluid, while the third may be either solid or fluid.

The process usually adopted is to place between two plates of different kinds of metal a liquid capable of exciting some chemical action on one of the plates, while it has no action, or a different action upon the other. A communication is then formed between the two plates.

226. **Galvanic Circuit.**—When two metals capable of exciting electricity are so arranged and connected that the positive and negative electricities can meet and flow in opposite directions, they are said to form a galvanic circuit, or circle. Such an arrangement is very generally termed, also, a simple galvanic battery.

A very simple, and at the same time an active galvanic circuit may be formed by an arrangement as represented in Fig. 64. C and Z are thin plates of copper and zinc immersed in a glass vessel containing a very weak solution of sulphuric acid and water. So long as the two metals do not touch each other, there will be but slight chemical action, and consequently little or no electricity evolved; but on bringing the two ends of the metal strips together, or by causing metallic contact by a connection of wires, X and W, a galvanic circuit will be formed, positive electricity passing from the zinc through the liquid to the copper, and from the copper along the conducting wires to the zinc, as indicated by the arrows in the figure. A current of negative electricity at the same time traverses the circuit also, from the copper to the zinc, in an opposite direction.

FIG. 64.

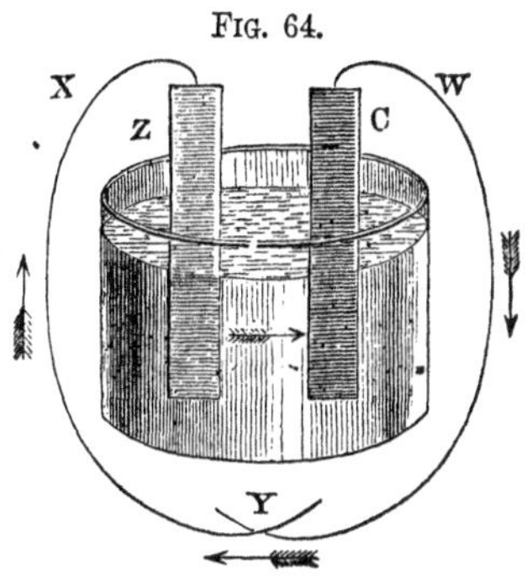

227. **Theory of a Simple Circuit.**—In the formation of a galvanic circuit, by the employment of two metals and a liquid, the chemical action which gives rise to the electricity takes place through a decomposition of the liquid.

When a plate of zinc and one of copper are immersed in water acidulated with sulphuric acid, the elements of the water, oxygen and hydrogen, are separated from each other, in consequence of the greater attraction which the oxygen has for the zinc. The oxygen, therefore, unites with the zinc, and by so doing

---

QUESTIONS.—What is a galvanic circuit, or simple galvanic battery? Describe the construction of such a circuit. What is the origin of the electricity evolved in a circuit composed of two metals and one liquid? Describe the theoretical action of such a circuit?

excites, or develops electricity in the metal. But as one kind of electricity can not be evolved without bringing an equal quantity of the other into activity, the act which develops negative electricity in the metal, instantaneously develops positive electricity in the liquid. It would naturally be supposed, that as the two opposite electricities have a strong attraction for each other, that they would again unite, and restore the equilibrium; such, however, from some unexplained reason, is not the case; but the electrical and chemical changes are so connected, that unless the equilibrium is restored, the action between the metal and the liquid will stop as soon as a certain quantity of electricity has accumulated. If, under these circumstances, the copper plate which is immersed in the liquid, but not acted upon by it, be brought in contact with the zinc, it will serve as a conductor, and will convey the positive electricity accumulated in the liquid to the zinc, restore the equilibrium of the two electricities, and cause the action between the liquid and the zinc to recommence. With the commencement of the flow of positive electricity from the liquid to the copper, and from the copper to the zinc, a current of negative electricity will tend to flow in the opposite direction, or from the zinc to the copper, and from the copper to the liquid.*

228. **Direction of the Current.**—In all cases, the direction of the current is dependent on the direction of the chemical action.

The positive electricity always sets out from the metal most acted upon by the exciting liquid, which may be, therefore, called the generating or positive plate. It traverses the liquid toward the less affected metal, which forms the negative, or conducting plate, and from this the force is transferred to the wire, or other conducting medium, between the two plates; thence it passes back again to the generating plate. In this way the circuit is completed, and unless this circulation can take place, all the phenomena of galvanic action will be suspended.

The electrical condition of the plates of copper and zinc as above described, it should be understood, applies only to those portions of the two metals which are immersed in the liquid. Those parts which are out of the liquid, and in the air, are in an exactly opposite condition. Thus the end of the zinc in the acid is +, or positive, while that in the air is —, or negative. The electrical state of the two ends of the copper is exactly the reverse.

If, in the arrangement above described, some liquid which acts upon the copper in preference to the zinc, as ammonia, had been used, the electrical

* In every voltaic current it is assumed that a quantity of negative electricity, equal to that of the positive set in motion, is proceeding along the conducting medium in a direction opposite to that in which the positive electricity is traveling; but in order to avoid confusion, whenever the direction of the current is mentioned, the direction of the positive electricity is alone referred to.

QUESTIONS.—What influences the direction of the current? What determines the electrical condition of the immersed metals?

condition of the two metals, and the direction of the flow of electricity, would have been reversed.

Although two metal plates are usually employed in a simple galvanic circuit, only one of them is active in the excitement of electricity, the other plate serving merely as a conductor to collect the force generated. A metal plate is generally used for this purpose, because metals conduct electricity much better than other substances exposing an equal surface to the fluids in which they are immersed; but other conductors may be used, and when a proportionately larger surface is exposed to compensate for inferior conducting power, they answer as well, and in some instances better, than metal plates. Thus charcoal is very often employed in the place of copper, and a very hard material obtained from the interior of gas retorts, "gas carbon," is considered one of the best conductors.

Two metals are not absolutely essential to the formation of a simple galvanic current. A current may be obtained from one metal and two liquids, provided the liquids are such that a stronger chemical action takes place on one side of the metal plate than on the other.

229. **Poles of a Galvanic Battery.**—The two metals forming the elements of the battery are generally connected by copper wires; the ends of these wires, or the terminal points of any other connecting medium used, are called the poles of the battery.

Thus, when zinc and copper plates are used, the end of the wire conveying positive electricity from the copper would be the positive pole, and the end of the wire conveying negative electricity from the zinc plate would be the negative pole. Faraday describes the poles of the battery as the doors by which electricity enters into or passes out of the substance suffering decomposition, and in accordance with this view he has given to the positive pole the name of *anode*, or ascending way, and to the negative pole the name of *cathode*, or descending way.

The manifestations of electricity will be most apparent at that point of the circuit where the two currents of positive and negative electricity meet.

When the two wires connecting the metal plates of a battery are brought in contact, the galvanic circuit is said to be closed. No sign of electrical excitement is then visible; the action, nevertheless, continues. The opposite electricities collected at the poles, in particular, neutralize each other perfectly on meeting; every trace of electricity must therefore vanish if a fresh quantity were not continually produced by the continuance of the chemical action.

QUESTIONS.—What is the necessity of two metals in a galvanic circuit? Under what circumstances can some other substance be substituted in place of the copper? What are the poles of a galvanic battery? What is the meaning of the terms anode and cathode? At what point of a galvanic circuit will the manifestation of electricity be most apparent? When is the galvanic circuit said to be closed?

230. **Compound Circuit.**—The electricity developed by a simple galvanic circuit, whether it be composed of two metals and a liquid, or any other combination, is exceedingly feeble. Its power can, however, be increased to any extent by a repetition of the simple combinations.

The discovery of this fact was first made by Volta, and applied by him in the voltaic pile before described.

Fig. 65.

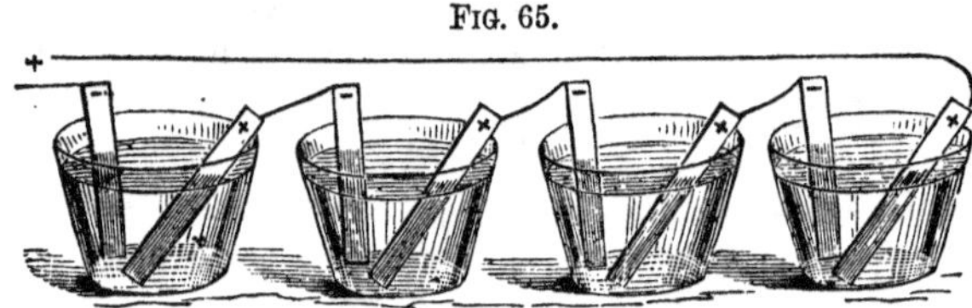

Fig. 65 represents, in its simplest form, the construction of a compound galvanic circuit, by the union of a number of simple circuits. Each glass contains one zinc and one copper plate, which are not immediately connected together as in a simple circuit; but every zinc plate is connected with the copper plate of the preceding glass by a copper wire or band. In the figure, the copper plate and the direction of the positive current is represented by the sign +, and the zinc plate and the negative current by the sign —.

In a compound galvanic circuit, like the one represented in Fig. 65, the positive electricity which the fluid in the first vessel acquires from the plate of zinc exposed to its action, is taken up by the copper plate and transferred to the second zinc plate in the second vessel, by means of its metallic connection. This transmits it, together with what itself generates, to the liquid of the second vessel. From this the double force is passed to the next copper, and by it to the third zinc, which it touches, and so on, every succeeding alternation being productive of a further increase in the quantity of the electricity developed. A current of negative electricity may in like manner be supposed to flow in an opposite direction, its quantity augmenting with each successive pair of plates. This action, however, would stop unless an outlet were given to the accumulated electricity by establishing a communication between the positive and negative poles of the battery, by means of wires attached to the extreme plate at each end. When these are brought into contact, the galvanic circuit is completed, and the electricities meet and neutralize each other, producing the various electrical phenomena. The electric current continues to flow uninterruptedly in the circuit so long as the chemical action lasts.

QUESTIONS.—What is the electrical power of a simple circuit? How may it be increased? Describe the construction of a compound circuit? In what manner does it accumulate electricity?

The simple and compound voltaic circuits in practical use, which in ordinary language are both designated as galvanic batteries, differ considerably in form and efficiency. The general principle of construction in all, however, is the same as that of the original voltaic pile.

231. **The Trough Battery.**—One of the earliest forms contrived is known as the Trough Battery, represented in Fig. 66. It consists of a trough of wood divided into water-tight cells, or partitions, each cell being arranged to receive a pair of zinc and copper plates. The plates are attached to a bar of wood, and connected with one another by metallic wires, in such a way that every copper plate is connected with the zinc plate of the next cell. The battery is excited by means of dilute sulphuric acid poured into the cells, and the current of electricity is directed by wires soldered to the extreme plates. When the battery is not in use the plates may be raised from the trough by means of the wooden bar.

Fig. 66.

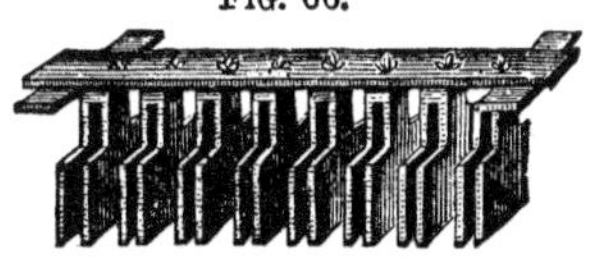

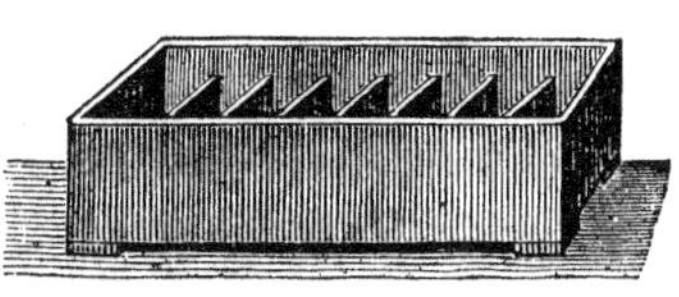

The battery by which Sir Humphrey Davy effected his splendid chemical discoveries was of this form, and consisted of two thousand double plates of copper and zinc, each plate having a surface of thirty-two square inches. Now, however, by improved arrangements, we can produce with ten or twenty pairs of plates, effects every way superior.

232. **Smee's Battery.**—The most easily managed form of galvanic battery at present used is that invented by Mr. Smee, and known as Smee's battery. (See Fig. 67.) It consists of a plate of silver coated with platinum, suspended between two plates of zinc, *z z*, the surfaces of which last have been coated with mercury, or amalgamated, as it is called. The three are attached to a wooden bar, which serves to support the whole in a tumbler, G, partially filled with a weak solution of sulphuric acid and water. The wires, or poles for directing the current of electricity are connected with the zinc and platinum plates by small screw-cups, S and A.

Fig. 67.

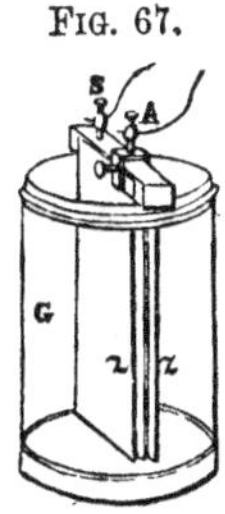

233. **Amalgamation of Zinc.**—The introduction of the process of amalgamating, or coating the zinc plates of a galvanic circuit with mercury, constituted an improvement of great value. In the original form of the galvanic battery, constructed of copper and ordinary metallic zinc, the waste of the latter metal by the action of the exciting acid upon it was very

---

QUESTIONS.—Describe the trough battery. What is the construction of Smee's battery? What is understood by the amalgamation of the zinc? What benefit results from this operation?

great; but by using amalgamated zinc this waste is diminished in an extraordinary degree, without at the same time diminishing the production of electricity. All improved batteries are, therefore, constructed with amalgamated zinc.

234. **Sulphate of Copper Battery.**—Another form of battery, called the sulphate of copper battery, from the fact that a solution of sulphate of copper (blue vitriol) is used as the exciting liquid, is represented by Fig. 68. It consists of two concentric cylinders of copper, C, tightly soldered to a copper bottom, and a zinc cylinder, Z, fitting in between them. Two screw-cups for holding the connecting wires are attached, one to the outer copper cylinder, and the other to the zinc.

Fig. 68.

The principal imperfection of the galvanic battery is the want of uniformity in its action. In all the various forms, the strength of the electric current excited constantly decreases from the moment the battery action commences. In the sulphate of copper battery, especially, the power is reduced in a comparately short time to almost nothing. This is chiefly owing to the circumstance, that the metallic plates soon become coated with the products of the chemical decomposition, the result of the chemical action whereby the electricity is developed.

This difficulty is obviated in a great degree by the use of a diaphragm, or a porous and permeable partition between the two metallic plates, which allows a free contact of the liquid on both sides within its pores, but prevents the solid products of the chemical action from passing from one metallic plate to the other. Bladder, leather, clay, porcelain, cloth, etc., have been used for this purpose.

Fig. 69.

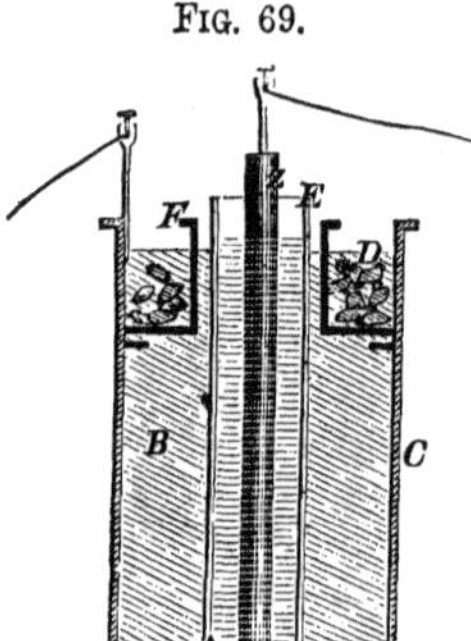

235. **Daniel's Constant Battery,** constructed according to the above described principle, and represented by Fig. 69, maintains an effective galvanic action longer than any other. The outer case, C, consists of a cell, or cylinder of copper, which is so constructed as to retain liquids, and is filled with a solution of sulphate of copper, B, acidulated with one eighth of its bulk of sulphuric acid. The solution is kept saturated with the salt by means of crystals of sulphate of copper, D, which rest upon the perforated shelf, F. In the center of the cell is placed a tube of porous earthen-ware, E, filled with

QUESTIONS.—Describe the sulphate of copper battery. What is the principal imperfection of the galvanic battery? How is it obviated? What is the construction of Daniel's battery?

an acid solution, A, which consists of one part of sulphuric acid diluted with seven parts of water. A rod of zinc, Z, is placed in this tube. On making a metallic communication between the zinc rod and the copper cell, a voltaic current is established.

236. **Grove's Battery.**—One of the most efficient batteries is that known as Grove's battery, from its inventor, and is the form generally used for telegraphing, and other purposes in which powerful galvanic action is required. It is constructed upon the same general principle as Daniel's battery, and consists of a plain glass tumbler, in which is placed a cylinder of amalgamated zinc, with an opening on one side to allow a free circulation of the liquid. Within this cylinder is placed a porous cup, or cell of earthenware, in which is suspended a strip of platinum fastened to the end of a zinc arm projecting from the adjoining zinc cylinder. The porous cup containing the platinum is filled with strong nitric acid, and the outer vessel containing the zinc with weak sulphuric acid. Fig. 70 represents a series of these cups, arranged to form a compound circuit, with their terminal poles, P and Z. This form of battery is objectionable on account of the corrosive character of the acids employed, and the deleterious vapors that arise from it when in action.

FIG. 70.

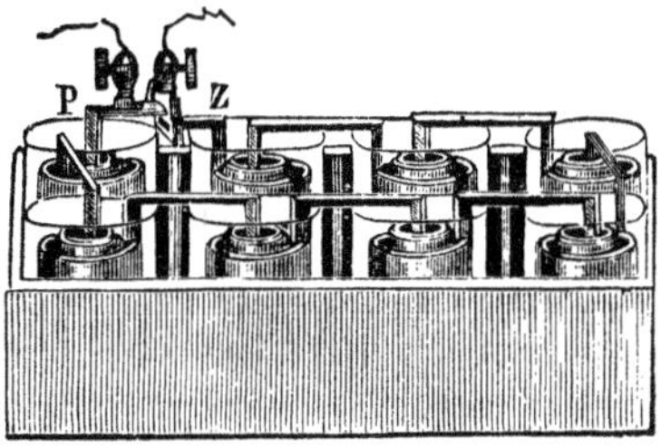

In what is known as Bunsen's Carbon Battery, a cylinder of carbon is substituted, on the ground of economy, in place of the platinum plates of Grove's battery.

237. **Resistances to the Circulation of the Galvanic Current.**—The amount of force or of electricity which circulates in a galvanic circuit does not depend wholly upon the energy of the chemical action which is exerted between the generating metal and the exciting liquid. "The current experiences a retardation or resistance from the very conductors by which its influence is transmitted; just as in the transmission of mechanical force in an arrangement of machinery, the intervention of the pivots and levers which are required for its conveyance introduces additional friction and additional weight, which are required to be overcome or moved, and which thus diminish the efficient power of the machine."—MILLER.

The resistances of the galvanic current arise from the imperfect conducting power of the liquid which is employed to excite it, and of the plates, wires, etc., the resistance offered by the liquid being the more considerable

QUESTIONS.—Describe Grove's battery. Is the electricity of a galvanic circuit always in proportion to the chemical action exerted? What are the resistances it experiences? To what are these resistances proportional?

of the two. The further the plates are removed from each other in the liquid, and the longer the column of imperfectly conducting matter which the electricity is obliged to traverse, the greater the resistance. The same thing is also true of the conducting wire. A wire one tenth of an inch in diameter, will for equal lengths offer four times the resistance of a wire two tenths, or one fifth of an inch thick.

238. **Characteristics of Ordinary and Galvanic Electricity.**—Electricity in its ordinary manifestations, as when developed by friction or by an electrical machine, exhibits itself in sudden and intermitted shocks, accompanied with a sort of explosion; galvanic electricity, or electricity produced by chemical action, is, on the contrary, a steady flowing current.

The electricity evolved by a single galvanic circle is great in quantity, but weak in intensity.

The electricity, on the contrary, produced by friction, or that of a thunder-cloud, is small in quantity, but of high tension,* or intensity.

These two qualities may be compared to heat of different temperatures. A gallon of water at a temperature of 100° has a greater quantity of heat than a pint at 200°; but the heat of the latter is more intense than that of the former. Again, in the phosphorescence of the sea, which often spreads over thousands of miles, we have an illustration of light very feeble in intensity, but enormous in quantity.

239. **Quantity and Intensity, how Measured.**—We measure the *quantity* of electricity in many ways; but most conveniently by the amount of any chemical compound which it can decompose. A machine or battery, for example, which, when arranged so as to decompose water, evolves from it four cubic inches of oxygen and hydrogen in one minute, is furnishing twice the quantity of electricity supplied by an apparatus which evolves only two cubic inches of the gases in the same time.

The *intensity* of electricity is less easily measured; but it is comparatively indicated by the ease with which it can travel through bad conductors; by

* "Tension is merely a synonyme for intensity, which originated in the hypothesis of electricity being an elastic fluid, which might be regarded as existing in a thunder-cloud, or on the conductor of a friction-machine in a state of condensation or compression, like high-pressure steam struggling to escape from a boiler, or air seeking to force its way out of the chamber of an air-gun. The word tension has been preferred to intensity, simply on account of its brevity, and its convenience in forming a double noun with electricity.

QUESTIONS.—What are the characteristic differences between galvanic and ordinary electricity? To what may quantity and intensity be compared? How are these two qualities measured? What is understood by the term tension?

its power to overcome energetic chemical affinity, such as that which binds together the elements of water; by the length of space across which it can pass through dry air (as in the case of a lightning flash striking a tree from a great distance); by the attractions and repulsions it produces in light bodies; and by the severity of the shock it occasions to living animals.

Galvanic electricity will traverse a circuit of 2,000 miles of wire, rather than make a short circuit by overleaping a space of resisting air not exceeding one hundredth part of an inch. Frictional electricity, on the other hand, will force a passage across a considerable interval, in preference to taking a long circuit through a conducting medium.

The assertion is within bounds, that the whole electricity of a destructive thunder-storm would not suffice for the electro-gilding of a single pin—so insignificant is its amount. A small copper wire, dipped into an acid along with a wire of zinc, would evolve more electricity in a few seconds than the largest friction electrical machine, kept constantly revolving, would furnish in many weeks. No shock, on the other hand, would be occasioned by the electricity from the immersed wires; nor would it produce a spark, or decompose water—so low is its intensity. A galvanic battery of many plates will, however, produce electricity of sufficient intensity to kill a large animal, and produce other effects analogous to lightning.

Electricity of intensity then, or tension-electricity, is electricity characterized by the greatness of its intensity—or whose intensity is greater than its quantity. Electricity of quantity, on the other hand, has its quantity greater than its intensity.

The intensity diminishes as the quantity increases; but the ratio which the one bears to the other differs through a very wide scale, so that a knowledge of the degree of the one does not often enable us to predicate the amount of the other. Practically, we have no difficulty in reducing both to a minimum, or in exalting the one whilst we reduce the other; but we can not at once exalt both intensity and quantity. The discovery of a method of effecting this will make a new era in the science; and admit of the most important applications to the useful arts.

240. **Practical Applications of Electricity of Quantity and Intensity.**—In the arts, it depends much upon the purpose to which electricity is to be applied whether it should be chosen great in quantity, or great in intensity. If the chemist desires to analyze a gaseous mixture by exploding it, he will use an electrical machine to supply a momentary spark of great intensity. But the electro-plater, who has constantly to decompose a compound of gold or silver, employs a small voltaic battery—which furnishes great quantities of electricity of considerable intensity. The

---

QUESTIONS.—Illustrate the differences between quantity and intensity. What definition may be given of the two? What relation exists between them? What are their practical applications?

electric light requires both quantity and intensity to be very great. The electric telegraph demands great quantity, but the intensity need not be very high.

241. Electro-chemical Decomposition.—When a current of galvanic electricity is made to pass through a compound liquid, composed of one conducting and one non-conducting substance, its tendency is to decompose and separate it into its constituent parts.

242. **Decomposition of Water.**—The most remarkable illustration of this power is to be found in the decomposition of water. This substance is composed of two gases, oxygen and hydrogen, united in the proportions of one measure of the former to two of the latter. When two gold or platinum wires, connected with the opposite ends of a galvanic battery, are placed in water at a short distance from each other, the water is decomposed, the hydrogen arising in bubbles from the negative pole of the battery, and the oxygen from the positive pole. When two glass tubes are placed over the platinum poles, as is represented in Fig. 71, we can collect the bubbles as they rise, the volume of the hydrogen being twice as great as that of the oxygen.

Fig. 71.

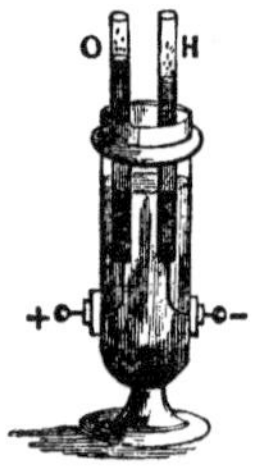

When copper wires, or the wires of metals which tend strongly to unite with oxygen are employed, gas escapes from one wire only; whilst if platinum or gold wires be used, gas is evolved from both. In the first case, the oxygen combines with the copper or other oxydizable metal, and forms an oxyd, which is dissolved by the liquid, and therefore hydrogen alone escapes; in the second case, both gases are evolved, since neither platinum nor gold has sufficient chemical affinity for oxygen to combine with it at the moment of its liberation.

243. **Electrodes.**—The term electrode is often used to designate the poles of a galvanic battery. It is especially applied in those cases in which the connecting wires of a circuit are terminated with strips of platinum, gold, charcoal, or some other good conducting, non-oxydizable substance.

244. **Theory of Electro-chemical Decomposition.**—Scientific men are not fully agreed upon the explanation of the phenomenon of chemical decomposition by means of the galvanic current. A general idea of what takes place may perhaps be best gained from what is called the electro-chemical theory. According to this, chemical attractions, which we distinguish by the name of affinity, and electrical attractions depend on the same cause, acting in one case on atoms, and in the other on masses of matter. Every atom of matter is regarded as charged in respect to all other

---

QUESTIONS.—What is the influence of the electric current in producing electro-chemical decomposition? How is this illustrated in the decomposition of water? What are electrodes? What is the theory of the decomposing action of galvanic electricity?

atoms, with either positive or negative electricity. In the case of water, hydrogen is the electro-positive element and oxygen the electro-negative element. It has been already shown that bodies in opposite electrical states are attracted by each other. Hence, when the poles of a galvanic battery are immersed in water, the negative pole will attract the positive hydrogen, and the positive pole the negative oxygen. If the attractive force of the two electricities generated by the battery is greater than the attractive force which unites the two elements, oxygen and hydrogen, together in the water, the compound will be decomposed. Upon the same principle other compound substances may be decomposed, by employing a greater or less amount of electricity. In this way Sir Humphrey Davy made the discovery that potash, soda, lime, and other bodies, were not simple in their nature, as had previously been supposed, but compounds of a metal with oxygen.

This theory, as presented, is not received as strictly in accordance with the fact. Recent experiments of Faraday have proved that the electricity which decomposes, and that which is evolved by the decomposition of a certain quantity of matter, are alike. Thus, water is composed of oxygen and hydrogen; now, if the electrical power which holds a grain of water in combination, or which causes a grain of oxygen and hydrogen to unite in the right proportions to form water, could be collected and thrown into a voltaic current, it would be exactly the quantity required to produce the decomposition of a grain of water or the liberation of its elements, oxygen and hydrogen.

The quantity of electricity, however, which is required to effect chemical decomposition is enormous. Faraday estimates the amount of electricity required to decompose a single grain of water to be equal to that evolved by a powerful flash of lightning.

245. **Limits of the Decomposing Action.**—Decomposition by the agency of the electric current takes place solely at those points where the electricity enters and leaves the liquid.

Thus, when a portion of water, for example, is subjected to decomposition in a glass vessel with parallel sides, oxygen is disengaged at the positive electrode, and hydrogen at the negative, the gases being perfectly pure and unmixed. If, while the decomposition is rapidly proceeding, the intervening water is carefully examined, not the slightest disturbance or movement of any kind will be perceived; nothing like currents in the liquid, or transfer of gas from one part to the other can be detected; and yet two portions of water, separated by an interval of four or five inches, may be respectively evolving pure hydrogen and oxygen. Now, since we know that every particle of water is composed of oxygen and hydrogen in the exact ratio of two

---

QUESTIONS.—Explain the decomposition of water. What fact has been proved by the experiments of Faraday? What is the relative quantity of electricity required to effect chemical decomposition? At what points of the galvanic circuit does the decomposing action take place? Illustrate this. In what respect is this action contrary to what might be naturally expected?

measures of the latter to one of the former, it would naturally be supposed that the electric current having separated the oxygen at one point, hydrogen would, having lost its combining element, also escape at the same point. This, however, is not the case, and great difficulty has been experienced in accounting for it.

The difficulty will be more evident, says Mr. Hunt, if we take the experiment on a larger scale; for example, if on one side of a wide river the positive pole is placed in the water, and the negative pole on the other, we shall still have—the battery being of sufficient power—oxygen given off on one side of the river, while hydrogen would be evolved at the other.

The following is the received explanation:—The arrangement of the particles constituting a line or layer of water between the poles of a galvanic circuit may be represented as follows, the positive atom, hydrogen, of each particle of water being turned by the influence of the electricity toward the negative pole, and the negative atom, oxygen, toward the positive pole—

Positive pole — O H, O H, O H, O H, O H, O H — Negative pole.

The same thing may be also illustrated in Fig. 72, where the particles of water are supposed to be spherical, the shaded portion of each sphere representing the hydrogen half of the particle, and the light portion the oxygen half.

Fig. 72.

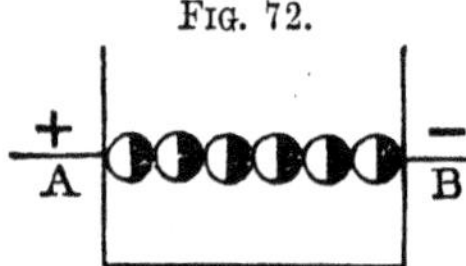

If the positive pole is placed on the left and the negative on the right, oxygen passes off from the first, and hydrogen from the last; if we reverse the poles, the order of the decomposition is changed also. It is not, however, to be supposed that when H. is liberated from O. at the negative pole, that the O. of that particle passes over along the line to the positive pole; but the view taken is, that as soon as the atom of oxygen loses its hydrogen, it combines with the atom of hydrogen of the next particle of water, and a new particle of water is reproduced. The oxygen of the second particle being thereby liberated, combines with the hydrogen of the next particle of water, and thus the decomposition and recomposition is continued on to the end of the series. Resorting again to symbols, No. 1 will represent the state of things before any change has been effected, and No. 2 the change after the circuit is complete—

No. 1. + O H, O H, O H, O H, O H −

No. 2. + O, H O, H O, H O, H O, H —

It should also be borne in mind, that the changes described are not successive, but simultaneous at each end of the series of particles, and at all intervening points in the line of the series.

246. Electrolysis and Electrolytes.—The process of resolving compounds into their constituents by electricity is

Questions.—What is supposed to actually occur in the decomposition of water? What is electrolysis?

termed Electrolysis, and a body susceptible of such decomposition is termed an Electrolyte.

No elementary substance can be an electrolyte; for from the nature of the process, compounds alone are susceptible of electrolysis. Electrolysis occurs only whilst the body is in the liquid state. The free mobility of the particles which form the body undergoing decomposition is a necessary condition of electrolysis, since the operation is always attended by a transfer of the component particles of the electrolyte in opposite directions.

The passage of a current of electricity through the liquid used in the cells or cups of a galvanic circuit depends upon the decomposition of its particles, in the same manner as in the case of water. No fluid, therefore, which is not an electrolyte, or in other words, which is not capable of being decomposed, is suitable for exciting a battery.

247. **Electro-chemical Order of the Elements.**—All the elementary substances, according as they appear at the positive or negative poles of a galvanic circuit, have been classified into electro-positive and electro-negative substances.

In the following table the most important of the elements are arranged in the order of their relative negative and positive properties, the most intensely negative element being placed at the top of the series, and the most intensely positive at the bottom:

ELECTRO-NEGATIVE.—Oxygen.
Sulphur.
Nitrogen.
Chlorine.
Fluorine.
Carbon.
Phosphorus.
Hydrogen.
Gold.
Platinum.
Mercury.
Silver.
Copper.
Tin.
Lead.
Iron.
Zinc.
Sodium.
Potassium.—ELECTRO-POSITIVE.

QUESTIONS.—What are electrolytes? Why can not an elementary substance be an electrolyte? What conditions are necessary for electrolysis? What fluids only are capable of exciting a galvanic battery? How may the elementary substances be classed as respects their electrical properties?

In this arrangement, each metal is positive as respects all that stand before it, and negative as respect those that succeed it. Oxygen is negative in every combination, and potassium appears to be uniformly positive. Hydrogen is highly positive when compared with oxygen and chlorine, but with metals it always exhibits negative electric energy.

248. **Electro-metallurgy**, or electrotyping, is the art or process of depositing, from a metallic solution, through the agency of galvanic electricity, a coating or film of metal upon some other substance.*

The process is based on the fact, that when a galvanic current is passed through a solution of some metal, as of sulphate of copper (sulphuric acid and oxyd of copper), decomposition takes place ; the metal, being electro-positive, attaches itself in a metallic state to the negative pole, or to any substance that may be attached to the negative pole ; while the oxygen, or other electro-negative element before in combination with the metal, goes to, and is deposited on the positive pole.

In this way a medal, a wood-engraving, or a plaster cast, if attached to the negative pole of a battery, and placed in a solution of copper opposite to the positive pole, will be covered with a coating of copper; if the solution contains gold or silver instead of copper, the substance will be covered with a coating of gold or silver in the place of copper.

The thickness of the deposit, provided the supply of the metallic solution be kept constant, will depend on the length of time the object is exposed to the influence of the battery.

In this way, a coating of gold thinner than the thinnest gold-leaf can be laid on, or it may be made several inches or feet in thickness, if desired.

The usual arrangement for conducting the electrotype process is represented by Fig. 73 It consists of a trough of wood, or an earthen vessel, containing the solution of the metal, the decomposition of which is desired—for example, sulphate of copper. Two wires, one connected with the positive, and the other with the negative pole of a battery, Q, are extended along the top of the trough, and supported on rods of dry wood, B and D. The medal, or other article to be coated, is attached to the extremity of the negative wire

* The general name of electro-metallurgy includes all the various processes and results which different inventors and manufacturers have designated as galvano-plastic, electro-plastic, galvano-type, electro-typing, and electro-plating and gilding.

QUESTIONS.—What substance is always negative? What one always positive? Define electro-metallurgy. Upon what is the process based? How is the thickness of the deposit regulated? Describe the arrangement for conducting the electrotype process.

and a plate of metallic copper to the end of the positive wire. When both of these are immersed in the liquid, the action commences—the sulphate of copper is decomposed—the copper being deposited on the medal attached to the negative pole, and the oxygen, before combined with it, on the copper plate attached to the positive pole, forming oxyd of copper. As the withdrawal of the metal from the solution goes on, the oxyd of copper thus formed

Fig. 73.

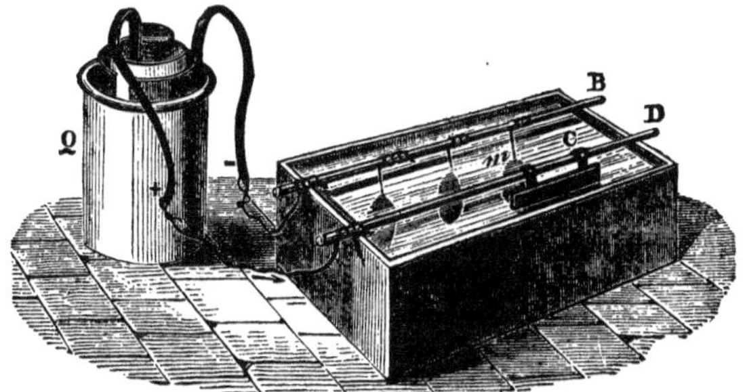

unites with the sulphuric acid which is liberated in the solution, and forms sulphate of copper. This dissolving in the liquid, maintains it at a constant strength.

The sole object of attaching a plate of metallic copper to the positive pole is to thus preserve the strength of the solution of sulphate of copper. If the positive pole had terminated with a plate of platinum or gold, the action would have commenced equally well, but the oxygen liberated from the copper, through its want of affinity to either the platinum or the gold, would have escaped as gas, and the solution gradually becoming weaker from the withdrawal of its elements, the electro-plating action would cease. When the operator judges that the deposit on the medal is sufficiently thick, he removes it from the trough, and detaches the coating. The deposit is prevented from adhering to the medal by rubbing its surface in the first instance with oil, or black-lead, and if it is desired that any part of the surface should be left uncoated, that portion is covered with wax, varnish, or some other non-conductor.

In this way a most perfect reversed copy of the medal is obtained—that is, the elevations and depressions of the original are reversed in the copy. To obtain a fac-simile of the original, the electrotype cast is subjected to a repetition of the process.

In general, it is found more convenient to mold the object to be reproduced in wax, or Plaster of Paris. The surface of this cast is then brushed over with black-lead to render it a conductor, and the metal deposited directly upon it. The deposit obtained will then exactly resemble the original object.

The pages and engravings in the book before the reader are illustrations of the perfection and practical application of the electrotype process. The engravings were first cut upon wood-blocks, and then, in combination with the ordinary type, formed into pages. Casts of the whole in wax were then made, and an electrotype coat of copper deposited upon them, and from the copper plates so formed the book was printed. The great advantage of this is, that the copper being harder than the ordinary type metal, is more durable, and resists the wear of printing from its surface for a longer period.

The improvement effected by electro-metallurgy in engraving is very great. When a copper plate is engraved, and impressions printed off from it, only the first few, called "proof impressions," possess the fineness of the engraver's delineation. The plate rapidly wears and becomes deteriorated. But by the electrotype process, the original plate can at once be multiplied into a great many plates as good as itself, and an unlimited number of the finest impressions procured.

In this way the map plates of the Coast Survey of the United States, some of which require the labor of the engraver for years, and cost thousands of dollars, are reproduced—the original plate being never printed from.

The metals upon which an adherent coating of silver or gold is most readily deposited are brass, copper, bronze, and German silver. The articles to be plated or gilded must be carefully cleansed from all adhering greasy matters by boiling them in a weak alkaline solution, and then rubbing them with chalk, rotten-stone, etc. The articles are then carefully washed, attached to a clean copper wire, and immersed in the silvering solution. The deposit is hastened by keeping the solution moderately warm, especially at the commencement of the process. The articles, when plated, have a dead white, or chalky appearance, but by burnishing they assume the brilliant luster of polished silver.*

249. **Protection of Metals from Corrosion.**—When two metals which are positive and negative in their electrical relations to each other, are brought in contact, a galvanic action takes place which promotes chemical change in the positive metal, but opposes it in the negative metal.

Thus, when sheets of zinc and copper immersed in dilute acid touch each other, the zinc oxydizes or rusts more, and the copper less rapidly, than

* The teacher, for experiment, can best illustrate the deposition of metals by electro-chemical action in the following manner:—Put a piece of silver in a glass containing a solution of sulphate of copper, and into the same glass insert a piece of zinc. No change will take place in either metal so long as they are kept apart; but as soon as they touch, the copper will be deposited upon the silver, and if it be allowed to remain, the part immersed will be completely covered with copper, which will adhere so firmly that mere rubbing alone will not remove it.

QUESTIONS.—How has the electrotype process affected the art of engraving? What are the peculiarities of the process of electro-plating and gilding? Under what circumstances can metals be protected from chemical action? Illustrate this.

without contact. Iron nails, if used in fastening copper sheathing to vessels, rust much quicker than when in other situations, not in contact with the copper. The reason of this is, that the two metals, in consequence of the electricity developed by their union, are placed in opposite electrical conditions. The copper, which is ordinarily positive, is rendered negative by the contact of the zinc, or iron; it, therefore, is not only entirely wanting in attraction for the negative corroding oxygen of the air, or water, on the principle that bodies similarly electrified repel each other, but even has a tendency to abandon any oxygen with which it may have previously combined. The zinc and iron, on the contrary, in virtue of the exaltation of their naturally positive condition, combine with the negative oxygen most readily, on the principle that bodies in the opposite electrical condition attract each other. The positive metal, therefore, oxydizes most speedily, while the negative metal remains uninjured.

What is called galvanized iron, is iron covered entirely, or in part, with a coating of zinc. The galvanic action between the two oxydizes the zinc, but protects the iron from rust. Sir Humphrey Davy attempted to apply this principle to the protection of the copper sheathing of ships (which wastes rapidly through the action of the oxygen in sea-water), by placing at intervals over the copper small strips of zinc. The experiment was tried, and a piece of zinc as large as a pea was found adequate to preserve forty or fifty square inches of copper; and this wherever it was placed, whether at the top, bottom, or middle of the sheet, or under whatever form it was used. The value of the application was, however, neutralized by a consequence which had not been foreseen; since the protected copper bottom rapidly acquired a coating of sea-weeds and shell-fish, whose friction on the water became a serious resistance to the motion of the vessel. The adhesion of these, under ordinary circumstances, is prevented by the corrosion of the copper by oxygen, and by the poisonous action of the compounds of copper and oxygen which are thereby formed.

The principle, however, has been applied with success for the protection of iron pans used in evaporating sea-water, and in other similar apparatus.

Questions.—How is this action accounted for? What is galvanized iron? What practical application of this principle was attempted by Sir Humphrey Davy?

INORGANIC CHEMISTRY.

THAT department of Chemistry which treats of inorganic, or unorganized bodies, is termed Inorganic Chemistry.

It includes the doctrines of affinity, the laws of combination, the chemical history of the elementary bodies, and of those compounds of the elements which are not the product, either directly or indirectly, of living, organized bodies.

CHAPTER V.

THE GENERAL PRINCIPLES OF CHEMICAL PHILOSOPHY.

250. **Elements.**—A chemical element is a material substance not yet analyzed or taken apart—not yet resolved by any process into two or more bodies differing from itself.

No one substance within the reach of man is, however, *positively* known to be elementary; and the student should distinctly understand, that it can not rightly be inferred, because a body has not yet by any known process been decomposed, that it never will be.

251. **Number of the Elements.**—The number of elements at present *fully* recognized by chemists is sixty-two. Of these only twenty-nine were known at the commencement of the present century.*

* This fact will illustrate to the general student one great feature in the progress of modern chemistry; but to the chemist, the discovery of thirty-three new elementary

QUESTIONS.—What is inorganic chemistry? What is a chemical element? Is any substance positively known to be elementary? What is the number of the elements?

252. Classification of the Elements.—The elements are usually divided into two great classes, the metallic and non-metallic substances, or the Metals and the Metalloids. The substances comprised in the first class are the more numerous, but those in the latter are the more abundantly distributed.*

Of the sixty-two elements, five are gases, viz., oxygen, hydrogen, nitrogen, chlorine, and fluorine; two are simple liquids, mercury and bromine; the remainder are solids, at common temperatures. Only fourteen of the elements are of common occurrence, and of these the great mass of the earth, with its atmosphere and water, are composed. The remainder occur only in comparatively small quantities, and fully one third of the whole number are so rare as not to admit of any useful application.

A very few only of the elements are found naturally in a free or uncombined state; of such we may mention oxygen and nitrogen, existing in the atmosphere; sulphur, carbon, and a few of the metals, as gold, platinum, copper, etc., distributed throughout the earth. The majority exist only in

---

bodies implies an amount of laborious and protracted research, preceding and following each discovery, of which words can convey to the uninitiated no adequate idea.

* The alchemists regarded the metals, the only elementary bodies with which they were acquainted, as compound substances. The baser metals, as lead, iron, copper, etc., they believed to contain the same elements as gold, from which they differed on account of their association with impurities; these impurities being separated, it was imagined that gold would remain.

The problem, known as the "transmutation of metals," which they sought to solve, and labored for centuries to effect, was not to generate or create metals, but to change the proportion of the elementary substances which composed them. "For a century or more," says Professor Faraday, in a recent lecture, "it has been the custom to spurn the doctrines of the alchemists as devoid even of the semblance of philosophic truth. The time has, however, past for this opinion to be maintained, and within the last few years a series of manifestations have been noticed which go far to vindicate many of their opinions." At a meeting of the British Association for the Promotion of Science in 1851, M. Dumas and Professor Faraday both avowed their belief in the possibility of transmutation, and the latter stated that he had even experimented with a view of producing this result, and should continue to do so. It is not, however, to be understood that chemists expect transmutation will be effected in exactly the sense of the old alchemical philosophy. There is no evidence that lead can be converted into silver, or copper into gold. M. Dumas suggests that the first successful transmutation as regards metals will be to effect a change of physical state merely, without touching chemical composition; thus, already we have carbon, which, as the diamond and as charcoal, manifests two widely different states. Sulphur also assumes two forms, as also phosphorus, silicon, and boron. Then why not a metal?

Within a very recent period (1857), a series of experiments have been published by Dr. Draper of New York, which seem to indicate that silver is capable of transmutation into another metal, possessing some of the properties and characteristics of gold. "It is hard

---

QUESTIONS.—Into what two great classes are the elements usually divided? How many of the elements are gaseous? How many liquid? How are the elements distributed in nature? In what condition are they generally found?

combination with each other, and in this condition they are so completely disguised as to manifest few or none of their characteristic properties.

253. **Compound Bodies.**—All compound bodies are formed by the chemical union of two or more of the elementary substances.

The compounds so resulting are, as might be supposed, almost innumerable, and the progress of research is continually adding to their number. Many of the compounds artificially formed by chemical action have no existence in nature. Some of them are of eminent utility to man, while others possess properties of a strange and fearful character. Happily, however, the majority of those compounds which are especially deleterious are, by the difficulty and expense of their preparation, placed far beyond the reach of the majority of mankind.

254. **Cause of Chemical Combination.**—In the early days of chemistry, chemical combination between different substances was supposed to take place through the agency and guidance of some spiritual or supernatural power which invested, or dwelt in every form of matter, both animate and inanimate. The popular names of many chemical substances at the present time, such as *spirit of wine*, *spirit of nitre*, etc., are evidences of the former general credence in this doctrine. Stahl, a noted chemist who died in 1685, taught that chemical combination proceeded from an approximation of the combining parts, somewhat after the manner of wedges. Modern chemistry explains chemical combination between different substances, as occurring through the agency of an attractive force, acting only between the atoms, or molecules of dissimilar substances, and only at insensible distances. This force, to distinguish it from other forms of attraction, is termed affinity. To

---

to think," says Sir David Brewster, " that the so-called elements are truly simple. The instinct of humanity revolts against believing that the Maker has departed from his wonted simplicity of procedure in this one part of creation, and flung such a number of unchangeable elements from his immediate hand. Many thoughtful and ingenuous men, indeed, have frankly supposed that it were more like the nature of Deity, as shown by his interpreted works, to pour forth the unreckonable variety of things from the bosom of one or two principles. Thales and the Greek physicists, Roger Bacon, Stahl, Lavoisier, Sir H. Davy, and Berzelius, have all given more or less expression to this idea. The greatest question in chemistry, or in plain earnest, the one question of the age then, is precisely this:—What is the interior nature of these elements? Science bids us ask, and perhaps nature is ready to answer it; but what shall be done, since no analytical power can move one of those steadfast natures from its propriety? Let synthesis be tried if analysis has failed; synthesis has never been tried. It is in the highest degree probable that all the present elements are equi-distant from simplicity, and all equally compound, if there be any truth in the unanimous testimony of chemical analogy. Their case is exactly like that of potassa, soda, lime, and their congeners, before the discovery of potassium;—that is to say, potassa once discovered to be metallic oxyd, all the rest were clearly metallic oxyds too, as experiment was not long of showing. In the same way, if the secret of one of these silent, tantalizing elements be discovered, the secret of them all is out."

---

[1] QUESTIONS.—How are compound bodies formed? Do all the compounds known to the chemist exist in nature? How did the early chemists explain chemical combination? How does modern chemistry explain it?

the question "What is the attractive force thus designated?" no satisfactory answer can be given. There are, however, some reasons for supposing it to be a modification of electrical force.

255. **Characters of Chemical Affinity.**—Chemical affinity is distinguished from all other kinds of attractive forces which act at minute distances, by certain peculiar characteristics. These are briefly as follows :—

I. It is exerted within its own limits with intense energy, but beyond those limits it is entirely powerless.

An iron wire which will support a weight of a thousand pounds without breaking, will in a few minutes yield to the almost noiseless action of a mixture of sulphuric acid and water. The tenacious metal will dissolve—particle by particle will be detached from the iron—and in the clear liquid which results, no vestige of the structure of the metal will remain. It is rarely possible by minute subdivision to cause the particles of different substances to approximate sufficiently near to produce chemical action. Tartaric acid and carbonate of soda may be incorporated by grinding for hours in a mortar, but they will not act chemically upon each other. If, however, we add a portion of water, which dissolves the particles of both and allows them mutually to approach closer, a chemical union, accompanied by an effervescence, immediately takes place.

The amount of power or work produced by the action of chemical affinity is in general very great, and in some instances we may approximately measure and compare it with other forces. For example, coal burns and produces heat solely in consequence of the affinity, or attractive force, which causes particles of oxygen in the air to unite with particles of coal. Now, a pound of the purest coal, burned under the proper circumstances, and its resulting heat applied to the production of steam, will generate a power capable of lifting a weight of 100 pounds to a height of 20 miles, or 1 pound 2,000 miles. This result, therefore, is a measure of the chemical force of affinity which operates between the particles of a pound of coal and the quantity of oxygen that unites with them.

II. It is only exerted between dissimilar substances.

No manifestations of this force can take place between two pieces of iron, two pieces of copper, or two pieces of sulphur; but between sulphur and copper, or sulphur and iron, chemical action of the most energetic kind may occur.

Were there but one kind of matter in the universe, the force of affinity could not exist; no chemical action could take place, and the science of chemistry would be unknown.

Questions.—What do we know of the nature of affinity? State and illustrate the first characteristic of affinity? What is said of the amount of work, or power which chemical affinity is capable of producing? Give an illustration. State and illustrate the second characteristic of affinity.

III. Generally speaking, the greater the difference in the properties of bodies, the greater is their tendency to enter into chemical combination. Between bodies of a similar character, the tendency to union is feeble.

IV. Chemical affinity occasions an entire change in the properties of the substances acted upon.

This change is most remarkable, and is of such a character as could not be predicted from any acquaintance with the substances in a separate condition. Thus, if we dissolve copper in sulphuric acid, we obtain a blue, semi-transparent substance; while iron treated in the same manner, yields a light green product.

Although in a combination, the properties of the constituents are changed, and, as far as ordinary observation is concerned, are destroyed, yet they really exist in the compound, and can be again reproduced by restoring the combining elements to their original condition.

V. The power of affinity is exerted between different kinds of matter with different, but definite degrees of force.

Nitric acid, for example, will combine with and dissolve most of the metals, as silver, mercury, copper, and lead; but it unites with them with very different degrees of intensity. With silver the combination is less powerful than with mercury, less so with mercury than with copper, and with copper less again than with lead.* Indeed, the different elements may be arranged in tables, in such a way as to indicate by their order the degree of affinity which they respectively have for some particular element.

Fig. 74.

* The difference in the strength of the affinity existing between different substances may be easily illustrated by the following experiment:—Dissolve a few crystals of acetate of lead (*sugar of lead*) in a small quantity of water, and fill a phial with the solution. If a piece of zinc be now suspended in the liquid, it will, after a little time, become covered with a gray coating, from which brilliant metallic spangles will gradually shoot forth (see Fig. 74) somewhat in the shape of a tree. These are pure lead, and the phenomenon is familiarly known as the lead tree. The effect thus produced is due to the superior affinity of the zinc for the acetic acid combined with the lead, which causes the two metals to interchange places—*i. e.*, the zinc combining with the acid and entering into solution, and the lead being deposited in a metallic state, in place of the zinc. If the action be kept up sufficiently long, every particle of lead may be in this way withdrawn from the liquid.

QUESTIONS.—What is the third characteristic of affinity? What is the fourth? Illustrate this. Is the force of affinity always uniform? How may this be shown?

VI. However much the properties and form of bodies may be changed by the action of chemical affinity, no destruction of matter ever ensues—the weight of the products of combination being always exactly equal to that of the component elements before combination.

By means of a simple experiment it may be shown that even although a substance may, through the action of chemical affinity, vanish from our sight, it still continues to exist as a gas which has the same weight as the visible solid which furnished it. Into a glass flask, A, Fig. 75, of about 250 cubic inches capacity, which is provided with a brass cap and stop-cock, 10 or 12 grains of gun-cotton are introduced. The air in the flask is then completely exhausted by means of an air-pump, and the flask weighed. The cotton is then ignited by means of two wires, *a* and *b*, proceeding from a galvanic battery, and passing through the cap of the flask. On the transmission of a voltaic current, the cotton entirely disappears with a brilliant flash, but the flask, if weighed again, will be found to be as heavy as before the cotton was fired.

FIG. 75.

VII. Chemical combination of substances may either occur instantly on mixture, or may be indefinitely postponed until some other force, as heat, for example, produces a commencement of the action.

In a large proportion of cases, chemical action will not commence spontaneously. A heap of charcoal will remain unaltered in the air for years; but if a few pieces be made red hot and then thrown upon the heap, chemical combination between the charcoal and the oxygen of the air is commenced by the heat, and continues until the whole mass is burned. In other instances chemical action commences without the application of any extraneous force. Phosphorus begins to burn slowly the instant it comes in contact with the atmosphere, and exposed to the heat of the sun, speedily bursts into a flame.

Ca-tal'y-sis.—The mere *presence of a third body* will sometimes awaken or excite the force of affinity between

QUESTIONS.—Is matter in its changes consequent on the action of affinity ever destroyed? What experiment illustrates this? Under what varying circumstances will chemical combination take place? Will chemical action between different substances generally commence spontaneously? Illustrate this. What is catalysis?

two other bodies to an extent sufficient to cause their union—without itself undergoing any alteration, either mechanical or chemical. Such an action is termed *Catalysis.* It is also sometimes called the *action of presence.*

Phenomena of this character are the most curious, and, in some respects, the most difficult of explanation of any in chemistry. A familiar example of this action is afforded us in the case of yeast, a most minute particle of which is able to excite fermentation in a large quantity of sugar in solution. Other examples will be noticed in the progress of this work.

Nascent State.—Chemists have long recognized the fact, that bodies, when in the act of liberation, or separation from other substances, display far more energetic affinities than under ordinary circumstances. This condition is termed the nascent (from the Latin *nascor, to be born,*) state. Thus, hydrogen and nitrogen gases, under ordinary circumstances, do not unite if mingled in the same vessel; but when these two gases are set free at the same time from the decomposition of some substance, they readily combine.

VIII. Chemical compounds may be formed either by the direct union of their ingredients, or by the displacement of one substance by a different one in a compound previously formed.

IX. Whenever the elements unite directly with each other, heat is generally evolved, and in many instances, light also; the amount of each being proportioned to the rapidity of the action.

256. **Laws of Chemical Combinations.**—It might naturally be supposed that chemical combination between the various elementary substances would take place in all proportions indifferently, in the same manner as unlike particles of matter can be mingled together mechanically. Such, however, is not the case, but the relative proportions in which different elements unite is determined by fixed laws.*

---

* It should be here remarked, that the views adopted in this work are those of Berzelius, Mitscherlich, Dumas, Hayes, and most of the leading chemists of the day—viz. that all mixtures of gases with gases, liquids with liquids, and all solutions *proper*, of solids in liquids, are not chemical combinations, unless they take place in definite proportions. Apparent combination in indefinite proportions, as of alcohol with water, may be

---

QUESTIONS.—What is understood by the nascent state? In what two ways may chemical compounds be formed? What phenomena of heat and light attend chemical combinations? Do substances enter into combination in all proportions? Enumerate the laws which regulate chemiual combination.

These laws, which are three in number, regulate the mode of combination of every known chemical compound, and are usually called the Law of Definite Proportions, the Law of Multiple Proportions, and the Law of Equivalent Proportions.

257. **Law of Definite Proportions.**—In every chemical compound the nature and proportion of its constituent elements are fixed, definite and invariable.

For instance, 100 parts of water contain 88·89 of oxygen and 11·11 hydrogen. It matters not in what condition the water may exist—in springs, or in the ocean, in the form of ice, dew, cloud, or steam, its composition is uniform and certain. When artificially prepared, by causing the gas hydrogen to unite chemically with the gas oxygen, the same proportions are required, that is, 11·11 grains, ounces, or pounds of hydrogen must be taken for every 88·89 grains, ounces, or pounds of oxygen. If either one of the constituents be in excess, combination will still take place, but the excess will be rejected. So also in the case of other simple compounds. A piece of flint, or of clear quartz crystal, come from whatever source it may, yields in every 100 parts, 48·2 of the element silicon, and 51·8 of oxygen.

The law of definite proportions may be proved in two ways: first, by analysis, that is, by taking the compound apart and comparing the products of decomposition; and, secondly, by synthesis, that is, by uniting the elements in definite proportions to form the required compound.

Although of great simplicity, it constitutes one of the fundamental principles upon which modern chemistry, as an exact science, rests. It enters into all the practical applications of chemistry to the arts, and is relied upon by the analyist as a means of verifying and classifying his results. It also enables us to draw a broad and clear distinction between a mechanical mixture and a chemical combination; between the force of affinity and the force of adhesion, which produces the solution of solids in liquids.

258. **Law of Multiple Proportions.**—It frequently happens that one elementary substance will unite with another in more than one proportion. The compounds so obtained differ greatly in their properties, but still preserve a simple relation to each other. The law which governs these relations, and which is known as the law of multiple proportions, may be stated as follows:—

If the elements A and B unite together in more proportions than one, the several quantities of B, which unite

---

explained by supposing that definite combination takes place between limited quantities of the combining substances, in the first instance, and that the compound thus formed is afterward mechanically mixed with the excess of either of the constituents.

QUESTIONS.—State the law of definite proportions. What are illustrations? How may this law be proved? What is its practical value? What is the law of multiple proportions?

with the same quantity of A, will bear a very simple relation to each other.

Thus we may have a series of compounds like the following:—A+B; A+2 B; A+3 B; A+4 B; A+5 B, etc., in which one part of the element A unites respectively with one, two, three, four and five parts of B, to form five different compounds, each possessing different properties. Such a simple series represents the five different compounds which nitrogen forms with oxygen—one, two, three, four and five parts (by weight) of oxygen uniting with one part of nitrogen. In some instances the relation is less simple, one or two proportions of one element combining with 3, 5, 7, etc., of another—the law simply requiring that the proportionals shall all be multiples of the smallest. Thus, compounds represented by the following formulas may exist:—

2 A+3 B: 2 A+5 B; 2 A+7 B, etc.

In this $1\frac{1}{2}$ is considered as the smallest combining proportional of B.

259. **Law of Equivalent Proportions.**—When an elementary body (A) unites with other bodies (B, C, D, etc.), the proportions in which B, C and D unite with A, will represent in numbers the proportions in which they will unite among themselves, in case such union takes place; in other words, the fixed proportions in which the elements unite among themselves, may be represented numerically.

Oxygen is an element that forms at least one definite compound with every other elementary substance, with a single exception. United with hydrogen it forms water, and 100 parts of water, as before stated, contain 88·89 parts of oxygen and 11·11 hydrogen. United with the element calcium, it forms lime, and 100 parts of pure lime, if examined, will be found to consist of 28·58 parts of oxygen and 71·42 of calcium. In like manner 100 parts of potash contain 17·02 of oxygen and 82·98 of the element potassium. It will be apparent, from these illustrations, that the quantity of oxygen is not the same in its compounds with the different elements, and the inquiry next arises, does any constant relation exist between the proportions of oxygen and the proportions of the different elements which unite with it to form compounds? The existence of such a relation may be shown in the following manner:—Having ascertained the proportions in 100 parts of the various compounds which each elementary body forms when it combines with oxygen, determine by calculation the proportion in which each element unites with the same fixed quantity of oxygen, as 8 parts, for example. A series of proportional numbers will thus be obtained, which will represent the ratios in which

---

QUESTIONS.—Illustrate the law of multiple proportions. What is the law of equivalent proportions? How is the law of equivalent proportions demonstrated?

each of the elements combines with oxygen. For example, in the case of water, it will be seen that for each 8 parts of oxygen, 1 part of hydrogen is present.

For 88·89 (the quantity of oxygen in 100 parts of water) : 11·11 (the quantity of hydrogen) : : 8 : 1.

So also in lime, for each 8 parts of oxygen, 20 of the element calcium are present.

For 28·58 : 71·42 : : 8 : 20.

And in potash, for every 8 of oxygen there are 39 of potassium.

For 17·02 : 82·96 : : 8 : 39.

In like manner it has, by careful and laborious investigation, been shown that the proportions which exist between oxygen and the other elements in their respective combinations, are capable of being represented numerically. Thus, 8 parts of oxygen unite with 14 of nitrogen, 16 of sulphur, 6 of carbon, 28 of iron, 32 of copper, 100 of mercury, 104 of lead, 108 of silver, and so on.

But further experiments have led to the very remarkable discovery, *that these numbers not only represent the quantities of the different elements which unite with 8 parts of oxygen, but they also indicate the simplest proportions in which the different elements can unite with each other.*

For example, not only does 1 part, by weight, of hydrogen, 16 of sulphur, 28 of iron, and 100 of mercury, severally unite with 8 parts of oxygen, but 1 part of hydrogen unites to form a compound with 16 parts of sulphur, and 16 of sulphur in turn unites to form different compounds with 28 parts of iron and 100 of mercury, or 39 of potassium.

260. **Law of Substitution.**—It very often happens also, that through the varying force of affinity, one element is able to expel and replace another in a compound previously formed. When such a substitution takes place, it always happens in the quantities indicated by their proportional numbers.

This principle may be illustrated as follows:—In mercantile transactions, 100 dollars in money will purchase 6 ounces of gold, or 12 ounces of platinum, or 100 ounces of silver, or 1,500 ounces of mercury; consequently, 6 ounces of gold have the same commercial value as 12 ounces of platinum, or 100 ounces of silver, etc. The same principle holds good in chemistry: 28 ounces, or 28 parts of any other denomination by weight, of iron, 100 of mercury, 108 of silver, or one of hydrogen, combine with 8 of oxygen. Accordingly 28 ounces of iron have same chemical value as 100 ounces of mercury, 108 of silver, or 1 ounce of hydrogen.—STOCKHARDT.

261. **Chemical Equivalents.**—The proportions, or quantities by weight, in which different substances unite

QUESTIONS.—What remarkable fact has been ascertained respecting the proportion of the elements which combine with oxygen? What are examples? What is understood by the law of substitutions? How is this illustrated? What is the meaning of chemical equivalents?

to form definite chemical compounds, are called Chemical Equivalents (from *æquus*, equal, and *valor*, value). They are also sometimes designated as combining, or equivalent weights. The numbers representing or expressing these proportions are termed equivalent numbers.

Thus, by 1 equivalent of oxygen is to be understood 8 parts of it by weight; by 1 equivalent of iron, 28 parts by weight; by 1 equivalent of mercury, 100 parts by weight.

It will be readily observed that the numbers used to designate equivalents merely express the *relative* quantities of the substances they represent; it is therefore a matter of little consequence what numbers are employed to express them, provided the relations between them are strictly observed. Thus we may represent the equivalent of hydrogen (which is the smallest of all the equivalent numbers) by 100, or 1,000 as well as by 1, provided all the other equivalent numbers are multiplied in an equal ratio; or hydrogen may be represented by .01 or .001, if all the other numbers are equally reduced. If hydrogen were represented by 100, oxygen would be 800, and iron 2,800. Or if hydrogen were 0·01, oxygen would be 0·08, and iron 0·28. It is the ratio, or relative proportion, which gives value to these numbers.

In England and the United States, the combining number of hydrogen is made the unit of comparison. The reason why this element is selected is because it combines with oxygen and other elements in a smaller proportion by weight than any other known substance, and the numbers representing the combining proportions of all the other elements, may also, with few exceptions, and without material error, be taken as multiples by whole numbers of the equivalent of hydrogen. The equivalent number of hydrogen in this scale is 1, and as one part of hydrogen is united in water with exactly 8 parts of oxygen, the equivalent number for oxygen is 8.

On the Continent of Europe, most chemists make oxygen the unit of comparison, and assume its equivalent number to be 100: the equivalent number of hydrogen will be, therefore, 8 times less, or 12·5, and the equivalent numbers of the other elements, calculated according to the hydrogen scale, will also be changed proportionally.

In the following table the elementary substances are arranged alphabetically, with the symbols used by chemists to designate them affixed to each. The numbers representing their equivalent or combining proportions, calculated according to the hydrogen scale, are placed opposite to each element.*

* The numbers on the hydrogen scale will be adopted in this work, and, generally speaking, fractional quantities will be omitted.

QUESTIONS.—What of equivalent numbers? May the numbers expressing equivalents be varied and changed? On what principle? What is the unit of the scale adopted in England and the United States for indicating the numerical relations of the equivalents? Why is hydrogen adopted? What is the unit adopted upon the Continent of Europe?

The names of the elements which from their rarity may be regarded as unimportant, are given in Italics.

TABLE OF THE ELEMENTARY SUBSTANCES, WITH THEIR EQUIVALENTS AND SYMBOLS.

Name.	Symbol.	H=1.	Name.	Symbol.	H=1.
Aluminum.................	Al	13·7	Molybdenum................	Mo	46·
Antimony (Stibium)........	Sb	129·	Nickel.......................	Ni	29·6
Arsenic.....................	As	75·	*Niobium*....................	Nb	
Barium......................	Ba	68·50	Nitrogen.....................	N	14·
Bismuth....................	Bi	212·	*Osmium*.....................	Os	99·6
Boron......................	B	10·9	Oxygen.......................	O	8·
Bromine...................	Br	80·	*Palladium*..................	Pd	53·3
Cadmium..................	Cd	56·	*Pelopium*....................	Pe	
Calcium....................	Ca	20·	Phosphorus..................	P	32·
Carbon.....................	C	6·	Platinum....................	Pt	98·7
Cerium.....................	Ce	47·	Potassium (Kalium)..........	K	39·2
Chlorine....................	Cl	35·50	*Rhodium*....................	R	52·2
Chromium..................	Cr	26·7	*Ruthenium*.................	Ru	52·2
Cobalt......................	Co	29·5	Selenium....................	Se	40·
Copper.....................	Cu	31·7	Silicium, or Silicon..........	Si	21·3
Didymium.................	D		Silver (Argentum)...........	Ag	108·
Erbium,....................	E		Sodium (Natrium)............	Na	23·
Fluorine....................	F	19·	Strontium....................	Sr	44·
Glucinium.................	G	6·9	Sulphur......................	S	16·
Gold (Aurum)..............	Au	98·	*Tantalum* (Columbium).....	Ta	92·
Hydrogen..................	H	1·	*Tellurium*...................	Te	64·
Ilmenium..................	Il		*Terbium*....................	Tb	
Iodine......................	I	127·	*Thorium*....................	Th	59·6
Iridium....................	Ir	99·	Tin (Stannum)...............	Sn	59·
Iron........................	Fe	28·	*Titanium*...................	Ti	25·
Lantanium................	La	36.	*Tungsten* (Wolfram).........	W	94·
Lead (Plumbum)...........	Pb	103·5	Uranium....................	U	60·
Lithium....................	Li	6·9	*Vanadium*..................	V	68·6
Magnesium................	Mg	12·2	*Yttrium*.....................	Y	32·2
Manganese.................	Mn	27·6	Zinc.........................	Zn	32·5
Mercury...................	Hg	100·	*Zirconium*..................	Zr	22·4

Three other substances discovered within the last few years, and designated as Aridium, Donarium, and Norium, are claimed to possess an elementary character. If their existence is fully established, the number of the elements must be considered as sixty-five.

The law of equivalents applies to compound substances equally with the elements—the equivalent of a combining number of a compound being always the sum of the equivalent of its components.

Thus, since water is composed of 1 equivalent, or 8 parts of oxygen, and 1 equivalent, or 1 part of hydrogen, its combinining proportion or equivalent is 9. The equivalent of sulphuric acid is in like manner 40, because it is a compound of 1 equivalent, or 16 parts of sulphur, and 3 equivalents of oxygen; ($3\times8=24$), and $16+24=40$. The equivalent number of potassium is 39,

QUESTION.—Does the law of combination by fixed equivalents extend to union of compound substances? Illustrate this.

and as this element combines with 8 of oxygen to form potash, the equivalent of the latter must be 39+8=47. Now, when these compounds unite, one equivalent of the one combines with one, two, three, or more equivalents of the other, precisely as the elementary substances do. For example, water unites with potash to form a compound, but it does so only in the proportion of 9 to 47; sulphuric acid also unites with potash to form a compound (sulphate of potash), but only in the proportion of 40 to 47.

To illustrate the advantage in practical operations of employing the scale of equivalents, we will suppose a person wishes to manufacture sulphate of potash, which is one of the ingredients which enter into the composition of alum. Having purchased in the market the necessary components of sulphate of potash, viz., sulphuric acid and potash, he mixes the two together, according to their equivalents, in the proportion of 40 parts (pounds, ounces, or tons) of sulphuric acid with 47 parts of potash. The result is, that *all* the sulphuric acid unites with *all* the potash, and the greatest product of the compound is obtained. If, on the other hand, he had mixed the sulphuric acid and the potash in any other than the above, or some multiple of the above proportions, there would have been an excess or deficiency of one of the ingredients, and consequently a loss of material. The sulphate of potash formed by the partial combination would also prove to be an imperfect article, from the mechanical mixture of the excess of one of the ingredients throughout its substance.

Previous to the discovery of this law of equivalents, at about the commencement of the present century, it could only be ascertained by laborious trials, how much of one chemical substance was required to combine with, or replace another. It is now only necessary to refer to the table of the proportional, or equivalent numbers to ascertain beforehand the quantity to be employed.

262. **Equivalent Volumes.**—When bodies are capable of assuming the form of a gas, or vapor, and in this condition act chemically upon and combine with each other, a very simple ratio prevails between the quantities which enter into combination, measured merely by their bulk or volume.

Thus, one volume of a gas, which may be distinguished as A, unites with one, two, or three volumes of B, or two of A may unite with three of B.

If when two gases capable of union by contact are brought together, the volume of one is greater than its combining proportion, the excess remains uncombined.

The volume of two gases, after combination, is often less than the sum of

---

QUESTIONS.—Show in what manner the law of equivalents is practically applied in chemical operations? What is understood by equivalent volumes? Does the volume of the gases always remain the same after combination?

of their volumes in their separate state; or in other words, the two gases or vapors, by the act of union, sometimes experience a condensation.

It is, however, a very curious fact, that when such a diminution of the volume occurs, it always takes place in a simple ratio to the volume of one or both of the combining gases. Thus, three volumes of hydrogen and one of nitrogen unite to form ammonia; but when the union takes place, the four volumes instantly contract to two, or one half their former bulk. The weight, however, of the ammonia formed is equal to the united weight of the hydrogen and nitrogen that have entered into its composition.

263. **Atomic Theory**.—A consideration of the facts set forth, naturally suggests the inquiry,—Why is it that all the different kinds of matter with which we are acquainted, in entering into chemical combination with each other, are constrained to do so according to certain fixed weights and volumes, and not otherwise? The response from every thinking mind will unhesitatingly be that the phenomena in question must originate in accordance with some great law or principle in nature, so extensive and general in its character as to affect all matter. Experiment and observation do not, and probably can not, enable us to say definitely what this law is; but a careful consideration and comparison of all the facts in the case, led Dr. John Dalton, an eminent English chemist, about the year 1808, to propose a theory which so satisfactorily explains the remarkable circumstances attending chemical combination, that scientific men of all countries receive it as substantially true. This theory is known as the "Atomic Theory," or the "Theory of Atoms."

The atomic theory supposes, in the first instance, that all matter is composed of ultimate particles, or atoms, which are incapable of subdivision. (*See* § 4, *page* 10.)

A belief in this hypothesis dates back to a very remote period. It was a doctrine taught by that sect of the Greek philosophers known as the Epicureans, and during the middle ages it formed a part of certain theological dogmas maintained by parties in the church. In more modern times, it received the sanction of many men of high scientific attainment, as Newton, Bacon, and others.* These opinions can not, however, be regarded in any other light than as mere speculations, and it was not until laborious study and

* "It seems to me," says Sir Isaac Newton, "that in the beginning, God formed matter in a solid mass of hard, impenetrable particles; and that these primitive particles being solids, are incomparably harder than any porous bodies compounded of them; so very hard as never to wear or break in pieces, no ordinary power being able to divide what God made one in the first creation."

QUESTIONS.—What inquiry naturally arises in the mind from a consideration of the facts stated? According to what theory is chemical combination explained? Who proposed this theory? What does the atomic theory suppose in the first instance? Is this supposition of recent origin?

research had elevated chemistry to the rank of an exact science, that any rational evidence upon the subject could be appealed to.

The atomic theory, as proposed by Dalton, further supposes, that the atoms of each separate elementary substance have all the same characteristic form and weight, and that when combination between two different elements takes place, one or more atoms of one substance arrange themselves in the most symmetrical manner possible by the side of one or more atoms of another substance, and thus form a compound atom.

In the simplest combination, one atom of one substance combines with one atom of another, but in other instances the proportion may be as 1 to 2, 3, 4, and 5, or as 2 to 3, 5, 7, etc. One atom of one kind can not combine with one half an atom of a different kind, or with any other fractional part of an atom, for the reason that no such quantities exist—the atoms being incapable of division. Hence the immutable nature of all compound bodies existing either in nature or art.

Furthermore, as combination of different substances takes place atom by atom, and as the atoms of each substance have a size and weight peculiar to themselves, we have an explanation of the circumstance that the chemical union of quantities of different kinds of matter only occurs in unchanging proportions by weight and volume—for what is true of all the atoms of a mass, must be true of the whole.

Again, a compound atom formed by the union of two dissimilar atoms, must, in uniting with other bodies, necessarily obey the same laws of combination as the elementary atoms, and be in turn incapable of division, since the very act of division would be its destruction, so far as its compound character is concerned.

A strong argument in favor of the truth of the atomic theory is, that no reasonable explanation of the facts pointed out can be given by the adoption of any other theory. If matter is infinitely divisible, and if atoms have no real existence, then there is no reason why bodies should not combine in all proportions. One grain, ounce, or pound, of one substance ought to combine with the half, quarter, tenth, hundredth, and every other proportion of a grain, ounce, or pound, of some other substance, so as to form an infinite number of compounds, all possessing different properties. But this, as has been already stated, never happens.

Dr. Dalton was also the first who conceived clearly the idea, that from the

QUESTIONS.—What does the atomic theory of Dalton further suppose? How does the immutable character of chemical compounds necessarily follow from the admission of these views? How is the doctrine of equivalent proportions explained by the atomic theory? What is a strong argument in favor of the atomic theory? Can the relative weights of the ultimate atoms be inferred from the relative actual weights of the elements?

relative actual weights of the elements which make up the mass of any compound, the relative weights of the ultimate atoms themselves might be ininferred, and represented numerically. The method of reasoning and deduction by which this result is arrived at is as follows:—

It is obvious that if we can, by any method, exactly fix the relative weights of the atoms of a few of the great elementary bodies, oxygen, hydrogen, nitrogen, carbon, etc., we can, by an extension of the process, solve the question for all other simple bodies, and for the most complex compounds into which they enter. Now, to attain this result, it is necessary to take one point as granted—the truth of which, although not susceptible of absolute demonstration, is yet rendered probable by many concurrent facts. This once allowed, the process becomes one of simple inductive reasoning. It is assumed that when two elementary substances unite in several proportions to form different compounds, that the combination takes place in the first or simplest compound in the proportion of one atom of the one to one of the other; in the second compound, of one atom to two atoms; in the third, of one to three, and so on.

Let us next examine the practical application of this supposition. Water, composed of oxygen and hydrogen, is found to contain these ingredients in the proportion of 8 to 1 by weight. Assuming, which many reasons make probable, that it is their simplest form of union, viz., of atom to atom, we obtain at once the relative weight of the ultimate atoms of oxygen and hydrogen—as 8 and 1 respectively.

Again, we have a series of five chemical compounds of oxygen and nitrogen, in which the proportion of oxygen increases uniformly in the ratio of the simple numbers, so that nitric acid, the fifth in order of these compounds, contains exactly five times the weight of that which exists in the protoxide of nitrogen, the first of the series. Concluding that the latter is the simplest form, and consists of a single atom or combining proportion of each of its elements, we obtain, by analysis of this gas, the relative weights of 8 and 14 for the atoms of oxygen and nitrogen composing it.*

Here then we have already a short scale of proportions fixed; in which hydrogen is the unit, oxygen 8, and nitrogen 14. The next step, in completing the circle of combination, furnishes a test of the truth of these results. Ammonia is a compound of hydrogen and nitrogen; and its analysis, exactly

* The student will perhaps be able to obtain a clearer idea of the relation of weights and proportions existing in the five compounds of oxygen and nitrogen from the following table.

	RELATIVE WEIGHTS.		RELATIVE PROPORTIONS.	
	Nitrogen.	Oxygen.	Nitrogen.	Oxygen.
Protoxyd of nitrogen, ...	14	8	1	1
Deutoxyd of nitrogen, ...	14	16	1	2
Nitrous acid, ...	14	24	1	3
Hyponitric acid, ...	14	32	1	4
Nitric acid, ...	14	40	1	5

QUESTION.—How is this conclusion arrived at?

made, gives proportions of the two which involve the same numbers as were obtained by the preceding methods.

This test obviously becomes more stringent and complete as we extend the number of bodies thus brought into conjunctions, and find the relative weight, so determined for each, strictly maintained in all their forms of combination. The atomic weight of sulphur, for instance, is found, by analysis of its compounds with oxygen, to be 16. Examining its simplest form of union with hydrogen, in sulphuretted hydrogen, the proportion is found to be exactly 16 to 1, or one atom of each, thus verifying the respective numbers before obtained. In a like manner all the other elementary bodies have been submitted, by experiment, to the same law, and have been found to furnish proofs precisely similar in kind. Thus the circle of demonstration has been continually enlarged; the evidence increasing in a geometrical ratio with the number of objects brought within the scope of inquiry. The conclusion is as certain and complete as any one of pure mathematics; or, if there be any exceptions, they are only such as may be ascribed to imperfect examination, or some other cause not infringing on the truth of the fundamental principle.

From what has been stated, it follows that the word atom may be used to express either an ultimate individual particle of a substance, or the simplest and smallest combining proportion of a substance. Indeed it is customary in chemical works to employ the word in both its significations—atom and atomic weights expressing the same thing as equivalent and equivalent weights.

Many other curious facts and relations have been discovered since the first announcement by Dalton of the atomic theory, which present strong additional evidence of the correctness of his views.

264. **Specific Heat of Atoms.**—For example, there appears to be a relation between the atomic weight of a body and its capacity for heat. Thus, the atomic weights of the metals, iron, copper, mercury, and lead, are respectively represented by the numbers 28, 32, 100, 104. Now if any of these four metals be taken in these relative proportions, it will require the same expenditure of heat to make them equally hot. 104 pounds of lead can be heated up to 212°, for example, by burning the same amount of alcohol which will heat 100 pounds of mercury, 32 of copper, or 28 of iron. A similar correspondence is also known to exist between the atomic weights and the capacity for heat of tin, zinc, nickel, cobalt, gold, platinum, sulphur, and tellurium, and according to some authorities, the correspondence extends to all the elements. If this last supposition is true (which is not proved), the determination of the specific heat of a substance would also afford the means of knowing its atomic weight and combining equivalent. Compound atoms

---

QUESTIONS.—Since the announcement of the atomic theory, have any circumstances confirmatory of its correctness been discovered? Is there a relation between the atomic weight of an element and its capacity for heat?

have also, in some instances, been proved to have the same relations to heat as the simple atoms composing them.

There has also an interesting relation been traced between the atomic weights, the specific gravities, and the combining measures or volumes of those elements which exist in the gaseous state, or are capable of assuming it. For example, a cubic foot of nitrogen weighs just 14 times as much as a cubic foot of hydrogen; a cubic foot of chlorine 35 times as much; of bromine, 80 times as much; of oxygen, 16 times as much; and the same measure of the vapor of iodine, 127 times as much. Now, these numbers respectively represent the density or specific gravity of these gases, compared with hydrogen as unity; and they also represent the atomic weights, or combining equivalents, of these several elements,—with the exception of oxygen, which is double.

It is important for the student, in the consideration of the whole subject, to clearly distinguish between the doctrine of chemical combination by equivalents, or, as it is often termed, "by atomic weight," and the atomic theory. The first is a truth independent of all theory, and rendered manifest to our comprehension by experiment and practical demonstration. The atomic constitution of matter, on which the law of combination by proportions is supposed to depend, can not, on the other hand, be proved by experiment, and still remains, and probably ever must remain, in the condition of a highly probable theory. The most subtile and refined analysis has never yet enabled any one to isolate an indivisible portion of matter, or even to adduce any direct evidence of the absolute existence of matter in this condition.*

---

* Experimental researches have, however, in some instances been made with a view of obtaining information on this subject. Dr. Thompson, of England, from certain assumed, but probable data, estimated an atom of lead, which, according to the table of equivalents, is 104 times larger than an atom of hydrogen, as only 1-310,000,000,000th of a grain. Ehrenberg, the eminent microscopist, has proved that the size of atoms, if they exist, must be less than 1-6,000,000 of a line in diameter, a line being assumed as 1-12th of an inch. More recently, Professor Faraday has endeavored, through the agency of light, to obtain some evidence of the existence of atoms. (See observations on divided gold, *London Phil. Mag.*, 1856-57, also *Annual of Scientific Discovery*, 1857-58.) The only positive result attained to was, to demonstrate that metallic gold, distributed mechanically throughout a liquid in particles so minute as to defy detection by the most powerful microscope, still retained its general physical properties.

Concerning the form of atoms two views are entertained. According to one hypothesis, atoms have the same form as the fragments obtained by splitting a crystallized body in the direction of its lines of cleavage. (See p. 55, § 73.) Antimony, which may be cleft in directions parallel to the faces of an acute rhombohedron, is resolved by this mode of division into similar rhombohedrons of continually smaller and smaller dimensions; and if we conceive the cleavage to be carried to the utmost possible limit, the smallest rhombohedrons thus obtained will be the atoms of antimony. Other substances, in like manner,

---

QUESTIONS.—Is there any relation between the atomic weight, the specific gravity, and combining volume of certain elements? What clear distinction should be made between the atomic theory and the law of equivalent proportions?

265. **Chemical Nomenclature and Symbols.**—Chemists recognize three great classes of substances, viz., Acids, Bases, and Salts.

Acids.—The common idea of an acid is, a substance soluble in water, which possesses the property of sourness, and which exerts such an action on vegetable blue colors as to change them to red. The chemist, however, disregards these properties, and considers all those substances to be acids which enter into combination with bases to form salts.

Vinegar, oil of vitriol or sulphuric acid, and aquafortis or nitric acid, are familiar examples of the class of acids.

Bases.—A substance which is capable of entering into combination with an acid, and by so doing destroys, or neutralizes its properties, is called a Base. The bases include those substances known as the alkalies, beside many other bodies of entirely different character.

Alkalies.—An alkali is a substance possessing many qualities exactly the reverse of those which belong to an acid. It dissolves in water, and produces a liquid, soapy to the touch. It has an acrid, nauseous taste, and restores the blue color to vegetable extracts which have been previously reddened by acid.

Potash, soda, and hartshorn or ammonia, are instances of well-known alkalies.

Salts.—Any compound produced by the union of an acid and a base is termed a Salt.

By the voltaic pile, salts are decomposed into acids and bases, the acids going to the positive pole, and the bases to the negative. We, therefore, call

---

admit of cleavage into cubes, prisms, etc. This view of the form of atoms offers the easiest explanation of the regular crystalline form, and the cleavage of simple substances.

The second hypothesis supposes that atoms have a spherical form; and that regular crystalline forms are occasioned by the peculiarity of their arrangement in varying numbers and angles. Thus, 4 spheres forming a base, and 4 placed perpendicularly over them, may form a cube; 2 or 4 layers of 3 each would give a prism, and so on.

QUESTIONS.—What three great classes of substances are recognized by chemists? What is an acid? What are examples of acids? What are bases? Define an alkali. What are examples of alkalies? What are salts? In the decomposition of a salt by the voltaic pile, how do its constituents distribute themselves?

the acid, in reference to its electrical character, the electro-negative constituent of a salt, and the base the electro-positive.

Some of the properties of acids and alkalies may be experimentally illustrated by means of a colored vegetable solution, such as the purple liquid prepared by slicing a red cabbage and boiling it in water. If a quantity of this infusion be divided into two portions, and to the one be added a little weak sulphuric acid, a red liquid will be obtained. If to the other a solution of an alkali be added, as potash or soda, a liquid of a green color is formed. On gradually adding the alkaline solution to the other, stirring the mixture constantly, the green color of the portions first added instantly disappears, and the whole liquid remains red; as more and more of the solution containing the alkali is added, the red by degrees passes into purple, and on continuing to add it, a point is reached when the original red liquid acquires a clear blue tint. At this moment there is neither free alkali nor free acid in the liquid, for the two have chemically united with each other, and have lost their characteristic properties. If the solution be now evaporated at a gentle heat, a solid crystalline substance is obtained, resulting from the combination of the sulphuric acid with the potash. This substance is a salt, and is called sulphate of potash.*

The acids and the alkalies are both remarkable for their great chemical activity. The acids dissolve all the metals, even the most compact. They also, except when very weak, destroy the skin and nearly all animal and vegetable substances. The action of the alkalies, especially potash and soda, is no less marked. They destroy the skin, if allowed to remain on it, and gradually remove the glaze from vessels of glass and earthen-ware which contain them. They also quickly remove paint from the surface of any object upon which their solutions fall. But the most remarkable property of acids and alkalies, is the power which they have of uniting with each other, and destroying, or neutralizing the chemical activity which distinguishes them when separate.

No simple or elementary substance has the properties of either an acid or alkali. Consequently, all acids and alkalies are compounds of two or more elements.

266. **Neutral Bodies.**—A substance which possesses neither the properties of an acid nor a base, is termed neutral.

* In practical chemistry, a blue substance, called "litmus," extracted from a species of lichen, is used extensively for determining the presence of an acid or alkali. Paper, colored blue with the tincture of litmus, is instantly changed to red by contact with the most minute quantity of an acid in solution; and the red color thus obtained is as quickly destroyed, and the original blue restored by the action of an alkali. Little strips of blue and red paper thus prepared, are kept constantly on hand in the laboratory, and are designated as "test papers."

QUESTIONS.—How may the properties of the acids and alkalies be illustrated? What are the characteristic properties of acids and alkalies? Does any simple substance possess the properties of an acid or alkali? What are the neutral bodies?

Water is the perfection of a neutral substance, although, in some instances, it may supply the place of an acid or a base.

267. **Origin of Chemical Nomenclature.**—The principles upon which chemical nomenclature is founded, were established by a committee of the French Academy in 1787. It was found that owing to the rapid progress of science, the number of new chemical substances increased so fast, that unless some uniform system of naming and classifying were adopted, the most inextricable confusion would result. The committee, therefore, devised a nomenclature which aims not merely to give a distinguishing name to the substances spoken of, but also to convey a knowledge of their components, and even of the proportions in which those components occur. This object was in a great degree attained to, and the system then instituted remains in use, so far as its essential features are concerned, to the present day.

268. **Nomenclature of the Elements.**—The elements which have been known from the most remote period retain their common names, and also their Latin names, to a considerable extent—as for example, Iron (Ferrum), Gold (Aurum), Copper (Cuprum), Mercury (Hydragyrum), Silver (Argentum), Lead (Plumbum), Tin (Stannum). If the element has been made known in modern times through chemical research, the name it bears generally indicates some distinguishing feature by which it is characterized: thus, Phosphorus (from the Greek φως, light, and φερω to bring), from its property of shining in the dark; Chlorine (from χλωρος, green), from its peculiar color; Bromine, from βρῶμος, a stench, etc. To the recently discovered metals, a common termination in *um* has been assigned, as Platinum, Palladium, Iridium, Potassium, Sodium, Aluminum, etc.

269. **Nomenclature of Compounds.**—When two elements unite, the product is called a *binary* compound (from *bis*, twice); thus, water, composed of oxygen and hydrogen, sulphuric acid, composed of oxygen and sulphur, and oxyd of iron, composed of oxygen and iron, are examples of binary compounds.

Compounds of binary combinations with each other, as sulphuric acid with oxyd of iron, are called *ternary* compounds (from *ter*, thrice), three elements being concerned. Most of the minerals are ternary compounds.

Combinations of salts with each other are named *quaternary* compounds, or double salts. Alum is an example, being a compound of sulphate of potash and sulphate of aluminum.

Compounds of oxygen are termed oxyds. Thus water is an oxyd of hydrogen, iron-rust an oxyd of iron.

The binary compounds of chlorine, bromine, iodine, fluorine, and several other elements which resemble oxygen in their mode of combination, are

QUESTIONS.—What was the origin of the chemical nomenclature now in use? What is the general nomenclature of the elements? What are binary compounds? What are examples? What are ternary compounds? Give examples. What are quaternary compounds? What are examples? What are compounds of oxygen called? What the compounds of chlorine, iodine, fluorine, etc.?

distinguished by the final termination *ide*. Thus chlorine forms chlorides; iodine, iodides; fluorine, fluorides; sulphur, sulphides, etc.*

When oxygen combines with the same element in more than one proportion, forming different oxyds, the several combinations are distinguished from each other by the use of prefixes. Thus, the first oxyd, or the one which contains but one equivalent of oxygen, is known as the *Protoxide* (from the Greek πρωτος, the first); the compound of two proportions is, in like manner, designated as the deutoxyd (δευτερος, double), and also as the binoxyd (βι, double); the compound of three proportions is also known as the tritoxyd (τριτος, third).

The oxyd, also, which contains the largest proportion of oxygen with which the body is known to unite, is termed the peroxyd. In like manner, the highest combinations of chlorine, sulphur, iodine, etc., are termed perchlorides, persulphides, periodides.

For example, oxygen unites with hydrogen in two proportions: the first combination is the protoxyd of hydrogen (water); the second and highest is the peroxyd. Again, with manganese, oxygen unites in three proportions: the first is termed the protoxyd, the second the deutoxyd, or binoxyd, and the third the peroxyd.

With some elements oxygen enters into combination in the proportion of 3 to 2, or in the ratio of 1½ of oxygen to 1 of the element. Such a compound is termed a sesquioxyd (from the numeral *sesqui*, once and a half). Certain other oxygen compounds are formed in the proportion of 2 of the element to 1 of oxygen; such are termed suboxyds, as the suboxyd of copper.

When the compounds formed by the union of oxygen with the different elements possess an acid character (as very many of them often do), a different plan is adopted to mark this peculiarity. The compound is then termed an acid, and its name is derived from the substance which combines with the oxygen, with the termination *ic* added. Thus, sulphur with oxygen gives sulphur*ic* acid; carbon with oxygen, carbon*ic* acid; and phosphorus with oxygen, phosphor*ic* acid. It frequently happens, however, that an element forms more than one acid with oxygen. When this is the case, the termination *ic* is applied to the strongest acid, and *ous* to the weaker. Thus we have sulphuric and sulphurous acids, nitric and nitrous acids.

The salts which these and other similar acids form by uniting with bases, are named in an equally simple manner, the acid supplying the generic, and the base the specific name: the *ous* termination of the acid is also changed into *ite*, and *ic* termination into *ate*. Thus, sulphite of soda, nitrite of potassa,

* **Binary compounds of sulphur, phosphorus and carbon, are also very generally known by the termination *uret*, as sulphuret of iron, carburetted hydrogen, etc.**

QUESTIONS.—**How are the first, second and third oxyds distinguished? What is a peroxyd? What is a protoxyd? What is a perchloride? What is a binoxyd? What are sesquioxyds? What are suboxyds? How are acid compounds of oxygen named?**

sulphate of soda, and nitrate of potassa, are salts respectively of sulphurous, nitrous, sulphuric and nitric acids.*

This nomenclature served to distinguish these acids and their salts until, as the science of chemistry advanced, a compound of oxygen and sulphur was discovered containing less oxygen than the sulphurous, and then a new name was required; it was therefore called *hyposulphurous* acid, and the salt formed with it is termed a *hyposulphite* (from the Greek ὑπὸ, under); so also, when an acid was discovered containing less oxygen than the sulphuric, but more than the sulphurous, it was called hyposulphuric, and its salt a hyposulphate. In some cases acids have been discovered containing more oxygen than those already named with terminations in *ic;* to these the prefix *hyper* (from the Greek ὑπὲρ, over) is attached. Thus chloric acid was for a very long time the highest oxygen compound with chlorine, but another still higher is now known. The last, therefore, is designated as hyperchloric, and sometimes as perchloric acid. Its salts are called hyperchlorates.

270. **Classification of Acids.**—It was once supposed that the presence of oxygen in a substance was essential to its acidity, but the progress of research has revealed the existence of acids which are entirely wanting in oxygen. Most of the acids which are wanting in oxygen contain hydrogen in its place. They are distinguished by prefixing to them the word *hydro*, as an abbreviation for hydrogen. Thus, chlorine and hydrogen form an acid, hydrochloric acid, often called muriatic acid; cyanogen and hydrogen form hydrocyanic acid, or prussic acid; sulphur and hydrogen form hydrosulphuric acid, etc.† Some chemists, especially the French, transpose these terms; they speak of chlorohydric, acid cyanhydric acid, sulphydric acid, etc. There is an advantage in this alteration, as it avoids any ambiguity which might arise from the use of the prefix *hydro*, which has sometimes been applied to compounds which contain water.

271. **Classification of Salts.**—In the early days of chemistry, the term salt was applied to all substances indifferently, which resembled common salt in appearance and properties. Subsequently, the use of the term was restricted to those compounds only which were formed by the union of an acid and a base: but when chemical knowledge had still further progressed,

* It may here be well to caution those who are just commencing the study of chemistry, of the necessity of distinguishing clearly between compounds such as the sulphites and the sulphates, or the sulphides and the sulphites. Sulphide of sodium is a binary compound of two elementary bodies, sodium and sulphur; sulphite of soda is a more complex compound, formed by the union of sulphurous acid and the oxyd of sodium (soda); sulphate of soda is formed by the union of sulphuric acid and soda.

† The acids formed by the union of sulphur and arsenic with hydrogen are also very commonly known as sulphuretted hydrogen, and arseniuretted hydrogen.

QUESTIONS.—How are the different acid compounds distinguished? How are salts named? What gives the generic and what the specific name to a salt? How do acids in forming salts change their terminations *ic* and *ous*? What do the prefixes hypo and hyper designate? Is the presence of oxygen essential to the existence of an acid? What element generally supplies the place of oxygen in acids wanting this element? How are hydrogen acids named?

it was found that if this definition was rigidly enforced, it would exclude from the class of salts a considerable number of compounds which possess the physical characteristics of a salt in a most eminent degree. Among these was common salt itself, which, although the type of all salts, is not a compound of an acid or a base, but a compound of two elements, chlorine and sodium. In like manner, the compounds of iodine, bromine, and fluorine with the metals, possess in a very high degree the saline character. To obviate, therefore, the somewhat startling proposition, that common salt is no salt at all, and to avoid doing violence to a long received and expressed common idea, two classes of salts were established.

The first class includes all those binary compounds which, like common salt, are formed by the direct union of a metal with some other substance, called a salt radical, as chlorine, fluorine, bromine, etc. Compounds of this character are termed Haloid Salts.

Radical.—The term radical in chemistry, is generally applied to any substance, simple or compound, which can unite with hydrogen to form an acid compound, and with a metal to form a salt.

The second class includes all those salts formed by the union of an acid and a base.* These are termed oxy-salts, or oxygen acid salts.

Many of the compounds of sulphur with the metals, as the compound of sulphur and potassium, also possess a saline character, and are termed sulphur salts.

Such in general are the principles of chemical nomenclature, as established by the Committee of the French Academy. As before said, the object of the inventors of this language was not only to give a distinguishing name to the substances spoken of, but also to convey a knowledge of its chemical composition. That this has been accomplished in a great degree, will be evident from one or two illustrations. Thus, the name *bi-chromate of potash* indicates by simple inspection that the substance is an oxygen acid salt, composed of chromic acid and potash, the prefix *bi* showing that the equivalent or proportion of acid to base is as two to one. Again, the name *permanganate of potash* indicates a compound of manganic acid and potash, and the prefix *per* shows that the acid in question is the highest oxygen compound of manganese known.

272. **Symbols.**—Although the chemical nomenclature in use is most convenient, and perhaps as perfect in principle as the nature of our language

* A beautiful illustration of the universality of the law, that bodies replace each other in combination in fixed equivalent quantities, is found in the combination of salts. Thus, when equivalents of two neutral salts, which are capable of decomposing each other, are brought into chemical contact with each other, the two bases exchange acids by an exact compensation; the original compounds are altogether lost, and two new salts evolved, without either loss or addition of any kind in the process.

QUESTIONS.—What two classes of salts have been recognized in chemistry? What are haloid salts? What are oxysalts? What are sulphur salts? Illustrate by example the manner in which the chemical name of a substance indicates its composition. What is the necessity of using symbols in chemistry?

will allow, yet the impracticability, in many cases, of contriving convenient names expressive of the constitution of many complex chemical compounds (the existence of some of which was not known or even anticipated by the inventors of chemical language), has led to the employment of symbols. These constitute a species of short-hand, which not only supplies all deficiencies of the nomenclature, but enables us to represent to the eye, and describe with mathematical accuracy and rapidity the known composition of every chemical substance, and the changes which it may undergo. The employment of symbols has now become universal, and is also indispensable to both teacher and student in the study of chemistry.

273. **Symbols of Elements.**—It has been agreed by all chemists to use, as symbols of the elements, the first letter of their Latin names. When two or more names commence with the same initial, a second distinguishing letter is added.

In the table of elementary bodies, the symbol of the several elements will be found opposite to their names.

The symbols, when used singly, represent not merely the element for which they stand, but one equivalent of that element. Thus, the symbol O stands not for oxygen in general, but for one equivalent of oxygen, or, hydrogen being unity, for the number 8. H, in like manner, stands for one equivalent of hydrogen, and the number 1; C for one equivalent of carbon, and the number 6; Pb for one equivalent of lead, and the number 104.

If more than one equivalent of a body has to be expressed, it is signified either by writing a small figure to the right of the symbol, and generally below the line. Thus—

O_2 stands for 2 equivalents, or 16 of oxygen.
O_5 " 5 " or 40 "

The same may be represented also by prefixing the number to the symbol, as 2O, 5O.

The symbol may also be considered as representing the atomic constitution of a body. For example, O stands for one atom of oxygen as well as for one equivalent; O_2 for two atoms; O_5 for five atoms.

274. **Symbols of Compounds.**—In order to form the symbol of a compound, we unite the symbols of the elements of which it consists, one after the other, indicating by means of figures the number of each which have entered into combination.

Thus, HO is the symbol of water, a compound consisting of one equivalent, or 1 of hydrogen, and of one equivalent, or 8 of oxygen; SO_3 is the symbol

---

QUESTIONS.—What symbols are used to designate the elements? What does a single symbol of an element represent? How are several equivalents of an element represented by symbols? How is the constitution of compounds represented by symbols?

of sulphuric acid, a compound consisting of one equivalent, or 16 of sulphur, and three equivalents, or 24 of oxygen. $C_{12} H_{11} O_{11}$ is the symbol of common sugar, a compound consisting of twelve equivalents of carbon, eleven equivalents of hydrogen, and eleven equivalents of oxygen.

A collection of symbols indicating the constitution of compounds, is called a formula.

Compounds united with compounds, such as salts, are expressed in a similar manner, the base of the salt, or the electro-positive element, being always placed first. Thus, sulphuric acid has the formula SO_3, and oxyd of iron, that of FeO, consequently the formula $FeO+SO_3$ will represent one equivalent of sulphate of the protoxyd of iron. Frequently a comma is placed between the two compounds instead of the algebraic sign +. Thus, sulphate of iron may be written FeO, SO_3. This mode is usually adopted to express a more intimate union than when the sign + is used. Thus, SO_3, HO,+2 HO indicates that an equivalent, or compound atom of sulphuric acid has united with three equivalents of water, two of which are loosely retained, and one very strongly.

Where it is necessary to indicate more than one equivalent of a compound, the whole formula of that compound is included in a bracket, and preceded by the indicating number. Thus, three equivalents of sulphate of iron would be written 3[FeO, SO_3]. The figure prefixed multiplies nothing beyond the symbols included within the bracket. Thus, in the formula for crystallized alum—

$$Al_2\,O_3,\ 3[SO_3]+KO,\ SO_3+24\ HO,$$

the 3 which precedes $S\,O_3$ only indicates that three equivalents of sulphuric acid are present. Frequently the employment of brackets is neglected, and then the figures multiply all the symbols included between them and the next comma or sign of addition.

275. Reactions and Reagents.—The various chemical changes, to which all matter is more or less liable, are termed, in the language of chemistry, *reactions* and the agents which cause these changes, *reagents.*

In addition to the information which symbols convey relative to the composition of the substances for which they stand, they can also be so combined in the form of equations, as to show in the most perfect manner the various products which result from chemical reactions. For this purpose, the symbols of the substances involved in the reactions are placed together, so as to form one side of the equation, and the symbols of the products resulting from the reactions on the other side. But as not the smallest particle of matter can be annihilated by any chemical action, it follows that the value of both sides of

---

Questions.—What are chemical formulæ? How is the composition of salts indicated by symbols? Which constituent of a salt is placed first? What does the sign + mean? What is to be understood by the terms reactions and reagents? How may symbols be arranged so as to indicate chemical reactions and their products?

the equation must be equal, or in other words, the sum of the weights of the products of every reaction must be always equal to the sum of the weights of the substances involved in the change. For example, the decomposition of carbonate of lime (marble) by sulphuric acid, and the liberation of carbonic acid gas may be represented by the following equation:

$$20+8,+6+16,+16+24=20+8,+16+24+6+16=90.$$
$$Ca\ O,\ C\ O_2+S\ O_3=Ca\ O,\ S\ O_3+C\ O_2.$$

The correctness of this equation may be proved by adding together the equivalents of both sides, when the sums will be found to be equal.

A very little practice will render the use of symbols familiar to all. To expedite the acquisition of this knowledge, the student will find it advantageous to exercise himself in the expression of chemical changes by symbols, whenever the opportunity occurs, until he is thoroughly acquainted with their signification and use.

276. **Isomerism.**—Until within a recent period, it was an acknowledged principle, that two bodies containing the same elements combined in exactly the same proportion, must of necessity possess the same properties, and be mutually convertible into each other. Such, however, is not the fact, and numerous substances are now known to exist, which are identical in chemical composition and yet exhibit totally distinct physical and chemical properties. Different bodies thus agreeing in composition but differing in properties, are said to be *isomeric* (from ἰσος, equal, and μερος, part), and the phenomenon in general is termed *Isomerism.*

A great class of bodies known as the volatile oils, oil of turpentine, oil of rosemary, oil of lemons, and many others, are examples of bodies which differ widely from each other in respect to odor, medicinal effects, boiling point, specific gravity, etc., and yet are exactly identical in composition—that is, they contain the same elements, carbon and hydrogen, in the same proportions.* "The crystallized part of the oil of roses, the delicious fragrance of which is so well known, a solid at ordinary temperatures, although read ly volatile, is a compound body containing exactly the same elements, and in the same proportion, as the gas we employ for lighting our streets."

The difference of properties in isomeric bodies is explained very simply by the atomic theory. "It is supposed that the atoms in each particular case are differently arranged, in the same way as the most manifold grouping may be produced on a chess-board by transposition of the white and black squares, as is shown in Fig. 76. Each figure is composed of eight white and eight black squares, but though the absolute number is the same, the grouping is different. In *a* one and one, in *b* two and two, in *c* and *d* four and four

* Two conditions of isomerism may be noted; one in which the absolute number of atoms, and consequently the atomic weight of the compound, is the same; the other where, though the relative proportions of the elements are the same, the absolute number of atoms of each is different.

QUESTIONS.—Illustrate this by example. What is isomerism? Give examples of isomeric bodies. How is isomerism explained?

squares are so joined as to present a different appearance. If we imagine these squares to be atoms, we obtain an idea of isomeric bodies, and it is thus rendered clear how there may be bodies of the same constitution and form,

FIG. 76.

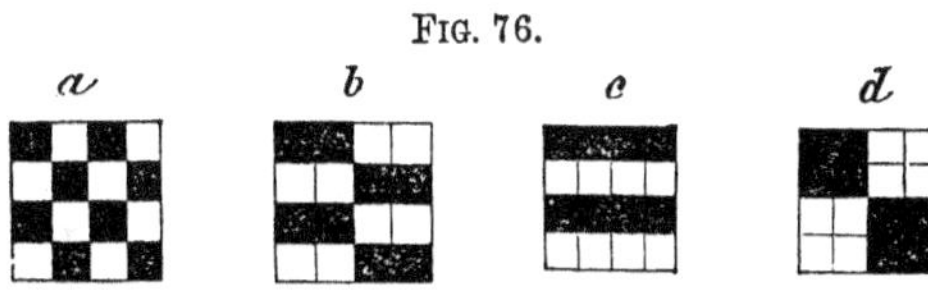

yet presenting an entirely different appearance and possessing different properties."—STOCKHARDT.

277. **Allotropism.**—Many of the elements are capable of existing in two or more different conditions, or forms, in each of which they manifest different, and often opposite properties. This principle is termed Allotropism, and bodies manifesting changes of such character are called Allotropic (from *ἀλλοτρόπος, different nature*).

One of the most striking illustrations of allotropism is to be found in the case of the element carbon, which exists in a pure state in the brilliant transparent diamond, in the opaque and black charcoal, and in the metallic-like body known as graphite, or black-lead. Sulphur, phosphorus, silicon, boron, oxygen, and other elements, are susceptible of similar changes.

Bodies in allotropic conditions differ in their chemical as well as in their physical properties. Carbon as the diamond is almost incombustible; carbon as lamp-black inflames at a low temperature, and sometimes ignites spontaneously. Phosphorus, in the ordinary condition, is soft, yellowish in color, has a powerful smell and taste, and can scarcely be handled with impunity, since it bursts into a flame at a temperature a little above that of the human body; allotropic phosphorus, on the contrary, is of a dark color, hard, devoid of both smell and taste, and may be handled without danger, and be even carried in one's pocket.

The explanation of allotropism is referred to difference in the arrangement of the particles or atoms constituting the body. Thus the same fibres of cotton, when closely matted together, constitute hard, tough paper; when simply carded, wadding; when twisted, yarn, or thread; and when intertwined, cloth.

QUESTIONS.—What is allotropism? What are examples of allotropism? How is this condition explained?

CHAPTER VI.

THE NON-METALLIC ELEMENTS.

278. The generally recognized division of the simple substances into Metallic and Non-metallic elements, or the Metals and Metalloids, (from μεταλλον, metal, and ειδος, appearance,) although most convenient for description, is not established in nature, and no strict line of separation, moreover, between the two classes can be indicated, since some of the elements possess, in a nearly equal degree, the characteristics of both.

Metalloids.—The number of the elements generally included in the class of metalloids is fourteen, which may be enumerated as follows:—Oxygen, Hydrogen, Nitrogen, Chlorine, Iodine, Bromine, Fluorine, Sulphur, Selenium, Tellurium, Phosphorus, Silicon, Boron, and Carbon.

Characteristics of the Metalloids.—The characteristics which serve in general to distinguish the metalloids from the metals are as follows:—They do not possess a metallic appearance, and are bad conductors of heat and electricity. When binary compounds of the metals and metalloids are decomposed by the agency of galvanism, the metalloids always separate at the positive pole (copper side), and the metals at the negative pole; as bodies endowed with opposite electricities only are attracted, the metalloids are, for this reason, termed electro-negative elements, and the metals electro-positive elements. Almost all the metalloids combine with hydrogen, but, as a general rule, the metals do not.

SECTION I.

OXYGEN.

Equivalent 8. *Symbol* O. *Density* 1·1. (*Air*=1.)

279. **History.**—Oxygen gas was discovered by Dr. Priestley, an English clergyman, in 1774. He called it dephogisticated air.

In the following year it was again discovered by Scheele, a Swedish chemist, and by Lavoisier, the illustrious French chemist, without cognizance of Priestley's discovery. The latter, supposing it to be the sole agent which imparted to bodies their acid properties, gave it its present name, oxygen, (from οξυς, acid, and γεναω, I give rise to).

Questions.—How are the elements divided? Is this division founded in nature? How many of the elements are generally included among the metalloids? Name them. What are the characteristics of the metalloids? When and by whom was oxygen discovered? From whom did oxygen derive its name?

280. **Natural History and Distribution.**—Oxygen is the most abundant of all the elementary substances, but is never met with in nature in a pure or isolated condition. It constitutes at least one third part of the solid crust of the globe, eight-ninths by weight of all the water upon its surface, more than one fifth of the atmosphere, and eight-ninths of the vapor contained in the atmosphere. It is also an essential constituent of all living structures, and is the immediate agent by which animal life and all the processes of combustion are sustained.

The meteoric masses which fall to the earth from the inter planetary spaces, have little or no oxygen in their composition, and in this respect they are unlike any of the compound substances which compose the crust of the globe. Hence the inference has been drawn, that in some of the great planetary masses of the solar system, from whence meteorites are undoubtedly derived, oxygen does not exist at all, or in much smaller proportions than upon the earth.

281. **Preparation.**—Many solid substances, which contain oxygen in combination, readily evolve it in a gaseous form when subjected to a sufficiently high temperature.

Fig. 77.

A very easy method of obtaining a small quantity of oxygen gas for experiment, which at the same time illustrates the original process by which Priestley discovered it, is to heat a little of the red oxyd of mercury in a thin glass tube (Fig. 77) over a spirit-lamp.* In this substance the affinity, or chemical attraction which holds together the mercury and the oxygen is so feeble, that a very slight degree of heat suffices to bring about decomposition;—the mercury collecting in small globules on the bottom and sides of the tube, and the oxygen escaping as a gas. The presence of the latter element may be demonstrated by holding an ignited substance over the mouth of the tube.

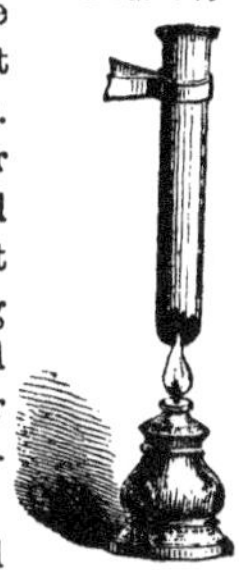

If it is desired to collect and preserve the oxygen liberated in this experiment, one end of a bent glass tube† is fitted by means of a per-

Fig. 78.

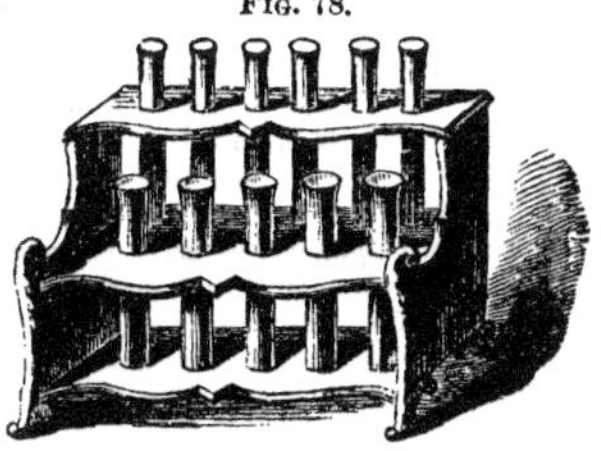

* Cylindrical glass tubes, with rounded bottoms, known as "test tubes," are generally used in chemical experimentation. A simple wooden rack, as in Fig. 78, serves as a convenient stand for them. Teachers will do well to furnish themselves with a supply of these tubes, as they are inexpensive, and can be used for a great variety of purposes.

† Glass tubing prepared expressly for chemical manipulations can be procured at a small expense of any dealer in chemical apparatus. By means of a Berzelius spirit-lamp, and with a little practice, an inexperienced person can, in a short time, learn to bend and adapt his tubing to his apparatus with ease and rapidity.

Questions.—What is said of the importance and distribution of oxygen? What inference has been drawn from the composition of meteoric stones? How is oxygen generally procured? By what simple method may a small quantity of oxygen be obtained?

forated cork into the mouth of the generating tube, and the other end is conducted into a vessel filled with water. The apparatus thus arranged may be supported by means of a piece of cord or wire, or by a sort of wooden vice (retort holder) constructed for chemical purposes, and represented in Fig. 79. The oxygen escaping in bubbles from the end of the tube under water is collected in a glass bottle or jar, which has been previously filled with water and inverted in the vessel; care being taken either to close the mouth of the jar, or else keep it continuously under water during the act of inversion. No water will escape from the jar until bubbles of gas from the tube are passed into it; but when this is permitted, the gas, by reason of its superior levity, ascends and displaces the water. As soon as one jar is filled it may be removed, and its mouth closed with a cork, or kept below the water level, and another substituted in its place. (See Fig. 79.)

FIG. 79.

For the production of oxygen gas in considerable quantity, materials less expensive than the red oxyd of mercury are used. The most convenient, and under ordinary circumstances the most economical method which can be adopted is, to expose to heat in a retort, or flask furnished with a bent tube, a perfectly dry mixture of equal parts of chlorate of potash and black oxyd of manganese. A common Florence flask will serve for this purpose, but a flask constructed of sheet copper and fitted with a small lead tube and screw-cap, is preferable.* A spirit-lamp affords sufficient heat to effect the chemical decomposition, and the gas liberated is collected in the manner before described. The salt chlorate of potash is very rich in oxygen—every 124 parts of it by weight containing 48 parts of this element united in the solid form with 36 parts of chlorine and 39 of the metal potassium. On the appli-

* Flasks, or generating bottles constructed of thin sheet copper, and furnished with a small leaden tube and a screw-cap, may be purchased of dealers in chemical apparatus, or can be easily manufactured by a coppersmith. For a continuous course of experiments their employment is strongly recommended, as they obviate entirely the annoyance and trouble arising from the fracture of glass, and the adjustment and preparation of the tubes.

QUESTIONS.—What is the most convenient and economical method of obtaining oxygen in moderate quantities? Describe the method of obtaining oxygen from chlorate of potash?

cation of heat, all this oxygen is driven off in a gaseous state, and chlorine, united with potassium, forming the chloride of potassium, remains. The reaction may be represented as follows:—

$$\begin{array}{l} 35+40+39+8=35+39+48=122 \\ Cl\ \ O_5\ \ K\ \ O=Cl\ \ K+O_6. \end{array}$$

Chlorate of potash and black oxyd of manganese both yield oxygen when heated separately, but under the conditions of heat and mixture above specified, the chlorate of potash alone disengages oxygen. The manganese, however, without taking any part in the chemical decomposition, exercises an important influence on the process, apparently by its mere presence, causing the oxygen to be liberated with the utmost facility and regularity, and at a much lower temperature than when the chlorate is used alone. The action of the manganese in producing this effect has been explained, by supposing that it mechanically separates the particles of the salt, and thus distributes the heat uniformly; but if this is true, clean sand, powdered glass, or any other similar material, ought to act equally well, which is not the case.

FIG. 80.

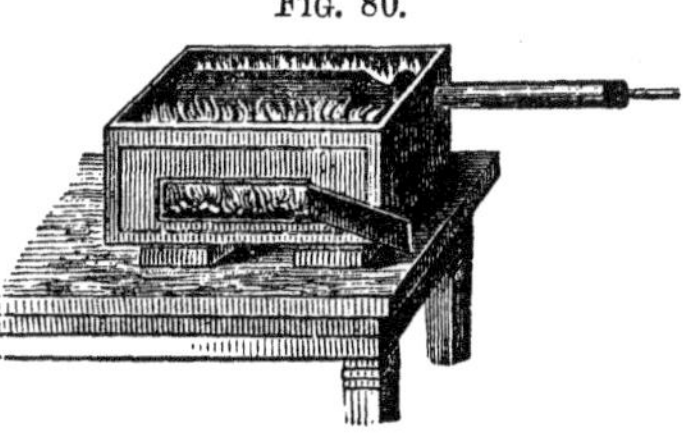

When very large quantities of oxygen are required, and perfect purity of product is not essential, an economical plan is generally adopted of heating the black, or peroxyd of manganese to redness in an iron retort, arranged in a suitable furnace. (See Fig. 80.) One pound of good oxyd of manganese thus heated, will yield seven gallons of oxygen, with some carbonic acid. This last may be entirely removed by causing the gas to pass through a solution of potash. In this process MnO^2 becomes converted in MnO+O.

Oxygen may be obtained from various other substances, but those already mentioned are the best, and the most frequently employed.* Red lead (oxyd of lead), and likewise saltpetre, when heated strongly, will furnish this

* A new method of preparing oxygen on an extensive scale for economic purposes, has recently been proposed by M. Boussingault. He states that caustic baryta, when heated to a particular temperature in the free presence of air, absorbs oxygen, and becomes peroxyd of barium, but on increasing the heat, the oxygen absorbed is given up. Thus the same quantity of baryta may be made to alternately absorb oxygen and evolve it into a reservoir.

QUESTIONS.—How much oxygen does this substance contain? What is the chemical reaction in this process? What is the object of mixing manganese with chlorate of potash? Is the action of the manganese understood? When large quantities of oxygen are required, what method is adopted? What is the chemical reaction in this process? From what other sources may oxygen be obtained?

gas. A mixture of strong sulphuric acid and one half its weight of black oxyd of manganese, or bichromate of potash, will liberate oxygen when heated.

All the green parts of plants evolve oxygen when exposed to the light of the sun. This fact may be readily demonstrated by placing a leafy branch, which is still connected with the parent plant, or a number of fresh leaves, under a jar filled with water, and then exposing them to the influence of solar light. After a short time small air-bubbles, consisting of pure oxygen, will collect in the upper part of the vessel. The minute bubbles, also, which may be often seen adhering to the leaves of aquatic plants under water, are generally pure oxygen.

282. **Properties of Oxygen.**—Oxygen, when pure, can not be distinguished from atmospheric air, being colorless, tasteless, and, under ordinary circumstances, destitute of odor. It is, however, somewhat heavier than atmospheric air; the density of the latter being represented by 1·00, that of oxygen would be 1·10.

One hundred cubic inches of dry oxygen weigh 34·20 grains. In its separate condition it is known only as a gas, all attempts to reduce it, by immense pressure and extreme low temperature acting conjointly, into a solid, or even liquid condition, having failed. Yet the learner will not fail to perceive, that oxygen when locked up in combination with the solid substances from whence we obtain it, must be itself a solid; and this consideration enables us to form some conception of the enormous force which, under the name of affinity, is capable of producing this effect.

Oxygen is very slightly soluble in water; a hundred volumes of this fluid, at ordinary temperatures, dissolving only four and one half volumes of the gas. Oxygen possesses a wider range of affinities than any other known substance, and combines in one or more proportions with all the elements except fluorine. The act of union of a substance with oxygen is termed oxydation, and the product of the union is called an oxyd. Oxyds are classified and divided, as has been before shown (§ 265), into acids, bases, alkalies, etc.

The tendency of oxygen to unite with other substances varies according to the circumstances under which the latter are presented to it, being greater under the influence of heat than of cold, and greater where there is an excess than where there is a deficiency of oxygen. Oxygen, at ordinary temperatures, enters slowly into combination with most of the metals. This action takes place much more rapidly in a moist than in a dry atmosphere. A bar of polished iron, in perfectly dry air at the ordinary temperature, will

---

QUESTIONS.—Do plants evolve oxygen? What experiment proves this? What are the properties of oxygen? Has oxygen ever been condensed into a liquid or solid substance? Is it known to exist in either of the latter conditions? What is said of its solubility in water? Of its range of affinity? What are the products of the union of oxygen with other substances called? How does the tendency of oxygen to unite with other substances vary? What is said of the oxydation of the metals? Will iron rust in dry air at ordinary temperatures?

remain unchanged for any length of time, but if moisture be present, it quickly becomes rusty. In the case of iron, the oxydation once commenced will spread through the entire mass of the metal; but in other instances, as in the case of lead and zinc, a superficial coat of the oxyd is formed, which adheres firmly to the surface, and protects the metal beneath from further change.

In order to commence and carry on oxydation, it is generally necessary to apply heat. An iron bar, when heated red hot, and exposed to the oxygen of the air, will rapidly become covered with a scale of oxyd, or rust. A stick of charcoal may be kept in oxygen at common temperatures for years without entering into combination with the gas, but the smallest spark upon the surface of the coal will cause the two elements to unite with great rapidity.

The direct union of oxygen with a substance is always attended with an evolution of heat.

In the ordinary rusting of iron, the disengagement of heat is too slow and feeble to be readily perceptible; but in some instances, where the union with oxygen at ordinary temperatures is rapid, the heat accumulates, and oftentimes rises sufficiently high to cause the materials to burst into a flame, producing what are called cases of "spontaneous combustion." This phenomenon is often exhibited when tow, "cotton-waste," or other fibrous materials that have been used in lubricating machinery, are laid aside in heaps. The oil upon them being spread over a large surface, absorbs oxygen with great rapidity, and the temperature of the mass continues to increase until the whole bursts into flame. Charcoal, reduced to a fine powder and exposed to the air, moist hay in stacks, and damp cloths in bales, frequently take fire under the same circumstances.

When the direct union of oxygen with a substance is attended with an evolution of both light and heat, the process is called Combustion, and the body is said to burn. On the other hand, the body which can combine with oxygen under such circumstances, is termed a Combustible, and the oxygen a supporter of combustion.

All the ordinary forms of combustion are simply processes of oxydation, and are accompanied by a withdrawal of free oxygen from the surrounding air; and in most instances the oxydation is commenced, or, as we express it, "the fire is kindled," by the application of some ignited substance, which raises the temperature of the combustible body sufficiently to enable it to at-

---

QUESTIONS.—In order to commence and carry on oxydation, what is generally necessary? What are examples? What phenomenon always accompanies the direct union of oxygen with a substance? What is spontaneous combustion? Give examples. What do you understand by the ordinary meaning of the term combustion? What is a combustible body? Why is it generally necessary to apply heat in order to cause combustion to commence?

tract the oxygen of the air, or commence burning; afterward, the heat which is liberated during the process is more than sufficient to carry it on, and thus the combination of one portion of oxygen with a burning body, causes the absorption of another portion.*

Bodies which will burn in the air, together with many substances which are generally considered as incombustible, burn in oxygen gas with great splendor. Experiments illustrative of these facts are among the most brilliant and interesting in the whole science of chemistry.

Fig. 81.

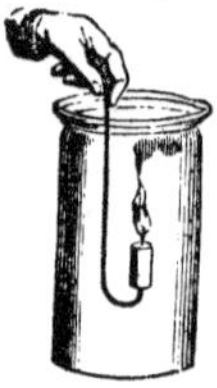

If we blow out a lighted candle in the air, the wick continues to glow for a few moments, but the flame does not spontaneously re-appear. If, on the contrary, the candle, still presenting some incandescent points, be plunged into a receiver containing oxygen (see Fig. 81), it inflames instantly, and burns with great brilliancy. This experiment, which may be repeated with a small narrow mouth jar of oxygen a great number of times, is characteristic of pure oxygen, and is the principal test used to detect its presence.

A glowing slip of wood introduced into oxygen, bursts into flame with a slight detonation. A bit of charcoal bark, slightly ignited, attached to a wire and lowered into a jar of oxygen, burns with great rapidity, sending off showers of brilliant scintillations in all directions. If a moistened slip of litmus paper (§ 266) be introduced into the jar after the combustion, it immediately turns red, a change not affected by atmospheric air, or pure oxygen; consequently an acid gas has been formed from the charcoal and the oxygen, which is called *carbonic acid.*

Fig 82.

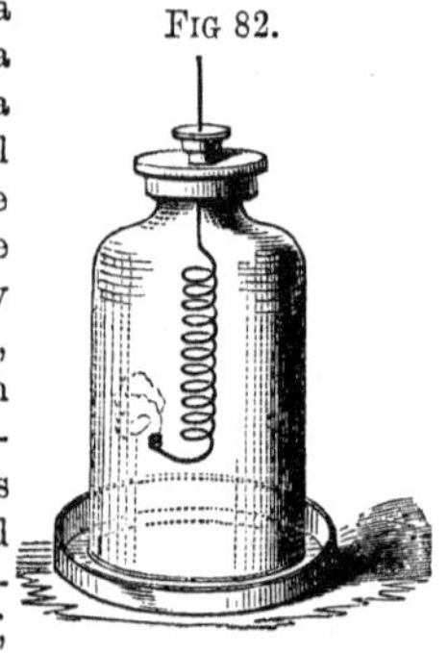

The combustion of iron in oxygen constitutes a most beautiful experiment. For this purpose a piece of fine iron wire, or, what is still better, a steel watch-spring, coiled in the form of a spiral (see Fig. 82) is employed. One end of the wire is tipped with a bit of sulphur, or tinder, and the other attached to a cork, so that the spiral may hang vertically. The sulphured end is then lighted, and the wire suspended in a jar of oxygen, open at the bottom, as is represented in the figure, supported upon an earthenware plate. The wire burns with an intense white light, the oxyd of iron formed darting out in brilliant coruscations in every direction. Melted globules of oxyd occasionally fall off, of so elevated a temperature, that they remain red hot for some time under

* For a particular consideration of combustion, see Chapter VII.

QUESTIONS.—How does pure oxygen act on combustible substances? Explain the experiments detailed.

the surface of water, and fuse deeply into the substance of the plate or glass upon which they strike.

FIG. 83.

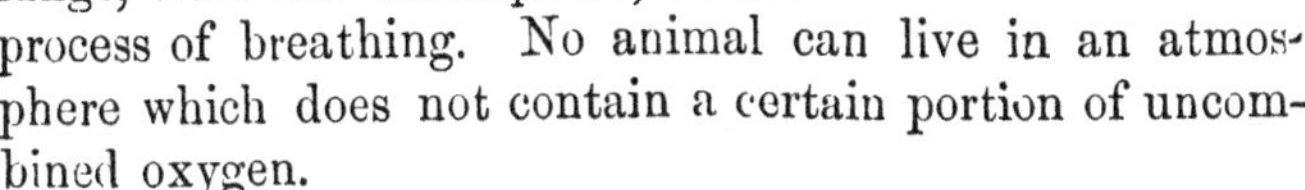

The light produced by phosphorus burned in oxygen is too brilliant and intense to be endured by the eye; and the jar, during combustion, becomes filled with a dense white vapor,—phosphoric acid, which is slowly absorbed by water. (See Fig. 83.)*

Kindled sulphur burns in oxygen with a beautiful blue light.

283. Oxygen and Respiration.—Oxygen is necessary to respiration, and is constantly taken into the lungs, from the atmosphere, in the process of breathing. No animal can live in an atmosphere which does not contain a certain portion of uncombined oxygen.

Oxygen, by the chemical action involved in the process of respiration, passes from a free state into a state of combination with other substances, and thereby becomes unfitted for the further support of animal life. If a bird be confined in a limited portion of atmospheric air, it will at first feel no inconvenience; but as a portion of oxygen is withdrawn from a free state at each inspiration, its quantity diminishes rapidly, so that respiration soon becomes laborious, and in a short time ceases altogether. Should another bird be then introduced into the same air, it will be almost immediately suffocated; or if a lighted candle be immersed in it, its flame will be extinguished. Respiration and combustion both produce the same effect, in causing free oxygen to be removed, or absorbed from the atmosphere. An animal can not live in

* This experiment should be performed with great care; otherwise the combustion-jar is liable to be broken, and the burning, liquid phosphorus dispersed in every direction. The combustion ladle should be deep—an iron cup or a piece of chalk scooped out and attached to a wire, the whole perfectly dry. The phosphorus should be divided under water, and afterward dried, not by wiping, but by contact with bibulous paper. It should not be allowed to project above the level of the deflagrating ladle, because during the act of combustion burning particles might disperse and stick against the sides of the jar, thus infallibly causing rupture of the glass. A similar result might be occasioned by employing wet phosphorus, the aqueous moisture from which, by expanding into steam, would scatter the melted phosphorus in all directions. One other point should be particularly attended to. The phosphorus placed in the ladle, and lowered into the jar, should be ignited on the surface by touching it with a hot wire, and not by holding the whole ladle over a flame. These directions being attended to will insure the success of the experiment, whereas by neglecting them, simple though they may appear, or any one of them, failure of the experiment is certain, and danger imminent.—FARADAY.

QUESTIONS.—Is oxygen necessary to respiration? What effect has the process of respiration on oxygen? Illustrate this. What analogy is there between respiration and combustion?

air unfitted to support combustion; and, under all ordinary circumstances, combustion will not continue in air containing too little oxygen for respiration.

Fermentation also acts like respiration and combustion in absorbing free oxygen from the atmosphere.

Although oxygen, as a constituent of the atmosphere, is necessary to respiration, it is destructive of animal life when breathed for any considerable length of time in a state of purity. When a rabbit, for example, is immersed in an atmosphere of pure oxygen, it at first experiences no inconvenience, but after an interval of an hour, or more, an unnatural excitement of the system is occasioned, accompanied by a rapid respiration and circulation of the blood; this is soon followed by insensibility, and death ensues in from six to ten hours.

284. **Magnetism of Oxygen.**—Oxygen is highly magnetic; that is, it sustains the same relations in degree to a magnet, that iron does. It has been further proved that, like iron, it loses its magnetism when strongly heated, but recovers it when the temperature falls. Faraday computes the magnetic effect of oxygen in the air to be equal to that of a metallic shell of iron, 1-250th of an inch in thickness surrounding the globe of the earth.

285. **Oxygen in Combination.**—The force which holds oxygen in combination varies extremely in different substances. In silica, (quartz, rock crystal, etc.), nearly one half the entire weight of which is oxygen, it is combined, or imprisoned, so to speak, with such force, that its liberation can only be effected by the most powerful agencies—heat alone failing to produce the slightest effect. In other solid oxygenized bodies, however, the affinities are so nicely balanced, that the slightest decomposing cause is sufficient to rend the elements, as we may say, from each other, and set the oxygen free. A very striking instance of this is furnished by chlorate of potash, the substance generally employed in the production of oxygen—every 124 parts of which, by weight, contain, as before stated, 48 of oxygen. A very slight degree of heat suffices to overcome the admirably poised balance of affinities, by which the combined elements of this salt are held together, and liberate every particle of oxygen. But this result can be effected by other agencies. For example, if we take a small quantity of sulphur, charcoal, phosphorus, sulphuret of antimony, or, to generalize, any other solid which has a strong attraction for oxygen, and mix either of them with a little chlorate of potash, carefully and with an avoidance of friction, the compound so obtained, when struck with a hammer upon an anvil, will explode violently. The experiment is best conducted by folding the mixture in a piece of paper. With phosphorus the explosive violence is greatest, with charcoal least, the variation being indicative of the respective tendency of these substances to combine with oxygen under the circumstances of the experiment.

Questions.—What effect does oxygen have on animal life when breathed pure? What is said respecting the magnetism of oxygen? Illustrate the various conditions under which oxygen exists in combination?

Gunpowder is another example of a substance holding a large amount of oxygen in combination, ready to spring into action with an almost irresistible violence.

286. **Active and Passive Condition of Oxygen.**—Oxygen, as hitherto considered, assumes two conditions, or states, widely different from each other. These may be termed its active and passive conditions. As locked up in rock-crystal, flint, clay, and other solids; as constituting eight ninths of the bland liquid, water; as an uncombined gas in the atmosphere, it is quiescent, inactive, waiting—retaining, however, all its forces in a latent state. This inactivity is one extremity of the scale of qualities possessed by oxygen. Intense violence characterizes its other extreme condition—"manifested," says Professor Faraday, "with tremendous energy in the phenomena of combustion and explosion—rushing with violence into other forms—displaying the most glorious exhibitions of light and heat—generating combinations of characters diametrically opposed, from the extreme of alkalinity on the one hand, to the most violent acidity on the other, and finally, having gone through its metamorphic phases, assuming its appointed place of rest in the world's economy."

287. **Ozone.**—In addition to these two extreme conditions, oxygen may assume another, in some respects still more extraordinary;—a state in which it is neither fully active or fully passive, but intermediate between the two former conditions—a state in which the activity possessed is not only less in amount, but different in quality. This condition of oxygen is characterized by the name of Ozone.

It has long been noticed that the working of an electric machine, especially in a close apartment, was accompanied by a peculiar sulphur-like odor, and also, that a similar odor pertained for some little time to places that had been struck by lightning. Beside recognizing these facts, and designating the odor in question as "*the electric smell,*" no explanation of the phenomenon was attempted by scientific men until within a very recent period; (since 1840). It was at last noticed, almost accidentally, that if a piece of paper moistened with a solution of starch, and a peculiar compound of iodine (iodide of potassium), was exposed in places pervaded by this odor, it was speedily turned blue. Now, this turning blue is an indication of the liberation of iodine from its combination; and the liberation of iodine is an indication of the agency of oxygen; so that in the determination of this additional fact, a connection was established between oxygen in an active state and—the electric smell.

The germ of knowledge thus obtained was expanded and generalized by Professor Schonbein of Bâle, who showed, by carefully conducted experiments, that the same smell and its corresponding action might be generated at pleasure, by various means—that the agent producing the odor occasioned other effects beside that of affecting the starch paper, such as bleaching, de-

QUESTIONS.—Under what two conditions does oxygen generally manifest itself? What is the third condition of oxygen? What is this condition termed? What circumstances led to the discovery of ozone? What discoveries were made by Schonbein?

odorizing, and corroding—and finally, that the mysterious gaseous agency itself was neither more nor less than oxygen—oxygen gas existing in a marked condition, or, as it is termed, in its allotropic form.—FARADAY.

Preparation.—Ozone may be obtained by passing a succession of electric sparks through a tube or vessel containing atmospheric air, or pure oxygen gas. It is also produced by the slow action of phosphorus upon oxygen, or atmospheric air. This latter reaction may be readily demonstrated as follows:

Take a quart glass bottle, and place in it a little water and a stick of phosphorus, first demonstrating the absence of ozone by testing it with iodine-starch paper.* Close the bottle, and allow the whole to remain for a little time. On again immersing the paper slip, it changes color, assuming a tint of blue. This result is not due to the vapors of phosphoric acid which may be noticed in the bottle, as they are readily absorbed by passing the gaseous contents of the bottle through water, while the ozone remains unaltered.

The formation of ozone may be also shown by another process still more simple. Take a glass jar, and first demonstrate by the iodine-starch paper the absence of ozone. Then pour into the jar a little ether, and there is still no ozone; but if we heat a glass rod in the flame of a spirit-lamp, and immerse it moderately hot (see Fig. 84), ozone will be abundantly produced.

FIG. 84.

Properties.—Ozone has never been obtained in a separate state, and appears to be entirely insoluble in all liquids. It has a peculiar odor, whilst ordinary oxygen is totally devoid of all smell. It possesses powerful bleaching properties, and if a solution of sulphate of indigo be poured into a vessel containing ozone, its deep blue color is destroyed with great rapidity. If the same experiment be tried with common oxygen, no bleaching action takes place. Ozone also exercises a remarkable influence over certain odors; thus, if a piece of tainted meat be immersed in this gas (see Fig. 85) the effluvium is instantly destroyed.

Ozone is perhaps the most powerful of all oxydizing agents. It corrodes even organic bodies, such as cork and India-rubber, while fragments of iron, copper, etc., rapidly absorb it, and become converted into oxyds. Silver,

* Iodine starch paper may be simply prepared by mixing a little starch with a solution of iodide of potassium—a salt obtained of any druggist—and imbuing unsized paper with the compound.

QUESTIONS.—How may ozone be obtained? What are the properties of ozone? What is said of the oxydizing influences of ozone?

under ordinary circumstances, is not affected by oxygen, and has hence been considered as one of the noble metals; but if a piece of silver-foil, moistened with water, be plunged into ozone, it rapidly crumbles into dust—oxyd of silver. Ozone displaces iodine from its combinations with the metals, setting the iodine free. This reaction is so easily produced, and is so sensitive, that it furnishes the readiest and most delicate method of detecting the presence of traces of ozone in the air. A slip of paper, as before stated, moistened with starch and iodide of potassium, and inserted in a vessel containing the slightest admixture of ozone, becomes blue from the action of the liberated iodine, which immediately unites with the starch, and forms the blue iodide of starch.

FIG. 85.

One of the most singular circumstances connected with ozone is the effect of heat upon it. A temperature not much higher than boiling water is sufficient to destroy it entirely. Advantage is taken of this fact to demonstrate the absolute chemical identity of ozone and oxygen. Ozone passed into one end of a red hot tube comes out ordinary oxygen at the other end.*

* Respecting this strange condition of allotropism, of which ozone is a particular example, Professor Faraday, in a recent publication, remarks:—"There was a time, and that not long ago, when it was held among the fundamental doctrines of chemistry, that the same body always manifested the same chemical qualities, excepting only such variations as might be due to the three conditions of solid, liquid, and gas. This was held to be a canon of chemical philosophy as distinguished from alchemy; and a belief in the possibility of transmutation was held to be impossible, because at variance with this fundamental tenet. But we are now conversant with many examples of the contrary; and, strange to say, no less than four of the non-metallic elements, namely, oxygen, sulphur, phosphorus, and carbon, are subject to this modification. The train of speculation which this contemplation awakens within us is extraordinary. If the condition of allotropism were alone confined to compound bodies, that is to say, bodies made up of two or more elements, we might easily frame a plausible hypothesis to account for it; we might assume that some variations had taken place in the arrangement of their particles. But when a simple body, such as oxygen, is concerned, this kind of hypothesis is no longer open to us; we have only one kind of particle to deal with, and the theory of altered position is no longer applicable. In short, it does not seem possible to imagine a rational hypothesis to explain the condition of allotropism as regards simple bodies. We can only accept it as a fact, not to be doubted, and add the discovery to that long list of truths which start up in the field of every science, in opposition to our most cherished theories and long-received convictions."

QUESTIONS.—What reaction takes place when ozone turns iodine-starch paper blue? What effect has heat upon ozone? How is ozone proved to be simply modified oxygen?

Ozone may be generally recognized in air which has swept over the ocean, although generally absent in air which has swept over land. It would appear that a moist state of the atmosphere is necessary to its development.* Mr. Wise, the celebrated æronaut states, that when on one occasion during an ascension, he became enveloped in a thunder-cloud, he found the surrounding air most powerfully impregnated with the peculiar odor of ozone.

It can not be doubted that so active an agent as ozone present in the atmosphere, must exercise an important influence in the economy of nature. What this influence is, is not definitely known. There can be but little doubt, however, that it acts as a purifying agent—oxydizing or burning up noxious products floating in the atmosphere. This supposition coincides with the opinion extensively entertained, that when ozone is in excess in the air, diseases of the lungs, influenza, etc., prevail (as would be expected from its irritating character): and that when it is deficient, fevers, etc., are common. Observers generally agree, that during those seasons in which cholera rages, the quantity of ozone in the atmosphere is greatly diminished.

288. **Daily Consumption of Oxygen.**—"It is not easy," says Professor Faraday, "to form an adequate idea of the aggregate results accomplished by oxygen in the economy of the world. For the respiration of human beings alone, it has been calculated that no less than one thousand millions of pounds of oxygen are *daily* required, and for the respiration of animals double that quantity; whilst the processes of combustion, fermentation, decay, and the like, continually going on, increase the daily sum total to eight thousand millions of pounds. Reduced to tons, we have the figures 7,142,847 as representing the daily consumption, and 2,609,285,714 the yearly consumption. Taken in connection with these statements, the fact that from one half to two thirds of the bulk of all the matter upon our planet consists of oxygen, does not seem wonderful.

SECTION II.

MANAGEMENT OF GASES.

289. **Pneumatic Trough.**—For collecting gases not absorbed to any considerable extent by water, an arrangement, known as the Pneumatic Trough, is always employed. For small operations this apparatus may be simply constructed by fixing a perforated shelf within a shallow dish, or wooden tub, in such a way, that when the vessel is filled with water to the proper height,

* Prof. Smallwood, of Montreal, in a communication to the American Association for the Advancement of Science, in 1857, stated that during the seven years ending in 1856, there were at Montreal, 918 days on which rain and snow fell; and during the like period, there were 816 days on which ozone was present in the air in appreciable quantity.

QUESTIONS.—Under what circumstances is ozone noticed in the atmosphere? What influence is ozone supposed to have in the economy of nature? What is said respecting the daily consumption of oxygen? How are gases not absorbed by water collected? Describe the pneumatic trough.

the shelf will be covered by it to the depth of about an inch. (See Fig. 86.) Another and more elegant arrangement, constructed of glass, and suitable for a lecture table, is represented by Fig. 87. The vessel intended for the reception of gas is filled with water, inverted and placed upon the shelf of the pneumatic trough, with its mouth directly over the perforation in it. The extremity of the gas-delivering tube, which dips into the water, is brought directly beneath the shelf, in such a way that the bubbles of gas escaping, ascend through the opening in the shelf into the vessel above.

Fig. 86.

Fig. 87.

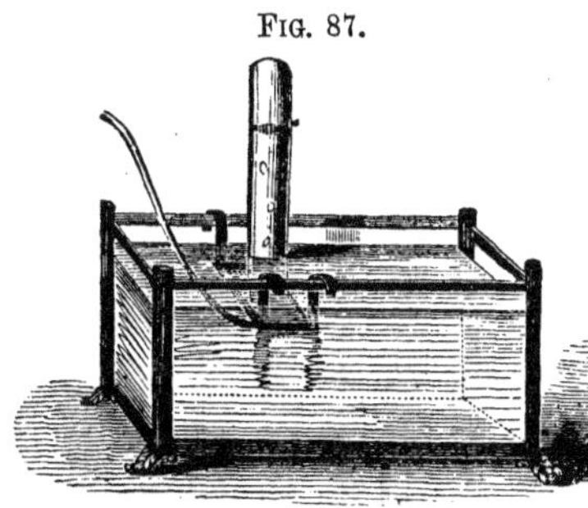

For permanent use, the pneumatic trough is usually constructed on a larger scale, of copper or tin plate, or of wood, and furnished with perforated shelves, arranged below the water level, of sufficient extent to accommodate a number of gas receivers at the same time. Fig. 88 represents the construction of such a pneumatic trough.

Fig. 88.

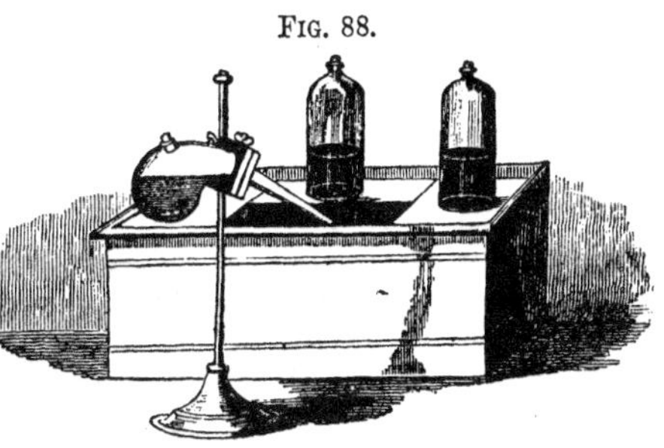

Water is supported in the gas-receivers above the level of the pneumatic trough by reason of the pressure of the atmosphere, on the same principle as mercury is sustained in the tube of a barometer.

In the collection of gases over the pneumatic trough, it should be observed that the gas which first comes over is mixed with the atmospheric air of the generating vessel, or retort; hence a volume of gas equal to about twice the volume of the retort should be allowed to escape, as impure. This precaution is especially to be attended to in the case of gases (such as hydrogen) which form explosive mixtures with atmospheric air. Gases may be transferred from one vessel to another, over the pneumatic trough, with the utmost facility, by first filling the vessel into which the gas is to be passed with water, inverting it, carefully retaining its mouth below the water-level, and then bringing

Questions.—What precaution should be observed in collecting gases over a pneumatic trough? How may gases be transferred from one vessel to another?

FIG. 89.

beneath it the mouth of the vessel containing the gas. (See Fig. 89.) On gently inclining the latter, the gas passes into the second vessel.

A jar, wholly or partially filled with gas at the pneumatic trough, may be removed by placing beneath it a common plate, deep enough to contain sufficient water to cover the edges of the jar. In this way gas, especially oxygen, may be preserved for a considerable length of time without admixture with the external air.

290. **Gasometers.**—In order to collect and preserve large quantities of gas, and to experiment with them more conveniently, capacious vessels of sheet-iron, or copper, called gasometers, are used. They consist in general of a cylindrical reservoir, suspended with its mouth downward, and fitting into an exterior and larger cylindrical vessel, or cistern, filled with water, as is shown in Fig. 90, which represents a pair of gasometers. The inner cylinder moves freely in the outer one, rising and falling as the gas is forced in or pressed out. The posts on each side of the cylinder are hollow, and contain weights, suspended to and balancing the inner moveable cylinder, so that it only presses on the gas as required. An upright rod of metal, shown in the engraving, rising from the inner cylinder, and passing through the supporting frame-work, keeps the cylinder steady in its place, as it rises or falls. Pressure, for forcing out the gas, is obtained by slipping on to this rod slit-weights of iron, as is seen in the figure. Gas is introduced into, and discharged from the gasometer, by means of a metal pipe, furnished with stop-cocks, and entering at the bottom of the stationary cylinder. For convenience, this pipe is carried up in front of the gasometer on the outside (as seen in the engraving), and by

FIG. 90.

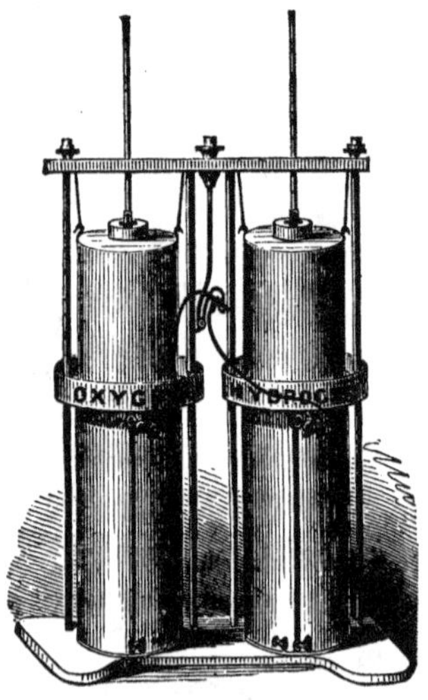

QUESTIONS.—What are gasometers? How are they constructed?

means of flexible tubes of India-rubber or gutta-percha, which screw on to its extremity, the gas can be conducted to any distance and in any direction.

The stop-cocks seen at the bottom of the gasometer are for the purpose of letting off the water, whenever this becomes necessary.

The large gasometers used for the collection and storage of illuminating gas are constructed upon precisely similar principles. Their general construction is represented in Fig. 91. The gas from the retorts is conducted by a pipe

Fig. 91.

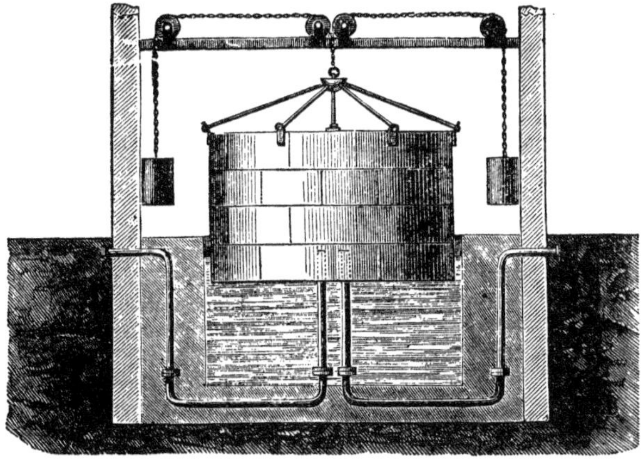

into the interior of the gasometer, and elevates it. Another pipe, opening also into the interior, is connected with the service-pipes which supply the gas. The gasometer is balanced by counter weights, supported by chains, which pass over pulleys, and just such a preponderance is allowed to it as is sufficient to give the enclosed gas the compression necessary to drive it through the pipes to the remotest part of the district to be illuminated.

SECTION III.

HYDROGEN.

Equivalent 1. *Symbol* H. *Density* 0·0692 (*Air*=1.)

291. **History.**—Hydrogen was first correctly described by Cavendish, an English chemist, in 1766. Before this it had been confounded with several of its compounds, under the designation of inflammable air. Its name is derived from ὑδωρ, water, and γεννάω, I give rise to, and refers to its production of water by uniting with oxygen.

Questions.—What is the history of hydrogen? What is its equivalent, symbol, and density?

292. Natural History and Distribution.—Hydrogen is never found in nature in a free state. The substance which contains it in the greatest abundance is water, of which it forms one ninth part by weight. As a constituent of other *inorganic* bodies, it is not very abundant in nature, but in the organic kingdom it enters largely into the composition of most animal and vegetable substances.

293. Preparation.—Hydrogen is always obtained for practical or experimental purposes from the decomposition of water.

It is liberated in the state of greatest purity through the agency of the voltaic current. When the wires connecting the poles of a galvanic battery in action are caused to terminate in water, decomposition is occasioned—hydrogen being evolved at the negative pole and oxygen at the positive. (See § 242, p. 148.) By placing tubes filled with water over the respective poles (see Fig. 92) the two gases may be collected in a separate state.

FIG. 92.

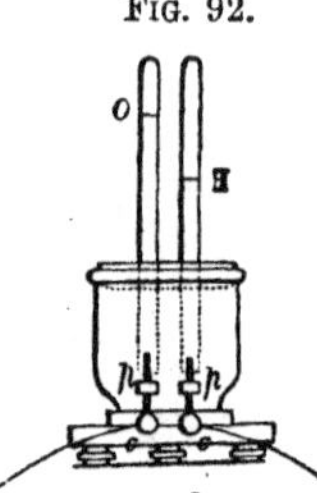

Water can not, under all ordinary circumstances, be decomposed by the action of heat alone.* Hydrogen may, however, be separated from water by heating this fluid in contact with substances which absorb its oxygen. Thus, if the vapor of water (steam) is passed over finely divided iron, heated to bright redness, the water is decomposed, oxygen uniting with the iron to form oxyd of iron, and hydrogen being set free.

This experiment, which was devised by Lavoisier, in order to prove that water is a compound substance, is easily performed by placing a quantity of iron filings in an iron tube (a gun-barrel, or better, a porcelain tube, protected by a covering of sheet-iron), arranged in a furnace, as is represented in Fig. 93; one end of the tube is connected with a retort, or flask, *a*, containing a small quantity of water, from which, by the heat obtained from

FIG. 93.

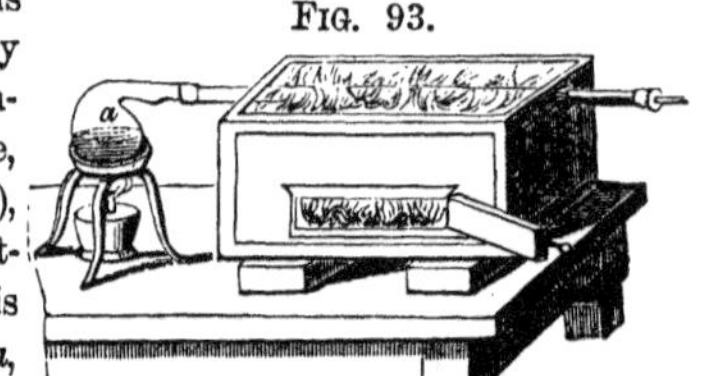

* Mr. Grove, the eminent English physicist, has recently shown that the vapor of water is decomposed to a small but sensible extent by an exceedingly high temperature, and resolved into its constituent gases.

QUESTIONS.—What is said of its natural history and distribution? How is hydrogen obtained? What process yields it in the greatest purity? Under what circumstances can water be decomposed by heat? Describe the experiment of Lavoisier.

a spirit lamp, a current of steam is driven through the tube, at the moment the metal has attained a full red-heat.

If the conditions of this experiment are reversed, and a stream of hydrogen be made to pass over oxyd of iron heated to redness, the hydrogen unites with and removes the oxygen of the oxyd of iron, thereby leaving metallic iron, and producing water.

If we sprinkle water in small quantity upon red-hot coals, a portion of it will be decomposed on the same principle as in the above experiment. The oxygen combines with the carbon and increases the intensity of the fire, while the liberated hydrogen burns and develops a very high degree of heat. Blacksmiths, it is well known, are accustomed to sprinkle their fires with water, in order to augment the heat, and too little water thrown upon a conflagration will often produce more injury than benefit.

Some of the metals, such as potassium and sodium, are capable of decomposing water (combining with the oxygen and liberating hydrogen), without the aid of heat. This may be shown by the following experiment:

FIG. 94.

Fill a glass tube with water, from which the air has been expelled by boiling, and invert it in a vessel of water. Pass into the mouth of this tube, by means of a wire, a small piece of sodium, as is represented in Fig. 93. This metal, being lighter than water, ascends to the surface, and absorbing oxygen from the water, rapidly liberates hydrogen.

Hydrogen gas is most conveniently obtained by putting pieces of zinc or iron into oil of vitriol, or strong sulphuric acid, diluted with six or eight times its bulk of water. Practically, this process may be conducted as follows:—Introduce into a suitable jar or bottle a small quantity of sheet zinc (or in the absence of zinc, scraps of iron, nails, etc.) cut into small pieces, together with water sufficient to more than cover the same. Then add a small quantity of strong sulphuric acid, and the evolution of gas immediately commences. By inserting into the opening of the flask, a perforated cork, to which a bent glass tube is fitted (see Fig. 95), the gas is easily collected over water in the usual way. Particular care should, however, be taken not to admit the gas into a receiver, until all the atmospheric air in the flask has been expelled.

FIG. 95.

An ounce of zinc is sufficient to liberate from water about two and a half gallons of

QUESTIONS.—Why does a blacksmith sprinkle his fires with water? Do any of the metals decompose water without the aid of heat? What experiment illustrates this fact? How is hydrogen obtained most conveniently? Describe the practical performance of this process?

hydrogen, and the evolution of the gas is regulated by the supply of acid. By means of a funnel-tube fitted into the cork of the generating vessel, and descending within the vessel to a point below the level of the contained liquid (see Fig. 96), the acid may be added from time to time in exactly the quantities necessary to produce the best effect. No gas can escape by this funnel-tube, as its extremity within the vessel is always covered by the fluid.

FIG. 96.

The theory of the liberation of hydrogen in this process is as follows: neither zinc nor iron is capable of uniting directly, as a metal with sulphuric acid; but oxyd of zinc and of iron combine readily with it. Thus a decomposition of water is determined. The zinc or iron takes oxygen from the water, and forms oxyds of these metals respectively, while the hydrogen before in combination with the oxygen passes off in the gaseous form. The oxyds of zinc and iron formed are insoluble in water, but are readily dissolved by the sulphuric acid, forming salts of sulphate of iron or zinc. The surface of the metal is thus left clean and exposed to the water, from which it attracts another portion of oxygen, which is dissolved as before. The reaction which takes place may be expressed by the following equation:—

$$Zn + SO_3 + HO = Zn\ O,\ SO_3 + H.$$

Sulphuric acid does not take any direct part in the decomposition of the water; but its presence seems to facilitate the processes by increasing the affinity between the metal and the oxygen of the water; it also dissolves the oxyd as fast as it is formed, which is essential to the continuance of the action.

294. **Properties.**—Hydrogen is a colorless gas, which has never been liquefied. When pure, it is without taste or odor, but as prepared in the way last described, it has a nauseous, disagreeable odor, arising from the presence of impurities contained in the materials used. It is slightly soluble in water, and does not support respiration: an animal plunged in it soon dies for want of oxygen. When mingled with a large quantity of air, it may be breathed for a time without inconvenience, and the voice of the person inhaling it, acquires a peculiar shrill squeak. Sounds produced in this gas are hardly perceptible.

Hydrogen is the lightest substance in nature, being sixteen times lighter than oxygen, and 14·4 lighter than air; 100 cubic inches of it weigh only 2·14 grains. Owing to its levity, it has been extensively used in filling balloons, which begin to rise when the weight of the material of which they are made and the hydrogen together, are less than the weight of an equal bulk of air. At the present time, coal gas, owing to the greater facility with which it can be obtained, is generally substituted in the place of hydrogen for

---

QUESTIONS.—What is the theory of the liberation of hydrogen under such circumstances? What is the chemical reaction? What part does the sulphuric acid sustain? What are the properties of hydrogen? What is said of the lightness of hydrogen?

ærostatic purposes—although of much greater density. Soap-bubbles inflated with hydrogen rise rapidly through the air. In order to obtain these bubbles, we fill a bladder, or gas-bag, provided with a stop-cock, with hydrogen gas, and attach to the stop-cock a common tobacco-pipe, or what is better, one of metal. (See Fig. 99.)* The extremity of the pipe is dipped into soap-suds, and the bubbles are blown by opening the stop-cock and gently pressing the bladder.

Hydrogen, beside being the lightest body in nature, possesses also the greatest tenuity, and there is reason for supposing that its atoms or molecules are smaller than those of any other known substance. No receptacle that is at all porous, as a bladder or India-rubber bag, can be used for storing hydrogen for any considerable length of time, the remarkable law of the diffusion of gases already explained (§ 52, p. 39) promoting its escape, and causing an interchange of the surrounding air. Faraday, in an attempt to liquefy hydrogen through the agency of cold and pressure, found that it would leak freely with a pressure of 28 atmospheres through stop-cocks which were perfectly tight with nitrogen at 60 atmospheres. A minute crack in a glass jar, quite too small to leak with water, will allow hydrogen to escape readily. Hydrogen also enters into combination in a smaller proportionate weight than any other element, and has hence been chosen as the unit of the scale of equivalents. Owing to the lightness of hydrogen, a jar may be filled with it by displacement, without using the pneumatic trough. Thus, if a bottle or jar be inverted over the extremity of an upright tube delivering the gas (see Fig. 97), the air it contains will be entirely displaced by the hydrogen rising into it. The gas may be retained for some minutes, even when removed from the source of supply, provided the jar be still held in an inverted position; but if its mouth be turned upward, the gas almost immediately escapes.

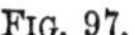
FIG. 97.

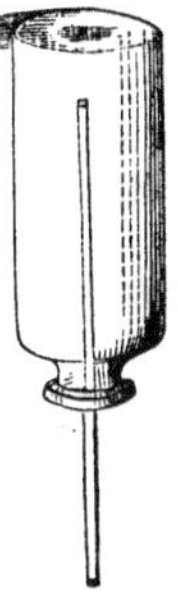

295. **Combustion of Hydrogen.**—Hydrogen is extremely inflammable; when a lighted taper is plunged into a jar of it, the gas takes fire, but the taper is extinguished, since there is no oxygen above the mouth of the jar to support combustion. This experiment is best shown by thrusting up a lighted bit of candle into an inverted jar, or bottle of hydrogen. The ignited gas burns quietly at the mouth of the jar, and the extinguished candle may be again relighted by it. If the bottle is suddenly reversed after the gas has burned awhile, the remaining gas will burst into flame with a slight explosion.

* India-rubber gas-bags, with metal pipes, stop-cocks, etc., are prepared especially for this purpose by dealers in chemical apparatus. A tobacco-pipe attached to the India-rubber delivery-tube of a gasometer may also be employed.

QUESTIONS.—What of its tenuity and smallness of particles? What are some illustrations of these properties? Why has hydrogen been chosen as the unit of the scale of equivalents? What is said of the inflammability of hydrogen?

A jet of hydrogen burns with a bluish white flame, and a feeble light. The experiment can be shown by adapting to the cork of a flask from which hydrogen is evolved, a piece of pipe-stem, or a small glass tube drawn out to a point. (See Fig. 98.)

Fig. 98.

If a dry, cold tumbler be held over a jet of burning hydrogen, its interior will rapidly become covered with a copious deposition of moisture. This results from a condensation of the vapor of water produced by the union of the hydrogen with the oxygen of the atmosphere.

296. **Explosion of Mixed Oxygen and Hydrogen.**—If the hydrogen before being kindled is mixed with air sufficient to burn it completely, or with between two and three times its volume, and then ignited, combustion takes place instantaneously throughout the whole mass, and is attended with a violent explosion. Hence particular caution is necessary in using hydrogen to avoid the slightest admixture of common air.

When pure oxygen is substituted in the place of air, the explosion is much more violent.

A mixture of oxygen and hydrogen will never unite under ordinary circumstances of temperature and pressure; but the passage of an electric spark, or the application of an intensely heated body, will cause instantaneous union, accompanied by an explosion. The product of such combination is always water.

In illustrating by experiment the explosive combination of oxygen and hydrogen, the proportions which produce the best effect are 2 of hydrogen to 5 of air, or 2 of hydrogen to 1 of oxygen. As the explosions are most violent, small quantities only of the mingled gases can be safely employed.

The experiments may be varied by inflating a soap-bubble with the gaseous mixture, and igniting it with a candle as it ascends; or by blowing up a quantity of bubbles in a shallow dish, as is represented in Fig. 99; or by filling a bladder with the mixed gases, and igniting it from a distance by means of a candle fixed to the end of a pole.

Fig. 99.

What is called the hydrogen-gun consists of a strong tin tube, about an inch in diameter and eight inches in length, open at one end and provided with a small vent hole at the other. In loading it, the vent is stopped by

Questions.—What are the peculiarities of the hydrogen flame? If a cold glass tumbler be held over the jet, what phenomenon is noticed? If hydrogen, before ignition, be mingled with air, what happens? Will oxygen and hydrogen unite of their own accord? What are the best explosive mixtures of oxygen and hydrogen? How may the explosive effects of mixed hydrogen and oxygen be illustrated? Explain the hydrogen-gun.

wax, the tube filled with water, and the proper mixture of gases introduced from a receiver under water. The tube thus filled is closed with a cork, and afterward fired at the vent. The explosion is sufficient to expel the cork with violence, and produce a loud report. The same experiment may be more simply performed by inverting a vial, or test tube over a jet of hydrogen, and allowing the escaping gas to mingle with, but not wholly displace the air. The mixture thus obtained may be exploded by applying flame to the mouth of the tube.

The loud, sharp report which attends the combination of oxygen and hydrogen under these circumstances, is explained as follows:—The steam, which is the resulting product of the union, suddenly expands from the high temperature attendant on the combustion, and immediately afterward condenses; great dilatation is first produced, followed by the formation of a partial vacuum; the surrounding air rushes in to fill the void, and by the collision of its particles produces the report.*

The inflammation of an explosive mixture of oxygen and hydrogen, or of hydrogen alone, in contact with air, is not only effected by a lighted taper, or the electric spark, but it likewise takes place in the cold by the action of certain substances, the principal of which is "platinum sponge," or platinum in a loosely coherent state.†

If we throw a piece of platinum sponge into a vessel containing a mixture of 2 parts of hydrogen to 1 of oxygen, a combination of the two gases, accompanied by an explosion immediately ensues. The same thing also takes place, but more slowly, when a thin plate of platinum, rendered chemically clean, is employed.

This phenomenon has been considered as one of catalysis (p. 161), or in other words, as due solely to the mere presence of the platinum; but it is now generally believed to be the result of adhesion (§ 48). The gases, it is supposed, by reason of a strong adhesion to the metal, are condensed upon its surface, and being thus brought within the sphere of each other's attraction,

* "The whole range of natural phenomena," says Professor Faraday, "does not present a more wonderful result than this violent combination of oxygen and hydrogen. Well known and familiar though it be—a fact standing on the very threshold of chemistry—it is one which I ponder over again and again with wonder and admiration. To think that these two violent elements, holding in their admixed parts a force of the most extraordinary kind—a force which, if we reduce it to a certain kind of comparison, will be found equal to the power of many thunder-storms—should wait indefinitely until some cause of union be applied, and then furiously rush into combination, and form the bland, unirritating liquid, water;—is to me, I confess, a phenomenon which continually awakens new feelings of wonder as often as I view it."

† Platinum sponge is easily prepared by soaking a small piece of bibulous paper in a solution of platinum (the bi-chloride of platinum) and afterward drying and igniting it. A little pellet of asbestos may be substituted with advantage in place of the paper. The sponge, after a little time, loses its peculiar property, but it can be again restored by being strongly ignited.

QUESTIONS.—What occasions the detonation? How may a mixture of oxygen and hydrogen be exploded without the direct application of an ignited substance? What is spongy platinum? What experiment illustrates its action?

unite. By the act of combination heat is evolved—the platinum becomes red hot—the remaining uncombined gases are ignited by it, and an explosion occurs.

Other finely divided substances beside platinum possess this property of favoring the combination of oxygen and hydrogen in an inferior degree. Even pounded glass, charcoal, pumice, rock-crystal, etc., if warmed to 600° F. produce this effect. Finely divided palladium, rhodium, and iridium act in the same manner as platinum.

If we project a jet of hydrogen alone upon platinum sponge, this substance becomes incandescent, and the gas inflames.

297. **Döbereiner's Inflammable Lamp.**—Advantage has been taken of this circumstance to construct a machine for obtaining fire instantly by means of hydrogen gas. It consists of a conical glass, Fig. 100, attached to a plate and stop-cock, and suspended in a receiver, *a*, containing sulphuric acid and water. Within the inner vessel a piece of zinc, *z*, is suspended, and this by contact with the dilute acid evolves hydrogen. The gas accumulating in the inner vessel forces the acid into the outer vessel, until it no longer touches the zinc, and thus stops the further evolution of hydrogen. By opening the stop-cock, *c*, the accumulated gas issues upon a ball of spongy platinum, *d*, and almost immediately takes fire. As fast as the gas escapes from the interior vessel, the sulphuric acid which has been displaced rises to take its place, and again coming in contact with the zinc, evolves a fresh supply of hydrogen.

Fig. 100.

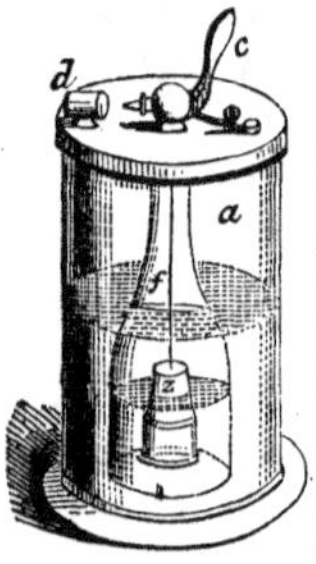

Fig. 101.

298. **Musical Tones.**—If a glass tube, open at both ends, be held over a jet of burning hydrogen (see Fig. 101), a rapid current of air is produced through the tube, which occasions a flickering of the flame, attended by a series of small explosions, that succeed each other so rapidly, and at such regular intervals, as to give rise to a musical note, or continuous sound, the pitch and quality of which varies with the length, thickness, and diameter of the tube. By sounding the same note with the voice, a tuning-fork, or musical instrument, the singing of the flame may be interrupted, or caused to cease entirely; or when silent, to recommence.

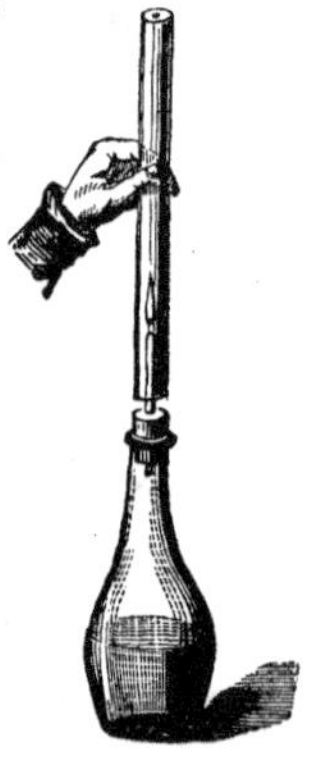

299. **Heat Generated by the Combustion of Hydrogen.**—The flame of hydrogen, although slightly luminous, produces a great degree of heat. When the combustion is assisted by oxygen gas, the heat gen-

Questions.—What other substances possess similar properties? When a jet of hydrogen is thrown upon spongy platinum, what ensues? What is the construction of Dobereiner's lamp? When hydrogen is burned from a jet in a tube, what phenomenon is noticed? What is said of the heating effects of the hydrogen flame?

erated is most intense, and is only exceeded by that produced by electrical agency.

300. **Oxyhydrogen Blow-pipe.**—The practical arrangement for effecting the combustion of hydrogen by oxygen, is known as the "Oxyhydrogen" or "Compound" Blow-pipe. As commonly constructed, it consists of two gasometers, containing, the one oxygen, and the other hydrogen. (See Fig. 102.) Tubes leading from these are brought together at their extremities, and the two gases delivered from apertures situated 1-30th of an inch apart, are burned in a single jet. The best result is attained by so arranging the stop-cocks of the gasometers, that the volume of hydrogen flowing out shall be double that of the oxygen.

FIG. 102.

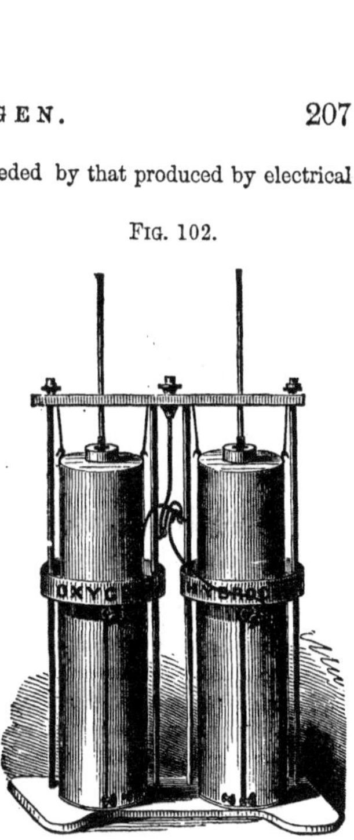

The effects of the compound blow-pipe may be produced in a degree by passing a stream of hydrogen through the flame of a spirit-lamp, as is represented in Fig. 103.

The effects of the oxyhydrogen blow-pipe are very remarkable. Substances that are infusible in the most intense blast furnaces, melt in the heat of its focus with the rapidity of wax. Iron, copper, zinc, and other metals, melt and burn in it readily; the first (when a watch-spring or steel file is employed) with beautiful scintillations, and the latter with characteristic colored flames. Thick platinum wire melts in it with ease, and may be even volatilized. Rock-crystal can be liquefied and drawn out into threads like glass, and the stem of a tobacco-pipe may be fused into an enamel-like bead.

FIG. 103.

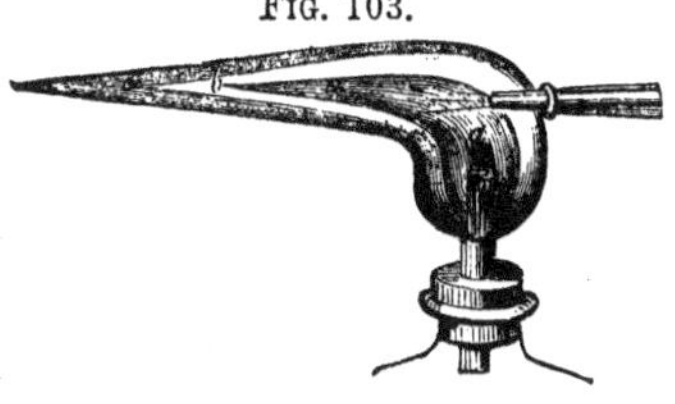

When the jet of the two gases, after being set on fire, is directed under water, it continues to burn beneath the surface of the liquid, in the form of a globe, and fuses and burns metallic wires held in it.

301. **Drummond Light.**—The flame of the oxyhydrogen blow-pipe is very pale in itself, but diffuses a dazzling light as soon as any solid body is introduced into it. By causing the flame to fall upon a cylinder of quick-

---

QUESTIONS.—Describe the oxyhydrogen blow-pipe. What are some of the effects produced by it? What is the Drummond light?

lime, an artificial light is produced, which for whiteness and brilliancy may be compared to the sun itself. With the requisite supply of gases this light may be maintained for hours, care being taken to expose to the flame fresh surfaces of the lime, by causing it to revolve by clock-work continually, but slowly. This light is generally known as the "Drummond Light," from the name of an English engineer, who first used it for signalizing at great distances; it is also often termed the "lime light."

The distances at which this light may be seen when its rays are concentrated by a parabolic mirror, are very great. In one instance, during the prosecution of the trigonometrical survey of Great Britain, it was seen by observers stationed upon a mountain peak, at a distance of 108 miles, during daylight.

The combination of hydrogen with other bodies is not attended with the development of light and heat, with the exception of oxygen and chlorine—two of the most highly electro-negative of all known substances.

302. **The Chemical Characteristics of Hydrogen** ally it very closely with the metals—particularly with zinc and copper—and there are some reasons for supposing that it is itself a metal, exceedingly volatile, and sustaining in this respect the same relation to mercury, that mercury does to platinum. The fact that it is wanting in luster, hardness, and brilliancy—qualities which are popularly considered as essential attributes of the metals—is no argument against this supposition, inasmuch as mercury, when vaporized through heat, is as transparent and colorless as hydrogen itself. The vapor of mercury and of other volatile metals is also, like hydrogen, a non-conductor of heat and electricity. Yet mercury, in the state of vapor, is no less a metal than in its ordinary condition.

Although hydrogen is the lightest and the most attenuated substance in nature, and combines in the smallest proportional quantity of all the elements, its active power, considered in relation to its combining weight, is very great. Thus, it combines with chlorine in the ratio of 1 part by weight to 36; with bromine 1 to 80; and with iodine as 1 to 125; yet in each case it abundantly satisfies the combining affinities of the other elements, generates by its union powerful and not easily decomposed acids, and in every other respect manifests an equality of force. This circumstance of so much power existing in connection with so little ponderable matter, is, regarded by Professor Faraday, as one of the most remarkable characteristics of hydrogen.

303. **Compounds of Hydrogen with Oxygen.**—But two compounds of hydrogen with oxygen are certainly known to exist*—the protoxyd of hydrogen, or water, whose chem-

* According to some authorities, there is a third compound—the suboxyd of hydrogen—formed by the gradual absorption of hydrogen by water.

QUESTIONS.—To what distance is this light visible? Are the combinations of hydrogen generally accompanied by evolutions of light and heat? What is said of the nature of hydrogen? What, according to Faraday, is one the most remarkable characteristics of hydrogen? What compounds does hydrogen form with oxygen?

ical symbol is HO, and the peroxyd or binoxyd, whose symbol is HO_2. Water is the only natural combination; the binoxyd being an artificial preparation.

304. **Water** is the most important, and at the same time the most remarkable of all chemical compounds. It is the most abundant substance existing in a separate condition upon the face of the earth, and covers to an unknown depth three fourths of its surface. Water enters largely into the composition of nearly all organized matter, and of every structure that possesses corporeal vitality, it is an essential element.*

305. **Composition of Water.**—Water, as has been already stated, is formed by the union of two volumes of hydrogen and one of oxygen, or by weight, of 8 parts of oxygen to 1 of hydrogen. The composition of water by measure and by weight, upon which, as a basis, the whole theory of atomic constitution and the doctrine of equivalent proportions rests, may be proved by a great variety of experiments, both by analysis and by synthesis.

By analysis, by decomposing water by the galvanic current (§ 242, p. 148), and by passing the vapor of water over red hot iron (§ 293). By synthesis, by uniting the two gases in proper proportions by combustion—by the action of spongy platinum—by the electric spark—and by passing a current of hydrogen over oxyd of copper, heated to bright redness.

The most reliable synthetical process is that last indicated. The hydrogen passing over the heated metallic oxyd, combines with its oxygen and forms water, which passes off as steam—the copper being left in a metallic state, the steam collected and condensed gives the weight of the water formed; the loss in weight which the metallic oxyd experiences gives the weight of the oxygen which has entered into the composition of the water; and the difference between these two, gives the weight of the hydrogen contained in the water.

FIG. 104.

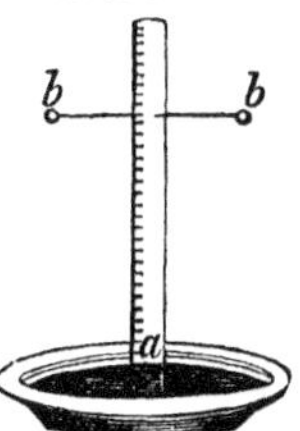

Eudiometer.—An apparatus by which a mixture of oxygen and hydrogen can be exploded by the electric spark, and the resulting product collected and examined, is termed an Eudiometer. It consists of a graduated glass tube usually placed over mercury, and so arranged that an electric spark can be passed into its interior. (See Fig. 104.) When a mixture of oxygen and hydrogen is exploded in such a tube over mercury, a vacuum is formed

* A man of 154 lbs. weight is made up of 116 lbs. of water and only 38 lbs. of dry matter: yet this proportion of water is small in comparison with the amount that enters into the economy of certain of the lower orders of animals. Of that class of sea-animals known as the medusæ, for example, it is estimated that at least 99-100ths of their whole structure by weight consists of water. They have, therefore, not inaptly been termed "living forms of water."

QUESTIONS.—What is said of water? What is the composition of water by measure and weight? How is the composition of water proved by analysis? How by synthesis? What is the most reliable synthetical method, and how is the composition of water calculated from the results obtained? What is an eudiometer?

by reason of their union and condensation, and the mercury rises to fill it. If the gases are mingled in the exact proportion to form water, the combination will be complete, and both will disappear entirely. If, however, one of the two elements is in excess, a gaseous residuum will remain. Thus, suppose we introduce into the eudiometer 100 measures of hydrogen and 75 of oxygen, we shall find after combustion 25 of oxygen remaining, but none of hydrogen. Therefore, 100 of hydrogen have combined with 50 of oxygen, or the union has taken place in the proportion of 2 volumes to 1. The graduations marked on the eudiometer tube enable us to proportion the quantities of the gases to be introduced, and also to estimate by the space unoccupied the volume of the residuum remaining after the combination.

306. **History.**—The history of water constitutes one of the most interesting portions of the whole record of physical philosophy. The old Greek philosopher Thales, in the earliest dawn of scientific speculation, taught that water was the "first and fontal" element of all material things—the earliest created substance. At a subsequent period, it was considered to be one of *four* primal elements; earth, air, and fire being the other three. This view of the elementary character of water remained unquestioned until nearly the close of the 18th century, or about the time of the first French revolution. Von Helmont, a contemporary of Galileo, and one of the most eminent scientific men of his day, maintained the doctrine that water was convertible into earth, and the following experimental results were appealed to as affording indisputable evidence of the fact, viz., that a tree when transferred from earth to water continues to develop itself and derive solid constituents from the liquid; and that when water was evaporated to dryness in a vessel, an earthy residuum always remained. The inference from these experiments was not, however, that water was a compound body, but rather that it possessed a generative character, and produced all the elements necessary for vegetable existence.*

Sir Isaac Newton, in 1704, in the course of his optical researches, remarked that water and the diamond both refracted light in the same way as substances of a highly inflammable character. He in consequence predicted the

* It is not a little singular that the compound character of water should have remained so long undetected by the Egyptians, Greeks, and Romans, who carried some branches of economical chemistry to a high degree of perfection, or in later times, by the Arabian chemists, or the mediæval alchemists. It would seem as if the phenomena of vegetation, and of animal life, if they had been watched with attention, would have shown that the elementary character of water was a most questionable doctrine. "Not a weed ever grew but what was possessed of the secret of its composite nature; not an animalcule ever lived but daily decomposed and changed the 'indivisible' into its own structure. No one, however, understood their language, or tried to interpret it, and hieroglyphics which seem to us pictures which tell their own story, revealed nothing to those who had already decided that they had no meaning."

QUESTIONS.—How may the composition of water be determined by the use of the endiometer? What opinions were formerly entertained respecting the nature of water? What doctrine respecting water was advanced by Von Helmont? Upon what did he base his conclusions? What facts were ascertained by Sir Isaac Newton?

future combustion of the diamond, and it is inferred that he anticipated, in a like manner, the combustibility of one of the elements of water.

Three quarters of a century after this, Lavoisier devised and carried out an experiment which is regarded as the commencement of the modern system of chemistry. He doubted the conclusions of Von Helmont, "and he asked nature if water could or could not be turned into stone, and asked in such a way that she granted an intelligible and unmistakable answer. He took an alembic, which may be described as an air-tight still or retort, in which the condensed steam or distilled liquor always flows back into the boiler—weighed it—put an ascertained quantity of water in it—made it air-tight—and set the water boiling; the steam or distilled liquor rising, became condensed, and continually trickled back through the tubular arms of the alembic into the original vessel. This arrangement was kept boiling for one hundred and one days and nights. At the end of that period, the whole apparatus had lost no weight; the alembic, however, had lost 17 grains, but the water had gained weight, and was muddy with earthy particles. When this muddied water was evaporated to dryness, there remained 20 grains of earth, 17 of which had clearly been worn out of the substance of the vessel; but where had the other 3 come from? Lavoisier at first assigned them to the incidental errors of the experiment, but it was afterward shown that they were derived from the water itself—from the saline and organic matter which it held in solution. Thus the earth, which Von Helmont traced to the transformation of water, was discovered to have come from the earthy vessel in which the water had been continuously boiled. Scheele, an eminent Swedish chemist, followed up the experiment, by analyzing the earth produced, and proved it to be the same as the material of the apparatus.

"The notable circumstance in this experiment is the use of the balance. Until this weighing of the alembic the balance had not been used in chemistry as an implement of research. Quality and not quantity was only regarded. But when Lavoisier ordered a balance with a view to its employment in research, the fate of old theories was sealed. The very thought of the balance implied the perception, by him that thought of it, of the central idea of all positive chemistry, namely, that every chemical operation ends in an equation; and that if 100 grains, ounces, or pounds of any substance whatsoever are burned, distilled, or in any way altered by a chemical process, then 100 pounds, ounces, or grains of material must be accounted for after the operation, for nothing is ever lost."—BREWSTER.

A few years after this experiment of Lavoisier, oxygen was discovered, and hydrogen first correctly described by Cavendish. Subsequently the composition of water was discovered almost simultaneously by James Watt, the inventor of the steam-engine, by Cavendish, and by Lavoisier; the first two by burning hydrogen in oxygen, and the last by decomposing the vapor of water.

QUESTIONS.—What experiment was instituted by Lavoisier? What were the results of this experiment? What was the most noticeable circumstance attending this experiment?

307. **Properties.**—The physical properties of water are so well known, and have been discussed to such an extent in the preceding departments of this work, that no lengthened description is necessary in the present connection.

In its ordinary condition as a liquid, and free from admixture, water is colorless, transparent, inodorous, and tasteless; it boils at 212° F., freezes at 32° F., and evaporates at all temperatures. It is 815 times heavier than an equal bulk of air.

308. **Coloration of Water.**—The peculiar colors which large bodies of water assume have not been satisfactorily accounted for. The color of the ocean "on soundings" is generally of a greenish hue; but off soundings it appears blue. It is maintained by some authorities that the blue tint of the ocean is only apparent, and is owing to a reflection of the most refrangible of the rays of solar light (the blue) in greater proportion than those which are less so. Sir Humphrey Davy attributed the blue color of the ocean to an admixture of iodine, and others have referred the very remarkable bright blue color of the Mediterranean to the presence of salts of copper; but although iodine exists in combination in all sea-water, and copper has been found in the waters of the Mediterranean, the quantities present do not appear to be sufficient to produce any perceptible coloration. The coloring matter of the Red Sea, which at particular seasons of the year is sufficiently intense to justify the appellation bestowed upon this body of water, has been proved to be owing to the presence of a prodigious quantity of microscopic plants.

309. **Transparency of the Sea.**—The transparency of the sea varies with the temperature. The maximum of visibility under water, under the most favorable circumstances does not exceed 25 fathoms, or 150 feet.

310. Purity of Water.—In nature, water is never found perfectly pure.

Rain-water collected in the country after a long continuance of wet weather is the purest natural water, but even this always contains atmospheric air, and the gases floating about in it, to the extent of about 2½ cubic inches of air in 100 of water. After rain-water, in the order of purity, comes river-water; next the water of lakes and ponds; next ordinary spring waters; and then the waters of mineral springs. Succeeding these are the waters of great arms of the ocean into which large rivers discharge their volumes, as the Black Sea, the water of which is only brackish; then the waters of the main ocean; then those of the Mediterranean and other inland seas; and last of all, the waters of those lakes which have no outlets, as the Dead Sea, Caspian, Great Salt Lake of Utah, etc.

311. **Spring Waters.**—Spring water, although it may be perfectly

---

Questions.—What are the physical properties of water? How much heavier than air is water? To what has the coloration of bodies of water been ascribed? What is said of the transparency of the ocean? Is water found pure in nature? What is the purest natural water? What is the relative purity of different waters? What is said of spring-waters?

transparent, always contains more or less of mineral matter dissolved in it. The nature of these substances will of course vary with the character of the soil through which the water percolates. The most usual impurities are carbonate of lime, common salt, sulphate of lime (gypsum), sulphate and carbonate of magnesia, and compounds of iron. Most spring waters also contain a proportion of carbonic acid gas.

312. **Mineral Springs.**—When the waters of springs retain in solution so large a proportion of mineral matter as to give them a decided taste, they are termed mineral waters, and are usually reputed to have some medicinal quality, varying with the nature of the substance in solution.

Waters which contain iron in quantity sufficient to impart to them an inky taste are termed *cha-lyb'e-ate;* the iron exists in the water most frequently in the state of carbonate, dissolved in carbonic acid, and rarely in a proportion exceeding one grain in a pound of water.

Waters impregnated with sulphuretted hydrogen gas are termed *sulphurous*, or *sulphuretted;* they may be readily recognized by their nauseous taste and odor. Remarkable springs of this character exist at Sharon, New York, and also in Virginia.

313. **Saline Springs.**—Springs whose waters contain a large proportion of earthy or alkaline salts, are called *saline*, although this term is generally applied to particularly designate springs containing common salt.

In some springs carbonic acid is very abundant, and imparts to the water an effervescent, sparkling character, like that noticed in the "Seltzer" and "Saratoga" waters.

314. **Thermal Springs.**—Many mineral springs are of a temperature considerably higher than that of the surface of the earth where they make their appearance, and not unfrequently discharge boiling water. The majority of hot springs occur either in the vicinity of volcanoes, or they rise from great depths in rocks of the oldest geological periods. With few exceptions, they discharge at all times the same quantity of water, and their temperature and chemical constituents remain constant. There is evidence to show that the temperature of some hot springs has not diminished for upward of a thousand years.

315. **River-water** is less fitted for drinking purposes than spring-water, although it often contains a smaller amount of dissolved salts. But river-water usually holds in solution or suspension large quantities of organic matter of vegetable origin, derived from the surface of the country drained by the stream. If the sewerage of large towns situated on its banks be allowed to pass into the stream, it is of course less fit for domestic purposes.

Water, however, which is contaminated with animal and vegetable matter,

---

QUESTIONS.—When are waters termed mineral? What are chalybeate waters? What are sulphurous, or sulphuretted waters? How may they be recognized? What are saline springs? What gives to Saratoga and Seltzer waters their sparkle? What are thermal springs? In what localities are they generally found? What is said of river-water? Can water purify itself?

is endowed with a self-purifying power of the utmost importance. The action of the oxygen of the air generates a species of fermentation, whereby the organic matters contained in the water become oxydated, deprived of both color and odor, and precipitated in part as sediment. The water of the river Thames, contaminated with the sewerage of London, is a remarkable illustration of this fact. Taken on board ships, it is at first nauseous, but after standing in casks for a few days, it becomes sweet and wholesome.

316. **Sea-waters.**—The most abundant substance in sea-water is common salt; next the chloride of magnesium and the sulphate of magnesia, which compounds give to the water its saline, bitter taste; then salts of calcium, potassium, with traces of iron, iodine, bromine, fluorine, silver, and some other of the metals. The specific gravity of sea-water varies slightly in different locations. The waters of the Baltic and of the Black Sea are less salt than the average, while those of the Mediterranean and some portions of the Gulf of Mexico, are more so. The whole amount of mineral constituents in the waters of the main ocean ranges from 3½ to 4 per cent.

The soluble earthy matters washed from the land by rains into the rivers, and by them carried into the ocean, remain there, since pure water alone evaporates from the surface of the ocean. The quantity of saline matter, therefore, in the ocean is continually accumulating. It is an error to attribute the saltness of the sea to the presence of vast beds of mineral salt; but the sea undoubtedly owes all its salts to *washings from the land.* The streams that have flowed into it for ages have been constantly adding to the quantity, until it has acquired its present briny and bitter condition. The evidence on this point is most conclusive; the saline condition of sea-water is but an exaggeration of that of all ordinary lakes, rivers, and springs. These all contain more or less of the mineral constituents of sea-water, but as their waters are continually changing and flowing into the sea, the salts in them do not accumulate.

Again, every lake into which rivers flow, and from which there is no outlet except by evaporation, is a salt lake; and it is extremely curious to observe that this condition disappears when an artificial outlet is provided. Examples of such lakes are the Dead Sea, the Caspian, the Sea of Aral, and the Great Salt Lake of Utah, the saltness of all of which greatly exceeds that of the ocean. Thus the waters of the ocean contain from 2 to 3,000 grains of saline matter in the gallon (70,000 grains); those of the Dead Sea, in some places, 11,000 grains, and those of the Salt Lake of Utah 22,000 grains, or nearly one third of their whole weight. In some instances, even this last proportion is exceeded.

317. **Relative Fitness of Waters for Use.**—Any water which contains less than 15 grains of ordinary mineral matter in a gallon is considered as comparatively pure, and may be employed for all domestic

---

QUESTIONS.—What is an illustration of this fact? What are the mineral constituents of sea-water? Why is the sea salt? What proof is there respecting the origin of salt in the ocean? What is said of the relative fitness of waters for use?

purposes, provided it does not contain too large a proportion of organic matter. Water, of which a gallon contains 60 grains of ordinary mineral matters, may be still good for drinking, but it is not fit for cooking vegetables or washing linen when it contains 8 grains to the gallon of either lime or magnesia. Waters which contain 6 grains of organic matter to the gallon are not fit for any domestic use; if this limit is exceeded, they act disastrously upon the animal economy, and may occasion dysentery and various other maladies. The presence of magnesia in considerable quantity in drinkable waters is undoubtedly injurious; the use of such waters in Switzerland is supposed to give rise to the frightful diseases known as "goitre" and "cretinism."* The disagreeable, earthy taste of certain well-waters, in most cases, arises from the presence of alumina, held in solution by carbonic acid.

One of the purest natural waters ever examined is that of the river Loka, in the north of Sweden, which flows mainly over granitic rocks, upon which water produces little impression. It contains only 1-20th of a grain (0·0566) of solid mineral matter per gallon. Such instances, however, are very rare; but water containing as little as 4 or 5 grains of solid matter to the gallon are not unfrequent. The quantity of organic matter in water is always greatest in summer, and disappears for the most part when the temperature of the water sinks to the freezing-point. Water, by filtration through finely powdered charcoal, may be almost entirely deprived of organic impurities.

318. **Hard and Soft Waters.**—Water is familiarly spoken of as *hard* or *soft*, according to its action on soap. Those waters which contain compounds of lime or magnesia occasion a *curdling* of the soap, as these earths produce with the fat of the soap a substance which is not soluble in water. Soft waters do not contain these earths, and dissolve the soap without difficulty. Many hard waters become softer by boiling, in which case the carbonic acid gas which holds the lime and magnesia for the most part in solution, is expelled by heat, and the mineral substances are deposited upon the interior of the boiler, causing a "*fur*," "scale," or incrustation.

Soft water, or that which is free from dissolved mineral matter, possesses a greater solvent power than hard water; therefore it is most suitable for washing and for the preparation of solutions. In culinary operations, where the object is mainly to soften the texture of animal or vegetable substances, or to extract from them and present in a

* Goitre is a swelling of the glands of the neck, and cretinism is a variety of idiotcy.

QUESTIONS.—How much organic matter in water will render it unsuitable for use? What effect is magnesia supposed to have in water? What in general is the cause of the earthy taste of certain waters? How does the organic matter contained in waters vary? When are waters said to be hard, or soft? What occasions the incrustation, or scale, upon the interior of boilers? What is said of the solvent action of hard and soft waters? What of their respective application for culinary operations?

liquid form some valuable constituent, as in the preparation of soups, tea, coffee, etc., soft water is the best. In other instances, in which it is desired to cook a substance, and not to dissolve it or extract its juice, hard water is preferable. To prevent the over-dissolving action of soft water in cooking, salt is frequently added, which hardens it.*

319. Much speculation has been occasioned by the circumstance that fresh water can generally be obtained by excavating for a few feet or inches on low sandy beaches, or islands in close proximity to the sea, and also by the occurrence, on many of the low coral islands of the Pacific, of fresh water springs which ebb with the tide. The explanation of these facts seems to be, that the fresh waters are derived from rains, and being lighter than the salt water of the ocean, remain suspended in the sands, resting upon the denser water beneath. They consequently rise and fall with the motion of the tides. It is also true that the water of the ocean, by filtration through sand, is deprived in part of its saline constituents.

320. All ordinary water contains in solution air, and generally a portion of carbonic acid gas. The quantity of these gases absorbed by water varies with its temperature, and also with the pressure of the atmosphere—cold water dissolving and retaining a larger quantity than warm or tepid water. When cold waters from springs or fountains are exposed to warm air, they become elevated in temperature, and the gases contained in them escape, rendering the water flat and insipid. The principal agent in imparting a sparkle and freshness to water is atmospheric air, and not carbonic acid gas, as is often supposed and taught.

Air and other gases existing in water may be expelled from it by raising the water to a boiling temperature, or by removing the pressure of the atmosphere. The presence of air in water may be beautifully illustrated by placing a vessel of spring-water beneath the receiver of an air-pump

---

* "These facts explain why it is impossible to correct and restore the flavor in vegetables that have been boiled in soft water by afterward salting them. It is also well known that peas and beans do not boil soft in hard water. This is owing to the effect which lime exerts in hardening or coagulating a peculiar substance ("casein"), which abounds in these seeds. Onions furnish a good example of the influence of *quality* in water. If boiled in pure soft water, they are almost entirely destitute of taste; though when cooked in salted water, they possess, in addition to the pleasant saline taste, a peculiar sweetness and a strong aroma; and they also contain more soluble matter than when cooked in pure water. The salt hinders the solution and evaporation of the soluble and flavoring principles."—*Youman's Household Science.*

QUESTIONS.—How is the presence of fresh water in close proximity to the sea accounted for? What is said of the presence of air in water? Why are waters which have been heated flat and insipid? How may air and the gases contained in water be expelled from it? How may the presence of air in water be demonstrated?

(Fig. 105), and gradually exhausting the air. As the exhaustion proceeds, the dissolved air escapes so rapidly as to occasion the appearance of ebullition.

FIG. 105.

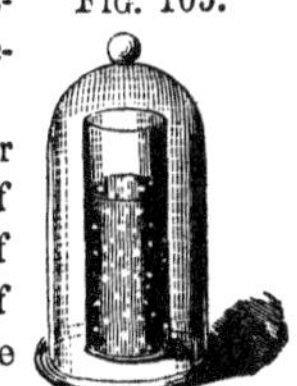

Fishes and other marine animals are dependent upon the air which water contains for their respiration and existence. If we place a fish in water which has been entirely deprived of air, it is almost immediately suffocated. This fact can, if desired, be demonstrated with the aid of an air-pump. The quantity of air retained by water, at an altitude of 6,000 or 8,000 feet, owing to a reduced atmospheric pressure, is two-thirds less than the usual proportion. Hence it is that fishes can not live in high mountain lakes—the amount of air contained in the waters being inadequate for their respiration.

A remarkable evidence of design on the part of Providence in supplying the wants of marine animals, which extract the oxygen they require for the support of life from the water in which they live, would appear to be found in the circumstance that water absorbs oxygen and nitrogen—the constituents of air—in proportions different from those existing in the atmosphere. Thus, ordinary air contains about 21 per cent. of oxygen, but air which exists in water contains from 30 to 33 per cent. Marine animals, therefore, can obtain more easily the necessary supply of oxygen from air which contains one-third of this gas, than from air containing but one-fifth.

It has also been recently discovered by Dr. Hayes, that the water of the ocean contains more oxygen near its surface than at a depth of one or two hundred feet. This fact has probably some connection with the comparative scarcity of animal life at great depths.

When water is in contact with an atmosphere of mixed gases, it dissolves of each a quantity precisely equal to that which it would have dissolved if in contact with an atmosphere of this gas alone.

Absolutely pure water can only be obtained by repeated distillations in clean vessels of hard glass.

321. Solvent Properties of Water.—The solvent properties of water far exceed those of any known liquid.

Most bodies are more soluble in hot than in cold water, the solubility increasing with the temperature. Among the few exceptions to this rule may be mentioned common salt, the solubility of which is nearly the same at all temperatures, and lime, which is more soluble in cold than in hot water.

322. Chemical Properties of Water.—Water is the perfection of a neutral substance, and enters into combi-

QUESTIONS.—What relation does air in water sustain to animal life? What are illustrations? What peculiarity characterizes the air contained in water? What is the condition of air at the surface and at the bottom of the ocean? In what manner does water absorb different gases? How may absolutely pure water be obtained? What is said of the solvent action of water in general? What of the chemical properties of water?

nation most extensively with acids, bases, with a large proportion of the salts, and, in short, with most bodies which contain oxygen.

A compound of water, in definite proportions with some other substance, is termed a *hydrate* ; and a body entirely free from water in combination is said to be *anhydrous*.

When a salt simply dissolves in water, the act of solution is uniformly attended with the production of cold; but when water* chemically combines with a salt, or forms a definite hydrate, the formation is always attended with heat; this circumstance indicates an essential difference between solution in water and chemical combination with water.

"Slacked lime" is a familiar example of a *hydrate.* When water is added to quick lime, it rapidly combines with it, producing great heat, and a chemical compound results, which is a "hydrate of lime." When water unites with potash and soda under the same circumstances the chemical union between the two substances is so strong, that no amount of heat alone is sufficient to separate them. So also when an acid has once been allowed to combine with water the entire separation of the two is seldom practicable, unless some base, for which the acid has a greater affinity than for water, be presented; in such a case the base displaces the water, and its expulsion by heat is then easily effected. For example, suppose that sulphuric acid has been freely diluted with water: upon the application of heat, the water at first passes off readily, leaving the less volatile acid behind. By degrees, however, it becomes necessary to increase the temperature in order to continue the distillation of the water, and at last the acid begins to evaporate also, and finally no further separation can be effected, as when the temperature rises to about 620° F., both water and acid distil over together. It is found on analyzing the liquid when it has reached this point, that the liquid contains one equivalent of acid and one of water, its composition being represented by the symbols SO_3, HO. If to this concentrated acid an equivalent of potash be added, the water is easily expelled, and an equivalent of anhydrous sulphate of potash (KO, SO_3) remains. Water, when it thus supplies the place of a base in combination with acids, is called basic water.

323. **Peroxyd, or Binoxyd of Hydrogen,** sometimes called oxygenated water, was discovered by Thenard, in 1818. It contains twice as much oxygen as water, and is a body characterized by most remarkable properties.

* For explanation of water, of crystallization, deliquescence, efflorescence, etc., see §§ 64, 65, 66, 67, pp. 48, 49.

QUESTIONS.—What is a hydrate? When is a body said to be anhydrous? What fact illustrates the difference between a solution in, and a combination with, water? What are illustrations of water in combination? When is water said to be basic? What is said of the second oxyd of hydrogen?

It is formed by decomposing peroxyd of barium by sulphuric or hydrofluoric acids. The process, however, is most difficult and complicated.

Peroxyd of Hydrogen is a syrupy liquid, of specific gravity 1·45, transparent, colorless, and almost inodorous, but possessed of a most nauseous and astringent taste. Although it differs from water only in containing an additional equivalent of oxygen, it is a powerful bleaching agent; and when applied to the skin for any length of time, whitens and destroys its texture. It can be preserved only at a temperature below 59° F. Heat rapidly decomposes it into water and oxygen gas, and at a temperature of 212° F., the evolution of gas is so rapid as to occasion an explosion. The mere contact of carbon, and of many of the metals and metallic oxyds also occasions its instantaneous decomposition, accompanied by an explosion and evolution of light.

The known properties of this substance render it highly probable that it would prove most valuable in its application to art—as a bleaching and oxydizing agent. The expense and difficulty attending its preparation have, however, thus far prevented its employment for any practical purpose.

SECTION IV.

NITROGEN, OR AZOTE.

Equivalent 14. *Symbol* N. *Density* 0·971.

324. History.—Nitrogen was first recognized as a distinct element by Dr. Rutherford, of England, in 1772.

Its name is derived from the Greek νίτρον, *niter*, and γενναω, *I form* (the generator of niter). Lavoisier, from its inability to support life, termed it Azote (from *a privative*, and ζωή, *life*).

325. Natural History.—Nitrogen is one of the most abundant of the elements.

As a constituent of the inorganic kingdom of nature, we find it in the atmosphere, of which it constitutes four-fifths; in ammonia; in bituminous coal; in the well-known salts, nitrate of potash and nitrate of soda (niter, saltpeter), and in many other mineral compounds. In the organic kingdom, nitrogen especially characterizes animal, in contradistinction to vegetable products; nevertheless it is found in the latter, but in small quantities. One-fifth of the weight of the dried flesh of animals is nitrogen. The plants which contain it in greatest quantity belong to the orders cruciferæ (turnips, cabbages, horseradish, mustard), and fungaceæ (mushrooms, etc.). Inasmuch as animals contain so much nitrogen, and vegetables so little, Berzelius imagined that nitrogen was generated in some unknown way by the animal functions. This idea, however, has been opposed by Liebig, who, with the majority of chemists, believes that the nitrogen existing in plants, upon which all animals directly

QUESTIONS.—How is it formed? What are its properties? Has the peroxyd of hydrogen been applied to any practical use? What is the history of nitrogen? What is said of its distribution in nature? What plants contain it in greatest abundance?

or indirectly feed, is sufficient to account for the large quantities of that element locked up in the tissues of animals. It is yet, however, one of the great unsettled questions in chemistry, and also in agriculture, whence plants derive their nitrogen;—whether from the soil (from manures and decaying organic matter), or from the air directly, or from the ammonia contained in the air.

326. **Preparation.**—The usual methods of obtaining a supply of nitrogen for the purpose of experiment are based upon the removal of oxygen from atmospheric air—leaving the nitrogen isolated or alone.

The simplest plan consists in placing a few fragments of phosphorus in a little metallic or porcelain cup, which is floated upon the surface of the water of a pneumatic trough. The phosphorus is ignited, and a glass jar or receiver, filled with air, is then inverted over it, with its lip in the water. (See Fig. 106.) The phosphorus burns at the expense of the oxygen in the confined air, and by reason of its great affinity for oxygen, it combines with every portion of this element contained in the receiver, leaving the nitrogen comparatively pure. As the combustion proceeds, the water of the pneumatic trough gradually rises in the jar to supply the place of the consumed oxygen. The product of the union of the phosphorus and the oxygen is phosphoric acid, which at first pervades the receiver as a dense white vapor, but after a little time is absorbed by the water.

FIG. 106.

Alcohol, ignited in a little cup, may be substituted in the place of phosphorus in this experiment, but it does not consume the oxygen entirely, and produces also a certain quantity of carbonic acid.

The removal of oxygen from the air may also be effected more slowly in various ways. A stick of phosphorus introduced into a jar of air standing over water, will slowly absorb the oxygen, and in two or three days about four fifths of the original bulk of the air, consisting of nitrogen nearly pure, will be left. Moistened iron filings produce a similar result, the metal gradually becoming oxydized, as is seen by the rusty appearance which it assumes.

Nitrogen may also be obtained by conducting chlorine gas into a solution of ammonia;* by exposing muscle (flesh) to the action of nitric acid in a retort to which heat is applied; and in a state of great purity by passing a current of air through a tube containing copper turnings heated to redness; the oxygen in this experiment being entirely absorbed by the copper to form oxyd of copper, while the nitrogen passes off.

327. **Properties.**—Nitrogen is a colorless, tasteless, and odorless gas,

* This experiment is a somewhat dangerous one. (See chloride of nitrogen.)

QUESTIONS.—Is it known in what manner plants obtain their nitrogen? How is nitrogen obtained? Enumerate some of the methods employed? What are the physical properties of nitrogen?

which as yet has resisted every effort to liquefy it. It is somewhat lighter than atmospheric air, having a specific gravity of 0·971 (air = 1·00).

One of the most distinguishing characteristics of nitrogen is its inertness, or "sluggishness;" it being, so far as chemical properties are concerned, in striking contrast with oxygen, which is one of the most energetic of the elements. It is neither acid or alkaline, and neither supports combustion, or burns. A burning taper is instantly extinguished in this gas, and an animal immersed in it quickly perishes; not because the gas is injurious, but for want of oxygen, which is required for both respiration and combustion. It, however, enters into the lungs with every act of inspiration; is a constituent of our most nutritious food, and a necessary component of the animal frame.

Nitrogen does not unite *directly* with any other single element; but its combination with the various elements all result from the agency "of indirect, oblique, or circuitous processes, which conditions being accorded, we frequently have whole classes of substances springing into existence; whereas, in the case of hydrogen, the combining tendency is satisfied with the formation of only one or two compounds."—FARADAY.

A striking illustration of the non-combining properties of nitrogen, is found in the fact, that no less than six tons of air pass through an average-sized iron blast-furnace every hour, during which transit the oxygen part of the air is most active in forming combinations, while the nitrogen, although subjected to precisely similar conditions of heat and contact, emerges as it entered, uncombined.

328. **Instability of Nitrogen in Composition.**—Nitrogen, of all ponderable substances, appears to have the greatest affinity for heat, and when in combination, constantly tends to unite with it, and resume its elementary condition of a gas. In consequence of this, and also by reason of its slight affinity for the other elements, the compounds of nitrogen are remarkably unstable. Many of them are decomposed with extreme suddenness by the slightest causes—the nitrogen being disengaged in the gaseous form, and often producing most violent explosions. Most of the explosive substances known are compounds containing nitrogen as an essential constituent; as, for example, gunpowder, gun-cotton, fulminating mercury (percussion-cap powder), fulminating silver, etc.

A substance known as the iodide of nitrogen strikingly illustrates *by its mode of preparation* the peculiarly indirect processes demanded by nitrogen for calling its powers of combination into play, and *by its character* when formed, the instability of the same element when forced into union with another body. Iodide of nitrogen is a simple compound of iodine and nitrogen. These two elements when mixed together directly manifest no disposition to unite, and may be preserved in contact unchanged for an indefinite period. But when nitrogen is brought in contact with iodine by an indirect process,

---

QUESTIONS.—What is one of the most distinguishing characteristics of nitrogen? Illustrate this. What is said of the combinations of nitrogen? What circumstance illustrates the non-combining properties of nitrogen? What peculiarity has nitrogen in composition?

as when a strong solution of ammonia is mingled in a glass vessel with a saturated solution of iodine in alcohol, combination almost immediately ensues, and a black powder, iodide of nitrogen, is formed. This, after standing for about a quarter of an hour, is separated from the liquid by filtration, washed in the filter with pure water, and dried by exposure to the air in a cool situation. As thus prepared, it is one of the most explosive substances known, the nitrogen being held by so slight an excess of force, that the merest friction between the particles of the compound is sufficient to shatter it into its elements. This result may be illustrated by a variety of experiments. A small quantity projected upon water explodes the instant it strikes its surface; the same result attends the dropping of a fragment from a slight elevation upon a hard surface, or by placing a small quantity upon one end of a counter and striking the other end with a hammer.*

Nitrogen, in some mysterious way, appears to be associated with all the higher forms of animal existence. The blood, the muscle, the brain, the nerves of animals, all contain it in large quantity, and these substances, of all organic compounds, are the ones most susceptible of decomposition.

Organic bodies which contain a large amount of nitrogen, generally emit a most offensive odor when they decay. The odor occasioned by the putrefaction of a dead human body, which is rich in nitrogen, is one of the most offensive in nature. Plants which contain this element in considerable quantity, as the cabbage and mushroom, putrefy with an animal odor. Substances containing nitrogen also emit an offensive and peculiar odor when burned; as for example, the smell of burnt hair, leather, flesh, bones, etc. This odor may be regarded as an invariable test of the presence of nitrogen.

Nitrogen constitutes an essential element of many of the most valuable and potent medicines, as quinia and morphia, and also of some of the most dangerous poisons, as prussic acid and strychnia.

A suspicion has always existed that nitrogen may be a compound body. One circumstance which has led to this idea, is its demeanor as respects electricity. Most of the binary compounds yield up their elements in obedience to the direction of this force, but electricity determines no liberation of nitrogen from any of its combinations. All attempts, however, to decompose it have failed, and its position among the elements must therefore remain undisputed.

---

* Iodide of nitrogen, prepared as above, is not liable to explode while moist, and in very small quantities may be used without danger. For the purpose of experiment minute portions of it should be taken upon the point of a penknife blade, or upon the end of a glass rod.

QUESTIONS.—Into what class of compounds does it particularly enter as a constituent? What characteristics of nitrogen are illustrated by the compound, iodide of nitrogen? What is said of nitrogen in the animal system? What circumstance characterizes the decay of bodies rich in nitrogen? What is one of the tests of the presence of nitrogen in a body? What is said of the elementary character of nitrogen?

THE ATMOSPHERE.

329. **History.**—The air was formerly supposed to be an element, but was not altogether regarded in the light of a material substance. The position which it held in the old systems of philosophy was similar to that assigned to light, heat, and electricity in some systems of the present day—a fluid substance without weight, form, color, or, in short, any of the ordinary attributes of matter.

It was not until 1673 that it was even suspected that airs, other than atmospheric air, might have an existence. About that time Robert Boyle, an English chemist, maintained "that some solid bodies do, in certain circumstances, as when heated, throw off artificial airs resembling atmospheric air in thinness and elasticity, as well as in dryness and permanency, but differing from it he could not well tell how."

In the beginning of the 17th century the workmen in certain German mines were molested (as miners still are) by certain agencies, some of which were liable to suffocate them silently but summarily (carbonic acid), while others burned, or exploding, blew them into fragments, (fire-damp, carburetted hydrogen). Von Helmont, the old alchemist, explained these phenomena by referring them to the agency of spirits, the guardians of the mineral treasures, whom he called *geists* (ghosts). From this originated the English word *gas*, which is still employed to designate aeriform substances.

Torricelli, a pupil of Galileo, first proved, in 1643, that atmospheric air possessed weight; and one hundred and fourteen years afterward, or in 1757, Joseph Black, a Scotch chemist of Edinburgh, first discovered and collected in a separate state a gas other than atmospheric air. He ascertained that limestone (chalk, marble, or oyster-shells) when burned in a kiln, or heated with a strong acid, parts with a kind of air in which no animal can breath and live. This gas (which we now call carbonic acid) Black termed *fixed air*, because it was imprisoned in the rock until the furnace or the acid extricated it from its fixture.

This discovery was one of the greatest that has ever been made in chemistry, since it for the first time clearly *proved* that there may exist different kinds of airy matter (just as there are different kinds of solid and liquid substances), differing as much from the gas of the atmosphere as oil or sulphuric acid differ from water, or as slate or marble from sandstone.

Shortly after this discovery by Black, Dr. Priestley devised the pneumatic trough (once known as the Priestleyan trough) and by so doing rendered easy the collection and handling of gaseous substances. He also discovered and isolated nine different gases, and among them oxygen. Scheele, working in an obscure Swedish town, with no other apparatus but phials and bladders, about the same time added two or three more to the list. Discoveries of the

---

QUESTIONS.—How was air regarded by the ancients? When was the existence of separate gases first suspected? What was the origin of the term "gas?" Who first demonstrated the weight of air? Who first collected and recognized a separate gas? What was the nature of Black's discovery? What is said of the importance of this discovery? What discoveries succeeded that made by Black?

same kind then took place in rapid succession all over Europe. Cavendish followed with hydrogen, Rutherford with nitrogen, while Lavoisier overthrew the great old doctrine of the elementary nature of air, by proving that it consisted of two gases mingled in unequal proportions.*

Within a comparatively recent period it has been admitted as a fundamental principle in physical science, that "gases are merely the steams of liquids which boil at immensely low points of temperature, these liquids being the liquefactions of solid bodies which melt at temperatures lower still, and that therefore there may be no end to the number of the kinds of gaseous matter, precisely as there is no known limit to the vast variety of liquids and solids."

330. **Atmospheric Air** consists essentially of nitrogen and oxygen mixed together in the proportion of four fifths by volume of the former to one fifth of the latter.

More correctly, the composition of air which has been freed from the presence of all foreign ingredients may be represented by measure and weight as follows:—

	By weight.	By measure.
Nitrogen	76·90	79·10
Oxygen	23·10	20·90
	100·00	100·00

In addition to oxygen and nitrogen, the atmosphere always contains small and variable proportions of carbonic acid gas and aqueous vapor; and very often, minute quantities of ammonia, nitric acid, the aroma of flowers, and various other organic and inorganic products;—in short, as the sea contains traces of almost every thing that is soluble, so the air contains traces of almost every thing that is volatile.

The oxygen and nitrogen existing in the air are merely intermingled, and not chemically combined with each other; but their relative proportions never vary. This has been proved by the analysis of air collected upon the summit of Mount Blanc, and upon the Andes; at an elevation of 21,000 feet by Guy Lussac in a balloon; over marshes; in hospitals; over deserts; and at the bottom of the deepest mines.

The quantity of carbonic acid, on the contrary, being much influenced by local causes, varies considerably. The average quantity is 4.9 volumes in 10,000 of air, but is observed to vary from 6.2 as a maximum to 3.7 as a minimum in 10,000 volumes. Its proportion near the surface of the earth is

* The experiment by which Lavoisier arrived at this result is described under the head of Combustion.

QUESTIONS.—What is now understood to be the true nature of gases? What is the composition of atmospheric air? In what condition do oxygen and nitrogen exist in the air? Are the proportions of those two gases variable? What is the proportion of carbonic acid in the air? Under what circumstances does it vary?

greater in summer than in winter, and during night than during day. It is also rather more abundant in elevated situations, as on the summits of high mountains, than in plains; this is probably owing to an absorption of the gas near the surface of the earth by plants and moist surfaces. An enormous quantity of carbonic acid gas is discharged from the elevated cones of the volcanoes of America, which may partially account for the high proportion of this gas in the upper regions of the atmosphere. The gas emitted from the volcanoes of the Old World is said to be principally nitrogen.

The quantity of watery vapor contained in the air varies with the temperature (§ 141, page 92). It seldom forms more than 1–60th or less than 1–200th of the bulk of the air.

Notwithstanding the difference in density between each of the principal constituents of the atmosphere—nitrogen, oxygen, carbonic acid, and the vapor of water—and notwithstanding, also, the absence of any chemical union between them, they are always, through the action of the law of the diffusion of gases (§ 51, page 39), found uniformly mingled together. The operations, also, of combustion, respiration, vegetation, and the like, continually going on upon the earth's surface, remove great quantities of oxygen from the air, and substitute a variety of other gases, the principal of which is carbonic acid; yet so beautifully adjusted is the balance of chemical action in nature, that no perceptible change in the composition of the atmosphere has been observed since accurate experiment on the subject was first commenced.

Ammonia seems to be an almost constant constituent of the atmosphere in exceedingly minute quantity. Recent most-carefully conducted experiments by M. Ville of France, fix the average quantity as 1 volume in 28,000,000 of air. Other experimenters have deduced a much greater result.

Nitric acid may be usually detected in the rain-water obtained during a thunder-shower. It is supposed to be formed by the union of the oxygen, nitrogen, and aqueous vapor of the air, through the agency of electricity.

Organic matter of some kind is almost always present in the atmosphere; but it not unfrequently happens that chemical tests fail to detect it, when the sense of smell and a peculiar effect upon the human constitution give abundant evidence of its presence. This is especially true of the odoriferous matters of flowers, and the miasmata of marshes. Dew collected over rice-fields often contains so much decomposing organic matter, as to become putrid after standing for a short time. Exposure to the night air of these localities in the hot season, invariably produces in the Caucasian race, malignant and almost incurable fevers.

The principal office which nitrogen appears to sustain in the atmosphere, is that of a diluent of the oxygen. If the quantity of oxygen in the air was increased much beyond its present proportion, the inflammability of most sub-

QUESTIONS.—How does the quantity of aqueous vapor vary? What is said of the uniformity of the condition of the atmosphere? What of the ammonia of the atmosphere? What of nitric acid? What of organic matter? What office does nitrogen appear to sustain in the atmosphere?

stances would be greatly augmented; and the functions of life would be called into such rapid action as to soon exhaust the powers of the system. Nitrogen being the most indifferent of all substances, and wanting in any poisonous qualities, dilutes the too active oxygen, and prolongs its action upon the system, in the same way as water dilutes and diminishes the stimulating action of spirituous liquors. Recent researches have also rendered it probable that the nitrogen of the air discharges an important office in respiration, by preserving the volume and tension of the cells and extreme tubes of the lungs.—PROF. MOULTRIE.

Oxygen is strikingly magnetic; nitrogen is singularly the reverse; and the atmosphere, a mixture of both, is nearly neutral as respects magnetism in all its relations to matter.

Another illustration of the adaptation of nitrogen to its atmospheric functions is to be found in its specific gravity, or density, which is nearly the same as that of its associated oxygen. Had there been any great difference in this respect, the tendency of the two gases would have been toward separation, and this, notwithstanding the influence of the law of diffusion. Again, as the atmosphere is now constituted, there exists a permanency of the pitch of sound: any tone being once generated, remains the same tone until it dies away. Its degree of loudness alters in proportion to the distance of the listener from the place where it originated, but its pitch—never. If the specific gravity of oxygen and nitrogen had, however, been widely dissimilar, there would have been a difference. No permanency of tone could then have been depended on, and the pitch of every original note would have varied continually. "All the studied arrangement of defined notes, which constitutes the art of music, would have been lost to us forever, had we been enveloped in such an atmosphere." These facts may be illustrated by striking a sonorous body in a receiver containing air, and afterward in one containing hydrogen, which is much lighter than air. (See Fig. 107.) The experiment may be varied by causing a tuning-fork in the key C to vibrate over a small glass jar, which, when made to resound, emits the same note, and is therefore in union with the fork. If the jar be now filled with hydrogen, and inverted, to prevent the escape of gas, and the fork be again caused to vibrate opposite its mouth, the unison is destroyed, and the sound is no longer responsive to the note C.

FIG. 107.

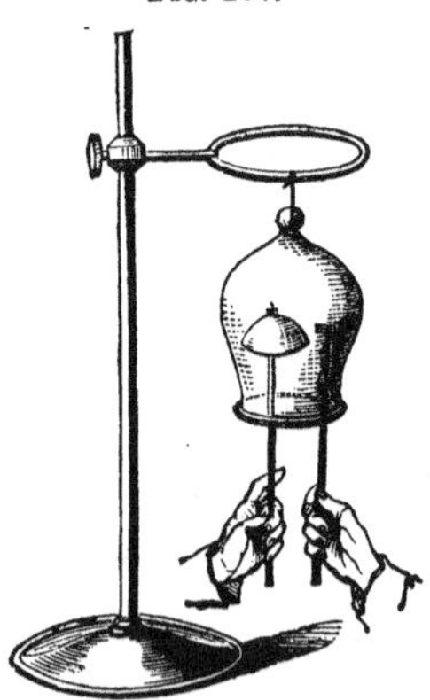

QUESTIONS.—What is said of the magnetic condition of the atmosphere? How does the specific gravity of nitrogen adapt it to its condition in the atmosphere? What experiments illustrate this?

331. **Analysis of Air.**—The proportions of oxygen and nitrogen in the atmosphere are determined by withdrawing the oxygen from a measured portion of perfectly dry air, through the agency of various substances which absorb it. (See § 326, page 220.) A stick of phosphorus introduced into a known measure of air in a graduated tube, the open end of which is beneath the surface of water (see Fig. 108), effects a complete absorption of the oxygen in about 24 hours. The water rising in the tube indicates a diminution of one fifth in the volume of the air—or what is the same thing, a withdrawal of from 20 to 21 per cent. of oxygen.

FIG. 108.

The carbonic acid, aqueous vapor, ammonia, and the occasional constituents of the atmosphere, are determined by passing a measured quantity of air through receptacles containing materials which absorb and retain them.

The arrangement by which this can be best effected is called an Aspirator. It consists simply of a tight cask of a known capacity, filled with water, and provided at the base with a stop-cock. At the top of the cask, a tube, or series of tubes, or other vessels are fitted, as is represented in Fig. 109; one filled, for example, with pumice stone drenched with strong sulphuric acid, and another with caustic potash. When the cock of the vessel is opened, and the water allowed to flow out, its place is supplied by an equal volume of air, which flows in through the tubes. The sulphuric acid absorbs all the moisture contained in the air which flows over it, and the potash all the carbonic acid. The quantity of air that passes through the tubes is known by the quantity of water that flows out of the cask, while the increased weight of the separate tubes gives the total amount of moisture and carbonic acid contained in such quantity.

FIG. 109.

332. Compounds of Nitrogen and Oxygen.—Nitrogen unites with oxygen to form five distinct compounds, containing, respectively, 1, 2, 3, 4, and 5 equivalents of oxygen, with 1 of nitrogen.

Their names and chemical constitution are thus expressed:

QUESTIONS.—How is air analyzed? How are the carbonic acid and aqueous vapor of the air determined? What is an aspirator? How many compounds of oxygen and nitrogen exist?

	Symbol.	Composed by weight of		
Protoxyd of nitrogen (nitrous oxyd)....	NO	14 nitrogen	+ 8	oxygen.
Deutoxyd of nitrogen (nitric oxyd)......	NO_2	14 "	+16	"
Nitrous acid...........................	NO_3	14 "	+24	"
Hyponitric acid (peroxyd of nitrogen)..	NO_4	14 "	+32	"
Nitric acid............................	NO_5	14 "	+40	"

Three of these compounds are acids; and all of them are endowed with qualities so marked, so powerful, and so well defined, that the original attributes of their elements are entirely lost.

333. **Nitric Acid, NO_5.**—Nitric acid is the most important of all the combinations of nitrogen and oxygen, and is the source from whence most of the compounds of nitrogen are generally obtained.

334. **History.**—It was known to the alchemists, but its true composition was first determined by Cavendish in 1785. The name formerly applied to it, and which is still used to some extent, was *aquafortis.*

335. **Distribution.**—Nitric acid occurs in nature usually in combination with potash, soda, or lime in the soil, especially in tropical countries, as in some parts of India and Peru. The compound formed with potash constitutes the nitre or saltpeter of commerce. In the desert of Atacama, in Chili and Peru, it exists in vast quantities combined with soda, forming nitrate of soda, which salt is also called " Chilian saltpetre," or cubic niter. Nitric acid, as already stated, also exists occasionally in the atmosphere, especially during and after the occurrence of thunder-storms.

336. **Preparation.**—When nitrogen is mixed with twelve or fourteen times its bulk of hydrogen, and a jet of the mixed gas is allowed to burn in air, or in oxygen, the water formed has a sour taste and an acid reaction from the formation of a small quantity of nitric acid. In this case the nitrogen burns by reason of the great heat developed during the combustion of the hydrogen, and the nitric acid combines at once with the water formed, which last substance, in some way by its presence, aids the operation. It was from noticing the acidity of water formed by the combustion of hydrogen in air, that Cavendish was led to institute an investigation which terminated in the discovery of nitric acid. He mixed together the two gases, oxygen and nitrogen, in a close tube, over a solution of potash, and then caused them slowly to combine by passing a series of electric sparks through the mixture for several successive days. At the conclusion of the experiment, the glass contained nitrate of potash (saltpeter). A similar result will be produced if a number of sparks be passed from an electrical machine, through air between two metallic points, over moistened litmus paper: a red spot will be produced upon the paper, owing to the formation of nitric acid in minute quantity by the combination of oxygen with nitrogen.

QUESTIONS.—Give the series. What is said of nitric acid? What of its history? What of its distribution in nature? How may nitric acid be formed? What circumstances led to its discovery?

For all practical purposes, nitric acid is always obtained by heating one of the natural compounds of nitric acid with potash or soda in a retort, with an equal weight of strong sulphuric acid. The nitric acid is displaced by the sulphuric acid, and distils over, being much more volatile than the sulphuric acid.

FIG. 110.

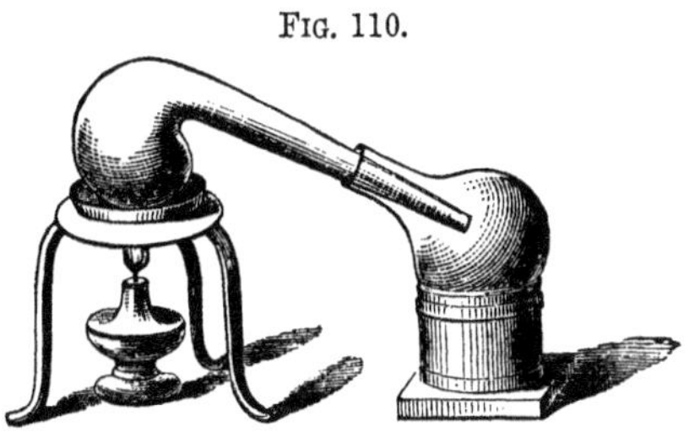

This process may be easily illustrated experimentally by introducing into a glass retort, Fig. 110, equal weights of powdered saltpeter and strong sulphuric acid. The retort should be supported upon a thin layer of sand contained in a tin or sheet-iron vessel (technically termed a sand-bath), and the heat supplied by an ordinary alcohol-lamp; a flask cooled by a wet cloth, or placed in a vessel of cold water, is adapted to the retort, and serves as a receiver. During the distillation red fumes appear in the retort, arising from a partial decomposition of the nitric acid formed, and a production of some of the lower oxyds of nitrogen.*

On a large scale, iron retorts coated on the inside with fire-clay are employed. The chemical reaction involved in this process may be represented as follows:

$$KO, NO_5 + SO_3 = KO, SO_3 + NO_5.$$

Or sulphuric acid and nitrate of potash give nitric acid and sulphate of potash.

337. **Properties.**—Nitric acid, when pure and in a concentrated state, is a colorless, limpid, fuming liquid, powerfully corrosive and intensely acid. As found in commerce, it is never pure, and is of a golden-yellow color. It is the highest oxyd of nitrogen known to exist, and has a specific gravity of 1·52 (water = 1). Anhydrous nitric acid, or nitric acid without water combined with it, can be prepared by a most carefully conducted chemical process; but under all ordinary circumstances it contains a certain proportion of water; its constitution being represented by the formula NO_5, HO. In the most concentrated state in which it can be used, it consists of 54 parts real acid and 9 of water.

Nitric acid is very readily decomposed, and mere distillation causes a partial decomposition. Exposure to light produces a similar result, oxygen and

* The retort generally breaks at the conclusion of this process from the crystallization of the sulphate of potash formed, but it may be saved by adding to it, when partially cooled, a small quantity of warm water.

QUESTIONS.—How is it practically prepared? What is the chemical reaction involved in the practical production of nitric acid? What are the properties of nitric acid? Does it exist apart from water? Is nitric acid easily decomposed? What effect has light upon it?

some of the lower oxygen compounds of nitrogen, which produce discoloration, being evolved—sometimes in quantity sufficient to expel the stopper of a bottle. In its concentrated form it begins to boil at 184° F., and freezes at about — 40° F.

338. **Chemical Action of Nitric Acid.**—Nitric acid is one of the very strongest acids, and ranks next to sulphuric acid. It attacks most inorganic substances, and all living tissues. It turns wool, feathers, the skin, and all animal matters containing albumen, a bright yellow color; the orange patterns upon woolen table-cloths are produced by means of it. In medicine it is not unfrequently used as a powerful cauterizing agent.

The effect of concentrated nitric acid upon animal tissues may be illustrated by applying a drop to a piece of parchment, which immediately becomes stained and shrivelled.*

The action of nitric acid on vegetable colors may also be illustrated by the following experiment:—Color some water blue in a test tube with a solution of indigo, and add to it on boiling, a drop of nitric acid; the blue color will almost immediately disappear.†

Nitric acid, when in its state of highest concentration, exerts no violent action upon certain organic substances, such as woody fibers, starch, etc., but unites with them to form most singular compounds. Cotton fibers immersed in it for a few moments and then carefully washed in water, are converted into a violently explosive substance. (See gun-cotton.)

Commercial nitric acid will completely dissolve, in the cold and without odor, a little less than its own weight of flesh and bone (beef), in a space of time varying from three to five hours. The action of nitric acid, however, upon organic substances and the metals is exceedingly different at different degrees of concentration.

Nitric acid very readily parts with a portion of its oxygen to the metals and to combustible bodies, and is therefore one of the principal agents made use of in chemistry for causing such substances to assume, or pass into a state of oxydation.

If nitric acid be dropped upon hot finely powdered charcoal, the charcoal burns vividly; if mixed with a little oil of vitriol, and poured upon oil of turpentine, it occasions an explosive combustion. Phosphorus is readily ignited

* It is an extraordinary, very cruel, and too common experiment made by physiologists to illustrate what they are pleased to call a power of vital contractility under the influence of a stimulus, by touching with a glass rod dipped in nitric acid, the heart of a living rabbit. In an instant the heart shrivels and contracts to one third its original size.—FARADAY.

† Indigo solution—a most useful chemical reagent—may be easily formed by pulverizing a small quantity of indigo, and forming a thin paste of it with strong sulphuric acid. After a few days add water, and a deep blue liquid, solution of indigo, is obtained.

QUESTIONS.—What are its freezing and boiling points? What is said of its chemical character? How may the action of nitric acid upon animal tissues be illustrated? How its action upon vegetable colors? How upon vegetable fibers? How is nitric acid able to produce oxydation?

by throwing it upon strong nitric acid. This experiment is a somewhat hazardous one, and particles of phosphorus scarcely larger than the head of a pin should alone be employed.

339. **The Action of Nitric Acid upon the Metals** is instructive, and serves to illustrate the manner in which metallic bodies combine with the acids generally. The metals will enter into direct combination with many of the simple non-metallic bodies. Thus antimony will unite with chlorine, iron with oxygen, and copper with sulphur; but no metal will unite directly with an acid. In order that combination between them should occur, it is necessary that the metal should be in the form of an oxyd. This oxydation may, however, be effected at the same time that the acid is presented to the metal, and the formation of the oxyd and its solution in the acid may appear to occur simultaneously. Zinc, for example, does not unite as zinc with sulphuric acid; but when this metal is placed in *dilute* sulphuric acid, the oxygen is supplied from the water contained in it, which is decomposed;—oxyd of zinc is produced and is immediately dissolved by the acid, whilst the hydrogen escapes in the gaseous form. When a metal, such as copper or silver, is dissolved by nitric acid, a preliminary oxydation is equally necessary; but owing to the facility with which nitric acid is decomposed, this oxydation is usually effected by depriving the acid of a portion of its oxygen, it being more readily decomposed than water. When this takes place, a part of the products of the decomposition of the acid escapes into the air in the form of deep red fumes (see hyponitric acid), while the compound of the metal with oxygen dissolves in another portion of the acid which has not undergone decomposition. It is through this peculiar action of nitric acid that it is rendered a most ready and powerful solvent of most of the metals.—MILLER.

340. **Salts of Nitric Acid.**—The salts formed by the union of nitric acid with the bases are termed *nitrates*, and are especially remarkable for the circumstance that they are soluble in water. When the nitrates are projected upon glowing coals they are decomposed; and by reason of the escape of oxygen, they deflagrate, or burn furiously with scintillations. If dissolved in water, and paper be moistened with the solution, allowed to dry, and then burned, the peculiar combustion characteristic of touch-paper will be produced. This property is, however, exhibited by the salts of some other acids.

Nitric acid is a substance much used in the laboratory, and in many of the operations of practical art.

341. **Protoxyd of Nitrogen, NO**;—*Nitrous Oxyd;—Exhilarating Gas.*—This gas was discovered by Priestley in 1776, but its properties remained unknown until investigated by Davy, in 1808. Since this period, a considerable degree of popular attention has always been bestowed upon it,

---

QUESTIONS.—Explain the action of nitric acid upon the metals, and the principle which such action illustrates. What are the salts of nitric acid termed? What are their distinguishing peculiarities? When and by whom was protoxyd of nitrogen discovered?

in consequence of the remarkable effects which it produces upon the animal system, when taken into the lungs.

342. **Preparation.**—Protoxyd of nitrogen is prepared by heating the salt known as nitrate of ammonia in a glass flask, furnished with a perforated cork and a bent glass tube, over a spirit lamp.* (See Fig. 111.)

Fig. 111.

Upon the application of a moderate temperature, the salt melts, and at about 400° F. apparently begins to boil; it is, however, in reality undergoing a process of decomposition, by which it is entirely resolved into gaseous protoxyd of nitrogen and steam (water). The temperature must be very carefully watched, and not allowed to rise so high as to occasion white vapors in the flask, as, in such case, some injurious products may be formed. The gas should be collected in a gasometer, or receiver filled with water of a temperature of about 90; cold water absorbing considerable quantities of it. It is also advisable to allow the gas to remain for a little time over water before attempting to respire it.

The reaction which takes place in the production of protoxyd of nitrogen may be explained as follows: Ammonia is a compound of nitrogen with hydrogen. When the nitrate of ammonia is heated, the hydrogen of the ammonia combines with a part of the oxygen of the nitric acid to form water, whilst the nitrogen of the ammonia at the same time becomes oxydized at the expense of another part of the oxygen of the nitric acid. The result is, that the whole of the nitrogen, both of the nitric acid and of the ammonia, is liberated in the form of protoxyd of nitrogen, thus:

$$\overbrace{NH_3,\ NO_5,\ HO}^{\text{Nitrate of ammonia.}} \text{ becomes } 2\ \overbrace{NO}^{\text{Protox. nitrog.}} + 4\ \overbrace{HO}^{\text{Water.}}$$

An ounce of nitrate of ammonia will furnish about 500 cubic inches of this gas.

343. **Properties.**—Protoxyd of nitrogen is a transparent, colorless gas, with a sweetish smell and taste. It is a heavy gas, its specific gravity being 1·52, or nearly the same as that of carbonic acid. It supports the combustion of many bodies with nearly the same energy and brilliancy as pure

* Nitrate of ammonia is a white crystalline salt, which can be cheaply purchased of dealers in chemicals, or can be easily made by neutralizing dilute nitric acid by carbonate of ammonia. In preparing exhilarating gas, not less than 6 or 8 ounces should be used.

Questions.—How is it prepared? What is the chemical reaction involved in the process? What are its properties?

oxygen; and when mixed with an equal bulk of hydrogen, forms an explosive mixture. It is, however, easily distinguished from oxygen by its ready solubility in cold water, which dissolves nearly its own volume of the protoxyd of nitrogen.

Under a pressure of 50 atmospheres at 45° F., it is reducible to a clear liquid, which, at a temperature of about 150 degrees below zero, freezes into a beautiful transparent crystalline solid. By mixing the liquid protoxyd with another very volatile substance, the bisulphide of carbon, and allowing the mixture to evaporate in vaccuo, M. Natterer, a few years since, obtained a reduction of temperature which he estimated at 220 degrees below zero;—a lower point than has been hitherto attained to by any other process.

Protoxyd of nitrogen, if quite pure, or merely mixed with atmospheric air, may be respired for a few minutes without inconvenience or danger. It then produces a singular species of transient intoxication, "attended in many instances with an irresistible propensity to muscular exertion, and often to uncontrollable laughter; hence the gas has acquired the popular name of exhilarating or *laughing*-gas. Different individuals are affected in different degrees and in various ways, according to the temperament of each. In plethoric persons, where there is any tendency to over-active circulation through the brain, the experiment is not a safe one. The intoxicating effects pass off in a few minutes, and frequently no recollection of what has passed is retained, and no lassitude is perceived after the extreme exertion."—MILLER. The gas should be inhaled from a large bladder or gas-bag, through a tube of an inch internal diameter.

An animal entirely immersed in this gas soon dies from the prolonged effects of the intoxication.

The idea that anæsthesis, or insensibility to pain during surgical operations, might be occasioned by the inhalation of gases, appears to have been first entertained by Dr. Horace Wells, of Hartford, Conn., from observing the action of protoxyd of nitrogen upon the animal system; and he succeeded in producing, by means of it, the same effects which are now accomplished by the agency of chloroform and ether.

FIG. 112.

344. **Deutoxyd of Nitrogen, NO_2:** *Binoxyd of Nitrogen, or Nitric Oxyd.*—This gas is easily prepared by pouring nitric acid upon clippings or turnings of copper, contained in a flask with a little water. As no heat is required, the double-tubed hydrogen gas apparatus may be employed. (See Fig. 112.) At the commencement of the action, the flask becomes filled with deep-red fumes, but if the gas be collected over water it will be found to be colorless.

The chemical action involved in the production of nitric-oxyd, by this process, is as follows: The copper takes oxygen from one por-

QUESTIONS.—How is it distinguished from oxygen? What effect has cold or pressure upon it? What effect does protoxyd of nitrogen produce upon the system when inhaled? What discovery was first suggested by the action of this gas on the system? How is nitric oxyd prepared? What is the chemical action involved?

tion of the nitr'c acid and becomes oxyd of copper, which combines with another portion of acid remaining undecomposed, and forms the nitrate of copper, the solution of which is of a blue color. That part of the nitric acid which is decomposed, loses three equivalents of oxygen, which are taken up by the copper; the remaining two equivalents of oxygen united with the nitrogen appear as the gas, thus:

$$\underbrace{3\ Cu.}_{\text{Copper.}} + \underbrace{4\ (NO_5)}_{\text{Nitric acid.}} = \underbrace{3\ (Cu.O,\ NO_5)}_{\text{Nitrate copper.}} + \underbrace{NO_2}_{\text{Nitric oxyd.}}$$

345. **Properties.**—Nitric oxyd is a colorless gas, which is but slightly absorbed by water. It is perfectly irrespirable, and excites a violent spasm of the throat when an attempt is made to respire it. Sir Humphrey Davy, in his experiments upon the respiration of the protoxyd of nitrogen, attempted to inhale this gas, but the result was nearly fatal, and would have been quite so, had not the glottis contracted spasmodically, and thus prevented its passage into the lungs.

Most burning bodies, when immersed in this gas, are extinguished by it, although it contains half its weight of oxygen. If phosphorus and charcoal, however, in a state of ignition, be introduced into it, the heat they evolve effects the decomposition of the gas, and the combustion continues with great brilliancy through the agency of the liberated oxygen.

The most remarkable property of nitric oxyd appears to be its great attraction for oxygen. When mixed with oxygen, or any gas containing oxygen (atmospheric air), dense red fumes are produced, which are soluble in water, and produce an acid liquid. In this way nitric oxyd may be used as a test to demonstrate the presence of uncombined oxygen in a gaseous mixture. Experimentally this action may be demonstrated as follows: Partially fill a tall glass jar or bottle with nitric oxyd, over a pneumatic trough; and then by lifting the end of the jar, admit a few bubbles of atmospheric air, or pure oxygen. In an instant, deep blood-red fumes will fill the vessel, and much heat will be generated. By agitating the contents of the jar with water, the red vapors are rapidly absorbed, and the experiment may be several times repeated, with the remaining portions of the gas.

Nitric-oxyd has never been liquefied.

346. **Nitrous Acid, NO_3.** *Hyponitrous Acid.*—The third compound of nitrogen with oxygen is a brownish red vapor at ordinary temperatures, and a volatile green liquid at a temperature 0° F. It is formed by mixing 4 volumes of nitrous oxyd with 1 volume of oxygen, both in a perfectly dry state. It unites with bases to form salts, which are called nitrites.

347. **Hyponitric Acid, NO_4.** *Peroxyd of Nitrogen.*—The red fumes which appear in mixing nitrous oxyd with oxygen, or atmospheric air, con-

---

QUESTIONS.—What are its properties? How does it act upon combustibles? What is the most remarkable characteristic of nitric oxyd? How may it be illustrated? What is the physical condition of nitrous acid? How is it prepared? What is said of hyponitric acid?

sist mainly of hyponitric acid. It may be formed in a state of purity, by mixing 4 volumes of Nitric-oxyd with 2 volumes of oxygen.

The compounds of nitrogen with hydrogen, carbon, and other non-metallic elements, will be considered in subsequent sections.

SECTION V.

CHLORINE.

Equivalent, 35·5. *Symbol*, *Cl*. *Density*, 2·47.

348. **History.**—This substance was discovered by Scheele in 1774, and called by him *dephlogisticated marine* acid.

It was universally regarded as a compound body until 1811, when Davy established its elementary character, and on account of its yellowish-green tint, gave it the appellation of chlorine (from χλωρὸς, green).

349. **Natural History and Distribution.**—Chlorine is a principal member of a small natural group of four closely-allied elementary bodies, viz., chlorine, iodine, bromine, and fluorine, which differ in many respects from all the other elements. They are characterized by a remarkable indifference for each other, and for an intense affinity for other substances at ordinary temperatures—an affinity so general as to preclude the possibility of any member of the class existing in a free and uncombined state in nature. Collectively they are termed the Halogens, from the circumstance of their forming with the metals saline compounds resembling common salt.—*Haloid salts*. (See § 271.)

FIG. 113.

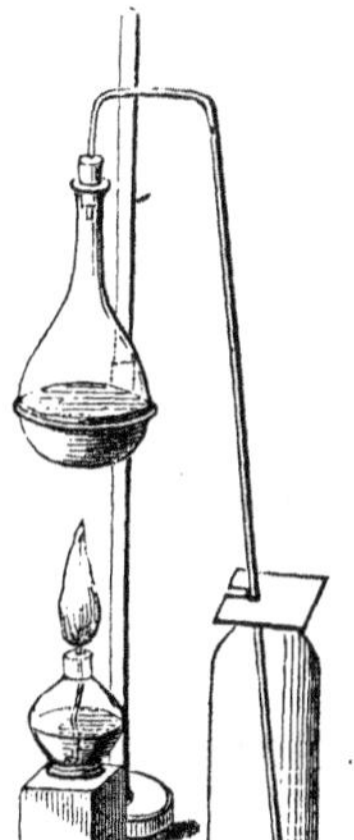

Chlorine united with other elements is a large constituent both of the inorganic and organic kingdoms. The great magazine of it in nature is rock, or common salt, which is a compound of chlorine and the metal sodium. Combinations of it also with other substances in the mineral kingdom are not unfrequent. In the organic kingdom it is found as a constituent of both animals and vegetables; existing in the greater number of animal liquids, and in various fluids and secretions of plants.

350. **Preparation.**—Chlorine is most easily prepared by pouring strong hydrochloric (muriatic) acid upon pulverized binoxyd of manganese contained in a glass retort or flask (arranged as in Fig. 113), and applying a gentle heat from a spirit-lamp. The proportions which give the best result are, one part by weight

QUESTIONS.—How is it prepared? Give the history of chlorine. To what other elements is chlorine allied? What are the characteristics of these associated elements? What designation is given to them as a class? What is said of the distribution of chlorine in nature? How is it prepared?

of binoxyd of manganese, and two parts by weight of hydrochloric acid. The gas may be collected over water, or more conveniently by the displacement of air in a dry, narrow-necked jar, as is represented in Fig. 113. The greenish color of the chlorine enables the operator to determine when the receiver is full. By closing the jars with glass stoppers, well smeared with tallow, the gas can be preserved unaltered for a considerable length of time.*

The chemical reaction which takes place in this process may be explained as follows: hydrochloric acid is a compound of hydrogen and chlorine; when mixed with the binoxyd of manganese in the proportion of 2 equivalents of the former to 1 of the latter, double decomposition ensues:—water, free chlorine, and a chloride of the metal being produced.

Thus,—$MnO_2+2HCl=MnCl+2HO+Cl$.

Three ounces of powdered binoxyd of manganese with half a pint of commercial muriatic acid diluted with 3 ounces of water, will yield between three and four gallons of the gas. Care should be taken not to use an acid more dilute than the one indicated, lest some explosive compound of chlorine should be generated.

Chlorine may also be readily obtained by distilling a mixture of 4 parts by weight of common salt, 1 part of binoxyd of manganese, 2 of sulphuric acid, and 2 of water. It is in this way that chlorine is prepared in enormous quantities for manufacturing purposes; but for the preparation of "chloride of lime," the first described method is followed, owing to the fact that the hy-

* The following memoranda respecting the preparation of chlorine are worthy of attention. The process should always be conducted in a well-ventilated apartment, altogether free from valuable furniture, and especially from colored curtains, paper-hangings, etc.—the action of the gas being most destructive of the color and texture of organic compounds.

Before applying heat to the generating flask, the operator should observe whether the interior of the glass has become thoroughly wetted by the acid, or whether a dry spot still remains. If the latter is the case, all heat should be withheld until the mass by agitation has become thoroughly incorporated, and the dry spot disappears. If this precaution is neglected, a fracture of the retort will probably take place on the application of heat. Most authorities recommend the collection of chlorine over warm water, inasmuch as cold water absorbs a considerable amount of the gas. This plan is attended with the serious disadvantage of causing chlorine to enter the bottle hot, and for that reason rarefied; so that when it cools and contracts, the stoppers of the bottles are found not unfrequently to be permanently fixed. Cold water should be employed, and except it be agitated whilst the gas is passing through it, so little of the chlorine is absorbed that the amount of loss is too small to be of consequence.—Faraday.

Every care should be taken in bottling up chlorine for preservation, to exclude water as much as possible, inasmuch as under the agency of light, water and chlorine react, forming hydrochloric acid, which is so violently absorbed by water, that the stoppers of the chlorine bottles become often irremediably fixed, owing to external pressure.—Ibid.

If the operator during the preparation of chlorine should inadvertently inhale a disagreeable quantity of the gas, the most effectual relief will be obtained from an immediate application of ammonia (smelling-salts) to the nostrils, or from inhaling the vapor of alcohol or ether.

Questions.—What precautions are to be observed in its preparation? What is the chemical action involved in this process? By what other process may chlorine be prepared? For what practical purposes are these two processes applied?

drochloric acid used is obtained as a waste product in the manufacture of carbonate of soda (soda-ash) from sea-salt.

351. **Properties.**—Chlorine is a dense, heavy gas, of a greenish-yellow color. It is characterized by a peculiar suffocating odor, almost intolerable to most persons even when greatly diluted with air, and occasioning a distressing irritation of the air-passages of the throat, attended with coughing. Any attempt to respire the gas in a pure form would probably be fatal, but when largely diluted with air, it is breathed without inconvenience by workmen in manufacturing establishments, and it has also been adopted as a remedial agent in this condition with benefit for pulmonary diseases. It should not, however, be used for this latter purpose without the sanction of a medical authority.

Chlorine is one of the heaviest of the gases, its specific gravity being 2·47 (air = 1). Under a pressure of 4 atmospheres, at 60° F., it condenses to a yellow liquid, of specific gravity 1·33, which remains unfrozen even at a cold of — 220° F.

Cold water absorbs about twice its bulk of chlorine gas. This solution, which is readily formed by agitating the gas and the water together, acquires the color, odor, and other properties of chlorine, and is much used for experimental and manufacturing purposes in preference to the pure gas. As it is slowly decomposed by the action of light, it should be preserved in bottles covered with paper, or in a dark place.

With water near its freezing point chlorine combines to form a definite hydrate, which contains 10 equivalents of water ($Cl+10HO$); this at a temperature of 32° F. freezes and forms beautiful yellow crystals. If these crystals be hermetically enclosed in a glass tube (see Fig. 114), they will, when exposed to a gentle heat, liberate free chlorine; this prevented from expanding, presses upon itself to such an extent that a portion of the gas liquefies, and may then be seen in the tube, floating upon the water which is present. This process furnishes the most ready way of obtaining liquid chlorine.

Fig. 114.

352. Chlorine is a supporter of combustion, but its effects are strikingly different from those manifested by oxygen. It does not combine directly with either oxygen or carbon, but has a most powerful affinity for hydrogen and the metals. Therefore, bodies rich in oxygen and carbon, either burn indifferently in chlorine or not at all, as in the case of charcoal; but on the contrary, bodies rich in hydrogen, together with many of the metals, burn in it with great brilliancy. The following experiments are illustrative of these facts:

Questions.—What are the general properties of chlorine? Is it at all respirable? What is the density of chlorine? Can it be liquefied? What is a solution of chlorine? What are its properties? What combination does chlorine form with water? How may liquid chlorine be prepared? What are the relations of chlorine to combustion?

A piece of flaming charcoal plunged into a vessel of chlorine is extinguished as instantly and as completely as if plunged into a vessel of water.

Wax and tallow are compounds of carbon and hydrogen. If a lighted taper be immersed in a jar of chlorine, its flame is extinguished; but the column of oily vapor rising from the wick is rekindled by the chlorine; the hydrogen part of the combustible burning with a dull reddish flame, while the carbon is separated in the form of a dense black smoke.

Another experiment illustrates the same action in a more remarkable manner. Oil of turpentine is a liquid exceedingly rich in hydrogen, and also in carbon. If a piece of paper soaked in it be fastened to the end of a rod and plunged into a jar of chlorine (see Fig 115), the chlorine unites with the hydrogen so readily as to instantly produce spontaneous combustion, while the carbon is separated and deposited as an abundant soot.

FIG. 115.

If a bit of ignited phosphorus be immersed in a jar of chlorine, as is represented in Fig. 116, the combustion continues, but the light evolved is hardly perceptible. If a piece of phosphorus be plunged into chlorine without ignition, it inflames spontaneously—a result which does not take place in oxygen. The feeble light which accompanies the combustion of phosphorus in chlorine, therefore, can not be explained by reason of any lack of affinity between these two substances, but it is due to the fact that the immediate products of the combustion are vaporous or gaseous, and are not rendered luminous by heating. (See Combustion.)

FIG. 116.

Antimony and many other metals finely powdered, and projected into a vessel of chlorine, take fire and produce a brilliant combustion. Thin sheets of copper leaf, attached to a copper wire, and dipped into chlorine, exhibit the same phenomenon.

353. The intense affinity which chlorine manifests for hydrogen is one of the most remarkable characteristics of this element, and is the property, above all others, which gives to chlorine its great value as an industrial agent. This affinity is, however, regulated, or rather called forth, by a most singular action of light. Thus, when chlorine and hydrogen, in the gaseous condition, are mixed together in equal volumes, they will remain for an indefinite period without action upon each other, if kept in the dark. If the mixture be exposed to diffused daylight, combination will take place gradually; but if the two gases

---

QUESTIONS.—What experiments illustrate its action in this respect? Why does phosphorus burn in chlorine with a feeble light? What are the relations of chlorine to hydrogen? What influence does light exert upon a mixture of these two elements?

are brought into direct sunlight, the union takes place instantly, accompanied with a powerful explosion.*

The following experiment is illustrative: Select a clear glass bottle (holding about a pint), and fill it over a pneumatic trough, to the extent of half its capacity, with chlorine gas; then carefully cover the bottle with a dark cloth, and add hydrogen from a gasometer sufficient to occupy the remaining space, or until all the water in the bottle is displaced; cork the bottle, keeping its mouth under water, and remove it from the trough carefully and entirely enveloped in the cloth. Then place the bottle in the direct light of the sun, and from a distance remove the cloth by means of a string or a long pole. If the preliminary conditions have been strictly complied with, the instant the rays of light fall upon the mixture a violent explosion will occur.

When chlorine in a free state, or in feeble combination with some other substance, is brought in contact with a body which contains hydrogen as one of its constituent elements, it manifests the same affinity for this element; and tends to " draw out," as it were, the hydrogen from its combination, and by uniting with it, to change or destroy the original compound. In this instance, also, light exercises a controlling influence.

For example: If a solution of chlorine in water (§ 351) be kept in the dark, no change takes place in it; but when exposed to the action of sunlight, it decomposes readily. This result is produced by the following reaction:—the chlorine contained in the solution, by reason of its intense affinity for hydrogen, withdraws this element from its combination with oxygen in the water, and uniting with it, forms an acid; the oxygen of the decomposed water, being no longer held in combination, escapes as a gas.

354. **Theory of Bleaching.**—It is this action of chlorine upon hydrogen which renders chlorine the most powerful of all known bleaching and deodorizing agents. Nearly all animal and vegetable coloring matters contain hydrogen as one of their essential constituents. When brought in contact with chlorine, the hydrogen they contain unites with it, and the original arrangement of particles, upon which the color of the body depended, being thus changed or broken up, the color itself is destroyed. Ozone, which is also a powerful bleaching agent, acts in a similar manner; the oxygen of which it consists, by reason of its highly active condition, withdraws hydrogen from its combination, unites with it to form water, and thus destroys the arrangement upon which the color depends. By withdrawing a single pillar of sup-

* It has also been shown by Dr. Draper, that pure and dry chlorine gas, when exposed for a time to the action of the sun's light, acquires and retains, for a considerable period, the power of forming an explosive union with hydrogen, even in the dark; while, on the other hand, chlorine prepared in the dark manifests no affinity for hydrogen until exposed to the light. This peculiar action of light is entirely confined to the chemical element of the solar ray.

QUESTIONS.—What experiment illustrates this? What are the relations which chlorine sustains to hydrogen in combination? Illustrate by example. What is the theory of bleaching by chlorine? By ozone? What is said of the permanency of the bleaching effect of of these agents?

port, the whole structure falls. Colors once removed by the action of chlorine or ozone can never be restored; and in this respect these two substances differ from most other bleaching agents.

The bleaching action of chlorine may be illustrated by a variety of experiments. For this purpose a solution of chlorine in water may be most conveniently used. If we pour a little of this solution upon red ink, red wine, the blue tincture of red cabbage, of litmus, on indigo solution, or on ordinary writing ink, their several colors almost immediately disappear. Paper, colored rags, and all varieties of cotton or linen fabrics immersed in a solution of chlorine, are bleached with great rapidity. The moist gas produces the same effect, but perfectly dry chlorine will not bleach. Fibres of wool are not bleached by the action of chlorine, neither is it usually employed for the bleaching of silk. It has no action upon "India," or printers' inks, for the reason that the coloring matter in these cases consists of minutely-divided carbon, which does not combine directly with chlorine.* By contact with chlorine for any considerable length of time, the texture of almost all organic substances is weakened and destroyed. This may be especially noticed in cases where cotton or linen fabrics have been wet with a chlorine solution, and then allowed to dry, without previous thorough washing.

355. **Disinfecting and Deodorizing Action of Chlorine.**—Chlorine acts upon noxious and odorous vapors and organic compounds to decompose and destroy them, in the same way as it does upon coloring agents. It differs essentially in its action from many substances used in fumigation, such as burnt paper, vinegar, pastiles, perfumes, and the like, inasmuch as, while the latter only disguise the ill odors, or mephitic atmosphere, by substituting one smell for another, the chlorine absolutely destroys the noxious matter itself.

The use of chlorine as a disinfectant, however, requires care. It should be used in the form of bleaching-powder ("chloride of lime"), mixed with water, and exposed to the air, in shallow vessels, if possible upon a high shelf. This compound is gradually decomposed by the carbonic acid of the atmosphere, and the chlorine being evolved falls slowly down, and is diffused through the

* A very elegant application of chlorine to bleaching purposes is made in the printing of bandanna handkerchiefs. The white spots, which constitute their peculiarity, are thus produced; First of all, the whole fabric is dyed of one uniform tint, and dried. Afterward many layers of these handkerchiefs are pressed together between lead plates, perforated with holes conformable to the pattern which is desired to appear. Chlorine solution is now poured upon the upper plate, and finds access to the interior through the perforations. By reason of the great pressure upon the mass, the solution can not, however, extend laterally further than the limits of the apertures, whence it follows that the bleaching agent is localized to the desired extent, and figures corresponding in shape and size to the perforations are bleached white upon a dark ground.—FARADAY.

QUESTIONS.—How may the bleaching action of chlorine be illustrated? What are exceptions to its action? What effect does continued contact with chlorine have upon organic substances? What is the disinfecting and deodorizing action of chlorine? How does chlorine differ in its action from many fumigating agents? How should chlorine be applied for disinfection and deodorizing?

room. If a more rapid action is required, a little dilute sulphuric or hydrochloric acid may be allowed to drop into the chloride of lime solution from a vessel suspended above it, by means of a piece of lamp-wick arranged in the form of a syphon. Another method is to suspend in the apartment, cloths steeped in a solution of bleaching-powder; and in the absence of bleaching-powder, the gas may be easily generated by one of the methods already dedescribed—care being taken to avoid excess.*

356. **Compounds of Chlorine.**—Chlorine combines with all the non-metallic elements, with perhaps a single exception. With many of them, however, it can not be made to unite directly. It enters into combination with all the metals; and with a large number of them directly, with an evolution of light and heat. The binary compounds of chlorine are termed *chlorides.* With the exception of the chlorides of silver and lead, and the subchlorides of copper and mercury, they are all more or less soluble in water, and in their taste and general physical character, resemble common salt.

It frequently happens that chlorine combines with the same metal in more proportions than one: for example, with iron, a protochloride ($Fe\ Cl$) and a sesquichloride (Fe_2Cl_3) may be formed; and with platinum a protochloride ($Pt\ Cl$) and a bichloride ($Pt\ Cl_2$); and, generally, for each oxyd of the metal that is capable of uniting with acids to form salts, a corresponding chloride exists.

Most of the chlorides of the metals are solid; but some few are liquid at ordinary temperatures; and one, the perchloride of manganese ($Mn_2\ Cl_7$) is gaseous. All of them are fusible at a moderate temperature, and many are readily volatilized in the operation; especially is this true of the chlorides of gold, copper, aluminum, magnesium, and several others. Geologists have taken advantage of this fact, in some instances, to explain the formation of mineral, or metallic veins in the rocky strata composing the crust

* "It must be particularly borne in mind, that chlorine in any form must only be used as an aid to proper ventilation. It is a necessary condition of health that our houses and rooms be properly ventilated. There is no substitute for ventilation any more than for washing or for general cleanliness. Chlorine, like medicine, ought in general to be used on special occasions, and under advice. In a sick-room, where ventilation is often difficult, chlorine, liberated in very minute quantities, will often be found singularly refreshing; but in this, as in all other cases of fumigation with chlorine, all metallic articles in the apartment ought to be removed, for these become speedily tarnished by the action of chlorine.

"For disinfecting the wards of hospitals and similar places, Prof. Faraday found that a mixture of 1 part of common salt, and 1 part of the binoxyd of manganese, when acted upon by 2 parts of oil of vitriol previously mixed with 1 part of water (all by weight), and left till cold, produced the best results. Such a mixture at 60° F., in shallow pans of earthen ware, liberated its chlorine gradually but perfectly in four days. The salt and the manganese were well mixed and used in charges of 3½ pounds of the mixture. The acid and water were mixed in a wooden tub, the water being put in first, and then about half the acid; after cooling the other half was added. The proportions of water and acid are 9 measures of the former to 10 of the latter."

QUESTIONS.—What is said of the compounds of chlorine? What are the compounds of chlorine termed? What are the general properties of the chlorides?

of our globe. It is supposed that the metals, in the form of chlorides, have been sublimed or volatilized by intense heat from the interior of the earth, and rising through openings and fissures in the rocks, have been deposited, as they cooled, in the situations in which they are now found.

Formerly, before the constitution of chlorine was fully understood, its compounds with the metals were termed *muriates*. The names, muriate of tin, muriate of soda, muriate of iron, have therefore the same signification as chloride of tin, chloride of soda, chloride of iron, etc.

357. Hydrochloric Acid, HCl. — *Chlorohydric Acid; Chloride of Hydrogen; Muriatic Acid.*—This substance, formed by the union of chlorine and hydrogen, is the most important of all the compounds which chlorine forms with the non-metallic elements.

It was first obtained by Priestley in its pure form of a gas, in 1772; and in a state of solution in water, it has long been known under the names of muriatic acid, and spirit of salt. In the latter condition it constitutes a strong, corrosive acid liquid.

358. **Preparation.**—When chlorine and hydrogen are mixed together in the proportion of equal volumes, and a chemical combination is effected between them, they unite, without condensation, to form hydrochloric acid gas. This union may be brought about by the action of light, in the manner before described (§ 353), by the application of an ignited match, or by the passage of the electric spark—the combination in the latter instances being always attended with an explosion.

Fig. 117.

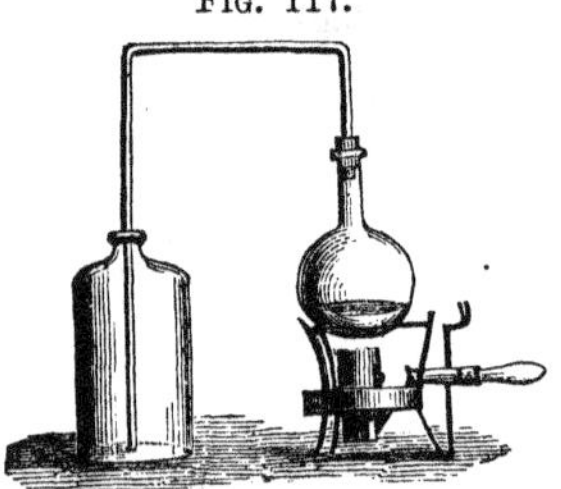

For experimental purposes, hydrochloric acid gas may be procured by heating in a glass flask, furnished with a perforated cork and tube, a quantity of strong commercial muriatic acid. The gas is readily given off by the application of a gentle heat, and may be collected by displacement of air in dry vessels. (See Fig. 117.)

For most practical purposes, hydrochloric acid is obtained by action of sulphuric acid upon common salt. When the process is conducted on a small scale, and in a glass retort, or an apparatus similar to that represented in Fig. 117, 3 parts of common salt, 5 of strong sulphuric acid, and 5 of water may be taken. The reaction in this case is as follows: common salt is composed of chlorine and sodium; when mixed with sulphuric acid and water, the water is decomposed; its hydrogen uniting with the chlorine of the common salt to form hydrochloric acid, and its oxygen with

---

Questions.—What theory has been proposed to account for the origin of mineral veins? What are muriates? What is hydrochloric acid? How may it be prepared? How is it prepared for practical purposes?

the sodium to form soda, which last unites with the sulphuric acid to form sulphate of soda. Expressed in symbols, we have—

$$\underbrace{NaCl}_{\text{Common salt.}} + \underbrace{SO_3}_{\text{Sulph. acid.}} + \underbrace{HO}_{\text{Water.}} = \underbrace{NaO,SO_3}_{\text{Sulphate of soda.}} + \underbrace{HCl}_{\text{Hydrochloric acid.}}$$

359. **Properties.**—Hydrochloric acid is a colorless acid gas, of a peculiar pungent odor, producing white fumes when allowed to escape, by uniting with and condensing the moisture of the atmosphere. It is quite unrespirable, but is much less nauseous and suffocating than chlorine. It produces coughing when breathed in even the most dilute condition. It is heavier than air, and has a specific gravity of 1·24. Under a pressure of 40 atmospheres at 50° F., it condenses to a colorless liquid, which has never been frozen. It is incombustible, extinguishes burning bodies and when brought in contact with dry and blue litmus paper reddens it.

Hydrochloric acid gas is especially characterized by a most intense attraction for water, which liquid at a temperature of 40° F. is capable of absorbing about 480 times its bulk of gas—increasing in volume thereby about one third, and acquiring a specific gravity of 1·21. Water of a higher temperature absorbs less. A piece of ice passed into a jar of hydrochloric acid gas standing over mercury is instantly liquefied by it; and the gas at the same moment disappearing by absorption, the mercury rises to fill the jar. By reason of its great solubility in water, it can only be collected over mercury, or by the displacement of air.

360. **Solution of Hydrochloric Acid**, which constitutes the liquid acid, or the muriatic acid of commerce, is prepared by generating the gas from a mixture of salt and dilute sulphuric acid, and allowing it to pass through and become absorbed by water.

The gas is conducted from the retort or generating vessel into a series of bottles or jars connected with each other and filled with water. When the water in the first vessel becomes saturated with hydrochloric acid, the gas passes over into the second, thence into the third, and so on, saturating each successively. Several contingencies, however, must be provided for in this operation; the evolution of gas may take place so rapidly as to rupture the receivers, or the gas delivered slowly may be absorbed so completely by the water as to produce a vacuum; in which case the whole liquid contents of the receivers flow back violently into the retort, and thus put an end to the process.

Questions.—Explain the chemical reaction in this case? What are the properties of hydrochloric acid gas? What is said of its attraction for water? What is the muriatic acid of commerce? How is it prepared? What precautions are to be observed in its preparation?

Woulfe's Apparatus.—To obviate these difficulties, a series of peculiar-shaped vessels, known as "Woulfe's bottles," are employed. These consist of glass jars, or bottles, provided with three necks, or apertures, (see Fig. 118), each of which is fitted with a perforated cork and tube. The manner in which the gas enters and is discharged from the vessel will be readily understood by an inspection of the figure. The middle aperture is fitted with a single upright tube, called the "safety tube," which dips beneath the surface of the liquid contained in the vessel. If the pressure of gas becomes

FIG. 118.

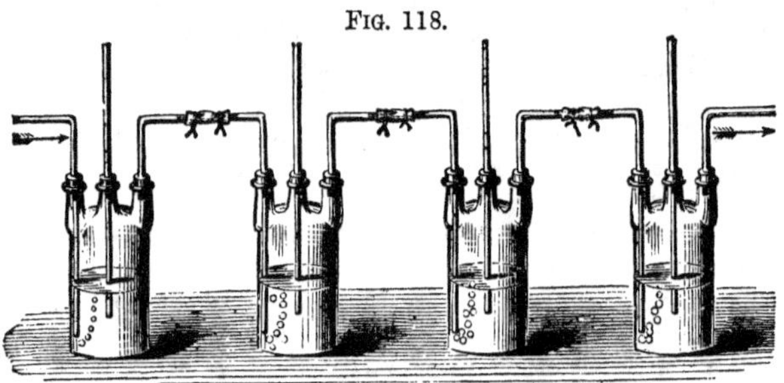

excessive, the water is forced up the center tube, and the pressure is relieved. If a vacuum is created, air enters from without to fill it. By the condensation of the hydrochloric acid gas, much latent heat is liberated, and the water which absorbs it soon becomes elevated in temperature; to obtain, therefore, the most concentrated solution of gas, it is necessary that the receivers should be immersed in cold water, or surrounded with ice. Connection between the separate Woulfe's bottles is effected by means of a flexible tube of India-rubber.

Hydrochloric acid solution, when pure, is a colorless liquid, fuming, when concentrated, on exposure to air. The commercial "muriatic" acid is generally of yellow, or straw color, owing to the presence of iron and other impurities. It constitutes one of the three great acids of commerce, and is extensively used as a reagent in chemical operations, and to some extent in medicine as a tonic. In the manufacture of "soda ash" (carbonate of soda), by the decomposition of common salt, hydrochloric acid gas is prepared as an incidental product in immense quantities; and in some of the great manufacturing establishments of Great Britain it is regarded as a waste product, the disposal of which is attended with difficulty and expense.*

* It was usual to allow the acid gas to escape into the air by means of a chimney, on emerging from the top of which it formed, in contact with moisture, white clouds of acid,

QUESTIONS.—Describe the construction and use of Woulfe's bottles. What are the physical properties of hydrochloric acid solution? What are its uses? Of what branch of manufacture is it an incidental product?

Free hydrochloric acid, derived from the salt contained in food, is found in the stomach, as a constituent of the gastric juice. Its presence, and that of the soluble chlorides in solution, is indicated by the formation of a white, curdy precipitate, when nitrate of silver in solution is added to the liquid. This precipitate—chloride of silver—is soluble in ammonia, but insoluble in nitric acid.

361. **Aqua Regia.**—*Nitro-Muriatic Acid.*—The name of aqua regia (*royal water*) was given by the alchemists to a mixture of nitric with hydrochloric acid, from the power that it possesses of dissolving gold, the "king of the metals."

Gold and platinum are insoluble in either acid separately; but when the two acids are mixed, they mutually decompose each other in the presence of the metals—free chlorine, and a compound of chlorine and an oxyd of nitrogen being liberated. The chlorine, in the moment of its extrication, or in its nascent state (page 162), acts upon the metals and dissolves them—the products formed being chlorides.

The best proportions of *aqua regia* are one of nitric acid by measure to two hydrochloric.

362. **Oxyds of Chlorine.**—Although chlorine and oxygen will not, under any circumstances, unite directly, several compounds of these elements may be obtained by indirect methods. The composition and names of the most important are indicated in the following table:—

	Symbol.	Composition by weight.	
Hypochlorous acid	ClO	35·5 chlorine.	8 oxygen.
Chlorous acid	ClO_3	35·5 "	24 "
Peroxyd of chlorine	ClO_4	35·5 "	32 "
Chloric acid	ClO_5	35·5 "	40 "
Hyperchloric acid	ClO_7	35·5 "	56 "

363. **Hypochlorous Acid.**—This compound may be produced by the action of chlorine upon red oxyd of mercury. It is a yellow gas, readily

---

which, wafted by the wind, produce a corrosive rain, most ruinous to the vegetation of the surrounding country. Many soda works in Great Britain were, therefore, indicted as nuisances on this account, and attempts were made to remedy the evil by discharging the fumes at great elevations, where it was supposed they would become so diluted by admixture with vapor as to be rendered harmless. To carry out this scheme, the most gigantic chimneys ever built were constructed. One near Liverpool is 495 feet high, 30 feet in diameter at the base, 11 feet at the top, and contains a million of bricks. Another at Glasgow is still larger. These costly structures have not, however, been found to answer the purpose for which they were intended, and it has become necessary to condense the gas as fast as it is liberated by bringing it into contact with cold water. But even this, taken in connection with the disposal of the great quantity of liquid acid formed, is a matter of great difficulty, and many arrangements have been patented to effect it.

QUESTIONS.—Does it exist in the animal economy? What is a test of its presence? What is aqua regia? Why so called? How is it enabled to dissolve gold? What is said of the oxyds of chlorine? What is hypochlorous acid?

absorbed by water, and condensed by the application of cold into an orange-yellow liquid.

364. **Bleaching Compounds.**—When chlorine gas is caused to pass through weak solutions of the alkalies, or over hydrate of lime (slacked lime), it is absorbed, and very peculiar compounds, possessed of remarkable bleaching properties, are produced. It is in this way that the bleaching agents so extensively used in the arts under the names of chloride of lime (bleaching powder) chloride of soda, and chloride of potash, together with what are called "disinfecting fluids," are prepared.

These compounds, according to the opinion of most chemists, are formed by the union of hypochlorous acid with an oxyd of a metal, and are termed *hypochlorites*. Thus the constitution of the so-called chloride of lime would be represented in symbols as follows: CaO, ClO. Other authorities deny the formation of hypochlorous acid, and regard the compounds in question as formed by the direct union of chlorine with an oxyd. According to this latter view, the constitution of chloride of lime, represented in symbols, would be as follows: CaO, Cl.

365. **Chloride of Lime**, or *Bleaching-Powder*, is the most important of all the bleaching compounds of chlorine, and is used in immense quantities for the bleaching of paper, cotton, and linen fabrics, and for disinfecting purposes. Its manufacture is almost a monopoly with Great Britain, and no attempt to prepare it on a large scale in this country has ever proved successful.* The process consists essentially in exposing fresh slacked lime, spread out upon shelves in large leaden or stone chambers, to the action of gaseous chlorine—the operation being continued until the lime has absorbed, or united with the largest possible amount of the gas. It is then withdrawn, and made ready for transportation by enclosure in tight casks. As thus prepared, it is a soft white powder, partially soluble in water, and possessing a chlorine-like odor. When exposed to the air it is readily decomposed, carbonic acid being absorbed, and chlorine liberated.

Ordinary bleaching-powders contain about 30 per cent. of available chlorine. The testing of their commercial value is termed *chlorimetry*, and the method adopted generally consists in ascertaining by experiment how many grains of a particular sample are required to destroy the color of a known weight of indigo in solution. The result, compared with the results of certain standard experiments, gives the percentage.

---

* The reason why the manufacture of bleaching-powder has not been introduced into the United States, is due doubtless to the fact, that it can only be economically conducted in connection with the manufacture of soda-ash, which process furnishes hydrochl ric acid, from whence chlorine is procured, at a mere nominal cost; and to carry on both operations advantageously requires the employment of great capital and skill.

QUESTIONS.—How are the ordinary bleaching compounds and "disinfecting fluids" formed? What is said of the composition of these compounds? What is said of chloride of lime? How is it prepared? What is chlorimetry? How is it conducted?

What is called "Labarraque's disinfecting liquid," is a solution of a compound of chlorine and soda, similar in composition to bleaching-powder. "Burnet's disinfecting fluid" is a compound of chlorine and zinc.

366. **Chloric Acid**, ClO_5.—This compound is not known in an isolated state, and is never obtained except in combination with water (ClO^5 HO). When a stream of chlorine gas is transmitted through a strong solution of caustic potash, the gas is absorbed, and a bleaching solution, as before described, is formed. This, by standing, or by the application of heat, loses its bleaching property, and becomes a mixture of chloride of potassium and chlorate of potash; the latter of which, being the least soluble, separates on concentrating the liquid, into shining tabular crystals. In this reaction, a part of the potash is decomposed; its oxygen combining with one portion of chlorine to form chloric acid, while the potassium is taken up by a second portion of the same substance; or in symbols:—

$$6\,Cl+6\,KO=KO,ClO_5+5\,KCl.$$

Chlorates of other bases are formed in a similar manner.

367. **Properties**.—All of the chlorates, when exposed to moderate heat, undergo decomposition, and liberate oxygen most abundantly; they are, therefore, generally used for the production of oxygen (§ 281). When thrown upon ignited charcoal, the chlorates deflagrate with scintillations, and when heated with substances which have a strong attraction for oxygen, such as phosphorus and sulphur, they explode violently (§ 285). Mere friction, also, with these elements is sufficient to cause a detonation; for example, if a half a grain of sulphur be triturated in a mortar with two or three grains of chlorate of potash, the friction is attended with a series of small explosions. A mixture of chlorate of potash, sulphur, and a little charcoal, was formerly used as a percussion powder for gun-caps; but its action upon the locks was found to be highly corrosive.

Paper soaked in a solution of a chlorate, burns in the same manner as touch-paper.

An attempt was made by the French government, toward the close of the last century, to substitute chlorate of potash in place of niter (saltpeter) in the manufacture of gunpowder; but the liability to accidental explosion was so greatly increased, that the enterprise was speedily abandoned. It is, however, still used to a very great extent in the manufacture of fire-works, and especially in the production of colored fires.

368. **Peroxyd of Chlorine**, ClO_4. *Hypochloric Acid.*—This substance, which can not be obtained in a state of purity without great danger, is prepared by distilling chlorate of potash with strong sulphuric acid. It is a gas of a yellow color, which is gradually decomposed by the influence of light, and at a temperature less than that of boiling water, its elements separate

QUESTIONS.—What are Labarraque's and Burnet's disinfecting fluids? What is said of chloric acid? How is chlorate of potash prepared? What are the properties of the chlorates? For what purposes are they practically employed? What is said of peroxyd of chlorine?

with a most violent explosion. Mere contact with many combustible matters also occasion an immediate explosion.

Some of the properties of this singular compound may be experimentally illustrated without danger. If a few grains of loaf sugar and chlorate of potash be separately pulverized, and mixed together in equal proportions, without friction, the addition of a single drop of sulphuric acid, let fall from the end of a glass rod, will produce instantaneous and brilliant deflagration. The chemical reaction which occasions this result is as follows: The sulphuric acid decomposes the chlorate of potash, and liberates peroxyd of chlorine; this, in turn, by contact with the sugar, is decomposed, and evolves heat sufficient to produce combustion.

Another very brilliant experiment consists in bringing phosphorus in contact with peroxyd of chlorine under water at the very instant of its development. A deep glass vessel being chosen (a conical wine-glass will answer), a few small pieces of phosphorus are first thrown in, and the glass two thirds filled with water. Crystals of chlorate of potash, about equal in quantity to the phosphorus employed, are then allowed to fall through the water and settle upon the phosphorus. All that now remains to be done is to bring sulphuric acid in contact with the two, which is easily accomplished by means of a dropping-tube, or small glass tube and funnel—the extremity of either of which being brought in contact with the mixture, the sulphuric acid is caused to touch the solids, without mixing with, and suffering dilution by, the water. (See Fig. 119.) Peroxyd of chlorine being rapidly evolved, the phosphorus reacts upon it, and flashes of a beautiful green light under water, attended with a crackling noise, are produced.

FIG. 119.

The two other principal compounds of chlorine with oxygen, chlorous and perchloric acid, although of scientific interest, are of no practical importance.

369. **Chloride of Nitrogen.**—The single compound which chlorine is known to form with nitrogen, is especially worthy of note as probably the most dangerous of all chemical combinations.

When a bottle of chlorine, perfectly free from greasy matter, is inverted over a leaden dish containing a solution of 1 part of sal-ammoniac (NH_4Cl) in 12 parts of water—the mouth of the bottle slightly dipping beneath the surface—drops of an oily-looking substance will gradually form upon the liquid and fall to the bottom of the dish,—chlorine slowly disappearing. The fluid substance thus generated is chloride of nitrogen. During the whole operation, the bottle must not be approached, unless the face is protected by a wire-gauze mask, and the hands by thick woollen gloves. The leaden dish containing the chloride of nitrogen, may, after a time, however, be withdrawn

QUESTIONS.—How may its properties be illustrated? What is said of chloride of nitrogen? How is it prepared?

from under the bottle, care being taken to avoid all agitation and contact with the glass.

As thus prepared, it is a volatile, oily liquid, with a peculiar, penetrating odor. When heated to about 200° F., or when merely touched with a greasy substance, with phosphorus or an alkali, or even when subjected to the slightest friction or jarring, it explodes with a flash of light and a violence that is difficult to conceive of. Glass and cast-iron in proximity to it, are shattered into fragments, and a single drop has been known to cause a perforation through a thick plank. A leaden vessel yields to its effects, and is merely indented.

The chemical constitution of this body is not certainly known; neither are the principles involved in its remarkable reactions at all understood. Similar compounds of nitrogen may also be formed with iodine, bromine, and cyanogen.

370. **History of Bleaching.**—The past history and present condition of the great industrial art of bleaching appropriately connects itself with the subject of chlorine. Before the discovery and application of this element, the operation of bleaching cotton and linen consisted essentially in washing and boiling the fabrics in hot water, with soap and alkalies, and subsequently exposing them for a lengthened period on the grass to the action of light and air. These operations were successively repeated, and the time required to render a piece of linen white and suitable for market, varied from four to eight months. During the 16th and 17th centuries, the Dutch enjoyed an almost complete monopoly of the business, and almost all the linen goods manufactured in Europe were sent to Holland to be bleached. The Dutch affixed their imprint on all goods bleached by them, which were afterward known as "Hollands," a term applied to linens even at the present day.

This method of bleaching was extremely expensive, not only on account of the time and labor required in the operation, but also from the great extent of grass-land necessary for the spreading of the cloths. Goods thus exposed out-of-doors served as a temptation to dishonest persons, and the old statute laws of England abound in severe penalties against trespassers upon bleach-fields.

The decolorizing action observed to take place when organic products are exposed to the action of light, air, and moisture, is explained on the same general principles, as in the case of chlorine (§ 354); viz., the coloring compound is broken up by the abstraction or union of its hydrogen constituent with the oxygen of the air, or with the oxygen contained in dew and aqueous vapors, it being a fact that "grass-bleaching" is most rapid at those seasons and times when the deposit of dew is most copious and abundant. It is also probable that the ozone present in the atmosphere exerts some effect, and the chemical action of light is known to be essential, inasmuch as the bleaching will not take place in the dark.

Questions.—What are its characteristic properties? What was the original method of bleaching? What is said of the early history of bleaching?

One of the improvements in bleaching introduced by the Dutch was that of "souring," which consisted in steeping the goods for a considerable length of time in sour milk; but about the year 1750, very dilute sulphuric acid was substituted in its place. This simple change was a most important discovery, inasmuch as it shortened the time required for bleaching linen by nearly three months, and greatly reduced the expense. In fact, the operation of "souring" by sulphuric acid still forms an essential feature of the modern processes of bleaching.

In 1785, Berthollet, a French chemist, while repeating some experiments on chlorine, which had been discovered by Scheele in 1774, ascertained that a solution of this gas in water was capable of destroying vegetable colors, and he was hence led to suggest its application to bleaching. About this time Berthollet was visited by James Watt, of England, celebrated from his connection with the steam-engine, to whom he related the results of his experiments. Watt, on his return to England, practically examined the subject, and made a successful trial of bleaching with the new agent, at an establishment near Glasgow, Scotland. From thence its use rapidly extended throughout Great Britain.

The introduction of chlorine as a bleaching agent, like all other great discoveries which tend to overturn old practices, encountered a most strenuous opposition.

The first method of using it, consisted in saturating cold water with the gas, and immersing the goods to be bleached in the solution. Heat being applied, the chlorine was evolved from the water and acted upon the coloring matters. The difficulties which attended this procedure were, that the gas was evolved so abundantly, that the workmen were unable to endure it, and the strength of the cloth also was impaired. A defect of the goods, becoming yellow after some days, led to the operation of boiling in alkaline leys, when it was discovered that solutions of the alkalies not only absorb a greater quantity of chlorine than water, but hold it with greater affinity—not allowing the gas to escape and affect the atmosphere, but at the same time imparting it regularly and effectively to fabrics in contact with them. The knowledge of these facts prepared the way for the further discovery, in 1798, by Mr. Tennant, of Scotland, of the compound known as "chloride of lime," or "bleaching-powder," the manufacture of which has been already described (§ 365).* During all this period the constitution of the bleaching agent in

* Chlorine, in its combination with lime, is completely under the control of the manufacturer, and can be used with any amount of violence, so to speak, within the limits of its powers. It may be developed at once, if desired, or its evolution can be effected by the slowest degrees. It is possible to so dilute bleaching-powder with water, that it shall exercise no bleaching effect of itself, but this effect shall be developed by the disturbing action of a third substance. This may be illustrated by making an exceedingly dilute

QUESTIONS.—What improvement was introduced by the Dutch? What was the next advance in the art? What did Berthollet discover? What followed Berthollet's discovery? What was the first method of bleaching by chlorine? What difficulties were encountered? How were they overcome?

question was unknown, it being regarded as a compound containing oxygen, termed "oxy-muriatic acid;" and it was not until 1811 that Sir Humphrey Davy demonstrated its true elementary character, and called it chlorine.

Some idea of the wonderful results which have flowed from the discovery and practical application of chlorine may be formed from the following facts: bleaching operations, which less than one hundred years ago, required from four to eight months, are now accomplished in comparatively few hours; the quantity of cloth bleached by several of the large establishments of England and the United States ranges from twenty to fifty thousand yards per day; and it has been further estimated that all the *available* labor of the civilized world would at the present time be insufficient to supply, by the old process, the present demand and consumption of bleached cottons and linens.

The operations of a modern bleachery for cotton fabrics may be briefly described as follows:—"All cotton fibers are covered with a resinous substance, which, to a certain extent, prevents the absorption of moisture, and also with a yellow coloring matter, which, in some kinds of cotton, is so marked as to give a distinctive character to the fabric made from it, as in nankeen, which is manufactured in China from a native brown cotton. In some varieties of cotton the quantity of coloring mater is so small that the fabric would not require bleaching were it not for the impurities acquired in spinning and weaving."

The first process of bleaching is called "singeing," and consists in passing the cloth with great rapidity over a red-hot copper cylinder. This burns, or "singes" off the fibrous down or "nap" from the surface of the cloth, rendering it smooth and more suitable for the reception of colors, in subsequent operations of dyeing and calico printing.

After singeing the goods are placed in large hollow wheels, each of which has four compartments, with openings upon its face. (See Fig. 120). Water being admitted into the compartments by means of a pipe concentric with the

solution of chloride of lime in water; so dilute that a solution of indigo poured into it is not perceptibly decolorized. If we now add a third or disturbing agency, in the form of a few drops of acid of any kind, chlorine is liberated, and decoloration takes place. A very beautiful application of this property is made in calico-printing. Suppose it is desired to produce a white pattern on a colored ground—white dots or leaves, for example, on a field of bright red, the red being a color removable by the agency of chlorine—this result is effected by the following course of manipulation: The whole fabric is first dyed of a uniform color, and then the form of the desired pattern is compressed upon the cloth with stamps coated with some substance containing a very weak acid. An acid known as citric acid (a crystalline solid), mixed with gum, is generally used for this purpose. The fabric is now dried, and still exhibits an uniform color. No sooner, however, is it dipped in a weak solution of chloride of lime, than the citric acid sets up just that amount of local decomposition necessary to affect the liberation of the chlorine, which immediately bleaches out the stamped pattern, leaving the unstamped portions of the fabric unchanged. The material which thus effects the liberation of chlorine, is termed a *mordant*, and the operation is called "*mordanting*."—FARADAY.

QUESTIONS.—Enumerate some of the results which have followed the discovery and application of chlorine. What is the natural state of cotton fibers? What are the successive operations of a modern bleachery?

axis, the wheel is caused to rotate rapidly, and the cloth, by agitation and dashing of the water, is speedily and thoroughly washed.

FIG. 120.

The next operation consists in boiling the cloth in an alkaline solution, which removes all the greasy and resinous matters. This is effected in a peculiar manner; the cloth is placed in large vats, on a grating, or perforated false bottom, through which, from a compartment below, rises a pipe, furnished on its extremity with a curved iron cover. (See Fig. 121.) A boiling solution of alkali is forced, by steam pressure, from the compartment below the vat up through this pipe, and striking against the cover, is reflected upon the goods in the form of a shower; thence filtering through the texture of the cloth, the liquor runs back into the lower compartment, to be again heated by steam, and forced up as before. This process is continued for about seven hours, and at its conclusion the color of the cloth is darker than at the outset. The cloth is then washed again in the wheels, and next steeped in a very dilute solution of chloride of lime, in large vats, for about six hours; it even then is not white, but of a gray appearance.

In the next process, the goods are steeped for four hours in very dilute sulphuric acid, when a minute disengagement of chlorine takes place throughout the substance of the cloth, and it immediately assumes a bleached appearance. The same operations of washing, boiling, bleaching, and souring, are then successively repeated, in less time, until at length the cloth is perfectly whitened.

The length of time required for all these operations is from 24 to 48 hours; one parcel of goods succeeding another in each successive stage of the process, so that all the departments of a bleachery are in full operation at the same time. The labor of handling the cloth, which may seem very great, is nearly all performed by machinery, with great rapidity, at a very slight expense—the average cost of bleaching cotton fabrics not exceeding one cent

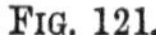
FIG. 121.

per yard. Cottons subjected to bleaching lose about 10 per cent. in weight. Wool is bleached by washing, and exposure to the vapor of burning sulphur.

SECTION VI.

IODINE.

Equivalent, 127. *Symbol*, I. *Specific gravity of vapor*, 8·7.

371. History.—Iodine was discovered in 1811 by M. Courtois, a chemical manufacturer of Paris.

He noticed that a dark-colored liquor, left after the preparation of soda from the ashes of sea-weeds, powerfully corroded his kettles, and that when sulphuric acid was added to the liquor, a brown substance separated, which on the application of heat was converted into a violet-colored vapor. A subsequent examination showed that the substance in question was a new element—Iodine.

***. **Natural History.**—Iodine is widely, but sparingly distributed in nature. In the inorganic kingdom it is a constituent of all sea-water, of many mineral springs (Saratoga, Carlsbad, etc.), and also of certain rare minerals. In the organic kingdom, it exists probably in all marine plants, but

---

QUESTIONS.—When and by whom was iodine discovered? What circumstances led to its discovery? What is said of its distribution in nature?

more abundantly in some species than in others; also in sponges, in the oyster and other sea-mollusks, and in some fishes (*i. e.*, cod-liver oil). It is always found in combination with other substances—generally as iodide of sodium, or magnesium.

372. **Preparation.**—The greater part of the iodine of commerce is manufactured at Glasgow, from "kelp," which is the ashes of sea-weeds collected and burned upon the coasts of Scotland and Ireland. The kelp is treated with water, which dissolves out a large quantity of soluble saline matters—carbonate of soda, common salt, chloride of magnesium, etc. When these substances are separated from the solution by partial evaporation and crystallization, there remains behind a dark-colored liquor, which contains iodine. This is heated with sulphuric acid and peroxyd of manganese, when the iodine distils over as a purple vapor, which is collected in receivers and condensed to a solid by cooling. A ton of kelp yields 9 pounds of iodine. The chief uses of iodine are for medicine, photography, and to some extent in dyeing.

373. **Properties.**—Iodine, at ordinary temperatures and pressures, is a solid, and is generally obtained by crystallization in the form of bluish-black scales, which possess a brilliant and somewhat metallic luster. Exposed to heat, it liquefies at 225° F., and boils at 350°, forming a magnificent purple vapor, from whence it derives its name (ιωδης, *violet-colored*). This property may be beautifully illustrated by heating a few grains of iodine in a glass flask, or test tube, over a spirit-lamp. (See Fig. 122.) On withdrawing the heat the vapor condenses, and forms brilliant crystals of solid iodine upon the sides of the flask.

FIG. 122.

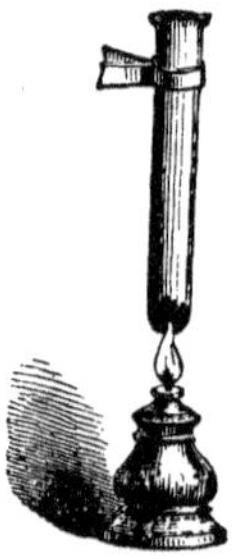

Dropped on a red hot surface iodine melts, and as a liquid assumes the spheroidal state (§ 154), forming a splendid experiment.

Iodine, when taken inwardly, acts in large doses as an irritant poison; but in small quantities it is a most valuable medicine. Long before its discovery, the ashes of a burnt sponge were often prescribed as a most efficacious remedy for certain diseases. Their effect is now known to have been due to the iodine contained in them. Iodine stains the skin and most organized substances of a brown color, and gradually corrodes them. Water forms with it a yellow solution, but dissolves it only in very small quantity—1 part in 7,000. Its bleaching properties are very feeble. Alcohol, ether, and solutions of the salts of iodine, dissolve it freely.

374. Iodine attacks the metals rapidly. Iron or zinc placed in water with it are readily dissolved, an iodide of the metal being formed. Some of the combinations of iodine with the metals are remarkable for their brilliant

---

QUESTIONS.—How is it prepared? What are its properties? From what circumstance does it derive its name? What are the effects of iodine upon the animal system? What is said of its combinations with the metals?

colors. An illustration of this forms an easy and striking experiment. Place in three test tubes a solution of iodide of potassium in water; if we add to the first a few drops of a solution of mercury (corrosive sublimate), we obtain a beautiful salmon-colored precipitate, which almost immediately changes to scarlet. A solution of sugar of lead added to the second, produces a bright yellow precipitate; and a solution of subnitrate of mercury added to the third, a green precipitate.

The distinctive test for iodine is a solution of starch, with which it strikes a deep blue color. The solution must, however, be cold, and no alkali must be present. This may be illustrated by experiment as follows:—Draw or paint upon a sheet of paper, figures in starch paste, and expose the paper to the vapor arising from iodine thrown upon a hot surface. The figures, which were before colorless, immediately become blue. If a little of the tincture of iodine be dropped upon flour, potatoes, etc., the presence of starch in these bodies will be indicated.

Iodine unites with hydrogen to form an acid, hydriodic acid, HI, and with oxygen, in several proportions, to form both oxyds and acids. Its principal oxygen compound is iodic acid, IO_5.

The most important compound of iodine is that which results from its union with potassium, forming a white crystallizable salt, the iodide of potassium, also termed the "hydriodate of potash" (KI). It is in this condition that iodine is chiefly used in medicine, and also in photographic operations.

SECTION VII.

BROMINE.

Equivalent, 80. *Symbol*, Br. *Specific gravity of vapor*, 5·3.

375. **History**—Bromine was discovered by M. Ballard, a French chemist, in 1826, in the "mother," or residual liquor left after the crystallization and separation of the salts of sea-water.

376. **Distribution**.—It exists in all sea-water in minute quantity, generally in the proportion of about one grain to the gallon; and for the most part in combination with magnesium, forming bromide of magnesium. It is also found in certain mineral springs, and in a few minerals.

377. **Preparation**.—Bromine is prepared by passing into the mother liquor of sea-water a stream of chlorine gas, and then agitating the liquor with ether. The chlorine sets the bromine free from its combinations, and the ether dissolves it. After standing for a little time, the etherial solution, having a fine red color, separates and floats at the top.

378. **Properties**.—Bromine, at ordinary temperatures and pressures,

---

QUESTIONS.—What is the distinctive test of iodine? What is the principal salt of iodine? When and by whom was bromine discovered? What is said of its distribution in nature? How is bromine obtained? What are its properties?

is a red liquid, so deep in color as to be nearly opaque. It is extremely volatile, and can only be preserved in very close vessels. A few drops slightly warmed in a glass flask, will fill the vessel with blood-red vapors. Its odor is somewhat like chlorine, but more offensive; hence the name Βρωμος, *bad odor.* When swallowed, it acts as a deadly poison, and a single drop upon the beak of a bird is said to occasion instant death. It rapidly destroys all organic tissues, and stains the skin permanently yellow. It boils at 145° F., and when exposed to a cold of —10° F., freezes into a crystalline solid. Bromine bleaches like chlorine, is slightly soluble in water, but dissolves freely in alcohol and ether. It combines directly with many of the metals, and forms bromides—the act of combination being often accompanied with an explosive evolution of light and heat. This may be experimentally shown by cautiously pouring a small quantity of powdered antimony or tin upon a few drops of bromine contained in a deep strong glass. In short, the properties of bromine greatly resemble those of chlorine, but they are less strongly developed.

Bromine is extensively used in photographic processes, and sometimes in medicine.

But one compound of bromine and oxygen is known, viz., bromic acid, BrO_5; it also unites with hydrogen to form an acid, hydrobromic acid, HBr.

SECTION VIII.

FLUORINE.

Equivalent, 19. *Symbol*, F. *Theoretical Density*, 1·31.

379. History.—Of this element but little is known except from its compounds.

Its affinities for the other elements are so powerful, and its action on the human system is so deleterious, that its isolation has been regarded as almost impossible. Within a comparatively recent period, however, several chemists have succeeded in separating it from all other bodies, in the form of a colorless gas. In its general nature and properties it undoubtedly resembles, and is closely allied to, chlorine, bromine, and iodine.

The only compound in which it exists in nature in any abundance, is a compound of lime, called fluor-spar, or fluoride of calcium. This mineral is found, in great quantity and beauty, in Derbyshire, England, and from it the well-known ornaments known as "Derbyshire spar," are manufactured. Fluorine is also found in a great variety of other minerals, and exists in minute quantities in the bones of animals, and in the enamel of the teeth.

Compounds containing fluorine can be decomposed without difficulty, and the fluorine transferred from one body to another; but so great is its affinity for the metals, and for silicon, a constituent of glass, that in passing out from

QUESTIONS.—How does bromine act upon the metals? What are its uses? What its compounds? What is known of fluorine? What is said of its distribution in nature? Why is it difficult to isolate fluorine?

a state of combination, it combines again immediately with the material of the vessel containing it.

379. Hydrofluoric Acid, HF.—Fluorine is not known to unite with oxygen under any circumstances, but with hydrogen it forms a very remarkable compound, "hydrofluoric acid."

This substance is formed by heating powdered fluor-spar with strong sulphuric acid, in a platinum or lead retort, furnished with a receiver of the same metal, which is kept cool by immersion in a freezing mixture. The chemical reaction which takes place may be expressed as follows:—

$$\overbrace{CaF}^{\text{Fluoride of calcium.}} + \overbrace{SO_3\,HO}^{\text{Sulphuric Acid.}} = \overbrace{CaO,\,SO_3}^{\text{Sulp. lime.}} + \overbrace{HF}^{\text{Hydrofluoric acid.}}$$

The acid thus obtained is a gas at ordinary temperatures, but is condensible by cold into a volatile, colorless liquid, which evolves white, suffocating fumes on exposure to the air; its attraction for water is very great, and when poured into it, it hisses like a red-hot iron. As vapor, and as an aqueous solution, it attacks and readily dissolves glass, and all compounds containing silica, together with some mineral substances that no other acid can affect. This property is often made available for etching upon glass.

In its most concentrated form, hydrofluoric acid is a most dangerous substance, and is more destructive of animal tissues than any other known agent. The most minute drop upon the skin occasions a deep and painful burn, often terminating in an ulcer difficult to cure. Its vapor is also in the highest degree corrosive.

The peculiar action of hydrofluoric acid vapor upon glass may be easily illustrated without danger, by the following experiment. Place in a small leaden dish, or an earthen cup, the interior of which has been slightly oiled, a little powdered fluor-spar, and add strong sulphuric acid, sufficient to form with it a thin paste. Cover the cup with a piece of window-glass which has received a coating of wax, and from some parts of which the wax has been removed, by scratching with a needle or other pointed instrument. (See Fig. 122.) After the lapse of some hours, remove the wax by melting and washing with oil of turpentine, when those parts of the glass left bare will be found to be deeply corroded. The same result can also be obtained in the course of a few minutes, by a gentle application of heat to the cup containing the mixture.

Fig. 122.

QUESTIONS.—What is its most remarkable compound? How is it prepared? What are its properties? How does it act upon organic substances? How may its action on glass be illustrated?

SECTION IX.

SULPHUR.

Equivalent, 16. *Symbol*, S. *Specific gravity*, 1·98, *in vapor*, 6.65.

380. **Natural History and Distribution.**—Sulphur is an element abundantly distributed in nature, most extensively as a mineral product, but widely and in small quantities as a constituent of animals and vegetables. It has been known from the most remote antiquity.

Sulphur is found in a native, or uncombined state, in all volcanic districts; and in Sicily and in some parts of South America, it exists in immense beds in the earth. Many of the compounds of sulphur with the metals occur as natural productions in great abundance, especially the sulphurets (sulphides) of iron, copper, lead, and zinc. The sulphuret of iron (iron-pyrites) is even employed as a source of sulphur. In an oxydized condition, as sulphuric acid, it is still more widely diffused in combination with various earths, as the sulphates of lime, magnesia, baryta, etc. Nearly one half the weight of sulphate of lime (gypsum, or plaster of Paris) is sulphur.

381. Most of the sulphur used in the arts is obtained from Sicily and the volcanic districts of southern Italy, the former exporting about 1,540,000 cwts. yearly. It is generally subjected, on the spot where it is dug from the earth, to a rough purification by fusion, and is brought into commerce in the form of amorphous, or semi-crystalline masses. Another commercial form is roll sulphur, or brimstone, which is generally the produce of roasting the sulphurets of iron and copper (pyrites), collecting the evolved fumes in condensing chambers, and subsequently fusing the sulphur into sticks. "Flowers of sulphur," a powder, is a third commercial state which this element is made to assume; and is produced by distilling sulphur and condensing the vapor.

382. **Properties.**—Sulphur in its ordinary condition is a yellow, brittle solid, which, by warmth and friction, emits a characteristic odor (brimstone odor). It is insoluble in water, and consequently tasteless; it is very slightly soluble in alcohol and ether; more so in oil of turpentine and some other oils; and readily in the bisulphide of carbon. It is a bad conductor of heat; and a roll of sulphur, when grasped by the warm hand, crackles and frequently falls in pieces from unequal expansion. It is a non-conductor of electricity, but when rubbed develops negative electricity abundantly.

Sulphur is highly inflammable, burning with a blue flame, and emitting suffocating fumes of sulphurous acid, (the familiar odor of a match). It has a powerful affinity for most of the other elements, and its act of combination

---

QUESTIONS.—What is the history of sulphur? What is said of its distribution in nature? From whence are supplies of sulphur chiefly derived? What are its commercial forms? What are the properties of sulphur? What is said of its solubility? What of its affinity for other elements?

with the metals to form sulphides, or sulphurets, is often attended with an evolution of light and heat. This fact may be experimentally illustrated by placing in a flask a few fragments of sulphur, and above them some copper turnings; on the application of heat from a spirit-lamp, vapor of sulphur rises, and coming in contact with the copper, enters into vivid combination with it.

383. Allotropism of Sulphur.—One of the most remarkable characteristics of sulphur is its allotropism, or power of existence in different states.

The first indication of this power is perhaps to be found in the fact, that it is capable of assuming two distinct crystalline forms. These are not merely modifications of one original primary figure (to which cause most crystalline variations can be referred), but they belong to two different, inconvertible, and incompatible systems of crystallization, viz., oblique rhombic prisms and right rectangular prisms. Examples of the first form, Fig. 123, (octohedrons derived from oblique rhombic prisms), occur in native sulphur, or in sulphur crystallized from a solution. Examples of the second form may be obtained by melting a quantity of sulphur in an earthen crucible, and allowing it to solidify on the surface; if the crust be then pierced with a hot wire, the fluid portion beneath may be poured off, when the interior of the crucible, on cooling, will be found to be lined with slender needles, or right rectangular prisms. (See Fig. 124.)

FIG. 123.

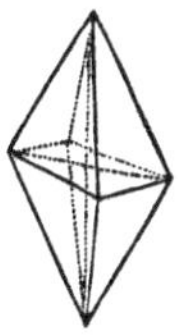

FIG. 124.

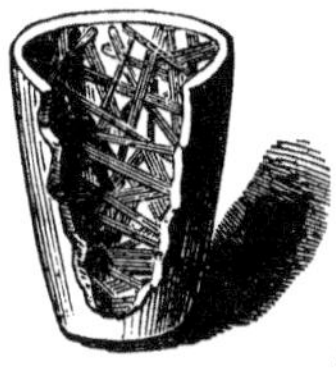

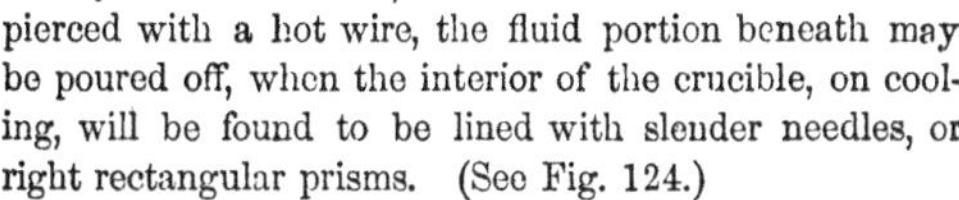

Both forms of crystals may be obtained by dissolving sulphur in boiling oil of turpentine; as the solution cools, the sulphur crystallizes out, first in the form of prisms; but afterward, as the temperature is reduced, octohedrons are formed.

The power possessed by sulphur of manifesting itself under two conditions, is, however, most strikingly illustrated by certain phenomena of its melting and subsequent cooling. Thus, if we heat a small quantity of sulphur in a glass flask over a spirit-lamp, it melts at a temperature of 250–280° F., into a clear, yellow liquid. If a portion of this liquid be poured into cold water, it immediately condenses into the state it had before melting—that is, into common, yellow, brittle sulphur. If to the portion remaining in the flask a stronger heat be applied (about 500° F.), the transparent fluid gradually thickens, becomes brown at first, and at last nearly black and opaque; in this condition the viscidity of the sulphur is such, that the flask may be inverted without escape of its contents. If the heat be still further increased, the black, tenacious sulphur once more liquefies, though it never be-

---

QUESTIONS.—What is said of the allotropism of sulphur? What is the first indication of this property? In what two forms does sulphur crystallize? What are examples? In what other way may the allotropic properties of sulphur be illustrated?

comes as fluid as when first melted, at the temperature of 226° F., and if suddenly cooled, by pouring it in a slender stream into cold water, it assumes a most singular state. It is no longer yellow and brittle, like ordinary sulphur, or like the product of pouring into water the first result of fusion, but it remains soft, tenacious, highly elastic, and of a brown color, resembling, in all its external characteristics, strips of India rubber or gutta percha. In this form it can be molded by the hand, and may be used to take impressions of seals, medallions, etc. After the lapse of a little time, it again becomes yellow, and returns to its original brittle condition, giving out in the transformation a quantity of latent heat.

384. **Milk of Sulphur.**—If we add to a strong boiling solution of potash or soda, a little of the flowers of sulphur, a part of the sulphur dissolves, and imparts to the liquor a yellowish-brown color. If a little of the clear solution be added to water, slightly acidulated, the acid will unite with the alkali holding the sulphur in solution, and cause the sulphur to be precipitated in the form of exceedingly minute particles, giving to the water a milky appearance. Sulphur in this form is nearly white in appearance, and is known as "Milk of Sulphur," or "Precipitated Sulphur."

In the organic kingdom sulphur is extensively, and perhaps universally diffused throughout animal substances, and exists in small quantities in most vegetables. The well-known blackening of a silver spoon immersed for some time in a boiled egg, is due to the presence of sulphur in the egg. The presence of sulphur also in a piece of flannel may be strikingly demonstrated by immersing the cloth in a mixture of oxyd of lead in a solution of potash; on applying heat, the flannel immediately turns black.

385. **Compounds of Sulphur and Oxygen.**—The compounds of sulphur with oxygen are numerous, but only two of them demand an extensive notice; these are Sulphurous acid, SO_2, formed by the union of one equivalent of sulphur with two of oxygen; and Sulphuric acid, SO_3, formed by the union of one of sulphur and three of oxygen.

Fig. 125.

386. **Sulphurous Acid,** SO_2 is formed when sulphur is burned in oxygen (See Fig. 125) or atmospheric air; and is the occasion of the well-known suffocating odor of an ignited match. It exists in nature in the vicinity of volcanoes, and is often evolved in immense quantities from their craters.

QUESTIONS.—What is milk of sulphur? What is said of sulphur in the organic kingdom of nature? What are illustrations of its presence in animal substances? What is said of the compounds of sulphur with oxygen? What of sulphurous acid?

When required in a pure state, it is best prepared by depriving oil of vitriol of a part of its oxygen. In order to effect this, two or three ounces of concentrated sulphuric acid are boiled in a glass retort or flask, with a half an ounce of copper turnings; pieces of charcoal may be substituted in place of the copper, but the gas evolved under such circumstances is not pure. In this process, a part of the acid gives up one equivalent of its oxygen to the metal, and sulphurous acid gas is liberated; the oxyd of the metal produced, unites with a portion of undecomposed acid to form a sulphate. Thus:—

FIG. 126.

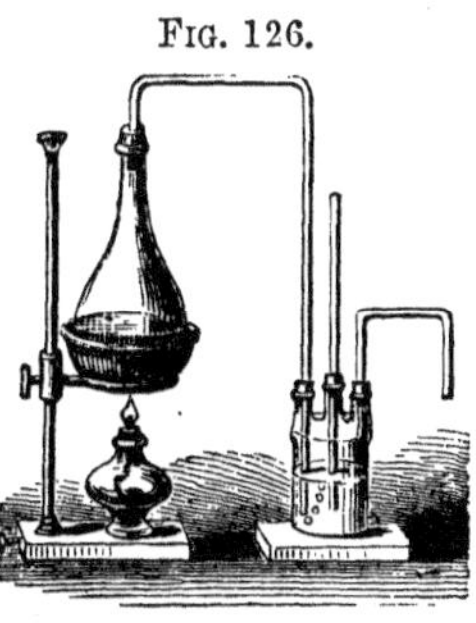

$$\underbrace{Cu}_{\text{Copper.}} + \underbrace{2SO_3}_{\text{Sulph. acid.}} = \underbrace{CuO, SO_3}_{\text{Sulph. copper.}} + \underbrace{SO_2}_{\text{Sulphurous acid.}}$$

By allowing the gas to bubble through water, a strong solution will be obtained, which may be used for illustrating the properties of sulphurous acid.

387. **Properties.**—Sulphurous acid is a colorless gas, with a characteristic odor, easily condensible by cold or by pressure, into a colorless, limpid liquid. Water, at 60° F., absorbs from 40 to 50 times its volume of sulphurous acid, and forms thereby a strongly acid liquid. Hence it is necessary to collect this gas over mercury or by the displacement of air from dry vessels. Its avidity for water is so great, that a piece of ice introduced into a jar of it, is instantly liquefied.

Sulphurous acid is not inflammable, and a lighted candle immersed in a jar of the gas, is immediately extinguished for the want of free oxygen. A most certain way of extinguishing a chimney on fire is to scatter flowers of sulphur on a pan of coals in a fireplace-opening beneath. The sulphurous acid gas formed by the combustion of the sulphur, ascends the flue, expels the atmospheric air present in it, and by depriving the burning soot of free oxygen, extinguishes it.

Sulphurous acid possesses bleaching properties, and is extensively employed in bleaching straw and wool. The articles are moistened and suspended in closed chambers in which sulphur is burned in an open dish; (an inverted barrel is often made to subserve the purpose of a bleaching chamber.) The sulphurous acid is absorbed by the damp goods, and discharges their color. The bleaching action appears to be due to the fact, that the gas unites with the coloring matters to form colorless compounds. It does not, like chlorine, decompose and destroy the coloring matter, since by the action of a stronger chemical agent, the colorless compound may be broken up and the original

---

QUESTIONS.—How is it usually prepared? Give the chemical reaction involved in its preparation. What are its properties? What are its relations to combustion? How is sulphurous acid employed in bleaching? What is the nature of its bleaching action?

color restored. This may be illustrated by holding a red rose, or any other red flower, over a bit of burning sulphur. The color is speedily discharged, but may be again restored by washing with dilute sulphuric acid. White flannel which has been bleached by sulphurous acid, when washed for the first time with an alkaline soap, has its original yellow color in part restored to it.

Sulphurous acid is also valuable as a disinfecting agent.

The compounds of sulphurous acid with the bases are termed sulphites. They are readily formed by transmitting a stream of gas through water in which the oxyd or the carbonate of the metal is dissolved or suspended, the carbonates being decomposed with effervescence. The sulphite of soda is known in commerce as *antichlorine;* since its solution in water is able to neutralize the chlorine which may remain in fabrics after bleaching, and thus prevents its destructive action.

388. **Sulphuric Acid,** SO_3.—This acid is one of the most important of all chemical reagents, and furnishes the means by which most other acids are prepared. Immense quantities of it are consumed in the manufacture of carbonate of soda, nitric and hydrochloric acids, chlorine, alum, sulphate of copper (blue vitriol), stearine, phosphorus, etc., and in dyeing, and in the refining of the precious metals. Its annual consumption in Great Britain alone is upward of twenty millions of pounds.

389. **Preparation.**—It has been already stated, that when sulphur is burned in air, or oxygen, the product is sulphurous acid. This gas, if made to combine with half as much oxygen again as it already contains, is converted into sulphuric acid; thus $SO_2+O=SO_3$. In other words, sulphurous acid must be oxydized in order to enable us to form sulphuric acid. Oxygen and sulphurous acid can not, however, be made to unite directly, but the intervention of some third substance is necessary. In the presence of water, the union takes place slowly, or if the two gases be mixed, and passed over spongy platinum, the union is effected immediately.

Neither of these processes can, however, be used with advantage in the arts; and the manufacture of sulphuric acid upon a large scale depends upon the fact, that when sulphurous acid mixed with oxygen is brought in contact with deutoxyd of nitrogen (NO_2), or any of the other higher oxyds of nitrogen, combination takes place with great rapidity; the presence of a very small proportion of deutoxyd of nitrogen being moreover sufficient to effect the combination of an almost indefinite amount of sulphurous acid and oxygen, provided that water is also present.

The following experiment will serve to illustrate the general principle upon which sulphuric acid is manufactured. Burn in a jar, containing a little water at the bottom, a piece of sulphur; as a consequence, the vessel becomes filled with sulphurous acid. If we now introduce into the gas a shaving moist-

QUESTIONS.—What experiments are illustrative? What is said of the compounds of sulphurous acid with the bases? What is antichlorine? What is said of sulphuric acid? What of its theoretical preparation? Upon what fact does its practical preparation depend? How may it be experimentally illustrated?

ened with nitric acid, reddish-colored fumes will immediately form around the wood, and gradually fill the whole vessel. (See Fig. 127.) These fumes are nitrous acid, and are produced by the action of the sulphurous acid, which decomposes the nitric acid, and by depriving it of 2 equivalents of oxygen, becomes sulphuric acid. Thus:

FIG. 127.

$$\underbrace{2SO_2}_{\text{Sulphurous acid.}} + \underbrace{NO_5}_{\text{Nitric acid.}} = \underbrace{2SO_3}_{\text{Sulphuric acid.}} + \underbrace{NO_3}_{\text{Nitrous acid.}}$$

The vapor of the sulphuric acid formed is absorbed by the water in the jar, and by repeating the experiment several times, a quantity of dilute sulphuric acid may be prepared.

On a large scale, the operation of manufacturing sulphuric acid is essentially the same in principle, and may be described as follows: immense chambers, lined with lead, are constructed; in some instances 300 feet long, 15 feet high, and 20 broad. (See Fig. 128.) The floor of these chambers is covered to the depth of a few inches with water, and at one extremity there is admitted by a suitable flue, B, sulphurous acid (from a furnace of burning sulphur), with atmospheric air; by another pipe, A, steam; and by a third, C, vapors of nitric acid (obtained by heating nitrate of soda with strong sulphuric acid). When these several substances meet within the chambers a most interesting and curious series of reactions take place;—the sulphurous acid withdraws oxygen from the nitric acid vapor, NO^5, and converts it into deutoxyd of nitrogen, NO_2, itself changing into sulphuric acid, SO_3. This last product then uniting with the steam, is precipitated to the bottom of the chamber, and is absorbed by the water. The deutoxyd of nitrogen does not remain unaltered, but in contact with the air admitted into the chambers, absorbs two equivalents of oxygen, and becomes converted into peroxyd of nitrogen, NO_4, forming red fumes (§ 347). These in turn, by contact with the sulphurous acid, give up their newly-acquired oxygen to form sulphuric acid, and are reconverted again thereby into deutoxyd of nitrogen. And this process is repeated over and over again, a small quantity of deutoxyd of nitrogen acting as the intermediate agent for withdrawing oxygen from the air, first to itself, and afterward giving it up to oxydate the sulphurous acid. The deutoxyd of nitrogen, together with the remaining nitrogen of the air, is finally allowed to escape at the further extremity of the chambers, and a fresh portion of nitric acid vapor is admitted to supply its place, and commence the reactions anew. The steam admitted into the chambers does not take any active part, but its presence is essential to the success of the opera-

FIG. 128.

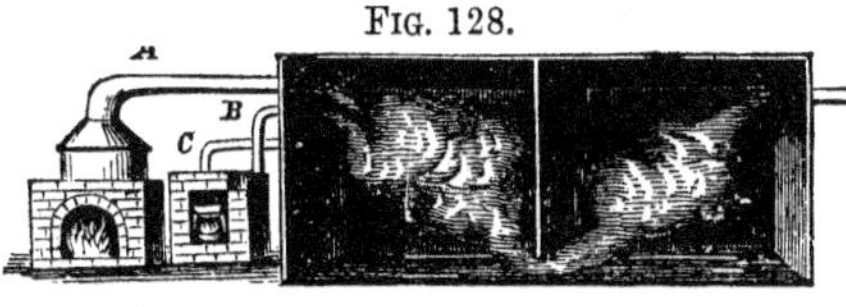

QUESTIONS.—How is the practical manufacture conducted? What reactions take place in the leaden chambers?

tion.* The chambers in which the acid is manufactured are usually divided into partitions, in order that the gases may mix together slowly and completely, before reaching an exit tube placed at the further extremity.

The sulphuric acid which collects in the water at the bottom of the chambers, is drawn off when it reaches a specific gravity of about 1·5; it is, however, in this state too dilute for sale, and is accordingly evaporated by heat in shallow lead pans, until it becomes strong enough to corrode the lead, when it is transferred into glass or platinum retorts,† and further heated until it attains a specific gravity of 1·84. In this condition it constitutes the concentrated oil of vitriol of commerce, and is transported in carboys, or large glass bottles packed in boxes. As thus produced, it is a definite hydrate, composed of 1 equivalent of acid, and 1 of water (SO_3, HO). This proportion of water, amounting to three ounces in every pound of acid, is held so firmly that it can not be driven off by heat. (See § 322.)

390. **Nordhausen Sulphuric Acid.**—In early times sulphuric acid was obtained by distilling dry sulphate of iron (green vitriol) in earthen retorts, at a high temperature. As thus prepared, it is a dark-brown, thick, oily liquid, and was originally called, from its derivation, "*oil of vitriol.*" It is the most concentrated form in which sulphuric acid can exist in a fluid condition, and contains less water than the ordinary concentrated sulphuric acid. When exposed to the air it fumes, and when dropped into water, hisses like a red hot iron. As acid in this state of concentration is required for certain processes in the arts, it is still prepared in the old way, especially at the town of Nordhausen, in Saxony, Germany;—hence its commercial name.

Sulphuric acid is known to combine with water in four proportions, forming four definite hydrates. Their composition may be illustrated as follows:—

Nordhausen acid, sp. gr.		1·9	$2(SO_3)HO$
Oil of vitriol,	"	1·84	SO_3,HO
Sulphuric acid of	"	1·78	SO_3,HO + HO
" "	"	1.63	SO_3,HO + 2HO (§ 274.)

* The description of the chemical changes involved in the manufacture of sulphuric acid in the leaden chambers, as thus given, is but an outline, embracing merely the fundamental principles. For the minute details, not suited for an elementary work, the student is referred to any of the modern encyclopedias of practical science.

† It was originally the custom to concentrate the sulphuric acid by boiling it in glass vessels, but the loss from breakage is so great, that in many manufacturing establishments platinum stills have been adopted, this metal resisting the action of the strongest acid at high temperatures. These stills are constructed in Paris of thin sheets of platinum soldered with gold. They are oval in form; and as a protection against the direct action of heat, they are inclosed in iron jackets. Their capacity varies from 500 to 2,000 pounds, and their cost from $8,000 to $13,000 apiece; and although one of these vessels only endures for a period of two or three years, their use has proved more economical than glass.

QUESTIONS.—What is the density of the acid thus formed? To what processes is it subjected? What is the composition of concentrated oil of vitriol? What is Nordhausen sulphuric acid? What are its properties? How many hydrates of sulphuric acid are known?

391. **Anhydrous Sulphuric Acid.**—When Nordhausen acid is carefully distilled in a retort furnished with a receiver kept cool by a freezing mixture, white fumes pass over, which may be condensed into a white, silky-looking, fibrous mass—anhydrous sulphuric acid. This substance possesses no acid properties, and may be handled without danger. When thrown into water, it hisses, and forms liquid sulphuric acid. It also liquefies on exposure to air, by the absorption of moisture.

392. **Properties.**—The oil of vitriol of commerce is a dense, oily-looking liquid, without odor, and of a brownish color. It is the strongest of all acids. It freezes at a temperature of —29° F., and boils at 620° F. Its affinity for moisture is most intense, and it abstracts it from every substance with which it is brought in contact. If a quantity of strong sulphuric acid be exposed in a shallow dish to the air, it frequently absorbs sufficient aqueous vapor from the atmosphere to double its weight. A piece of wood introduced into sulphuric acid becomes black and reduced to coal, the same as if it had been exposed to the action of fire. The explanation of this is as follows: the wood is a compound of oxygen, hydrogen, and carbon; the sulphuric acid abstracts the oxygen and hydrogen, which combine to form water, while the carbon remains behind. Gases containing aqueous vapor are deprived of it by causing them to bubble through strong sulphuric acid.

When concentrated sulphuric acid is mixed with water, great heat is evolved, and the mixture, when cold, occupies less bulk than the two liquids did separately. This fact may be strikingly illustrated by mixing 4 parts of oil of vitriol with 1 of water. Water in a test tube immersed in such a solution, may be caused to boil. (See Fig. 129.)

FIG. 129.

Sulphuric acid does not evaporate at the ordinary temperature of the air; but if a drop of dilute acid fall upon a cloth, the water gradually evaporates until the acid which is left behind acquires a considerable degree of strength, and then chars or destroys the cohesion of the fibers; hence the destructive action of sulphuric acid upon fabrics even when very much diluted.—MILLER.

Ordinary sulphuric acid is never pure, but always contains lead derived from the leaden chambers; when mixed with water, this lead is precipitated, and causes the solution to appear milky.

Sulphuric acid attacks all the metals except gold, platinum, iridium, and rodium.

393. **Hyposulphurous Acid,** $S_2 O_2$.—By digesting sulphur with

---

QUESTIONS.—What is anhydrous sulphuric acid? What are its properties? What are the properties of "oil of vitriol?" What is said of its attraction for moisture? What are illustrations of this? When concentrated sulphuric acid is mixed with water, what follows? What is said of the action of sulphuric acid on fibers? What of its purity? What of its action on metals? What is said of hyposulphurous acid and its compounds?

a solution of sulphate of soda, a portion of the sulphur is dissolved, and a salt containing hyposulphurous acid is formed—the hyposulphite of soda. The acid itself can not be isolated. Hyposulphite of soda is at present largely employed in photographic operations, owing to its property of dissolving certain salts of silver which are insoluble in water. The surface of the photograph is freed from them by immersion in a solution of it; after which, if well washed with water, it is no longer liable to alteration by exposure to light.

394. Sulphur and Hydrogen.

Hydrosulphuric Acid, HS.—*Sulphuretted Hydrogen, Sulphydric Acid.*—This gas is formed naturally during the putrefaction of many organic substances, and is also a constituent of many mineral springs. It is easily prepared by the action of dilute sulphuric acid upon protosulphide of iron, FeS.*

Fig. 130.

For this purpose an evolution flask (Fig. 130) is best adapted; but a common, open-mouthed bottle, fitted with a perforated cork and bent tube, will answer. (See Fig. 131.) Introduce into the flask protosulphide of iron in small quantities, with water sufficient to cover it; then add sulphuric acid until a copious disengagement of gas takes place. By introducing the evolution tube into cold water, a solution of the gas will be obtained, in which state its properties may be experimentally illustrated to the best advantage. The operation of preparing the gas should be conducted in a well-ventilated apartment, or in the open air.

Fig. 131.

The chemical reaction involved in this operation is as follows: water is decomposed; its oxygen uniting with the iron to form oxyd of iron, which dissolves in the acid to form sulphate of iron, while the hydrogen escapes, and takes with it the sulphur contained in the sulphide of iron. Thus:—

$$\underbrace{FeS}_{\text{Sulphide of iron.}} + \underbrace{SO_3,\ HO}_{\text{Sulphuric acid (dilute).}} = \underbrace{FeO,\ SO_3}_{\text{Sulphate of iron.}} + \underbrace{HS.}_{\text{Hydrosulph. acid.}}$$

395. **Properties.**—Hydrosulphuric acid is a transparent, colorless gas, of a disgusting odor, like that of rotten eggs. It is about one fifth heavier than common air, and burns with a blue flame, with a smell of sulphur. It is highly poisonous when respired in a concentrated form, and even when

* Protosulphuret of iron is prepared by heating 2 parts of iron filings with 1¼ parts of sulphur, to a red heat, in a covered earthen crucible.

Questions.—What of hydrosulphuric acid? How is it prepared? What chemical reactions are involved in its preparation? What is said of its properties? What of its poisonous effects?

present in the air in very minute proportions, it is rapidly fatal to the lower orders of animals. A single gallon of it, mixed with 1,200 of air, will render it poisonous to birds, and 1 in 100 will kill a dog. When inhaled it acts directly upon the blood, thickening it, and turning it black. It is this gas which makes an open or foul sewer so destructive of health to any district in which it may be situated. When present in the air of a room, it may be instantaneously destroyed by the action of a small quantity of free chlorine. A cloth moistened with alcohol, and held before the mouth, is a good protection also against its inhalation.

By pressure, sulphuretted hydrogen is reduced to a colorless liquid, which freezes at —122° F. into a crystalline, semi-transparent mass. Cold water dissolves between two and three times its bulk of this gas, producing a feebly acid liquid, which possesses the characteristic smell and taste of sulphuretted hydrogen, with all its properties. When exposed to the air, this solution becomes milky; the hydrogen being slowly oxydized to form water, while the sulphur separates. The solution, therefore, should be kept in well-stopped bottles, quite full.

Sulphuretted hydrogen is formed naturally under a variety of circumstances. Its chemical proportions being 1 equivalent of hydrogen (1) to 1 of sulphur (16), it follows that 100 parts of the gas contain only about 6 parts of hydrogen; so that a very small proportion of hydrogen causes a large amount of sulphur to assume with it an aeriform condition, and exhibit the fœtid odor and poisonous properties of the gas in question. In volcanic countries sulphuretted hydrogen is often evolved from fissures in the rocks, mixed with steam and other gases; in sewers and cesspools it is produced in large quantities by the decay of organic matter, and in marshes, where vegetable matter alone is undergoing decay, in the presence of water containing sulphate of lime (gypsum), its presence may be often detected. The waters of mineral springs, as those of Avon and Sharon, N. Y., and the sulphur springs of Virginia, often contain sulphuretted hydrogen, though rarely in a proportion exceeding 1½ per cent. of their volume; and the gas in solution in this small quantity, when taken into the stomach, acts as a valuable medicinal remedy for various diseases.

Hydrosulphuric acid, though a feeble acid, combines readily with bases to form sulphides, or sulphurets. Thus, if we place a drop of sulphuretted hydrogen water upon a bright silver or copper coin, or upon a piece of lead, a black spot will be quickly produced, owing to the formation of a black compound of the metal and sulphur (a sulphide). The black sulphide of lead formed when hydrosulphuric acid is brought in contact with the salts of lead, is particularly noticeable, and may be exhibited by exposing a piece of paper moistened with acetate of lead to air impregnated with this gas. This test is so delicate, that 1 part of sulphuretted hydrogen in 20,000 of air is said to

---

Questions.—What is said of the solubility of this gas? What of its natural formation and proportional composition? What is said of its presence in mineral springs? What of its combinations with the metals?

be sufficient to occasion a blackening of the paper. For the same reason, surfaces covered with lead paints, in the vicinity of sewers, cesspools, or the bilge-water of vessels, etc., soon become discolored. Sulphur unites with zinc in the same manner as with lead, but the resulting compound, sulphide of zinc, is white, and not dark colored like the sulphide of lead. Hence zinc paints, for many locations, are more suitable than lead paints.

When hydrosulphuric acid, either in the form of gas or solution, is added to a solution containing copper, silver, gold, lead, tin, antimony, or arsenic, these metals are precipitated as insoluble sulphides, and may be collected and separated from the solution by filtration. If iron, zinc, manganese, cobalt, and nickel are contained in the same solution, they are not precipitated until a stronger reagent is added. Hence sulphuretted hydrogen may be used to separate one class of metals from another; and in fact is employed extensively for this purpose in chemical analysis.

SECTION X.

SELENIUM AND TELLURIUM.

396. **Selenium, Se.**—This element was discovered by Berzelius in 1817, and was named by him Selenium, from Σεληνή, the moon. It is one of the least abundant of the elements, and always occurs in combination, generally in ores of iron, copper, and silver, forming selenides of these metals. The principal localities in which it exists are in Norway, Sweden, and the Hartz mountains of Germany. It is a dark-brown, brittle solid, opaque, and possessing a metallic luster somewhat like lead. It closely resembles sulphur in its properties, and forms acid compounds with oxygen (selenious and selenic acids) analogous to sulphurous and sulphuric acids. When heated strongly it gives out a powerful odor, like putrid horse-radish, by means of which the smallest trace of this element may be detected in minerals, when heated before the blow-pipe.

397. **Tellurium, Te,** is a rare substance, found chiefly in the mines of Hungary and Transylvania; sometimes native and nearly pure, but generally combined with various metals, such as gold, silver, bismuth and copper. It is a silver-white, brittle solid, possessing a strong metallic luster, and by some authorities is classed among the metals. It is, however, closely allied to sulphur and selenium in all its properties and combinations.

Selenium and tellurium both unite with hydrogen to form gaseous compounds, of singularly offensive and noxious properties. A single bubble of seleniuretted hydrogen allowed to escape into a room, produces on those who

---

QUESTIONS.—Why do surfaces painted with lead blacken on exposure to this gas? Why are zinc paints, for many situations, preferable to lead? Explain the manner in which hydrosulphuric acid is used in chemical analysis. What is said respecting selenium? What are its characteristic properties? What is tellurium? What are its properties? What is said of the compounds of selenium and tellurium with hydrogen? What effect has tellurium upon the animal system?

breathe it, all the usual symptoms of a severe cold and irritation of the throat, which continue for several days. But the most singular fact connected with tellurium is, "that when certain odorless preparations of this element are taken internally, they form compounds within the animal organization which impart to the breath and the perspiration so fœtid an odor as to render the person taking it a kind of horror to every one he approaches; and this lasts sometimes for weeks, though the dose of tellurium administered may not exceed a quarter of a grain."—JOHNSON.

SECTION XI.

PHOSPHORUS.

Equivalent, 32. *Symbol*, P. *Density*, 1·863.

398. **History.**—The credit of the discovery of phosphorus is ascribed to Brandt, an alchemist of Hamburg, who first recognized it while searching for the philosopher's stone in human urine, in the year 1669. Its method of preparation was, however, for a long time kept secret.

399. **Natural History and Distribution.**—Phosphorus is never found in nature in a free state, but exists in small quantities, widely diffused, in the mineral kingdom, principally in combination with lime. It is a constituent of most of the primitive and volcanic rocks, and by the decay of these it passes into the soil; from the soil it is extracted by plants, which accumulate it, particularly in their seeds (wheat, corn, oats, etc.). Man and animals deriving their support directly or indirectly from plants, in turn collect it in their systems—in such quantities that animal products furnish almost the only source from which phosphorus is artificially prepared. United with oxygen and with lime, it forms the principal mineral constituent of the bones. Thus the body of an adult man contains from 9 to 12 pounds of bones, which contain from 5 to 7 pounds of phosphate of lime (phosphoric acid and lime), or from 1 to 2 pounds of pure phosphorus.* "Phosphorus

* No seed suitable to become food for man and animals can be formed in any plant without the presence and coöperation of the phosphates, and a field in which phosphate of lime, or the alkaline phosphates form no part of the soil, is totally incapable of producing grain, peas, or beans.

Animals which are fed on food which contains no phosphate of lime, gradually lose their nervous irritability, and sink into a state of inanition and torpor, which is speedily followed by death. A *deficiency* of phosphate of lime in the food of young children, is also liable to produce a disease known as the *rickets*. As animals derive the phosphate of lime necessary for their support either directly or indirectly from plants, and as these in turn extract it from the soil, it is evident that the fertility of a soil can only be sustained by restoring to it the constituents thus abstracted from it. Hence the value of bones and animal products which contain phosphate of lime (as guano) as manures for wheat and plants of like character.

QUESTIONS.—What is the history of phosphorus? What is said of its distribution in nature? In what condition does it exist most abundantly?

also appears to be essential to the exercise of the higher functions of the animal, since it exists as a never-failing ingredient in the substance of which the brain and nerves are composed."

400. **Preparation.**—Phosphorus was formerly extracted from urine, but at the present time it is obtained almost exclusively from bones, from which immense quantities are prepared for the manufacture of matches and other uses.

The details of the process of preparation are briefly as follows:—The bones are first burned to whiteness and then reduced to a fine powder, which powder, being a phosphate of lime, insoluble in water, is technically known in chemistry and the arts as "bone-ash." So much sulphuric acid and water is then added to a suitable quantity of bone-ash as will, in the course of a few days, partially decompose it—two thirds of the lime uniting with the sulphuric acid to form an insoluble sulphate of lime, while the remaining one third continues in combination with the whole of the phosphoric acid to form a new compound, which is readily soluble in water. This new compound is called *superphosphate of lime*, and of late years has been extensively introduced into agriculture, as a ready means of supplying exhausted soils with the phosphorus needed for the production of crops. The chemical reaction which takes place may be expressed in symbols as follows:—

$$\overbrace{3CaO,PO_5}^{\text{Bone-ash.}} + \overbrace{2(SO_3,HO)}^{\text{Sulph. acid.}} = \overbrace{2HO,CaO,PO_5}^{\text{Superphosph. lime.}} + \overbrace{2(CaO,SO_3)}^{\text{Sulph. lime.}}$$

The insoluble sulphate of lime and the superphosphate of lime dissolved in the acid solution, are then separated from each other by filtration, and the latter, evaporated to a syrup, is mixed with charcoal, and heated in an iron, or earthen retort. Under these circumstances the charcoal decomposes the superphosphate of lime;—phosphorus rises as a vapor, and passing into cold water, is collected and condensed into a solid. The crude phosphorus thus obtained is purified by melting under water, and is then cast into sticks, in which form it is sold.

401. **Properties.**—Phosphorus exists in two conditions, viz.: in an ordinary state, and in an allotropic state. In its ordinary state it is a soft, semi-transparent, almost colorless, waxy-looking solid. It is insoluble in water, but readily soluble in ether, alcohol, and in various oils.

At all temperatures above 32° F., phosphorus, when exposed to the air, slowly combines with oxygen, and emits a feeble light, readily perceptible in the dark (hence its name, from φως, *light*, and φερειν, *to bear*). Exposed to a temperature of about 60° F. it bursts into a flame. This extreme combustibility of phosphorus renders it necessary to keep it continually under water, from which it should be taken, for the purpose of experiment, with great caution, and be held with a pair of forceps, or upon the point of a knife. When-

QUESTIONS.—How is phosphorus obtained? What is superphosphate of lime? What is the chemical reaction involved in its manufacture? What are the properties of ordinary phosphorus? What is said of its solubility? What of its inflammability?

ever, also, it is desirable to cut it into fragments, the operation should be performed under water. The burns occasioned by melted phosphorus are extremely severe, from the difficulty of extinguishing the flame.

Phosphorus is also easily ignited by friction, and for this reason is employed in the manufacture of matches. It burns in the air with a brilliant flame, and in pure oxygen gas with a light so dazzling that the eye can hardly sustain it. (§ 282.)

At a temperature of 111° F., air being excluded, phosphorus melts; and when fused under water, it can be molded as readily as wax. At 550° F., in close vessels, it boils, giving off a colorless gas. A solution of phosphorus in naphtha, by cooling and evaporation, yields crystals of phosphorus. Very fine crystals of phosphorus may be also obtained by exposing phosphorus to sunlight in a tube either exhausted of air, or filled with a gas which can not oxydize it.

The following experiments illustrate some of the characteristics of this element:—

Place in a glass flask about a quarter of an ounce of ether and a piece of phosphorus of the size of a pea. Cork the flask and allow it to stand some days, frequently agitating it. In this way an ethereal solution of phosphorus will be obtained, which, when rubbed upon the hands, renders them luminous in the dark. The explanation of this phenomenon is, that the ether evaporates, and leaves the phosphorus which it held in solution upon the hands in a state of minute subdivision. In this condition it combines with the oxygen of the air, or undergoes a slow combustion, diffusing a white smoke and a pale greenish light. Heat is at the same time evolved, but not sufficient to occasion ignition. By rubbing the hands, the light is rendered more vivid, as a fresh surface of phosphorus is thus continually presented to the oxygen of the air.

If we moisten a lump of white sugar with an ethereal solution of phosphorus, and throw it into hot water, the heat of the water will volatilize both the ether and the phosphorus; and the vapors, in rising to the surface of the water, and coming in contact with the oxygen of the air, will inflame spontaneously.

If we pour an ethereal solution of phosphorus upon fine blotting-paper, the latter will ignite spontaneously after the ether has evaporated.

If we place a piece of phosphorus of the size of a pea upon blotting-paper, and sprinkle over it some soot or finely-pulverized charcoal, the phosphorus, after a little time, melts, and at length spontaneously inflames. The finely-pulverized charcoal causes this combustion, owing to its porosity, which enables it to readily absorb oxygen from the air. This oxygen is in turn imparted to the phosphorus, and by uniting with it, occasions heat, which, prevented by the non-conducting properties of the charcoal from escaping, accumulates, and occasions combustion.

QUESTIONS.—What property renders phosphorus available for the manufacture of matches? What experiments illustrate the characteristics of phosphorus?

Phosphorus when taken internally is a most violent poison, and in combination with other substances, is frequently used for the destruction of rats and vermin. The so-called rat-exterminating poison is composed of 1 dram of phosphorus, 8 ounces of hot water, and 8 ounces of flour.

402. **Allotropic or Amorphous Phosphorus.**—It has long been noticed, when phosphorus is exposed to the action of light for a considerable length of time, that its exterior becomes coated with a red powder, and that the same product is formed when phosphorus is burned with a limited supply of air. This red powder was always supposed to be an oxyd of phosphorus, but within a recent period, Prof. Schrotter of Vienna has succeeded in demonstrating that the substance in question is merely an allotropic state of ordinary phosphorus. He has shown that if ordinary phosphorus be submitted to the action of a prolonged heat, within certain limits, and under circumstances involving an entire exclusion of oxygen, it becomes converted into a brick-red substance;—"not soluble in any of the ordinary solvents of phosphorus—not igniting by ordinary friction—not luminous at ordinary temperatures—not poisonous; distinguished, in fact, for negative properties, as common phosphorus is for active ones; and yet this wonderful change is only molecular; that is, the phosphorus is not converted into a compound: it has combined with nothing, it has lost nothing, but its particles have probably arranged themselves with respect to each other, in a manner different from that of the particles of common phosphorus." Common phosphorus we are obliged to keep in water, for the purpose of guarding against spontaneous combustion; allotropic phosphorus, however, may be kept unchanged in atmospheric air, and may be handled or even carried in the pocket with impunity. Exposed to a temperature of about 480° F., it melts, and returns to the condition of ordinary phosphorus; and at a temperature of 500° it bursts into flame with a sort of explosion. The identity of the two substances is proved by their ready conversion into each other, and by the fact that the compounds which they form with other bodies are the same.

403. **Matches.**—Some notice of the history and manufacture of matches is appropriate in connection with the subject of phosphorus.

The comparatively low temperature at which sulphur ignites, early suggested its application to the end of a strip of dry wood, as a means of procuring flame. The old sulphur match was chiefly used in connection with a flint and steel, and a box for holding tinder. The tinder, formed by the partial combustion of a linen or cotton rag, was first ignited by means of a spark resulting from a collision of a flint and steel, and this in turn communicated the fire to the match. Fifty years ago, a "tinder-box" was as much an indispensable article of household economy as a paper of matches is at the present day.

Soon after the discovery of phosphorus, attempts were made to use it as a

QUESTIONS.—What is said of the poisonous properties of phosphorus? What is rat-poison? What is said of allotropic phosphorus? In what respects does allotropic differ from ordinary phosphorus? How can we prove that allotropic and common phosphorus are the same? What is said of the history and origin of matches?

method of procuring fire, but its costliness prevented its general introduction and use for this purpose, for nearly one hundred and fifty years. One of the first methods of applying it was to put a piece of phosphorus in a phial, and then to stir it with a hot iron wire; the phosphorus was partially burnt in the confined portion of air, and the interior of the bottle became covered with an oxyd of phosphorus; on removing the wire, the phial was corked tighfly for use. When a light was wanted, a common sulphur match was dipped into the bottle, and a small portion of the phosphorus adhering to the tip, flame was produced by the energetic chemical action of the sulphur and the phosphorus. Various other inventions were employed for procuring fire;—such as the sudden condensing of air in a syringe furnished with a piston and an arrangement for holding tinder—apparatus for igniting tinder by an electric spark—Doberciner's Lamp (§ 297), etc., etc. In fact, during the whole of the last century, and even later, the invention of a safe, convenient, and reliable agent for kindling a fire or light, was regarded as one of the great wants of the age.

The next important step taken in perfecting the match, was the employment of chlorate of potash. The match stick was tipped with a mixture of chlorate of potash and sugar, and ignited by immersion in a little bottle containing asbestos soaked in sulphuric acid. (For explanation of this phenomenon see § 368.) Matches thus prepared were put up in cases, which contained in one compartment a small bottle of acid. Their price, when first introduced, was $4 75 for a case of 100; but subsequently was reduced to 50 cents. These matches continued in use until within a very recent period.

The next important invention was that of the so-called "Lucifer Matches," which were tipped with a paste of chlorate of potash and sulphuret of antimony mixed with starch, and were ignited by drawing the match between two surfaces of sand-paper. These were the first friction matches. In 1834, phosphorus was substituted in the place of antimony, and the match was ignited by friction upon any rough surface. Subsequently, saltpeter was substituted in the place of chlorate of potash, which produced quiet ignition instead of detonation.

The details of the manufacture of matches at the present time are generally as follows: The ends of the wooden match-splints, which are sawed by machinery, are first sulphured, by immersion in a pot of melted sulphur. When dried, they are next dipped in the phosphorus composition, which is a paste prepared by mixing together in a hot solution of glue, or gum, in water, phosphorus, saltpeter, and generally red-lead and some coloring ingredients;—if the tips of the matches are to be red, vermillion is added; or if blue, Prussian blue.

The various reactions which take place when a match is fired are as follows: the phosphorus contained in the composition is first ignited by the heat

QUESTIONS.—What were some of the early methods resorted to for the purpose of obtaining a light? When was phosphorus first applied to the manufacture of matches? What were the first friction matches? How are matches manufactured? What chemical reactions are involved in the firing of a match?

evolved by friction or compression; and the heat occasioned by its combustion decomposes the saltpeter and the red-lead; these substances, in their decomposition, evolve oxygen, which supports the flame, adds to its heat, and enables it to ignite the sulphur, which in turn inflames the wood. The odor of a burning match is occasioned by the combustion of the sulphur, and in some recent inventions, has been obviated by the substitution of stearine in the place of sulphur. The temperature required for kindling matches varies from 150 to 160° F.*

The manufacture of matches is attended with danger, not only from the highly inflammable nature of the ingredients used, but also from the fact, that a continued exposure to the vapor of phosphorus, produces a disorganization of the jaw-bones, causing excruciating suffering, and usually terminating in death. The phosphorus, in the first instance, attacks a little spot of decay upon a tooth, and from this ulceration spreads with great rapidity. Of these evils the first is greatly lessened, and the second altogether avoided, by the use of the amorphous or allotropic phosphorus, before described.

404. **Compounds of Phosphorus with Oxygen.**—Phosphorus unites with oxygen to form four compounds, viz.:—

		Composed by weight of	
Phosphoric acid....................	PO_5	32 phosphorus	40 oxygen.
Phosphorous acid....................	PO_3	32 "	24 "
Hypophosphorous acid..............	PO	32 "	8 "
Oxyd of Phosphorus................	P_2O	64 "	8 "

405. Phosphoric Acid, PO^5.—This acid, which is the most important of the oxyds of phosphorus, is the sole product of the rapid combustion of phosphorus in oxygen, or atmospheric air.

It appears as a dense white vapor, which condenses on cooling into a white powder. It may be easily collected by burning phosphorus in air under a dry bell glass. As thus prepared, it has so great an avidity for water, that when brought in contact with it, it hisses like a hot iron. Exposed to the air for a few moments, it absorbs moisture, and deliquesces to a liquid. When once converted into a hydrate, water can not be entirely separated from it. Its solution is intensely acid, and when evaporated to dryness, yields, on cooling, a glassy, transparent solid, known as *glacial phosphoric acid.*

Phosphoric acid may also be prepared by the action of nitric acid on phosphorus, and also from bones, by the action of sulphuric acid. It combines

* Some idea of the importance of the manufacture of matches as a branch of industrial art, may be formed from the following statistics of materials consumed in Austria in one year, 1849, for this purpose—125,000 lbs. of saltpeter, 32,500 lbs. of phosphorus, 1,500,000 lbs. of sulphur.

QUESTIONS.—What is the temperature required for kindling a match? What effect has the vapor of phosphorus upon the animal system? What compounds does phosphorus form with oxygen? How is phosphoric acid prepared? What are its properties?

with water in three proportions, to form three distinct hydrates, which unite with bases to form three classes of salts. The nomenclature and composition of these hydrates, which are of great scientific interest, may be represented as follows:—

	Acids.
Monobasic or metaphosphoric acid................	$HO.PO_5$,
Bibasic or pyrophosphoric acid..................	$2HO.PO_5$,
Tribasic or common phosphoric acid.............	$3HO.PO_5$.

It is in the form of phosphoric acid, united with some base, generally lime or magnesia, that phosphorus exists in the bones, in the seeds and tissues of plants, and in the soil.

406. **Phosphorus Acid**, PO_3 is the principal product which results from the slow combustion which occurs when phosphorus is exposed to the oxygen of the atmosphere. It may also be formed by burning phosphorus with a limited supply of air.

The other oxyds of phosphorus are comparatively unimportant.

407. **Phosphorus and Hydrogen.—Phosphuretted Hydrogen**, PH_3.—Phosphorus unites with hydrogen in three proportions to form three compounds; one of which, a gas, phosphuretted hydrogen, possesses the property, under certain circumstances, of inflaming spontaneously on exposure to air, or oxygen gas.

This substance is conveniently prepared by heating fragments of phosphorus in a retort, with a strong solution of caustic potash, or cream of lime, prepared from lime recently slacked.* On the application of a gentle heat to the retort, the beak of which is caused to dip slightly beneath the surface of water, the gas is evolved, and the bubbles, as they rise and come in contact with the air, spontaneously inflame. (See Fig. 132.) Each bubble, as it breaks and ignites, produces a beautiful white wreath of smoke (vapor of phosphoric acid), composed of a number of concentric rings, revolving around the axis of the wreath, as it ascends (see Fig. 133); thus tracing before the eye, with perfect distinctness, the peculiar gyratory move-

FIG. 132.

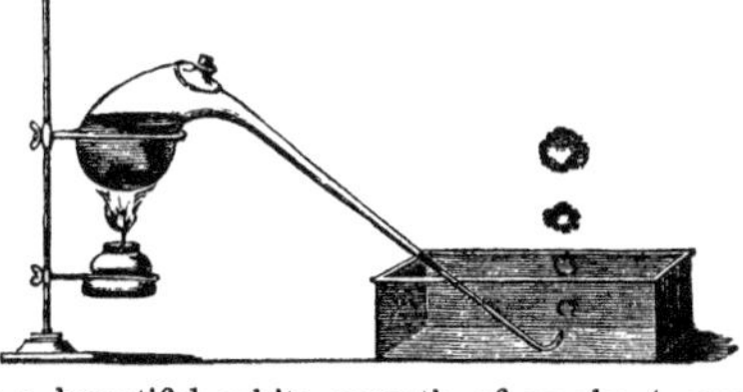

FIG. 133.

* In this experiment it is best to employ a very small flask or retort, and in order to avoid the presence of atmospheric air, it is advisable to fill it full to the neck with the cream of lime, or potash solution. For an ounce flask, a piece of phosphorus of the size of a pea is sufficient. It is best, also, not to apply heat to the glass directly, but to place it in a basin containing a solution of salt, which is then heated to a boiling temperature by a spirit lamp.

QUESTIONS.—What is said of its combinations with water? In what state does phosphorus generally exist in nature? What is said of phosphorus acid? What is said of phosphuretted hydrogen? How is it prepared? What phenomenon attends its evolution in air?

FIG. 134.

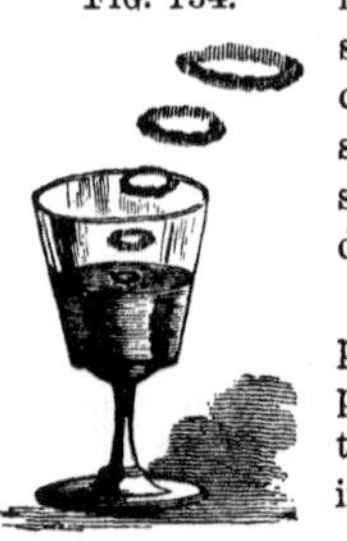

ments imparted to air by the impulse of a force acting suddenly upon a mass of air in all directions, from a center. The same phenomenon is also seen in the rays of smoke produced by the mouth of a skillful tobacco-smoker, and frequently also, upon a much larger scale, during the discharge of cannon on a still day.

Phosphuretted hydrogen may be more simply prepared by throwing into a glass of water a few pieces of phosphuret of calcium. This substance, by contact with the water, is decomposed, and evolves the spontaneously inflammable gas. (See Fig. 134.)

408. **Properties.**—Phosphuretted hydrogen is a colorless, transparent gas, possessing an offensive, fœtid odor, and producing a poisonous action upon the system, when inhaled. It loses its spontaneous inflammability by standing for a time over water, and also by the addition of the vapor of some inflammable bodies, such as ether, oil of turpentine, etc. By varying the conditions of its preparation, it may also be evolved without the self-lighting power.

The production of this gas, by the decay of bones and other organic products in wet, swampy places, and its subsequent ignition in contact with the air, is supposed to have originated the popular superstition known as the "Ignis Fatuus," or "Will-o'-the-wisp."*

SECTION XII.

BORON.

Equivalent, 10·9. *Symbol*, B.

409. **History and Distribution.**—Boron is an element that is always found in nature in composition with oxygen, forming boracic acid. The latter substance is found only in few localities, and in comparatively small quantities. United with soda it forms a salt, borax, which is a well-known article of commerce.

Until within a very recent period (1856–7), comparatively little has been known respecting the nature of the pure element, boron. It has been recently ascertained, however, that it is closely allied to carbon, and that it exists in

* It is generally taken for granted that luminous appearances in the air are often seen in the vicinity of swamps, grave-yards, or other receptacles of decaying organic matter. Such, however, is not the fact; and it is extremely doubtful whether any well authenticated instance of such an appearance can be cited. The generally-received account of the "Ignis fatuus" must therefore be regarded as a fiction.

QUESTIONS.—What are the properties of phosphuretted hydrogen? What popular superstition is it supposed to have originated? What is said of boron?

three allotropic conditions, viz., as a chocolate-brown amorphous substance; as an opaque, semi crystalline body, occurring in thin plates, with a black-lead luster; and, lastly, in a crystalline condition, resembling the diamond in luster, hardness, and refractive power. As yet, chemists have been only able to obtain it in very minute crystals; but if larger crystals can be prepared, it will undoubtedly take rank as one of the most valuable of gems. Its method of preparation consists essentially in fusing boracic acid with the metal aluminum.

410. **Boracic Acid,** BO_3 is found in small quantities in Thibet and in South America, but the principal supply is from volcanic districts of Tuscany, in Italy, called *lagoons*, where jets of vapor and of boiling water, charged with boracic acid, are continually issuing from fissures in the earth.*

The manner in which the boracic acid is collected is as follows: A locality is chosen, where the soil is observed to possess a high temperature, and a basin of moderate depth (A, Fig. 135) is excavated, and walled up with

FIG. 135.

a masonry—openings, *v*, being left in the bottom for the admission of the steam escaping from the earth.† Water from adjacent springs is then conducted into the basin, which absorbs the boracic acid brought up by the ascending vapor, and at the same time becomes heated to the boiling temperature. After the lapse of twenty-four hours, the solution is drawn off into a similar-constructed basin, B, at a lower level, from thence into a third, C, and

* "As you approach the lagoons, the earth seems to pour out boiling water, as if from volcanoes of various sizes, in a variety of soils, but chiefly of chalk and sand. The heat in the immediate neighborhood is intolerable, and you are drenched with vapor, which impregnates the atmosphere with a strong and somewhat sulphurous smell. The whole scene is one of terrible violence and confusion:—the noisy outbreak of the boiling water; the rugged and blasted surface; the volumes of vapor; the impregnated atmosphere. The ground burns and shakes beneath your feet, and the whole surface is covered with beautiful crystallizations of sulphur and other minerals."—DR. BOWRING.

† The dimensions of these basins vary from 100 feet in circumference and 7 feet deep, to 500 and 1000 feet in circumference and 15 to 20 feet deep.

QUESTIONS.—What are its properties? What is said of boracic acid? How is it collected?

so on, until the water, having absorbed the greatest possible amount of boracic acid, is transferred into shallow tanks, E, for purification. The solution thus obtained is evaporated in leaden pans heated by the volcanic steam, until the boracic acid contained in it is deposited in white, scaley crystals. The annual production of boracic acid from these sources is at present about three million pounds.

Boracic acid has a white, pearly luster and a greasy feeling. It is a feeble acid, sparingly soluble in cold water, but dissolving in three times its weight of boiling water. Its solution in alcohol burns with a beautiful green flame, which constitutes a test of the presence of boron. This property may be illustrated by igniting a solution of borax in alcohol in a shallow cup, and stirring the liquid with a glass rod while burning.

411. **Borax, or Biborate of Soda,** is formed by adding carbonate of soda to a solution of boracic acid. This salt is composed of two equivalents of acid, one of base, and ten of water—its constitution being represented as follows, $NaO, 2BO_3+10\ HO$. Borax is obtained naturally in small quantities and in an impure state, by the evaporation of the waters of certain lakes in Thibet, and is exported under the name of *tincal.*

Borax is chiefly used in the arts as a flux in the welding, soldering, and refining of metals.

The term *flux* is applied in metallurgy to those substances which assist fusion, either by expediting the process, or by protecting the substance melted from oxydation.

Borax, when heated, bubbles up, loses its water of crystallization, and at a temperature below red-heat, melts into a transparent glass. The property which this glass possesses of dissolving the metallic oxyds, gives to borax its value as a flux. For example: in the welding of iron, a union between two surfaces can not be effected unless both are clean and perfectly free from oxydation; but a piece of iron can not be strongly heated without the formation of a layer of oxyd upon its surface. This difficulty is obviated by sprinkling the hot surfaces with powdered borax, which, as it melts, not only dissolves off the oxyd, or scale already present, but keeps the metal bright by preventing all further oxydation.

Borax is also much used as a test before the blow-pipe, for recognizing the presence of certain metallic oxyds. For this purpose, a small crystal of borax is fused upon the end of a bent platinum wire, and a minute quantity of the substance to be tested is melted with the salt in the flame of the blow-pipe. The peculiar color which the borax glass receives, indicates the character of the coloring substance: thus, with an oxyd of chromium, the borax forms an emerald-green glass; with oxyd of cobalt, a blue; with manganese, a violet; with iron, a yellow, and so on.

QUESTIONS.—What are the properties of boracic acid? What is borax? For what purpose is it used in the arts? What is a flux? What gives to borax its value as a flux? Illustrate this. How does borax serve as a blow-pipe reagent?

SECTION XIII.

SILICON, or SILICIUM.

Equivalent, 21·2. *Symbol*, Si.

412. **Distribution.**—Silicon, in combination with oxygen, is the most abundant of all the solid substances which compose the crust of our globe. All rocks which are not calcareous (lime) are silicious.

It is only within a very recent period (1855-7) that chemists have been enabled to obtain any very definite knowledge respecting the nature and properties of pure silicon. It is now known to exist, like carbon and boron, in three allotropic conditions; in an amorphous nut-brown powder; in a condition resembling graphite (black-lead); and in a crystalline condition. It has most of the characteristics of the metals, and by the most recent authorities is classed with them. As prepared by a somewhat complicated process, it is easily fusible, and may be run into ingots and alloyed with copper and iron. At a meeting of the French Academy in 1857, two small cannon composed of an alloy of copper and silicon were exhibited.

413. **Silicic Acid, or Silica, SiO_3,** is the principal oxyd of silicon, and the most important of all its compounds. In fact, it is in this condition only that silicon is found in nature.

FIG. 136.

When pure, or merely colored by small quantities of different oxyds, it is very generally termed quartz. It is frequently found crystallized, its ordinary form being a six-sided prism, terminated by six-sided pyramids, as in rock-crystal. (See Fig. 136.) Sometimes the prism is very short and disappears entirely, and the pyramid only is seen, as in common quartz. In transparent and colorless rock-crystal, silica is almost absolutely pure, and in this condition is not unfrequently used in jewelry. Amethyst is crystallized quartz, colored purple by the presence of protoxyd of manganese. Common flint, agate, carnelian, chalcedony, jasper, and opal, are other varieties of nearly pure silica, their colors being occasioned by the presence of different metallic oxyds. Common sand is mainly composed of silica, colored yellow or brown by the presence of oxyd of iron; sand cemented into rock-masses, through the agency mainly of silica, is termed "sandstone."

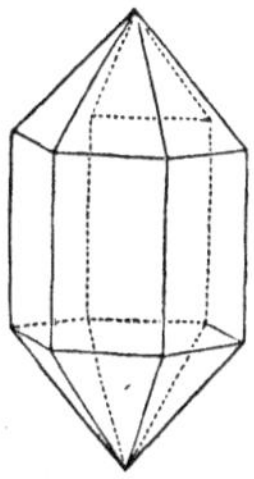

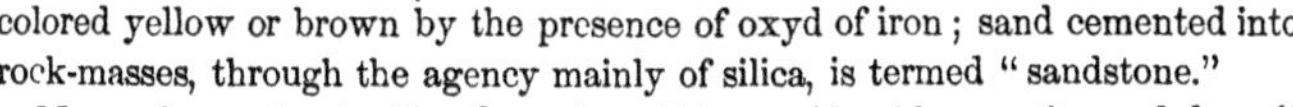

Many plants absorb silica from the soil in considerable quantity, and deposit

QUESTIONS.—What is the natural history of silicon? What is known respecting the pure element? What is silica? What is quartz? In what minerals does silica nearly pure exist? What is amethyst? To what are the colors of agate, chalcedony, opal, etc., due? What is common sand? What is sandstone? Does silica exist in plants?

it upon the exterior of their stalks, or stems. Examples of this may be seen in the glossy coating which invests the outside of straw, cane, rattan, bamboo, etc. In these instances, the silica subserves the same purpose in the structure of the plant that bones do in the structure of men and animals—that is, it gives to the stalk firmness and stiffness. The straw of wheat grown upon soils deficient in "soluble silica," is so weak as to be hardly capable of supporting the weight of the seed.

In the animal kingdom, silica exists in the feathers and hair of animals, and recent researches have also detected it in the blood.

414. **Properties.**—Pure silica is not affected by the heat of the strongest wind furnace, but before the flame of the oxyhydrogen blow-pipe it melts into a transparent glass. In its native state it is insoluble in pure water, and in all acids except hydrofluoric. In hardness it approaches the precious gems, and it scratches glass easily.

Silica, although it presents the characters of an earth, is in reality an acid, and a most powerful one. Under all ordinary circumstances, however, its acid properties are not manifested by reason of its almost entire insolubility.

When silica is digested in solutions of the alkalies it gradually unites with them, and forms salts—silicates of potash or soda—which are readily soluble. Even flints in their unground condition, or fragments of quartz when placed in strong solutions of caustic potash or soda, at a high temperature, are readily caused to pass into solution. When solutions of silica in an excess of alkali are concentrated, a semi-fluid mass closely resembling a solution of starch is produced. This product is known as *soluble glass*, and is readily soluble in hot water, and can be applied as a varnish for rendering surfaces of wood or cloth fire-proof. It has also been used to some extent as a substitute for starch or gum in the stiffening of fibrous substances. Ancient monuments or buildings constructed of soft and friable stone may be preserved in a great measure from decay and the action of the weather by a coating of soluble glass. For practical purposes, soluble glass is formed by fusing together 8 parts of carbonate of soda (or 10 of carbonate of potash) with 15 parts of pure sand, and 1 of charcoal. The product, when pure, resembles ordinary glass, but dissolves in boiling water without residue.

When a solution of soluble glass is rendered acid by the addition of hydrochloric acid, the silica after a little time separates as a transparent, tremulous jelly. This is a hydrate of silica, which once precipitated in this manner, is no longer soluble in either water or acids. By preventing the escape of moisture, it may be preserved in a gelatinous condition; but if once allowed to dry, it forms a white, gritty powder—white silicious sand.

Most natural waters contain a little soluble silica, which can be only separated by evaporating the water to dryness. Waters which contain alkaline

---

QUESTIONS.—What are illustrations? What is said of silica in the animal kingdom? What are the properties of silica? Is silica an acid? Under what circumstances does it pass into solution? What is soluble glass? What are its properties and uses? How may silica be separated from its solution in alkalies? Does silica exist in natural waters?

carbonates dissolve it more freely, and when the action of the alkaline liquid is aided by that of a high temperature, as is the case with the Geysers, or hot springs of Iceland, very large quantities of silica are dissolved. As the liquid cools, the silica is deposited, in an insoluble form, on the surrounding objects in contact with the waters, forming "petrifactions." Agates, chalcedony, carnelian, and onyx, have undoubtedly been thus formed by the slow deposition of silica from its solution in water.

The acid character of silica is especially exhibited when it is exposed, in contact with other salts, to a high temperature. It then displaces the most powerful acids from their combinations, and uniting with their bases, forms silicates. Thus when carbonate or sulphate of potash, soda, or lime, are mixed with silica and fused, the silicic acid displaces the carbonic and sulphuric acids from their combinations, and forms silicates of potash, soda, or lime. All the forms of clay, feldspar, mica, hornblende, and a great number of our most common minerals, are the salts of silicic acid.*

415. **Fluoride of Silicon, Si Fl.**—*Fluosilicic Acid.*—In order to prepare this gas, equal parts of finely-powdered fluor-spar and silicious sand, or powdered glass, are mixed in a capacious flask, with six parts of concentrated sulphuric acid. On the application of heat, hydrofluoric acid is liberated, and this immediately attacking the silica, produces a colorless gas, of which silicon is a constituent. When passed into water, the gas is decomposed, silicon is precipitated in the form of gelatinous silica, and the water becomes a solution of hydrofluosilicic acid. This reaction, which constitutes a very interesting experiment, may be easily exhibited by an arrangement of apparatus as represented in Fig. 137.

FIG. 137.

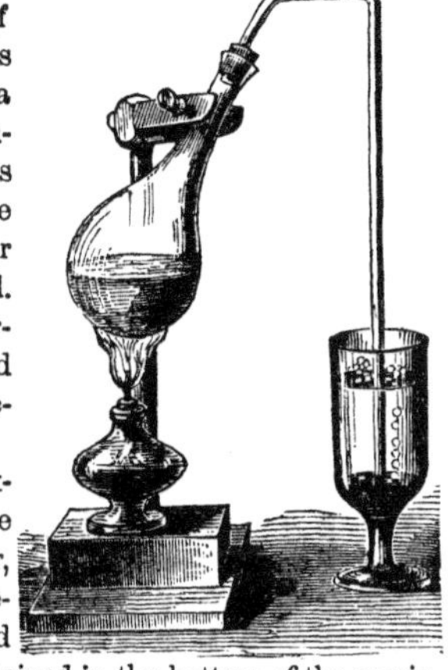

In transmitting the gas into water, the extremity of the evolution tube should not be brought into direct contact with the water, lest it become at once obstructed by the deposited silica; but it should be plunged beneath the surface of a little mercury contained in the bottom of the receiv-

* The composition of many of the silicious minerals is extremely complex, and in a scientific point of view, extremely interesting. Upon one group alone, the zeolites—hydrated silicates of alumina, with lime, potash and soda—an immense amount of labor has been expended by many of the most eminent chemists of the present century, and yet their chemical formula and most natural relations are still open to question.

QUESTIONS.—Explain the circumstances. What is the supposed origin of agates, carnelians, etc.? When is the acid character of silica especially manifested? Illustrate. What are examples of natural silicates? What is said of fluosilicic acid? What occurs when this gas is passed into water?

ing vessel, as is represented in Fig. 137. As the gas ascends through the mercury, and enters the water, it exhibits a most curious phenomenon; each bubble becoming invested with a white bag of silica, and rising, like a miniature balloon, to the surface; it often happens, also, in the course of the experiment, that the gas forms tubes, or conduit pipes of silica in the water, through which it gains the surface without decomposition.

SECTION XIV.

CARBON.

Equivalent, 6. *Symbol*, C. *Specific gravity as diamond*, 3·3 to 3·5.

416. **History.**—Carbon is one of the most abundant and important of the elementary bodies. In the inorganic kingdom of nature it exists chiefly as mineral coal; in the state of carbonic acid diffused throughout the atmosphere; and as a constituent of the great rock masses—carbonates of lime and magnesia. In the organic kingdom, it is the characteristic ingredient of all substances which are produced directly or indirectly from animal or vegetable organisms.

Carbon is found pure in nature in three allotropic forms or conditions, each of which, although possessed of identically the same chemical composition, exhibits properties singularly different from the others, and peculiar to itself. These are, 1. *The Diamond;* 2. *Graphite*, or *Plumbago;* 3. *Mineral Coal* and *Charcoal.*

417. **The Diamond** is pure carbon, crystallized.

It is found throughout a wide extent of country in India, but chiefly at Golconda, and in certain districts of Borneo and Brazil. It has also been found associated with gold and platinum in the Ural mountains, and in a few instances in the United States, principally in the gold districts of North Carolina.* In only a few instances has the diamond ever been found imbedded in rock masses, but it is usually associated with materials transported by water from a distance, such as loose sand and rolled gravel. In their natural

* The largest diamonds come from Golconda, but Brazil furnishes the greatest quantity. The yearly produce of the Brazilian mines at the present time is estimated at from 10 to 13 lbs., a large proportion of which, however, are unfit for jewelry.

QUESTIONS.—What is said of the distribution of carbon in the two great kingdoms of nature? In what conditions is carbon found pure naturally? What is the diamond? Under what circumstances is it found in nature?

condition, diamonds have usually the appearance of semi-transparent, rounded pebbles, and are covered by a thin, opaque crust; on removing this crust, their exceeding brilliancy becomes apparent.

The diamond is generally colorless, and such specimens possess the greatest value; but it is not unfrequently found of a blue, yellow, or rose color, and sometimes green or black.

FIG. 138.

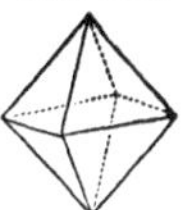

The primitive form of the diamond is that of an octohedron (see Fig. 138), but its faces are often convex, and its edges rounded. It is cut for jewelry in three forms, known as brilliants, Fig. 139, roses, Figs. 140, 141, and tables, Figs. 142, 143.*

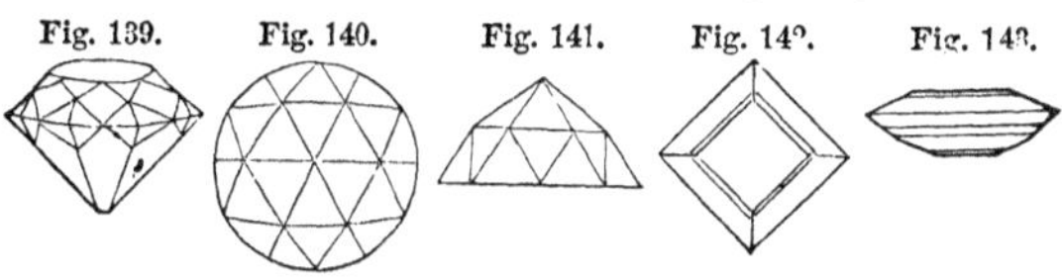

Fig. 139. Fig. 140. Fig. 141. Fig. 142. Fig. 143.

The diamond is the hardest of all known substances, and can be only cut or abraded by means of its own powder—inferior and imperfect stones being broken down for this purpose. The process of cutting is effected by a horizontal disc of steel, covered with diamond dust and oil, and revolving with a velocity of two or three thousand times per minute. The gem is fixed in a mass of lead, which is fitted to an arm, one end of which rests upon a table over which the plate revolves, while the other, sustaining the diamond, is pressed upon the plate by movable weights, at the discretion of the operator. The gem, however, cannot be ground into any form at pleasure, but only in directions parallel to its lines of cleavage. (§ 73.)†

* The f rm of the brilliant shows the gem to the best advantage, and may be recognized by its flat summit; the surface of a rose diamond is covered with equilateral triangles, terminating in a sharp point. The table form is only given to plates, laminæ, or slabs of diamonds, which have a small depth compared to their superficial extent. The brilliant and the rose lose in cutting and polishing somewhat less than half their weight, so that the value of a cut stone is double that of an uncut one, without reckoning the expense of the process.

† The method of cutting diamonds was discovered in 1456, and is still unknown in its perfection among Eastern nations. The business in Europe is carried on almost exclusively in Amsterdam, Holland. The heat developed in the cutting is frequently so great as to melt the lead in which the diamond is imbedded, and the time occupied in cutting a single face varies from 3 to 30 hours.

The weight of diamonds is estimated in carats—150 of which are equal to 1 ounce Troy, or 480 grains. "The rule for estimating the value of diamonds is peculiar, and supposing the gems under comparison to be equal in quality, may be expressed as being in the ratio of the squares of their respective weights. Thus, supposing three diamonds to exist, weighing respectively 1, 2, and 3 carats; their respective values would be as one, four, and nine. This rule, however, can only be considered as applying to gems of a moderate size; as very large diamonds, if estimated according to this mode of calculation, would become expensive beyond the means of the richest to command."

QUESTIONS.—What is its primitive form? In what three forms is it cut for jewelry? What is said of its hardness? How is it cut?

The diamond is remarkably indestructible, and is not acted upon by any solvent, neither is it affected by heat alone—since it may be heated, when removed from the access of air, to a white heat without injury. In the open air it burns at about the melting point of silver, and is converted into coal, or carbonic acid gas.

Many attempts have been made to fuse or crystallize some pure form of carbon, or, in other words, to manufacture diamonds, but they have all failed. In 1853, M. Desprétz of Paris succeeded, after long-continued voltaic action, in depositing at one of the terminal poles of a galvanic battery a quantity of carbon in the form of minute *microscropic* grains; these grains appeared to be octohedral crystals, and were capable of cutting and polishing diamonds and rubies; hence it has been inferred that they were actually themselves diamonds.

The origin of the diamond has been a subject of much curious speculation, inasmuch as the circumstances under which it is found in nature afford us no clue to the process of its formation. The structure of the diamond itself, however, furnishes us with some positive information on the subject, and indicates that it is a product, either directly or indirectly, of the vegetable kingdom.* Sir David Brewster, who has given much attention to the subject, is inclined to the opinion, that the diamond is a drop of fossilized gum, analogous in some respects to amber.

418. The largest known diamond is an uncut gem belonging to the crown jewels of Portugal. It was found in Brazil about the year 1808, and weighs 1,680 carats, or about 11 ounces. About the middle of the 16th century a diamond was found at Golconda in India, which had the form of half a hen's egg, and weighed nearly 6 ounces. This diamond, which was long known as the Great Mogul from its possessor, has disappeared, and is supposed to have been broken up;—the separate pieces, according to this theory, now constituting three of the largest existing diamonds, viz., 1, the great diamond in the possession of Russia, weighing 196 carats: 2d, the Koh i-noor, in the possession of the Queen of England, which weighed before cutting 186 carats, and after cutting 103 carats; and 3d, a diamond belonging to the Shah of Persia, of the weight of 130 carats. The value of the Russian diamond has been estimated at 20 millions of dollars, and that of the Koh-i-noor at from 3 to 10 millions.

The other large diamonds most worthy of notice are the following:—A yellow diamond belonging to the crown of Austria, which weighs 139 carats. The size and form of this diamond, which was once sold as a bit of colored

* The evidence on this point is principally as follows; diamonds have been found inclosing vegetable matter, and when the diamond is burned a minute yellowish ash is left, which generally possesses a cellular structure. Some other proof is also afforded by the action of refracted and polarized light.

QUESTIONS.—What is said of its indestructibility? Have any attempts been made to manufacture diamonds? What is said of the origin of the diamond? What evidence have we on the subject? How large a diamond has ever been found? What are some of the most valuable diamonds?

Fig. 144.

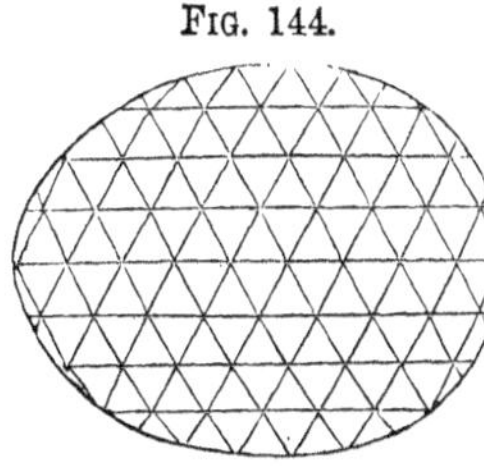

Fig. 145.

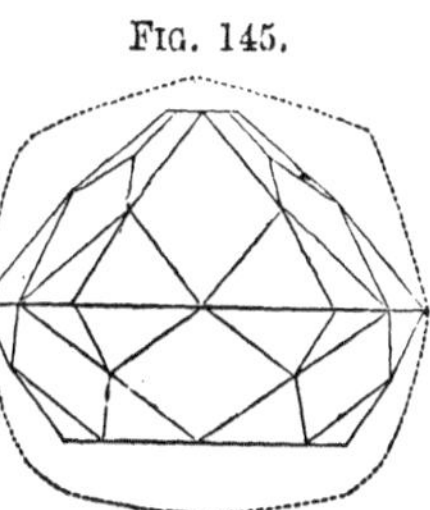

glass, are represented in Fig. 144. The Pitt or Regent diamond belonging to France, is represented in Fig. 145. the dotted line being the outline of the stone before cutting. This diamond, which is a light blue color, is allowed to be the finest in existence, and weighs 131 carats. It was brought from India by a Mr. Pitt, and sold to the Regent of France in 1717 for about $700,000. Its value, as estimated by a commission of Parisian jewelers, is about $3,000,000.

Fig. 146.

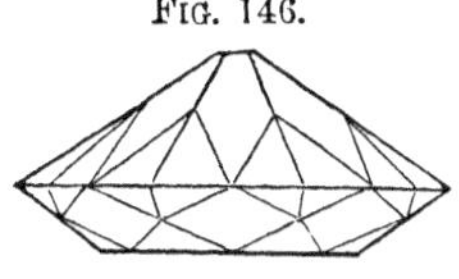

Fig. 146 represents a very beautiful diamond known as the Pigott diamond, which weighed 47 carats, and was sold for about $120,000.*

419. **Graphite, or Plumbago,** is the second allotropic form in which carbon occurs uncombined in nature. It has a metallic, leaden-gray luster, feels unctuous to the touch, and is generally known as "black-lead," although it has no trace of lead in its composition.

It is found chiefly in the older rocks (in many localities in the United States), chiefly in beds or rounded masses, but sometimes crystallized in flat six-sided prisms. It is never found perfectly pure, but usually contains a little iron and some other accidental impurities. Like the diamond, it can not be fused or volatilized by the action of the most intense heat; it burns, however, in oxygen gas, forming carbonic acid.

The principal use to which plumbago is practically applied is for the manufacture of "lead pencils." Most of the ordinary pencils now used are manufactured from a factitious paste, made of powdered plumbago, antimony, and sulphur fused together, and cast into blocks. These blocks are then sawed into small rectangular prisms, which are subsequently inclosed in cylinders of cedar wood. The best drawing-pencils are, however, made, by reducing the plumbago to a fine powder, freeing it from impurities, and then subjecting it to enormous hydrostatic pressure, simultaneously with the abstraction of all remaining traces of air by means of an air-pump. A coherent block is thus

* This diamond is not in existence, but was destroyed by a Turkish pasha in order to prevent it from falling into the hands of his enemies.

So rare are diamonds of large size, that it is stated that the whole number known to exceed 33 carats in weight does not exceed nineteen.

Questions.—What is graphite? In what conditions does it occur in nature? What is said of its infusibility? What of its practical applications?

obtained, which is subsequently sawed into bars. The particles of plumbago, although apparently very soft, are in reality extremely hard, and the steel saws employed to cut it rapidly wear out. Plumbago is also used for the manufacture of melting pots or crucibles, for the lubrication of the bearing surfaces of machinery, and for imparting a luster to iron.

Several modifications of graphite may be procured artificially. When cast iron is melted with an excess of charcoal, it dissolves a portion of the carbon. This carbon, when the iron is allowed to cool slowly, crystallizes out in the form of large and beautiful leaflets of graphite.

420. **Gas Carbon.**—Another exceedingly interesting variety of graphite is formed in the interior of the retorts used for the production of coal-gas. This substance (which may be procured in abundance at all gas-works) is known as "gas carbon." It possesses a luster resembling that of a metal, a hardness sufficient to enable it to scratch glass, and is one of the purest forms of carbon.

421. **Coal.**—The third allotropic modification of carbon includes all the varieties of mineral coal, wood, charcoal, lamp-black, soot, animal charcoal, etc., etc.

422. **Mineral Coal** is the product of an accumulated vegetation, which flourished mainly during a particular period of the earth's history, known in geology as the "carboniferous epoch."

It occurs on the earth in veins, or strata, enclosed between other strata of limestone, clay-slate, or iron ore.

We know that coal is of vegetable origin, because in every coal-mine we find leaves, trunks, and fruits of trees in immense numbers, many of them in the most perfect state of preservation; so much so, that the botany of the coal period can be studied with nearly as much certainty as the botany of any given section of the present surface of the earth; and, furthermore, whenever coal has not been too much changed by heat and pressure, a thin layer of it exhibits, under the microscope, all the ducts and vessels of the plant to which it originally belonged.

Coal consists, like vegetable matter in general, of carbon, hydrogen, and oxygen, with a small proportion of nitrogen. It contains, in addition, variable quantities of saline and earthy substances, which always enter into the composition of plants. These matters, when coal is burnt, are left unconsumed, and, together with some impurities, constitute its ashes.

423. **Anthracite Coal** differs from bituminous in this respect—that its original volatile constituents, oxygen, hydrogen, etc., have been mainly driven off by the agency of heat, leaving carbon in a dense and nearly pure

QUESTIONS.—How may graphite be formed artificially? What is gas carbon? What are its properties? What is the third allotropic form of carbon? What is mineral coal? What proof have we of its vegetable origin? What is the constitution of coal? What occasions the difference between anthracite and bituminous coal?

condition behind; bituminous coal, on the contrary, not having been exposed to the same degree of heat, retains its original vegetable constitution in a great degree unaltered.* When bituminous coal is ignited, its volatile constituents are expelled by heat, and burn with flame and smoke; while anthracite, from its previous deprivation of these substances, burns without flame or smoke.

424. **Coke** is bituminous coal heated apart from air, until its volatile constituents are in a great measure expelled. It produces a more steady and intense heat than the coal from which it is derived, and evolves no smoke.

425. **Charcoal** is that form of carbon which results from depriving animal and vegetable substances of their volatile constituents.

This is usually effected by the agency of heat; but the application of heat is not essential, since wood immersed in sulphuric acid, or buried for a long period in the earth, becomes converted into charcoal.

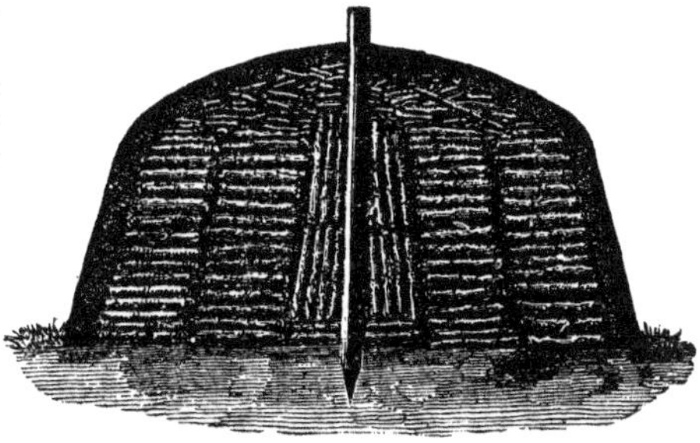

Charcoal is usually prepared by firing wood in mounds or pits, covered with turf or soil in such a way as to exclude in a great degree the admission of air, and thus prevent complete combustion. Fig. 147 represents the arrangement and construction of a "charcoal mound or heap." If the diameter of the heap be 30 feet or more, the operation is not complete in less than a month, and the slower the combustion the greater the product of charcoal. When the wood is thoroughly charred, the admission of air is entirely cut off, and the combustion ceases. The charcoal produced retains the form of the wood, but is much reduced in size; generally not amounting to more than three fourths of the bulk of the wood, and never exceeding one fourth of its weight. The nicest kinds of charcoal, such as are used in the manufacture of gunpowder, are prepared by heating wood in close iron cylinders.

FIG. 147.

426. **Soot** is coal in a state of minute division resulting from the imper-

* Wherever the strata inclosing coal have been disturbed and altered through the agency of subterranean heat, the coal is generally anthracite; but where the strata remain undisturbed, the coal is generally bituminous. Thus in Pennsylvania, the great coal-fields which are adjacent to the line along which the Appalachian chain of mountains have been elevated, furnish only anthracite; but as we recede from the mountains and go west, the coal becomes bituminous.

QUESTIONS.—What is coke? What is charcoal? How may it be prepared? What is the ordinary process of preparing charcoal? What is soot?

fect combustion of carbonaceous gases. Lamp-black is generally applied to designate the soot produced by the imperfect combustion of tar and resinous matters; it is much used in the manufacture of printers' ink and of paint.

Animal charcoal, bone-black, and ivory-black, are names given to the products produced by heating bones, ivory shavings, and like animal substances, in close vessels. The charcoal thus obtained is mixed with ten times its weight of phosphate of lime.

427. **Properties.**—Carbon in the form of charcoal is a black, brittle, insoluble, inodorous, tasteless substance. At ordinary temperatures it has little or no affinity for the other elements, and is, consequently, one of the most unchangeable of all substances. Grains of wheat charred at Herculaneum nearly 2,000 years ago, still retain their form. Wooden posts, if charred at the end before being set in the ground, are rendered far more durable. For the same reason, it is a common practice to char the interior of tubs and casks intended to hold liquids.

Charcoal, when subjected to the action of the most intense heat, is infusible, and if air be excluded, it remains unchangeable.*

At high temperatures, however, carbon surpasses all other bodies in its affinity for oxygen, and is, consequently, more suitable than any other substance for depriving the metallic ores or oxyds of their oxygen, and reducing them to a metallic state—an operation termed smelting.

The compounds of carbon with the other elements are termed *carburets*, or *carbides.*

Newly prepared charcoal possesses the remarkable power of absorbing and condensing within its pores, large quantities of certain gases and aqueous vapor. (The explanation of this phenomenon has been already given, § 48.) Charcoal from hard wood, or that which possesses fine pores, exhibits this property in the highest degree, and the gases which are absorbed most abundantly are those which are most readily liquefied by cold and pressure; thus of ammoniacal gas it absorbs 90 times its volume, of carbonic acid, 35 times; of oxygen, 9 times; of hydrogen, 1·75 volumes.

Charcoal in a finely-divided state has also the power of absorbing odoriferous effluvia, and the coloring principles of most animal and vegetable substances. Animal matter, in an advanced state of putrefaction, loses all offensive odor when covered with a layer of charcoal; it continues to decay, but without emitting any ill odor.

* An illustration of this is found in the fact, that charcoal thrown into a blast-furnace, and its access to air being cut off by an envelope of molten metal, will not unfrequently pass through the furnace unconsumed and unaltered.

QUESTIONS.—What is lamp-black? What is animal charcoal? What are the properties of charcoal? What is said of its indestructibility? What of its affinities? Why is carbon uniformly used in the reduction of metallic ores? What are the compounds of carbon with the metals called? What is said of the absorbing power of charcoal? What gases are absorbed most abundantly? What is said of its deodorizing and decolorizing agency? What are illustrations of its deodorizing action?

Advantage has been taken of this property of charcoal to construct a respirator for protection against the inhalation of malarious and infected air. It consists of a hollow case of wire-gauze filled with coarsely-powdered charcoal, and fitted over the mouth and nostrils by straps, as is represented in Fig. 148. All the air that enters the lungs must pass through this charcoal seive, and in so passing, is deprived of the noxious vapors or gases it may contain. For persons engaged in hospitals, dissecting-rooms, the holds of ships, or in the vicinity of sewers, this device is most valuable. Foul water filtered through a layer of powdered charcoal, is decolorized and purified. This action of charcoal may be illustrated by agitating water containing sulphuretted hydrogen in solution, with a small quantity of freshly-burned powdered charcoal; the offensive odor will completely disappear. Sugar-refiners render brown sugar white by passing it in solution through animal charcoal. Ale and porter, subjected to the same treatment, are not only decolorized, but deprived of their bitter principles. In case of poisoning with vegetable poisons, such as opium, morphia, strychnia, etc., one of the best immediate antidotes which can be given is powdered charcoal in water: this absorbs the poisonous principle, and renders it inactive. The decolorizing action of charcoal may be illustrated by filtering porter, port-wine, or water colored with ink, through a small quantity of animal charcoal. (See Fig. 149.) The filtered liquor will be deprived of smell, taste, and color.

FIG. 148.

FIG. 149.

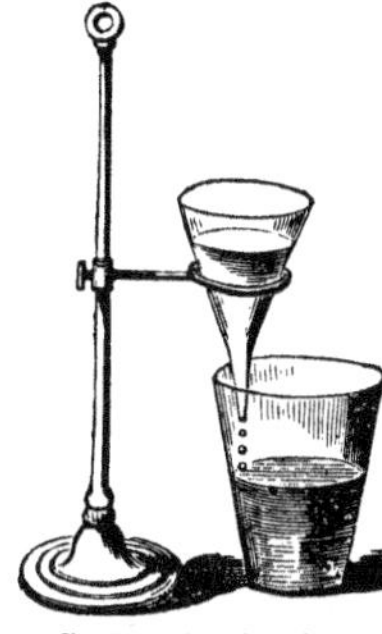

Charcoal loses its absorptive and decolorizing properties by use; but on heating it afresh, it regains them.

Carbon in the form of the diamond is a non-conductor of electricity; but in all its other forms it is an excellent conductor, ranking next to the metals in this respect. In a state of fine subdivision, carbon is a bad conductor of heat, but its conducting power increases with its density.

428. **Compounds of Carbon and Oxygen.**—The compounds of carbon with oxygen and hydrogen, and with oxygen, hydrogen, and nitrogen, are innumerable, and constitute the great bulk of the substance of all vegetable and animal products. The consideration of these compounds be-

QUESTIONS.—What advantage has been taken of this property? What are illustrations of the decolorizing action of charcoal? Under what circumstances may charcoal act as an antidote for poisons? What is said of the conducting powers of carbon for heat and electricity? What is said of the compounds of carbon with oxygen?

longs mainly to organic chemistry. With oxygen alone carbon unites directly to form only two compounds—carbonic oxyd and carbonic acid. Their composition may be represented as follows:—

	Symbol.	Composition by weight.
Carbonic oxyd........................	CO	6 carbon. + 8 oxygen.
Carbonic acid........................	CO_2	6 " + 16 "

429. **Carbonic Acid, CO_2** is the sole product of the combustion of pure carbon in oxygen gas or atmospheric air. It is also produced abundantly by all the ordinary processes of combustion, by respiration, fermentation, and by the decay of animal and vegetable products. It exists in a free state in the atmosphere, and in the earth in immense quantities, chiefly in combination with lime, forming carbonate of lime (marble, chalk, etc., etc.).

For an account of its discovery see § 329.

430. **Preparation.**—Carbonic acid may be prepared by burning charcoal in oxygen gas (p. 190); or by allowing a candle to burn as long as it will in a closed bottle or jar filled with air. Practically, however, it is obtained in a pure state, much more conveniently. It being a feeble acid, almost every other acid, which dissolves freely in water, is able to expel it from its compounds; it is, therefore, easily separated from its compounds by the addition of any of the common acids. Thus, fragments of chalk or marble, with a little water, are placed in an open-mouth bottle, or in an evolution flask (see Fig. 150, also Fig. 95), and dilute sulphuric or hydrochloric acid added. The acid seizes upon the lime, and displaces the carbonic acid, which escapes with an effervescence. It may be collected in the usual way over water, or in dry bottles, by the displacement of air.

Fig. 150.

Fig. 151.

431. **Properties.**—At ordinary temperatures and pressures, carbonic acid is a colorless, transparent gas, of a pungent odor, and acidulous taste. It is more than half as heavy again as atmospheric air, its specific gravity being 1·529 (air = 1·000); by reason of its great density, it may be poured from one vessel into another like water. (See Fig. 151.)

Carbonic acid is not inflammable, and extinguishes the flame of burning bodies, even when largely diluted with air, for a candle will not burn in a

QUESTIONS.—What is the composition of carbonic acid? What is said of its formation and distribution? How may it be prepared? How is it obtained practically? What are its properties? What is said of its density? What of its relation to combustion?

mixture of 4 volumes atmospheric air, and 1 volume of carbonic acid.* This property may be strikingly illustrated by placing a lighted candle at the bottom of a deep jar, and then pouring carbonic acid from another vessel upon it, as is represented in Fig. 151. The light will be extinguished as soon as the gas reaches the flame.

Carbonic acid in its pure state is irrespirable, producing, the moment it is inhaled, a spasm of the glottis, which closes at once the air passages of the lungs: an animal immersed in it, therefore, dies of suffocation. When diluted with air, it may be breathed without difficulty, but if the proportion in which it exists in the air exceeds 4 per cent., it acts as a narcotic poison.† A proportion of 10 to 12 per cent. is speedily destructive to animal life, and even so small a quantity as 1 or 2 per cent. is deleterious and depressing. The drowsiness and headache experienced in crowded and ill-ventilated apartments are chiefly due to the accumulation of carbonic acid as the resulting product of respiration.

Many persons have lost their lives, either intentionally or by accident, by sleeping in a confined room with a pan of burning charcoal; also from descending into wells, mines, vats, and sewers in which carbonic acid has accumulated. Accidents of the latter character may be prevented by taking the precaution to lower a lighted candle into the well or vat suspected to contain this gas, before descending into it; if the light remains undiminished, all may be considered safe; but if the flame be extinguished, or even sensibly impaired, there is evident danger. Wells, pits, etc., containing carbonic acid may be freed from it by lowering into them pans of recently-burned pulver-

---

* This property of carbonic acid has been practically applied for the extinguishment of fires in coal-mines—a stream of carbonic acid, generated by passing air through a furnace of coal, being blown into the mine until all its passages were filled with it, and the combustion arrested. In this way, a coal-mine in England that had been on fire for thirty years, and had extended over twenty-six acres, was extinguished in 1851. About 8,000,000 cubic feet of gas were required to fill the mine, and a continuous stream of impure carbonic acid was forced in by the agency of a steam-jet, day and night, for about three weeks. The difficulty lay not so much in putting out the fire, as in cooling down the ignited mass, so that it should not again burst into a flame on the readmission of air, and in order to effect the necessary reduction of temperature, water was blown in along with the carbonic acid, in the form of a fine spray, or mist. Subsequently, cold air mixed with the spray was thrown in; and in a month from the commencement of operations, the fire was found to be completely extinguished.

A portable arrangement for extinguishing fires, termed the "Fire Annihilator," embodies the same principles. It consists, essentially, of a tin or sheet-iron case, containing a substance holding carbonic acid in combination, together with a bottle of sulphuric acid. By means of a simple arrangement, this bottle of acid may be broken, when its contents, mixing with the solid, evolve carbonic acid; and this, flowing out from apertures in the case, fills the apartment, and extinguishes the fire.

† By a narcotic poison we understand one which produces sleep and insensibility, terminating, if taken in sufficient quantity, in death. Opium and morphia are examples.

QUESTIONS.—What of its relation to respiration? What are illustrations of the poisonous influence of carbonic acid? What precautions should be taken before descending into wells, sewers, etc.?

ized charcoal, or fresh slacked lime, or by showering down cold water—all of which substances absorb the gas freely.

To resuscitate those who have been exposed to the poisonous action of carbonic acid, dash cold water upon them freely, and assist the circulation by friction of the extremities.

432. Water at ordinary temperatures and pressures absorbs about two thirds of its bulk of carbonic acid; but it will take up much more if the pressure be increased. The quantity absorbed is in exact ratio with the compressing force, the water dissolving twice its volume when the pressure is doubled, and three times its volume when the pressure is trebled. On removing the pressure the greater part of the gas escapes, and produces that effervescence which we see when a bottle of ginger-beer, soda-water, cider, or champagne is opened.

Most of the beverage sold under the name of soda-water does not contain a particle of soda, but is merely water impregnated, by mechanical pressure, with about eight times its bulk of carbonic acid. In fermenting liquors inclosed in bottles, on the contrary, the carbonic acid is gradually evolved by the process of fermentation in the interior of the bottle. As fast as it is set free, the liquor dissolves it, the pressure of the gas upon the inner surface of the bottle increasing at the same time. The pressure thus generated is enormous, and beyond a certain limit the cork will either be forced out, or the bottle will burst. If the cork be withdrawn, the confined gas will drive out the liquor in its own eagerness to escape. The manufacture of champagne is always carried on in vaults far below the surface of the earth, in order to secure a low, and at the same time a uniform temperature. The reason of this is, that the absorption of carbonic acid by the liquor is greatly assisted by a reduction of temperature, and a rise of a few degrees of the thermometer in the vault is sometimes accompanied by the breakage of thousands of bottles.

Fermented liquors, by the escape of their carbonic acid, are rendered flat and insipid. A thick, viscid, or glutinous liquor, like porter or ale, retains the little bubbles of carbonic acid as they rise through it, and is thereby caused to froth; but a thin liquor, like champagne or cider, which allows the bubbles to escape freely, only sparkles.

A solution of carbonic acid in water has a pleasant, acid taste, and temporarily reddens blue litmus paper. The solvent powers of such a solution are far more extensive than those of pure water; and the hardest rocks and minerals are gradually disintegrated and broken down by the long-continued action of water charged with a small proportion of this gas.

433. **Solidification of Carbonic Acid.**—When carbonic acid at 32° F. is subjected to a pressure of 36 to 38 at-

QUESTIONS.—What is the antidote against poisoning with carbonic acid? What is said of the absorption of carbonic acid by water? What is ordinary "soda-water?" What is the source of carbonic acid in fermenting liquors? What takes place when a fermenting liquor is bottled? When does a liquor froth, and when sparkle? What is said of the solvent power of carbonic acid in solution? What is said of the solidification of carbonic acid?

mospheres, it condenses into a liquid as transparent and colorless as water. If a stream of liquefied acid be allowed to escape into the air, it freezes by its own evaporation into a white, snow-like solid.*

* The compressing force used to effect the liquefaction of carbonic acid is that of the elasticity of the gas itself. The experiment may be performed by generating carbonic acid in a closed glass tube, as has been previously explained (see § 178); but usually an apparatus constructed for this particular purpose is employed. This consists of two cylindrical vessels, Fig. 152, each of wrought iron, and each sufficiently strong to withstand

Fig. 152.

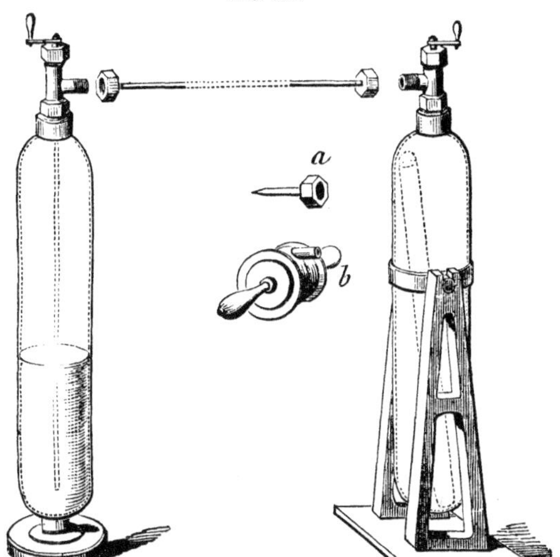

a pressure of 4,000 lbs. per square inch. One of these vessels serves as a generator, and the other as a receiver, and both are furnished with stop-cocks of a peculiar construction. The generator is furnished with an axis, and is mounted upon an iron frame, so that it may revolve in a vertical plane. The receiver is supplied with a tube, which goes nearly to the bottom, and the generator with a cylindrical copper vessel which admits of being filled with oil of vitriol.

The operation is conducted by charging the generator with a solution of bi-carbonate of soda, and the copper vessel with sulphuric acid. The stop-cock of the generator being now firmly closed, the generator itself is revolved upon its axis, by which means the oil of vitriol contained in the copper vessel runs out upon the carbonate of soda, and occasions a liberation of carbonic acid. After a time, when the action is complete, the receiver, which is immersed in a freezing mixture, is connected by means of a metallic tube with the generator, and the stop-cocks being opened, the carbonic acid contained in the generator rushes over into the cold and empty receiver, and becomes in part condensed.

Question.—Give a general description of the process.

In this condition it wastes away slowly, and may be handled and molded with ease. If suffered to remain in contact with the skin, however, it burns like a red-hot iron.

When a little mercury is placed in a porcelain cup and covered with solid carbonic acid, the addition of a few drops of ether occasions so rapid an evaporation that the mercury is immediately frozen, and may then be hammered and drawn out like lead. In this way ten pounds of mercury may be frozen in less than eight minutes.

434. Lime-water brought in contact with carbonic acid gas rapidly absorbs it, and becomes milky from the formation of carbonate of lime (chalk). This reaction, therefore, constitutes a test for the presence of carbonic acid. Thus if we expose fresh lime-water to the air, a pellicle of carbonate of lime soon forms upon its surface, proving the presence of carbonic acid in the atmosphere. In like manner, by blowing through a tube (see Fig. 153) into a vessel of lime-water, we can demonstrate the abundant presence of carbonic acid in the air expelled from the lungs. The milkiness occasioned by the contact of carbonic acid with lime-water, disappears when an additional quantity of acid is taken up by the solution—carbonate of lime being soluble in an excess of carbonic acid. Many natural waters, by virtue of an excess of carbonic acid contained in them, hold very considerable quantities of lime in solution, and are thereby rendered "hard." When such waters are heated or agitated with air, a portion of the carbonic acid escapes, and the lime is precipitated—forming in boilers and tea-kettles, and in the channels of streams, incrustations of lime.

FIG. 153.

435. **Petrifactions.**—It often happens, when an organic substance is placed in water holding lime in solution by virtue of an excess of carbonic acid, or other mineral matter, that its particles, as they decay, are replaced by particles of mineral matter, until at last all the organic particles disappear, and a stony mass is substituted, which resembles the original substance in

---

The stop-cocks are now closed, the vessels disconnected, and the generator opened and freed of its contents. It is then charged afresh, and the operation repeated as before; five or six repetitions being necessary before any very considerable quantity of liquefied acid becomes accumulated in the receiver.

The liquefied gas can be drawn off from the receiver by means of a jet, *a*, screwed on to its stop-cock. When a portion is discharged by means of this jet into a metallic box, *b*, fitted with perforated wooden handles, a part of the liquid gas assumes a solid condition in consequence of the intense cold developed by the evaporation of another portion, and the box becomes filled with a white solid, like dry snow.

QUESTIONS.—What are the properties of the solidified gas? What is a test for the presence of carbonic acid? What are illustrations? Under what circumstances does carbonate of lime dissolve in water? When is it deposited? What are petrifactions?

form and structure, and not unfrequently in color. This result is termed petrifaction. It is a mistake, however, to suppose that the original particles are converted into stone; for the process of petrifaction is one of replacement, and not of conversion, *i. e.*, a particle of mineral matter of the same form being substituted for each organic particle.

436. The presence of carbon in carbonic acid may be demonstrated by dropping a piece of ignited potassium into a small flask filled with the dry gas. The potassium, by depriving the carbonic acid of its oxygen to form potash, liberates carbon, which is deposited in the form of black particles upon the walls of the glass. This experiment, which is a very striking one, may also be performed by igniting a bit of potassium in a glass tube, through which a current of dry carbonic acid is at the same time transmitted.

437. Carbonic acid is evolved from the earth in many localities, particularly in volcanic districts. At one locality near Vesuvius in Italy, it is estimated that 600 lbs. weight are discharged every twenty-four hours.

438. The salts formed by the union of carbonic acid with the protoxyds of the metals, are numerous and important, and are termed CARBONATES. They are easily decomposed by contact with the stronger acids, and, with the exception of the carbonates of the alkalies, they are for the most part insoluble in water.

439. **Carbonic Oxyd, CO.**—When carbonic acid is passed over red hot coal, or metallic iron, it loses half of its oxygen, and becomes converted into carbonic oxyd.

This reaction is often witnessed in coal fires. The fuel in the lower part of the grate, which has free access to air, generates by its combustion carbonic acid. This passing up through the interior of the fire, where the supply of air is limited, is deprived of half of its oxygen, and becomes carbonic oxyd, while at the same time the carbon of the heated fuel which has entered into combination with the removed oxygen furnishes another equal quantity of the same gas. On coming in contact with the air at the top of the fire, the carbonic oxyd ignites, and burns with a flickering, pale-blue flame. This phenomenon may be particularly noticed in a charcoal fire, when fresh coal has been recently added.

Carbonic oxyd is a transparent, colorless gas, which is much more poisonous than carbonic acid; and the inhalation of air containing one two hundredths of it, for any considerable length of time, is said to be fatal. Carbonic oxyd may be obtained with facility by heating crystallized oxalic acid with five or six times its weight of concentrated sulphuric acid in a glass retort, and collecting over water. As thus prepared, it contains carbonic acid, from which it may be separated by allowing the mixed gases to bubble through milk of lime, or solution of potash.

QUESTIONS.—How may the presence of carbon in carbonic acid be demonstrated? What is said of the natural production of carbonic acid? What of its salts? What is carbonic oxyd? What is a familiar example of its production? What are the properties of carbonic oxyd? How is it prepared?

By generating the carbonic oxyd in the same manner in a test tube fitted with a perforated cork and jet, Fig. 154, the gas may be ignited as it is evolved, and its peculiar blue flame exhibited.

FIG. 154.

440. **Carbon and Sulphur.**

Bi-Sulphide of Carbon, CS_2.—When fragments of sulphur are dropped upon ignited charcoal contained in a peculiarly arranged earthen retort, the sulphur in the form of vapor unites with the carbon, and the product of the combination distilling over, may be condensed, in cooled receivers, into a colorless, transparent liquid—bi-sulphide of carbon.

This compound is highly volatile and inflammable, and is characterized by a most fœtid and peculiar odor. It possesses the power of refracting light in a remarkable manner, and as the most ready solvent known of gutta-percha, India-rubber, and various greasy and resinous substances, it is somewhat extensively applied to manufacturing purposes. It also dissolves sulphur, phosphorus, and iodine—these bodies being deposited again in beautiful crystals by the evaporation of their solvent.

441. **Carbon and Nitrogen.**

Cyanogen, NC_2 or Cy.—This substance, which is one of the most interesting compounds of carbon, strikingly resembles an element, and was the first compound body which was distinctly proved to be capable of entering into combination with the elements in a manner similar to that in which the elements combine with each other.

This discovery, made in 1814 by Guy Lussac, formed an epoch in chemical science, and by originating new views of chemical composition, revolutionized the whole subject of organic chemistry. Since then, numerous other bodies have been discovered, which deport themselves in respect to the elements exactly as cyanogen does—or in other words, as if they themselves were elements. Such compound bodies are known in chemistry as compound or organic radicals;—the elements being simple radicals. (See § 271.)

The name cyanogen (blue-producer, from the Greek *κύανος*, *blue*) is derived from the circumstance that this body forms an essential ingredient in the pigment, "Prussian Blue."

Cyanogen consists of 2 equivalents of carbon, and 1 of nitrogen; but no direct union of these elements can be effected.

For experimental purposes on a small scale, it may be obtained by heating in a small retort, or test tube (see Fig. 155), the salt known as cyanide of mercury, previously reduced to a fine powder, and well dried. The cyanide

QUESTIONS.—What is bisulphide of carbon? What is its method of preparation? What are its properties? What its practical applications? What is said of cyanogen? What are compound or organic radicals? What in chemistry is understood by a radical? What is the chemical constitution of cyanogen? How may it be prepared?

undergoes decomposition, like the oxyd of mercury under the same circumstances (§ 281), yielding metallic mercury and gaseous cyanogen, which should be collected over mercury.

Fig. 155.

442. **Properties.**—Cyanogen is a transparent, colorless gas, with a pungent, peculiar odor, somewhat resembling that of peach kernels; it is nearly twice as heavy as atmospheric air, and when inhaled, is poisonous. It is inflammable, and burns with a beautiful and characteristic purple flame. At a temperature —4° F., it liquefies, and forms a colorless, limpid liquid, which freezes at —30° F. into a transparent solid.

Cyanogen in many of its properties closely resembles chlorine, and like it unites with hydrogen to form an acid, and with the metals to form salts, termed cyanides, which latter possess the characteristic properties of the haloid salts.

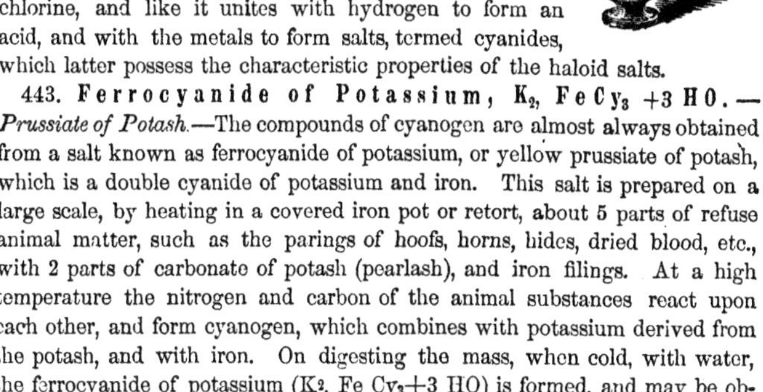

443. **Ferrocyanide of Potassium, K_2, $FeCy_3$ +3 HO.**—*Prussiate of Potash.*—The compounds of cyanogen are almost always obtained from a salt known as ferrocyanide of potassium, or yellow prussiate of potash, which is a double cyanide of potassium and iron. This salt is prepared on a large scale, by heating in a covered iron pot or retort, about 5 parts of refuse animal matter, such as the parings of hoofs, horns, hides, dried blood, etc., with 2 parts of carbonate of potash (pearlash), and iron filings. At a high temperature the nitrogen and carbon of the animal substances react upon each other, and form cyanogen, which combines with potassium derived from the potash, and with iron. On digesting the mass, when cold, with water, the ferrocyanide of potassium (K_2, Fe Cy_3+3 HO) is formed, and may be obtained, by filtering and evaporating the solution, in splendid, yellow, flat crystals. In this condition it forms an important article of commerce.

444. **Prussian Blue.**—When a solution of ferrocyanide of potassium is added to a solution of peroxyd of iron,* a beautiful, deep-blue, bulky precipitate is obtained, which, when washed and dried, constitutes the well-known pigment, Prussian, or Berlin blue—so called from its discovery at Berlin, in Prussia, in 1710. This substance is largely used in painting, in calico-printing, and dyeing, in staining wood and paper, and for concealing or neutralizing the yellow color of linen (an operation termed blueing). Cloth may be dyed blue by first immersing it in a solution of peroxyd of iron, and then in one of ferrocyanide of potassium; the two substances thus meeting in the structure of the cloth, precipitate or produce the color in the very interior of the fibers.

* A solution of peroxyd of iron may be readily obtained by dissolving a few crystals of copperas (green vitriol) in water, adding a little nitric acid, and heating the solution.

QUESTIONS.—What are its properties? What is said of its affinities and compounds? What is ferrocyanide of potassium? How is it prepared? How is Prussian blue prepared? What are its uses? How is cloth dyed of this color?

445. **Blue Ink.**—Prussian blue is insoluble in water and in dilute acids, with the exception of oxalic acid. The blue liquid obtained from its solution in this acid, thickened with gum, constitutes the well-known blue-ink, or writing fluid.

The color of Prussian blue is not very permanent, and is instantly destroyed by the action of the alkalies. The substance itself is formed by the union of cyanogen with iron; and its composition, which is somewhat complex, may be represented by the formula ($3\ Fe\ Cy+2\ Fe_2\ Cy_3$). Although containing cyanogen, a poison, Prussian blue is not poisonous, and is used by the Chinese in large quantities for the coloration of "green tea."*

When ferrocyanide of potassium is added to a solution of protoxyd of iron (green vitriol), it occasions a greenish-white precipitate, which, by exposure to air, rapidly becomes blue.

Ferridcyanide of Potassium.—When chlorine gas is passed through a solution of ferrocyanide of potassium, a salt crystallizing in ruby red crystals is obtained, which contains a larger proportion of cyanogen than the ferrocyanide of potassium; and is known as the ferridcyanide of potassium, or the red Prussiate of potash. When added to a solution of the protoxyd of iron, it produces a dark-blue precipitate, but with solutions of the peroxyd it forms no precipitate. By the use, therefore, of the ferro and ferrid cyanides of potassium, chemists are easily able to distinguish between salts of the peroxyd and salts of the protoxyd of iron.

446. **Cyanide of Potassium, KCy.**—When 8 parts of ferrocyanide of potassium, 3 of carbonate of potash, and 1 of charcoal, are exposed to a strong red-heat in an iron crucible, a compound of cyanogen and potassium is obtained—the cyanide of potassium. This salt, when pure, somewhat resembles white porcelain in appearance; it is freely soluble in water, and when taken into the stomach, is a deadly poison. The hands of the workmen who use this salt are also liable to ulceration.

The solution of cyanide of potassium in water possesses the property of dissolving most of the metallic oxyds, especially those of the precious metals; it is, on this account, therefore, extensively used for the preparation of the gold and silver solutions employed in electro-gilding and plating. A solution of cyanide of potassium will dissolve out the black marks of "indelible ink," which is a solution of the oxyd of silver.

447. **Hydrocyanic Acid, HCy.**—*Prussic Acid.*— This compound, so remarkable for its poisonous properties, is

* The progress of chemical science is strikingly illustrated by the fluctuations in the price of this pigment;—thus in 1770, its price was $10 perpound; in 1815, $3; in 1825, 60 cents; and at the present time, about 30 cents.

QUESTIONS.—What is blue ink? What is said of the permanency of the color of Prussian blue? What is its composition? What is the reaction of ferrocyanide of potassium and protoxyd of iron? What is ferridcyanide of potassium? What are its reactions with the solutions of the oxyds of iron? How is cyanide of potassium formed? What are its properties? What its practical applications? What is said of the formation of prussic acid?

formed by the indirect union of cyanogen and hydrogen. It is easily obtained by distilling cyanide of potassium with dilute sulphuric acid.

The reaction is similar to that involved in the production of hydrochloric acid from common salt and sulphuric acid (§ 358), thus:—

$$\underbrace{KCy}_{\text{Cyanide of potassium.}} + \underbrace{SO_3, HO}_{\text{Sulph. acid.}} = \underbrace{KO, SO_3}_{\text{Sulph. potash.}} + \underbrace{HCy.}_{\text{Hydrocyanic acid.}}$$

In its pure state, hydrocyanic acid is a colorless, transparent liquid, with a feeble acid reaction. It is lighter than water, and so extremely volatile, that if a drop be allowed to fall upon a glass plate, a part of the acid will be frozen by its own evaporation. Its vapor has an odor of peach-blossoms or bitter almonds, and both of these substances owe their peculiar flavor in part to the presence of this acid in their composition.

Hydrocyanic acid is the most fatal of all the poisons known to the chemist. A single drop of the concentrated acid upon the tongue of a large dog produces immediate death; and a slight inhalation of its vapor occasions very disagreeable sensations. When largely diluted with water, it is sometimes given in medicine in very minute doses. Ammonia, brandy, and chlorine are its best antidotes. A suspension of animation occasioned by an over-dose of it does not always result in death, if proper remedies are employed.

Physiologists are not fully agreed as to the cause of the almost instantaneous death occasioned by this acid. By some it is supposed to act upon the vital organs by reason of a sympathetic shock transmitted to the nerves; and by others the effect is ascribed to an almost immediate absorption of the poison into the system.

Various parts of many plants belonging to the order *Rosaceæ*, such as bitter almonds, the kernels of plums and peaches, the leaves of the cherry-laurel, etc., yield, on distillation with water, a sweet-smelling liquid containing hydrocyanic acid.

448. **Cyanogen and Oxygen.**—Cyanogen further resembles an element in the circumstance that it is capable of uniting with oxygen, in several proportions, to form acids, which, in turn, unite with bases to produce salts. The two best known of these oxyds, cyanic and fulminic acids, have an identity of chemical constitution, but entirely different properties.

449. **Cyanic Acid, CyO** is a highly volatile liquid, which decomposes quietly, but so readily, that it is exceedingly difficult to preserve it in unaltered condition; its salts are termed cyanates.

450. **Fulminic Acid, Cy_2O_2**, which, like cyanic acid, is composed of equal atoms of cyanogen and oxygen, is not known in a separate state. Its compounds with the metallic oxyds are termed fulminates, since they explode,

QUESTIONS.—What chemical reaction is involved in its preparation? What are the properties of prussic acid? What is said of its poisonous qualities? What are its antidotes? How is it supposed to occasion death? From what vegetable productions may prussic acid be obtained? In what other respects does cyanogen resemble an element? What is said of cyanic acid? What of fulminic acid?

from the slightest disturbing causes, with fearful violence. The compound with mercury, termed "fulminating mercury," is prepared by dissolving 1 part of mercury in 12 parts of nitric acid, and mixing the solution with an equal quantity of alcohol; on the application of gentle heat, chemical action ensues, accompanied by the evolution of copious white fumes, and the fulminate separates in white crystalline grains. As thus obtained, it constitutes, when mixed with six times its weight of saltpeter, and made into a paste with water, the composition used for filling percussion caps.

Besides the fulminate of mercury, analagous compounds may be formed in a similar manner with oxyds of silver, copper, zinc, and other of the elements. All of them are exceedingly dangerous to handle, and the fulminate of silver ranks next to the chloride of nitrogen in explosive character; thus, it explodes under water when heated to 212° F., and also when in a moist state by friction with a hard body; when dry, the touch of a feather, or the vibration of the house occasioned by the rolling of a carriage, is also sufficient to cause its violent decomposition. The fulminic acid separates, on exploding, into nitrogen, carbonic oxyd, and the vapor of water, the metal being set free.*

451. **Compounds of Carbon and Hydrogen.**—The compounds of carbon with hydrogen are numerous, and are all derived from the decomposition of bodies of an organic origin. Some of these are liquid, some solid, and others are gaseous.

The consideration of two of them only properly pertains to the department of inorganic chemistry. These are, light carburetted hydrogen gas, C_2H_4, and heavy carburetted hydrogen, or Olefiant gas, C_4H_4.

452. **Light Carburetted Hydrogen,** C_2H_4.—*Marsh Gas; Fire-damp.*—This gas occurs abundantly in nature. It is evolved from rock-fissures in coal mines, and forms, in connection with atmospheric air, an explosive compound, known to the miners as "fire-damp."† It is also a constant product of the putrefactive decomposition of wood and other carbonaceous bodies under water, and may be obtained from this source by stirring the mud at the bottom of stagnant pools, and collecting the gas as it rises by means of an inverted bottle and tunnel. (See Fig. 155.) At Kanawha, in Virginia, this gas rises in immense quantities in connection with salt-water from Artesian

* The detonating bombs with which the life of Napoleon III. was attempted in 1858, were filled with fulminating mercury.

† In this condition it accumulates in the galleries of coal mines in large quantities, and when fired by accidental contact with flame, explodes with fearful violence. The product of the explosive combustion is mainly carbonic acid, so that the workmen in the mine who escape death by burning, are almost certain to be afterward suffocated. By an explosion of this character at the Felling colliery in England, in 1812, 92 persons lost their lives. These accidents have now been in a great measure prevented by the use of the "safety lamp."

QUESTIONS.—How is fulminating mercury prepared? To what use is it applied? What is said of the other fulminates? What is said of the compounds of carbon and hydrogen? What is said of light carburetted hydrogen? What is "fire-damp?"

wells, and being conducted by an arrangement of pipes under the salt-boilers, furnishes sufficient heat by its combustion to evaporate the brine. A similar natural supply of this gas in the town of Fredonia, in New York, has for many years past been extensively applied for illuminating purposes.

Fig. 155.

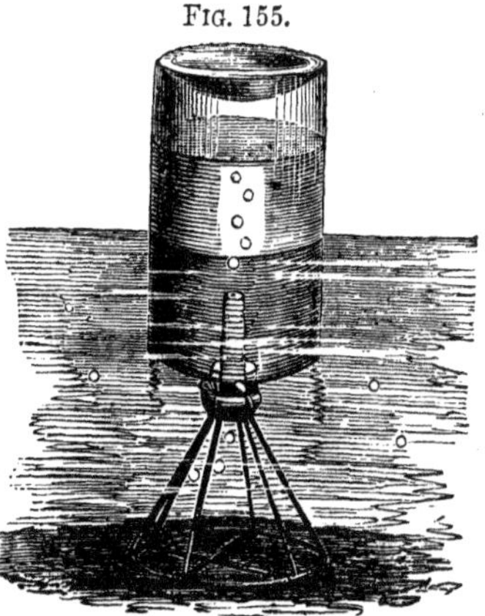

453. **Properties.**—Light carburetted hydrogen is a colorless, inodorous, tasteless gas, slightly soluble in water, and when diluted with common air may be inhaled without injury. Its weight is about half that of air. It does not support combustion, but is itself inflammable, and burns with a yellow, luminous flame. When mingled with air or oxygen gas it forms explosive compounds. 100 parts of this gas by weight, consist of 75 carbon and 25 hydrogen.

454. **Heavy Carburetted Hydrogen. Olefiant Gas,** C_4H_4.—This gas was discovered in 1796 by an association of Dutch chemists, who gave it the name of "olefiant" (oil-producer), from its formation with chlorine of a compound having the appearance of oil. It does not occur naturally, but is obtained by the destructive distillation* of oil, and also in connection with light carburetted hydrogen and some other substances, when coal, resin, tar, asphaltum, fat, animal refuse, and similar inflammable matters are distilled for the purpose of obtaining gas for artificial illumination.

Fig. 156.

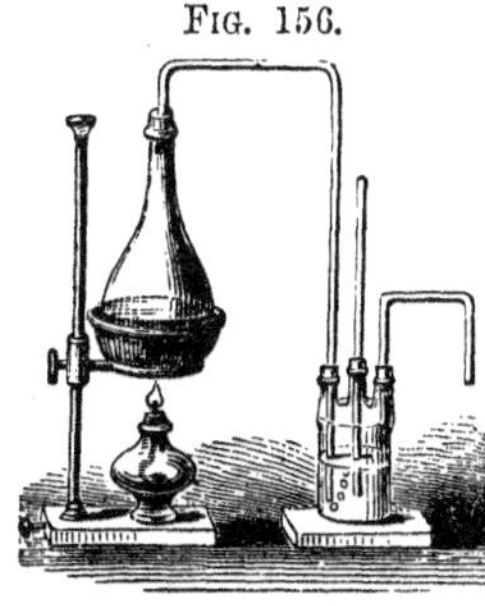

Olefiant gas is easily prepared by heating together 1 measure of strong alcohol with 2 measures of oil of vitriol in a retort or flask capable of holding at least four times the bulk of the liquid introduced. (See Fig. 156.) The gas comes off freely at first, but by degrees the mixture blackens and froths up, so that a careful regulation of the heat is necessary. It is accompanied by the vapor of ether, and toward the close of the process by sulphurous acid in large quantities; but it may be purified by causing it to pass, first through a Woulfe's bottle containing a solution of potash, then through oil of vitriol, and finally collecting over water.

* By destructive distillation, we understand the decomposition of a body subjected to heat in a retort, accompanied by a partial or entire volatilization of its products.

Questions.—What are its properties? What is said of olefiant gas? How is it obtained? What is understood by destructive distillation?

455. **Properties.**—Olefiant gas, as thus prepared, is a colorless gas, with a faint, sweetish odor. It is slightly soluble in water, and has about the same density as air. It was liquefied by Faraday under great cold and pressure, but remained unfrozen at —166° F.

Olefiant gas does not support life or combustion, but is itself very inflammable, and burns with a splendid white light, far surpassing in brilliancy that produced by light carburetted hydrogen. When mixed with oxygen and fired, it explodes with great noise and violence. This may be illustrated by passing bubbles of the mixed gases through water, and igniting them at the surface, care being taken not to communicate fire to the vessel containing the mixture.

The composition of olefiant gas is 2 volumes of hydrogen and 2 of carbon vapor condensed into 1 volume.

When olefiant gas is mixed over water with an equal volume of chlorine, the two gases gradually unite, and form a heavy, sweetish, aromatic liquid. This substance, which collects in oily-looking drops in the water, is commonly known as "Dutch liquid," from its discoverers.

An instructive experiment may be also performed by mixing in a tall jar 2 measures of chlorine and 1 of olefiant gas. On applying a light to the mouth of the vessel, the mixture burns quietly—the chlorine uniting with the hydrogen to form hydrochloric acid, while the carbon is deposited in the form of a dense black smoke.

456. **Illuminating Gas,** prepared by distilling in close vessels bodies rich in hydrogen and carbon, but deficient in oxygen, is always a mixture of olefiant gas, light carburetted hydrogen, carbonic oxyd, and hydrogen in variable proportions, depending upon the nature of the substance, and of the process of manufacture.

The most valuable constituent of all illuminating gases, is olefiant gas; and if this gas could be procured sufficiently cheap, it would be used alone in preference to all others; but as this is not the case, we are obliged, from motives of economy, to be content with a mixture of olefiant and other gases, such as is yielded by the decomposition of oils, fats, resins, coals, and the like substances. Oils and fats, when distilled, yield a product very rich in olefiant gas, which has double the illuminating power of the best coal gas, and three times that of ordinary coal gas. Resins also yield a highly illuminating gas. The first cost, however, of oil and resin is so much greater than that of coal, that the former are not able in an economical point of view, to compete with the latter, although the product of gas from coal is every way in-

---

QUESTIONS.—What are its properties? What is said of the illuminating properties of olefiant gas? What compound does it form with chlorine? What phenomena attend its combustion with chlorine? What is illuminating gas? What is its most valuable constituent? What are the comparative values of oils, resins, and coals for the manufacture of gas?

ferior to that from oil and resin. Thus a pound of coal yields from 3 to 4 cubic feet of gas; a pound of oil, 15 cubic feet; of tar, 12; and of resin, 10.

457. **Coal Gas** is only produced from the bituminous varieties of coal, but all bituminous coals are not fitted for gas manufacture. The production of gas depends upon the application of a high temperature to the coal. At a moderate heat, such as 400° F., the volatile constituents of the coal separate mainly as liquids —oil and tar—with little or no admixture of permanent gas; but at a cherry-red heat, or a little higher, there is an abundant production of gas, with only a small production of tar, etc. That variety of coal known as "cannel," is far superior to all others for the production of gas.

Fig. 157.

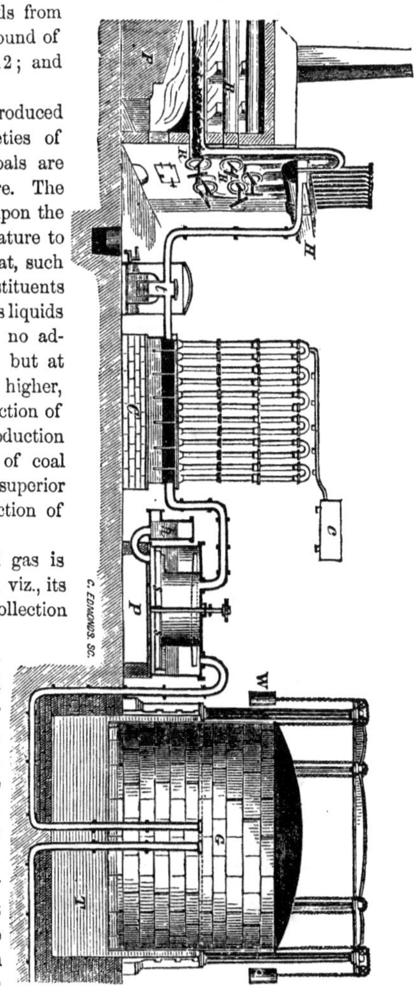

The manufacture of coal gas is divided into three processes, viz., its formation, purification, collection and distribution.

Its formation is always effected in semi-cylindrical tubes of cast-iron, called retorts, arranged in furnaces, as is represented at R F, Fig. 157. The cylinders are closed at the posterior end, and open in front, each being provided with a door, which is made to fit air-tight by means of screws and moist clay. In very extensive gas-works there are from 400 to 500 retorts, of which from 200 to 300 are worked night and day—each retort being charged with about 120 lbs. of coal every 6 hours. The residue left in the retorts after all

Questions.—What coals are used for gas manufacture? Upon what does the production of gas from coal depend? Into what three processes is gas manufacture divided? Describe the formation of coal gas.

the volatile products of the coal have been evolved, is coke, which is raked out, cooled, and used for fuel. It is worth, for heating purposes, as much or more than the coal originally was from which it is derived, and, therefore, the cost of the coal used in the retorts is, theoretically, nothing. Fig. 158 represents the manner of charging and clearing the retorts, and the general appearance of the furnaces of large gas-works.

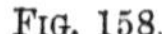

Fig. 158.

The volatile products evolved by heat from the coal are light and heavy carburetted hydrogen, carbonic oxyd, hydrogen, oily vapors, sulphurous acid, sulphuretted hydrogen, ammonia, carbonic acid, aqueous vapor, nitrogen, and small quantities of many other substances. This mixture is totally unfit for illuminating purposes until purified, which is accomplished as follows:—

Question.—What are the volatile products evolved from the coal?

The gases and vapors, as they are evolved from the coal, flow out of the retorts through iron pipes into a receiver half filled with water, which is called the hydraulic main, H, Fig. 157—the extremities of the pipes dipping beneath the surface of the water, in order to prevent the gas from returning into the retorts when the doors are opened. In the hydraulic main a considerable quantity of the matters volatilized with the gas are deposited, viz., ammonia and the oily vapors, which condense into a black, semi-liquid mass, known as "coal-tar." The gaseous products, however, being still hot, retain various other matters in a vaporous state, which, unless separated, would in cooling condense in distant parts of the apparatus, and stop up the pipes. The hot gas is therefore made to pass from the hydraulic main into large upright iron tubes, surrounded with cold water, which are called condensers, in which the remaining vapors condense into a liquid, and trickle down into reservoirs provided for their reception, C, Fig. 157. From the condensers, the gas passes through a cylindrical vessel, P, Fig. 157, filled with cream of lime, kept in a state of constant agitation by means of a wheel, or stirrer, *s s*. This lime removes the carbonic acid, the sulphur compounds, and the remaining ammonia from the gas, which is then discharged into the gasometer, and is ready for distribution.

Dry lime arranged upon a series of shelves, over which the gas is made to pass, is also used for purification. As the gas leaves the lime-purifiers, the aqueous vapor which it always contains in a greater or less quantity, takes up mechanically certain portions of the lime; each little particle being inclosed in a microscopic vesicle or bubble of vapor, which floats in the gas with its burden like a miniature balloon. In the combustion of the gas these vesicles of vapor burst, and their inclosed particles of lime being liberated, occasion the sparkling which may be generally observed in the flame of coal gas.

In the beginning of the distillation, the olefiant gas forms about one fifth of the entire volume, but toward the end of the process, or by too strong a red-heat, its quantity considerably diminishes, while that of hydrogen increases. The great bulk of ordinary coal gas is light carburetted hydrogen; the gas first given off from good coals consisting of 13 of olefiant gas, 82·5 carburetted hydrogen, 3·2 carbonic oxyd, and 1·3 nitrogen. After the lapse of 5 hours the product consists of 7 olefiant gas, 56 carburetted hydrogen, 11 carbonic oxyd, 21·3 hydrogen, and 4·7 nitrogen. The free hydrogen and carbonic oxyd present in coal gas give no light, and are positively injurious, by diluting the illuminating gases.

Gas is sold by the cubic foot, or by the thousand cubic feet; and an ordinary gas-flame is generally estimated to consume from 1 to 1½ cubic feet per hour.

458. **Gas Meters.**—Gas is measured by means of a self-acting instru-

QUESTIONS.—How is the gas purified? What proportion of coal-gas is olefiant? What proportion is light carburetted hydrogen? How is gas sold? How much gas will an ordinary burner consume in an hour? How is gas measured? Describe the construction of the meter.

ment called a meter. Its principle of construction and working may be illustrated as follows: When a number of vessels, of known capacity, are so arranged that (without loss of gas in the interval) one after the other shall be filled by gas in passing—and for this purpose, are inverted in water, into which the gas enters, as in the case of an ordinary gasometer—it follows, that just as many cubic feet will have passed as there are vessels that have been filled. If all these vessels are attached to a common axis and revolve with it, as each in succession fills and rises, the axis will be turned once round, thereby indicating the passage of 4 cubic feet of gas. Now, in the ordinary gas-meter (see Fig. 159), instead of four separate vessels, there is an outer, cylindrical case, A A, more than half filled with water, and a cylindrical drum, divided into four compartments, B B B B, revolving in it. The gas enters into the revolving inner drum, by a pipe at its center, and discharges its gas into the compartment which may happen to be over it, causing the compartment to rise, and the drum to perform a portion of a revolution. When the compartment becomes entirely filled, its edge, D, is lifted so far out of the water that the gas contained in it escapes (passing in the direction of the arrows) into the space between the two drums, and is conveyed away by a tube not shown in the figure. The revolving drum is connected with clock-work, which shows by an index the number of revolutions made, and the capacity of the compartments being known, the quantity of the gas passing through is accurately determined. The meter described is known as the "wet meter," and is the one in most general use. Other arrangements employed for measuring gas, dispense with the water, and are termed "dry meters."*

FIG. 159.

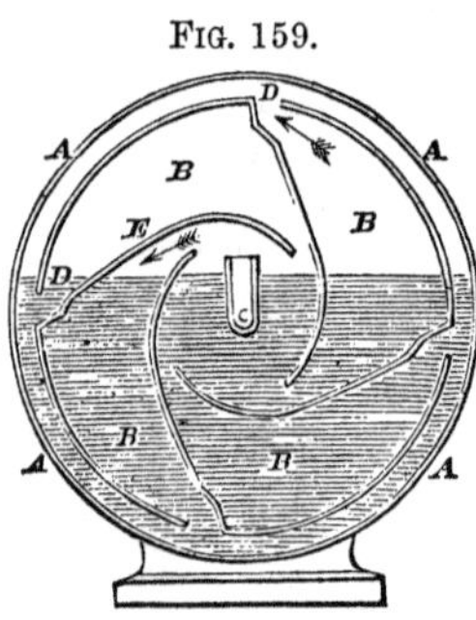

459. Illuminating gas of all kinds, when mixed with air in certain proportions, forms explosive mixtures; care, therefore, should be taken, not to enter an apartment pervaded with a strong odor of gas, with a light, until a thorough ventilation has been effected.

* The gas-meter, when properly constructed, is an exceedingly accurate instrument, though frequent differences arise on this subject between gas companies and their customers. These discrepancies, occurring between one period of consumption and another, and which are always attributed to the meter, arise most frequently from differences in the quality of the gas furnished; for it is a fact not sufficiently known, that the poorer the gas, the faster it will flow through the burners; and, though the meter has registered correctly the volume of gas delivered, it does not follow that the consumer has received an equivalent amount of light. A desirable improvement in this direction would be a meter registering the time or duration of light, rather than the *volume* of gas. Until that is accomplished, gas companies have no inducement to furnish good gas. The worst article with which consumers can be satisfied will be more likely to be manufactured, since it is the cheapest to *produce*, and the dearest to *sell*.

QUESTION.—What is said of the explosive compounds of illuminating gas?

460. **History.**—The fact that a combustible, illuminating gas, is produced during the decomposition of coal by heat, was first noticed in 1664, but it is only within the present century that any general, practical application of this knowledge has been made. Gas was first employed for street illumination in London in 1812, and in Paris in 1815. The majority of householders in London were opposed to its introduction into the streets of that city, and for many years the advocates of the use of gas for general illumination, encountered a great amount of opposition and ridicule.*

461. Gas is manufactured from oil, resins, grease, etc., by causing them to trickle into a retort containing fragments of coke, or bricks heated to redness. Decomposition of the oily substances immediately takes place, and the gas evolved needs only to be cooled to adapt it to immediate use.

CHAPTER VII.

COMBUSTION.

462. **History.**—Fire, in the opinion of the ancients, was one of the four elements of nature—earth, air, and water being the other three.

This doctrine was generally received until the middle of the 17th century (1650), when a new theory, accounting for the various phenomena of combustion, was proposed by Beecher, an eminent German physician and chemist,—which was afterward, toward the latter part of the same century, still further elaborated and explained by Stahl, also a German physician, and one of the most eminent scientific men of his age. This theory, which remained undisputed until after the discovery of oxygen in 1774, was known as the "*Phlogistic Theory.*"

It started with the assumption that there existed in nature a distinct substance, or agency, constituting the principle of fire, called *Phlogiston* (from the Greek φλογιζω, *to burn*). Phlogiston, although never isolated, was believed to exist in all combustible bodies, and to constitute a part of their

* At the present time it is estimated that 6,000,000 tons of coal are annually employed in England for the manufacture of gas, and from 60 to 75 millions of dollars are expended in its production. In London alone, 500,000 tons of coal are annually used, producing five thousand million cubic feet of gas, and yielding an amount of light equal to that which would be evolved from the combustion of ten thousand million of tallow candles, of six to the pound.

QUESTIONS.—What is said of the history and first introduction of gas? How is gas manufactured from oils and resins? What was the original supposition concerning fire? What theory succeeded? Explain the general principles of the phlogistic theory?

structure, and its presence in such bodies was supposed to endow them with the property of burning. When a body burned, phlogiston was liberated, and the light and heat which accompany combustion were attributed to the rapidity which which the phlogiston passed out. When a body was wanting in phlogiston, or had once lost it, it ceased to be combustible, and was said to be dephlogisticated.

For example, according to this theory, a lighted candle burns because it is a compound of candle-matter and phlogiston, which compound, in the action of burning, is decomposed, and the phlogiston, set free, appears, in escaping, in its natural character, as flame. The pure, dephlogisticated candle-matter is also liberated, little by little, as the candle burns away, and when collected, proves to be water and carbonic acid; so that, according to the phlogistic theory, tallow should be regarded as a compound of fire, with water and carbonic acid. Furthermore, "a stick of brimstone burns away with a blue flame and a suffocating vapor, and the residue of its combustion is sulphurous acid. In the language of the phlogistians, brimstone is a compound of two things, sulphurous acid and phlogiston; and when it is suffered to burn, it gives out its phlogiston, or flame of fire, and there remains its dephlogisticated sulphur, or sulphurous acid, in the separated state. Phosphorus, according to the same hypothesis, contains a white, deliquescent acid (§ 405) and phlogiston—the two so loosely united as to be kindled or decomposed by a little friction, or by a slight elevation of temperature; when burned, it sheds its phlogiston, and the phosphoric acid is reproduced."

It had been long before observed, that the metals, with the exception of gold and silver, were changed into rusts, or "calxes," resembling chalk, brick dust, or other highly colored earthy bodies,* when heated to a high temperature in the air. We now know these calxes to be simply oxyds; but the phlogistians, recognizing the only identity of this alteration of the metals with what is undergone by sulphur, phosphorus, or any common combustible when it is burnt in the air, explained the change as follows: they said that each metal was composed of its own rust, or calx, and phlogiston, and that when it was burned in the fire, it gave out its fiery principle, while its ashes or rust remained." Thus, iron was composed of iron-rust and fire; dephlogisticate it, that is, burn it to a cinder, and you have rust.

"Such bodies as wood, coal, and especially charcoal, which give out much heat, and leave apparently little dephlogisticated matter when burnt, were regarded as substances overcharged with phlogiston, and therefore capable of imparting it largely to others. Now, it always was, and still is, desirable to transform ores, such as iron rust in the various iron-stones, into metals, such as iron; and it has long been understood that the best way of doing so, consists in mingling those ores with carbon in some form or other, and heating them in a furnace; a thing but too easily explained by the phlogistic theory, for the carbon had only to pour its phlogiston into the ores to convert them into metallic natures, solid and bright. In the substance of silver and gold,

* Iron-rust (oxyd of iron), oxyd of lead, etc.

however, the phlogiston (fire) was so compacted and inherent, that nothing could take it out of them; and hence they remained fixed in the furnace under all ordinary circumstances."

The phlogiston, once liberated from a metal or combustible, could not, like the dephlogisticated matter—the phosphoric, or sulphurous acid, or the iron-rust—be caught and measured. In the opinion of the ancients, it ascended at once into a boundless space of pure fire, called the "empyrean," which was supposed to inclose the air as the air inclosed the earth; but according to the phlogistians, it was no sooner liberated from a combustible, than it passed into combination with the surrounding atmosphere. It could not, in their opinion, be emancipated from its union with one body, unless another was ready to take it without delay, and the appearance called fire, was the almost instantaneous glance of phlogiston in its passage from one engagement to another. Hence the necessity of the presence of air to the continuance of combustion; and hence Priestley, when he discovered oxygen, supposed it to be common air deprived of phlogiston; since it did not burn of itself, but powerfully supported combustion, by reason of its supposed attraction for the phlogiston contained in combustibles. He therefore called it dephlogisticated air.

Although the phlogistic theory ingeniously explained a great variety of phenomena, there were certain circumstances connected with combustion which could not well be accounted for. Thus it was observed that certain metals were heavier after heating than before: ten grains or ounces of lead weigh more than ten after having been burnt to calx; and ten ounces of iron increase in weight by conversion into rust;—in other words, the metals lead and iron, supposed to be compound bodies, gave off by heating, one of their ingredients, phlogiston, and were thereby converted into elements; and yet the product—the calx—was heavier than the original metal; whereas, if phlogiston was really a material substance, and had escaped from the lead or the iron, the products, after heating, ought to have weighed less. This difficulty was explained by assuming that phlogiston, alone of all substances, was endowed with the specific property of lightness, or levity, so that it buoyed up, or made lighter, every body with which it combined. "This singular evasion of the question of weight only introduced another perplexity; but the good old chemists were equal to the emergency. If the calx or rust of lead, or of any other metal, became lighter, in common balance-weight, by combining with phlogiston—that agent of positive levity—how was it that it also became *specifically* heavier? The calx was comparatively a light stone; but the lead into which it was converted by union with light phlogiston, was a comparatively heavy metal; a cubic inch of the metal being twice as heavy as a cubic inch of the stone. If the particles of an ounce of calx had buoys of fire attached to them, so as at once to change them into particles of lead, and to make them lighter in the aggregate, how should such enlarged and lightened particles produce a metal of so much greater a specific gravity than the unphlogisticated rust?" To this it was replied, "that the phlogisticated particles of calx were not enlarged, but only lightened; the fiery particles were not stuck on to the calx ones like so many vesicles; but they penetrated them, and then compressed them, so that a greater number of the fire-pierced earthy particles (thereby rendered metallic) packed into the same space, and therefore the metal was specifically heavier, though absolutely lighter, than the calx from which it was made."*—BREWSTER.

463. Such is a brief outline of the celebrated phlogistic theory which during the greater part of the last century received the sanction and support of all the chemists and scientific men of Europe. The honor of its overthrow and the establishment of correct views, belongs to Lavoisier, whose decisive experiments were instituted about the year 1780.

He took a glass flask, added to it a certain known weight of metallic mercury, filled the flask with oxygen gas (which had been discovered some years previously), and hermetically sealed it. The weight of the whole was then carefully ascertained. The mercury contained in the flask was then heated to about 600° F., at which temperature it entered into combination with the gas, and formed a calx, or oxyd of mercury. Lavoisier then weighed the flask and contents, and found that it had gained nothing and lost nothing; the phlogiston, therefore, if it had been driven out from the metallic mercury, still remained in or incorporated with the flask and its contents.

The flask being next carefully opened, the air from without was heard to rush into it, indicating the existence of a vacuum in its interior. The mercury, therefore, had not by heating imparted any thing to the gas of the flask, but had really abstracted something from it, and when taken out and weighed separately, was found to have increased in weight. That this increase in weight was due to the abstraction of oxygen, and to its incorporation with the substance of the mercury, he further proved, by decomposing the calx (or oxyd) of mercury (formed in the first experiment) into oxygen gas and metallic mercury, by heating it in a suitable apparatus to a temperature of about 900° F. The two products being carefully collected, their joint weight was found to be the same as that of the calx of mercury employed. These

* "How catholic, elastic, and satisfactory this venerable hypothesis must have been. It was all wrong, indeed, as a substantive doctrine. In one particular it was a sort of reverse of truth. It is not the calxes (ores and rusts) and acids that are simple; it is not the combustibles and metals that are compound; it is exactly the reverse. Sulphur, phosphorus, carbon, and the combustibles, on the one hand, with lead, iron, and the metals on the other, are elementary; the respective acids and calxes of these principles are the compounds. The phlogistians may, therefore, be said to have perceived the relation subsisting between these two classes of bodies upside down, like the figures in a camera obscura. As to the generic idea of phlogiston, erroneous though it was and is, it is extant in science yet; for it is impossible to see wherein caloric differs from it as a scientific conception, although elaborated with immensely greater precision, except that caloric is the matter of heat, while phlogiston is the matter of fire. Both phlogiston and caloric are substances which have no existence whatever in the external world; they have both been convenient, though fictitious representatives of natural realities, and they have both been eminently useful in standing for certain phenomena in their several days, but the latter creation of the materializing tendency of unripe science is not a whit better in essence than the former."—SIR DAVID BREWSTER.

QUESTIONS.—Who overthrew the phlogistic theory? By what experiments was its falsity demonstrated?

experiments, therefore, proved unmistakably that the calx, or red rust of mercury, was a compound of oxygen and mercury, and not an element, as had long been supposed; and that metallic mercury was not a compound of its own calx and the positively light phlogiston, but the real element.

Lavoisier also burned phosphorus in a jar of oxygen, and observed that much of the gas disappeared, and that the phosphorus gained in weight; and that the increase of the one was in the exact ratio of the decrease of the other. Iron wire, also, burned in oxygen, gave a result equal to the weight of the wire employed, plus the weight of the oxygen that had disappeared.

Observing also that the results of combustion in atmospheric air were the same in degree as those in pure oxygen, he next investigated the nature of air, and found that it consisted in part, of oxygen which supported and occasioned combustion, and of another gas which possessed properties entirely opposite, and which we now know to be nitrogen.

The results of the experiments of Lavoisier, therefore, demonstrated that there was no such substance as phlogiston, or the matter of fire; and that when a body, compound or elementary, was burned, it did not give off *imaginary* buoyant phlogiston, but took in *real* weighty oxygen.

Lavoisier commenced his investigations in 1772, and fully announced them in 1784. For eleven years he encountered the opposition of the whole scientific world, with but a single supporter—Laplace, the astronomer. Gradually, however, the new doctrines gained ground, and before the close of the 18th century were generally received.*

From this point discovery rapidly succeeded discovery, until it became at last understood that oxygen was not only the great agent in combustion, but that the respiration of all animals, the processes of vegetation, and the growth, sustenance, and decay of all organic beings were dependent upon it as a constituent of the atmosphere. The true idea of a chemical element was then first arrived at,—affinity or chemical attraction was recognized as an independent force, and the nomenclature of chemistry at present in use was established. In short, the whole science of modern chemistry may be said to date its origin from the epoch of the labors and investigations of Lavoisier.†

* The two great chemists of that day in England, Cavendish and Priestley, never, however, abandoned the doctrines of phlogiston. The former, when it became evident that the new theory of chemistry had won the day, gave up the science in disgust; the latter, becoming involved in theological difficulties, emigrated to Pennsylvania, where he afterward died—maintaining in his correspondence to the last, a defence of his favorite theory.

† LAVOISIER.—No attempt to sketch the history of chemistry can be considered complete without some notice of the life of this celebrated man. He was the son of a rich merchant of Paris, and was born in 1743. He early devoted himself to the study of chemistry, as it was then understood, was made a member of the French Academy at the age of 25, and was put at the head of the national powder and saltpeter works at 33. His great investigations on combustion, the composition of water, atmospheric air, etc., were carried on during the years 1772-83, during which period he filled the office of a receiver, or "farmer-general" of the public revenues. In 1790 he was a prominent member of the famous commission which originated the French system of weights and measures, now generally recognized as the true standard by most scientific men. His labors in other departments of science were also numerous and important. In the common course of

464. **Combustion**, in the strict chemical acceptation of the term, is a chemical process in which at least two elements enter into combination, producing heat and a new compound.

Combustion, in the ordinary sense, is the rapid chemical union of oxygen with a combustible body, attended with an evolution of light and heat.

Every species of combustion with which we are familiarly acquainted is simply a process of oxydation; but combustion may occur without the presence of oxygen, or in oxygen without the sensible evolution of either heat or light. For example, when antimony in powder or copper in the form of thin leaf is presented to chlorine, a combination is instantly effected between these bodies—a chloride of copper or antimony being produced, with an evolution of vivid light and heat; and on the other hand, the decay of wood, or the rust of metals in air—changes effected by union of these substances with oxygen—are true examples of combustion—heat and a new compound being produced without the evolution of light.

465. All bodies may, with reference to combustion, be arranged under one of three classes, viz, supporters of combustion, combustibles, and burnt bodies.

Supporters of Combustion are those bodies which, like oxygen, allow other substances to burn in them freely, but which can not themselves, in ordinary language, be set on fire. It is usual to reckon five supporters of combustion, viz., oxygen, chlorine, iodine, bromine and fluorine.

Combustibles are bodies which, like charcoal, actually burn when sufficiently heated in the presence of a free supporter of combustion.

Burnt bodies are those which will neither burn themselves nor support the combustion of others. They may be made red hot, but do not burn; sand, iron-rust, and earthy bodies are examples of this kind. They are for the most part compounds that have at some time or other been produced by combustion; or in other words, they are bodies which have been already burned, and are no longer fitted to undergo this change. Chemists further

---

events, it might have been expected that the latter years of his life would have been passed amid the admiration and reverence which naturally wait upon the originator of a new system of acknowledged truths. Such, however, was not his fate. He was arrested during the "reign of terror," and thrown into prison, on the wretched charge of having, in his capacity of a public officer, authorized the adulteration of the tobacco of the Republic. When brought before the revolutionary tribunal, he asked for a respite of a few days, in order to complete some researches, the results of which, he said, were important for the interests of humanity. The reply of the judge was, that the Republic wanted no scientific men, and forthwith condemned him to the guillotine, to which he was dragged the next day, May 8th, 1794, in the 52d year of his age.

QUESTIONS.—Define combustion. Is oxygen necessary for combustion? Into what three classes may all bodies be divided in respect to combustion? What are supporters of combustion? What are combustibles? What are burnt bodies?

distinguish and classify burnt bodies under the names of acids, alkalies, oxyds, salts, etc.—MILLER.

466. **Combustion and Explosion.**—Explosion in most, and perhaps all cases, is a species of combustion, differing from ordinary combustion simply in the rapidity of action; thus in combustion, the combustible and the supporter of combustion are brought together by degrees, as in the flame of a candle; but in an explosion the whole action occurs at once.

467. The origin of the heat which accompanies combustion has not been satisfactorily accounted for. Every change in the state of a body we know is accompanied by a change in temperature; but why the union of carbon with oxygen to form a gas, or oxygen with hydrogen to form a vapor, should produce a heat sufficient to melt the most refractory substances, still remains unexplained.

468. In all ordinary cases of combustion, the heat evolved does not depend upon the combustible, but upon the amount of oxygen that enters into combination; or in other words, that combustible will evolve the greatest quantity of heat which is capable, with a given weight, of combining with the most oxygen.

For example, a pound of hydrogen in burning consumes or unites with 8 pounds of oxygen; while a pound of carbon unites with but 2 2-3 pounds of oxygen. A given weight of hydrogen in burning will produce, therefore, three times as much heat as the same weight of carbon.

469. The quantity of heat which a combustible body evolves in combining with oxygen, is the same, whether the combustion takes place slowly or quickly, provided only that the relative quantities of the combining bodies are the same in both instances.

Thus, as much heat is given out in the decay (slow combustion) of a given quantity of wood in the air, as in its quick combustion in a furnace; but in the former case, the heat is much less intense, and often becomes insensible, because, during the long time occupied in the combination with oxygen, the greater part of it is carried away by conduction.

The temperature required to induce combustion, or the combination of any substance with oxygen, is different not only for different substances, but even for the same substance, according as the combustion is to take place rapidly or slowly. Thus phosphorus combines slowly, with oxygen, or exhibits slow combustion, at 77° F., but does not enter into rapid combustion till raised to

---

QUESTIONS.—What is the difference between combustion and explosion? What is the origin of the heat evolved in combustion? To what is the heat evolved by the combustion of a body proportioned? Illustrate this principle. Is the quantity of heat increased by the rapidity of the combustion? Illustrate this. Is the temperature at which combustion occurs constant for the same substance? What are examples of slow combustion?

140° F. Tallow thrown upon an iron-plate not visibly red-hot, melts and undergoes oxydation, diffusing a pale, lambent flame, only visible in the dark. When a coil of thin platinum wire is first heated to redness, and suspended in a glass containing a few drops of ether or alcohol (see Fig. 160), the vapors of these substances, mixed with air, oxydate upon the hot metallic surface, and sustain the wire at a red heat, so long as the supply of mixed vapor and air is kept up, without the occurrence of combustion with flame. The product of the oxydation thus effected, is a pungent, irritating vapor, which affects the nose and eyes unpleasantly.

Fig. 160.

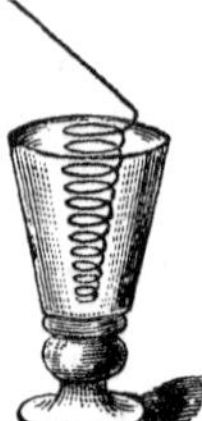

This experiment may be modified by suspending a coil of thin platinum wire, or a ball of spongy platinum, over the wick of a spirit-lamp, supplied with alcoholic ether, (see Fig. 161); on lighting the lamp, and then blowing it out as soon as the metal appears red-hot, slow combustion of the spirit vapor supplied by the capillary action of the wick, will take place, and the platinum will continue to glow for hours.

Fig. 161.

470. In combustion, no loss whatever of ponderable matter occurs—nothing is annihilated; but the products of combustion, when collected and weighed, always exceed the weight of the original substance burned, by an amount equal to the weight of the oxygen absorbed during combustion.

The most simple illustrations of this fact are obtained in the combustion of those bodies which afford a solid residue. Thus, when two grains of phosphorus are burned in a measured volume of oxygen gas, they are found converted, after combustion, into a white powder (phosphoric acid), which weighs 4½ grains, or the phosphorus acquires 2½ grains; at the same time, 7½ cubic inches of oxygen disappear, which weigh exactly 2½ grains.—Graham.

471. The constituents of all ordinary combustible substances—wood, coal, oils, fats, etc.—which give to them their value as fuel, are carbon and hydrogen. These substances also contain some oxygen; but this element contributes nothing whatever to their value as fuel, and the larger the proportion of oxygen in a combustible, the less adapted is it for fuel.

472. **Products of Combustion.**—When combustion takes place with a free supply of air, oxygen unites with the carbon of the fuel to form carbonic acid, and with the hydrogen to form vapor of water. These products being volatile, rise in the atmosphere, and disappear, forming part of the aërial column that ascends from a burning body.

473. The activity of combustion is greatly increased by increasing the num-

---

Questions.—Is any matter lost during combustion? How may this be illustrated? What are the valuable constituents of ordinary combustibles? What influence has oxygen as a constituent of fuel? What are the ordinary products of combustion? How may the activity of combustion be increased?

ber of particles of oxygen which are brought in a given time in contact with the combustible, and by carrying away the gaseous products of combustion, which are no longer capable either of burning or supporting combustion, and which, if allowed to accumulate, would cut off the supply of fresh oxygen. Hence the benefit of blowing a fire, or forcing a stream of fresh air upon it, from a bellows, in order to revive it, or increase its intensity. The influence of a long chimney, in producing a powerful heat in a furnace at its base, by increasing the draft, is similar; while the effects of diminishing the supply of air, by closing the damper, or shutting the door of the ash-pit, is seen in the diminished temperature, and reduced consumption of fuel which occurs under such circumstances.—MILLER.

474. The weight of the air required for the combustion of fuel far exceeds that of the fuel itself; and as the space occupied by a given weight of air is much greater than that of an equal weight of fuel, the bulk of the air employed to effect combustion is immense. For example, it requires 11·45 pounds of air to consume 1 pound of pure charcoal; and as 1 pound of air occupies about 13 cubic feet of space, the pound of charcoal will require for its combustion at least 150 cubic feet of air. As fuel is burned, however, a much larger quantity is employed; thus, anthracite coal requires theoretically 136 cubic feet per pound, but in practice, under steam boilers, 276 cubic feet are necessary.

The amount of heat which a pound of pure charcoal is capable of producing, through its union with oxygen in the process of combustion, is sufficient to convert 13 pounds of water at 60° F. into steam at 212° F. The ingenuity of man can not generate from the combustion of a pound of coal a greater amount of heat than this, or when generated, compel it to evaporate a greater quantity of water.

The quantity of heat which is obtained from fuel in practical operations, falls very far short of its theoretical value. In some of the Cornish steam-engines, of England, which are the best in the world, it is stated that the utmost theoretical quantity has been rendered available; but this statement is doubtful. Under ordinary steam-boilers not more than two thirds of the available heat is ever utilized, and in a majority of cases the proportion does not probably exceed one half.

The reason of this loss of heat in practice, is due mainly to two causes, viz., the heated air escapes up the chimney before it has surrendered to the boiler or heating apparatus, the full amount of heat it is capable of relinquishing; and, secondly, through want of a perfect combustion, the full amount of heat is not evolved from the fuel. The remedy for the first difficulty is to be sought for in improved mechanical arrangements of boiler and furnace; the

---

QUESTIONS.—How is it benefited by blowing it? Why is the temperature and consumption of fuel reduced by closing the draft? What is said of the amount of air required to produce combustion? How much air is absolutely required to burn a pound of charcoal? How much heat will a pound of charcoal in burning evolve? Is the largest possible amount of heat from fuel ever wholly utilized? Why is it in practice that we fail to utilize the full amount of heat derivable from fuel.

remedy for the second pertains to chemistry, and is to be found in perfecting the supply of air. When the supply of air is insufficient, carbonic oxyd becomes in great part the resulting product of combustion, instead of carbonic acid; but for the formation of the first-named gas, only one half the quantity of oxygen is required as for the production of carbonic acid, so that coal may be dissipated in vapor, and may be apparently wholly consumed by one half the amount of air that is usually required in an open fire, under circumstances where the full amount of heat is given out. In such cases a pound of charcoal, instead of emitting heat enough to convert 13 lbs. of water into steam, will only give out one fifth of the heat, and will therefore convert but little more than 2½ lbs. of water into steam.* That so great an amount of loss as this is ever practically experienced, is not probable; but in all furnaces of ordinary construction, the waste of fuel from this source is very great. Owing to the fact that carbonic oxyd is a colorless gas, and as the operations of the furnace appear to go on uninterruptedly, the loss of heat occasioned in this manner is very apt to remain unsuspected.

By admitting, in a proper manner, an adequate supply of air, all the carbon in burning is converted into carbonic acid, and the maximum of heat capable of being evolved from the combustion is generated.

475. **Light of Combustion.**—The light emitted by burning bodies is a direct consequence of the heat evolved in the process of combustion. All solids and liquids (as melted metals), when elevated to a sufficiently high temperature, (977° F.), become luminous.

The color of the light emitted from an ignited substance, depends upon the degree of temperature to which it has been elevated. As the temperature rises, the colored rays appear in the order of their refrangibility; first red, then orange, yellow, green, blue, indigo, and violet are emitted in succession. At about 2100° F., all these colors are produced, and from their admixture, white light results, and the ignited body is then said to be "white-hot."

In all luminous flames, the light is emitted from solid particles highly ignited.

A flame containing no such particles emits but a feeble light, even if its temperature is the highest possible. For example, the flame produced by burning a mingled jet of oxygen and hydrogen, although one of the most in-

---

* The great loss of heat involved in the production of carbonic oxyd, is due not merely to the fact that carbonic oxyd requires less oxygen for its formation than carbonic acid, but the former gas occupies twice the bulk of the latter, and, consequently, renders latent a greater amount of heat.

QUESTIONS.—Under what circumstances will fuel be burned to the best advantage? Upon what does the light which accompanies combustion depend? What relation is there between the light of an ignited substance and its temperature? What is flame? Upon what does the luminosity of flame depend? Illustrate this.

tense sources of heat at our command, is so little luminous as to be barely visible in clear day-light; if, however, we introduce into it a solid body, like lime, the light becomes so augmented that the eye can scarcely support it. When phosphorus is burned in oxygen, the light is most dazzling, but when burned in chlorine, it is extremely feeble; the reason for the difference in these two cases is, that in the first instance the product of combustion is solid, non-volatile phosphoric acid, the particles of which, becoming highly heated by the combustion, are highly luminous; in the second case, the product of combustion is a gas, and the heat which its particles acquire in combustion not being sufficient to render them luminous, little or no light is evolved.

476. **Materials for Illumination.**—The materials ordinarily employed for effecting artificial illumination, are solid or liquid compounds of carbon and hydrogen—coal, oils, tallow, etc.—which are generically termed hydrocarbons.

By heat we decompose them into gaseous compounds of carbon and hydrogen, and in this state only are they available for purposes of illumination. In the combustion of these two elements in the flame of a candle, the oxygen of the air combines with both, but by reason of a superior affinity, it unites first with the hydrogen to form vapor of water, producing, thereby, a most intense heat, but an almost imperceptible light. The hydrogen, in combining with oxygen, abandons the carbon, which, being thus set free in the form of minute solid particles in the midst of the heated space, becomes white-hot, and imparts luminosity to the flame. The moment, however, the incandescent, floating carbon comes to the edge of the flame, it finds the oxygen of the air, unites with it, and becomes converted into invisible gas—carbonic acid—while its place is immediately occupied by another particle of solid carbon.

Between the flame of a candle and the flame of gas-light there is no difference; in the case of the candle, however, the gas is generated and burned at the same time and place—the heat that produces it serving also to inflame it. In the case of a gas-light, on the contrary, the inflammable gas is distilled by heat from the illuminating substances in close vessels in one place, and conveyed by pipes to be burned at a place more or less distant where the illumination is required.

The fact that the combustion of a candle generates gas of the same nature as that produced in ordinary gas manufacture, may be demonstrated by introducing one end of a small tube of glass, *p*, Fig. 162, into the interior of the flame of a large candle, when a portion of the inflammable gas existing there may be drawn off and ignited at the upper extremity of the tube.

QUESTIONS.—What are the materials ordinarily used for artificial illumination? What are hydrocarbons? In what condition only are they available for illuminating purposes? What takes place during their combustion? What difference is there between the flame of a gas-burner and that of a candle? How may the production of inflammable gas in a candle be demonstrated?

FIG. 162.

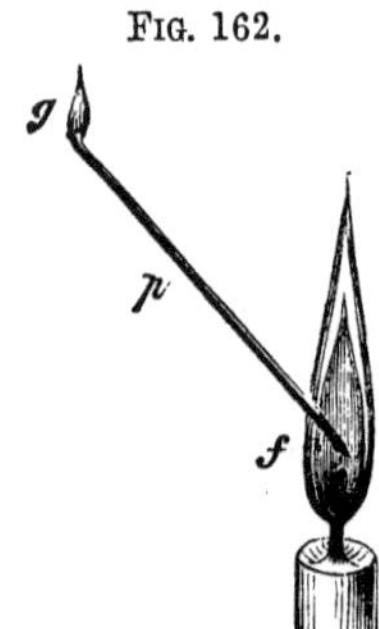

The existence of solid particles in every illuminating flame may be also demonstrated by introducing a cold body into the flame, which so interrupts the progress of combustion that the solid particles are no longer consumed, but are deposited as soot. When we say a lamp smokes, we mean that the carbon contained in the flame is passing off in an unconsumed state.

477. **Combustion of a Candle.**—A candle is an ingenious contrivance for supplying flame with as much melted fat as it can consume without smoking. It is easy to conceive that it would by no means be an impossibility to ignite a stick of wax or tallow by itself; it would, however, be a matter of difficulty, inasmuch as the material would melt away long before it could inflame. Supposing, nevertheless, it could be ignited, then a larger amount of combustible would be on fire than the air could consume, and a large, thick, smoky flame would result. By the use of a wick, this difficulty is avoided.

When the end of the wick which protrudes from the center of the candle is ignited, it radiates sufficient heat downward to melt a portion of the material of the candle, and form a hollow cup filled with liquid combustible around the wick-fibers. The spaces between the fibers of the wick, acting like a series of small tubes, convey the fluid fat by capillary attraction up to the flame, where it is decomposed into gaseous compounds of hydrogen and carbon.

478. **Structure of Flame.**—The flame of every lamp or candle consists of three distinct portions, or rather, cones concentric with one another. The innermost cone, *a*, Fig. 163, is formed entirely of combustible gases, produced by the decomposition of the illuminating material. This is at a temperature below redness, and is consequently non-luminous. Around this is the luminous cone (*b*), the flame proper, where the hydrogen is uniting with the oxygen of the air, and the particles of carbon not having yet done so, are floating about in an incandescent state and radiating light. Beyond the second cone is another film, or casing (*c*), where the oxygen of the air unites with the carbon, and in a properly adjusted flame is entirely consumed. In the flame of a gas-jet the same parts may be also recognized. At the base of every flame a pale blue line of light may be observed; at this point, the supply of oxygen from the air is insufficient to completely and simultaneously consume both the hydrogen and the carbon supplied from the interior of the flame, and there being no solid carbon eliminated, there is consequently but a feeble light.

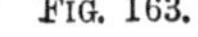
FIG. 163.

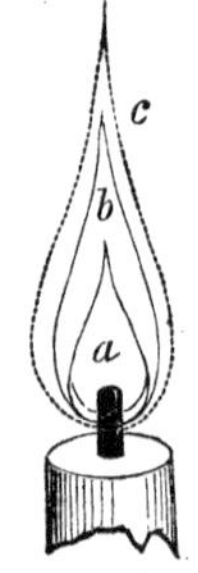

QUESTIONS.—How may the presence of solid particles in a flame be demonstrated? When we say a lamp smokes, what is understood? What is the necessity of the wick in a candle? Describe the structure of the flame of a candle.

The portion of wick within the interior of a candle flame is charred and blackened by the heat, but not consumed, owing to the fact that the burning envelop which surrounds it effectually cuts off all access of air, and thus prevents combustion. For the same reason, also, the interior cone of combustible gases, in every luminous flame, remains unignited.

The tapering and conical form which flames assume, is due to the ascending currents of rarefied air which are produced in the atmosphere by the heat attendant on the combustion.

479. That the combustion of a candle is superficial, and that the flame is a film of white-hot vapor, inclosing an interior portion which can not burn for want of oxygen, may be demonstrated by bringing down upon the flame a piece of thin glass, so as to make a transverse section of the flame; we shall then observe a ring of light surrounding the dark interior part of the flame. This experiment may be still better performed by means of a piece of fine wire gauze. When this is brought down upon the flame of a large and steadily burning candle, the flame will be cut off where it touches the gauze, and the exterior luminous circle will be well defined. (See Fig. 164.)

FIG. 164.

That no combustion can go on in the center of flame, may be shown in various ways; as for example, if we ignite a small quantity of strong alcohol in a saucer, and place a rod of white wood across it for a few seconds (see Fig. 165), it will be found on removing the stick, that it is burned or blackened at only two points, viz., where the flame was in contact with the air. The same thing may also be shown by holding a match stick for a few seconds across the middle of the flame of a spirit-lamp with a large wick. If a fragment of phosphorus be placed in a small circular spoon, ignited, and then introduced into the middle of a large flame, it will be extinguished, but will be rekindled the moment that the spoon is withdrawn from the flame.

FIG. 165.

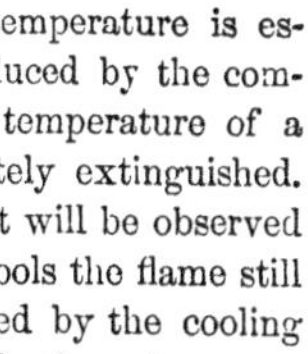

480. In order that a flame should exist, a very high temperature is essential. This is particularly the case with the flames produced by the combustion of the hydrocarbons; and if in any manner the temperature of a flame is reduced beyond a certain limit, it is immediately extinguished. Thus, if a stout copper wire be introduced into a flame, it will be observed that a dark space is produced around it; a second wire cools the flame still further; and a small flame may be completely extinguished by the cooling effect produced by bringing down a coil of wire upon it. If a fine wire-gauze be brought over a flame, the inflammable gases will be so far cooled by passing through its meshes (their heat being conducted off), that they no longer continue in a state of inflammation. (See Fig. 164.) If the meshes are very

QUESTIONS.—Why is not the wick of a candle consumed? Why are flames tapering and conical? What experiments prove that the flame of a candle is superficial? What that no combustion goes on in the interior of the flame? What is essential to the existence of flame? Illustrate this? Why can not a flame pass through a wire-gauze?

FIG. 166.

fine, the conducting power of the metal is sufficient to cool the flame below the point of ignition, even though the wire itself may be red-hot. The inflammable vapor which passes through the gauze may, however, again be kindled by the direct application of flame.

These experiments are well illustrated with a jet of gas issuing under low pressure. If the gauze be held over the jet before it is lighted, and a flame applied above, it will take fire there, but the flame will not pass through to the gas below. (See Fig. 167.) If we place a piece of camphor on the center of the wire-gauze, and apply a flame below, the camphor will melt and pass through the meshes, but will burn only on the under side. (See Fig. 168.)

FIG. 167.

FIG. 168.

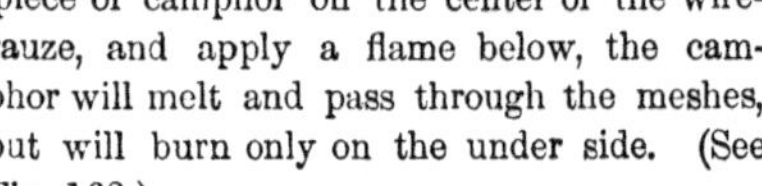

481. **Safety-Lamp.** — These facts, discovered by Sir Humphrey Davy, were beautifully applied by him in the construction of the "Safety-Lamp," which allows the miner to work in safety in an atmosphere pervaded with an explosive mixture of light carburetted hydrogen (fire-damp, see § 452). It consists merely of a common oil-lamp, the flame of which is completely inclosed within a cylinder of wire-gauze. (See Fig. 169.) This completely arrests the passage of the flame; so that, although the lamp be introduced into an explosive mixture, the flame will not pass through the gauze to ignite it.

FIG. 169.

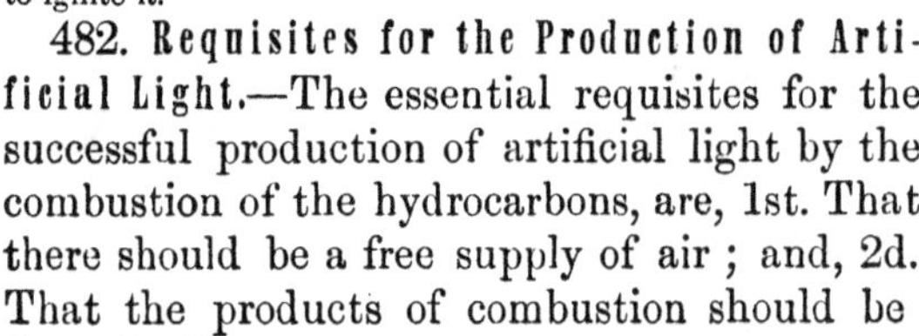

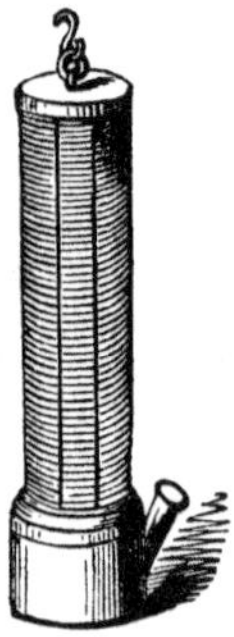

482. **Requisites for the Production of Artificial Light.**—The essential requisites for the successful production of artificial light by the combustion of the hydrocarbons, are, 1st. That there should be a free supply of air; and, 2d. That the products of combustion should be freely conducted off.

These two facts may be illustrated by placing a glass cylinder over a lighted candle, in such a way as to cut off its connection with air *from below;* the flame, in this case, will be extinguished for want of a free supply of air. If the cylinder be now *closed at the top*, but held over the candle in such a way that the air can gain admittance from below, the flame will also be extinguished, since the burnt gases, the products of combustion, are unable to escape, and by their accumulation, prevent combustion. If the cylinder be

QUESTIONS.—To what invention has this principle been applied? Describe the safety-lamp. What are the essential requisites for the production of artificial light? How may these be illustrated?

placed in such a way that the air can gain free admittance below, and escape freely at the top, bearing with it the products of combustion (see Fig. 170), the candle will not only continue to burn uninterruptedly, but its combustion will be more perfect, than when it is allowed to burn openly in the air. The reason of this is, that the ascent of the air, heated by the combustion, creates a rapid current of fresh air from below up through the cylinder—thus supplying more oxygen within a given time and space, which occasions more perfect combustion, and a stronger illuminating flame. Hence the benefit of surrounding a lamp-flame with a glass chimney, open at the bottom and top.

Fig. 170.

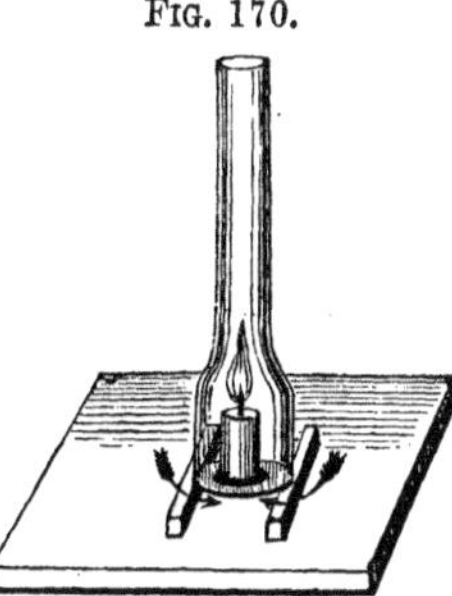

If too much air be supplied to a flame, the inflammable gases burn with a blue and feeble light, an effect which may be seen by blowing upon a common gas-flame, or by watching the exposed gas-lights of shops upon a windy night. In these cases, the gas becomes immediately mixed with the oxygen of the air, which burns up the solid particles of carbon before they are sufficiently heated to afford light.

The necessity of air for the support of flame, is also strikingly shown by the fact, that it is impossible to light a lamp or candle with a match, so long as the sulphur on the end of it is burning freely; since the sulphurous vapor abstracts the oxygen from the air around the wick, in order to form sulphurous acid.

Fig. 171.

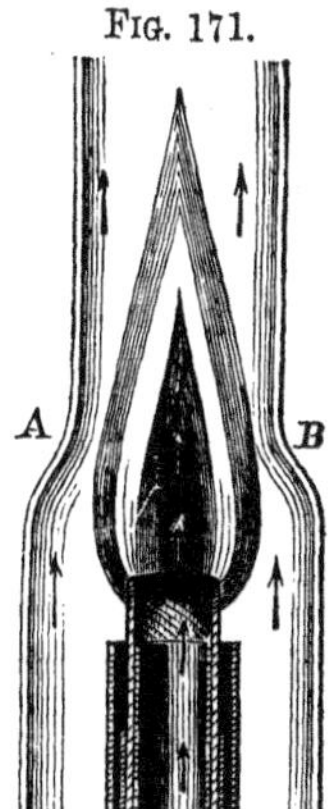

483. **Argand Lamps.**—In an ordinary lamp or candle-flame, the combustion goes on only at those points where the air has free access, viz., upon the outside of the flame, as is indicated by the existence of a dark central portion. If, however, air be introduced into the interior of the flame, combustion is effected both at the center and at the circumference, and the light is increased. This arrangement is practically carried out in those lamps which are fitted with hollow or circular wicks, and which are known as "Argand" lamps, from their inventor. In these, a current of air rushes up, through the hollow wick, into the center of the flame, as shown by the central arrows, Fig. 171, causing it to burn in the form of a hollow ring. The combustion is also made more powerful, by surrounding the flame with a glass chimney, which is usually made conical, or is

QUESTIONS.—What is the effect of admitting too much air to a flame? What is an Argand lamp? Describe its construction?

caused to contract at a certain height above the burner, so as to form a shoulder, A B, in order to deflect the ascending outer current of air, and throw it in upon the flame at an angle. In this way the temperature of the separated carbon particles of the flame is enormously increased, and the greatest quantity of light is produced, from a given amount of fuel.

FIG. 172.

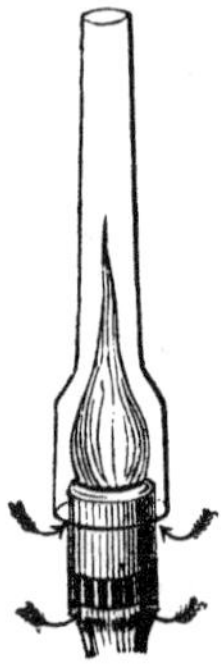

The effect of the draft through the interior of the wick, on the combustion of the inflammable gases, may be readily made apparent by closing, with a piece of paper, the openings in the base of the lamp, through which the air gains admission. The flame will immediately become impaired in brilliancy, burning with a red light, and the evolution of much smoke.

In an Argand lamp we are able to burn the poorer and cheaper oils (those which contain an excess of carbon) without the production of smoke; inasmuch as the greater supply of air effects the entire combustion of carbon; whereas, in an ordinary lamp, by reason of the limited supply of air, we can use only the best oils, or those which contain a large proportion of hydrogen. Fig. 172 exhibits the external construction of an Argand burner, and the direction of the currents of air.

FIG. 173.

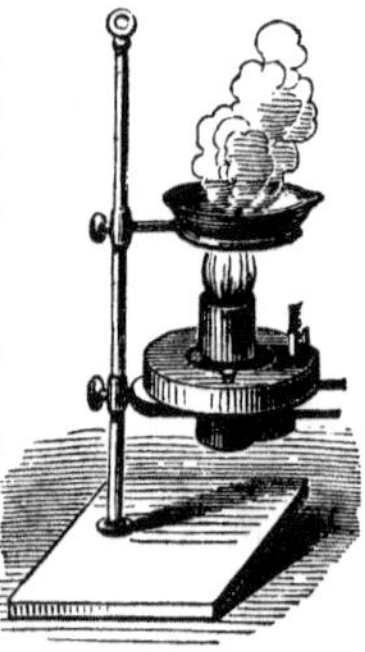

FIG. 174.

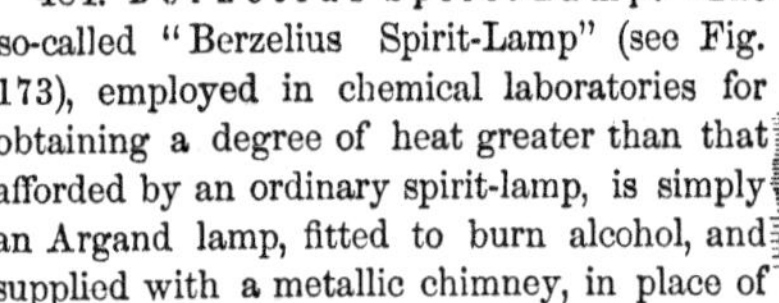

484. **Berzelius Spirit-Lamp.**—The so-called "Berzelius Spirit-Lamp" (see Fig. 173), employed in chemical laboratories for obtaining a degree of heat greater than that afforded by an ordinary spirit-lamp, is simply an Argand lamp, fitted to burn alcohol, and supplied with a metallic chimney, in place of one of glass. The standard to which it is attached is provided with several rings of various sizes, for sustaining crucibles, porcelain dishes, etc., which are to be heated.

485. **The Blow-Pipe.**—The principles upon which the blow-pipe operates are essentially the same as those involved in the construction of the Argand lamp: a jet of air or oxygen is thrown into the interior of a flame, by which the rapidity of combustion is increased, and the heat of the flame powerfully augmented.

The mouth blow pipe consists essentially of a bent tube, generally of brass, terminating in a fine uniform jet. (See Fig. 174). It is usually also constructed with a chamber, or enlargement of the tube, near its small extremity, which serves to collect the moisture which condenses from the breath. When the jet of the blow-pipe

QUESTIONS.—What is the effect of closing the inner draft of this lamp? How is an Argand lamp enabled to burn cheap oil? What is a Berzelius spirit-lamp? What is the theory of the blow-pipe? What is the construction of a blow-pipe?

is inserted into the flame of a candle, and a current of air forced from it, the flame loses its luminosity, and is projected laterally in the form of a beautiful, pointed cone, in which two parts are distinctly discernible, viz., a small, blue interior cone, *a b*, and a larger exterior cone of a yellowish appearance, *c*. The different parts of this flame possess very different properties.

FIG. 175.

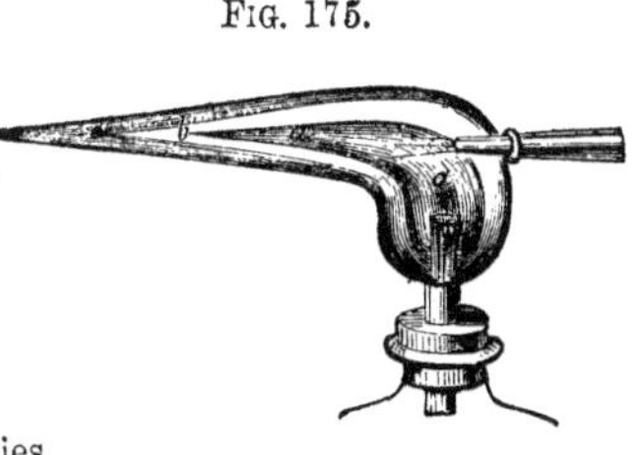

The blue cone is formed by the admixture of air with the combustible gases rising from the wick; in this part of the flame the combustion is complete, and the heat greatest. In front of the blue cone is the luminous portion, consisting of unburnt combustible gases at a high temperature, which of course have a powerful tendency to combine with oxygen. If a fragment of some metallic oxyd, such as oxyd of copper, be introduced into this part of the flame, the oxyd will be deprived of its oxygen, in consequence of the superior affinity of the hot gases for this element, and will be reduced to a metallic state: hence this portion of the flame of the blow-pipe is termed the "*reducing flame.*" At the apex, or extreme point of the outer flame, these effects are reversed. Here atmospheric oxygen at a high temperature exists, and its tendency is to unite with any substance with which it may be brought in contact. Hence if a fragment of metal, such as lead, tin, copper, etc., be placed at this point, it will quickly become covered with oxyd; and this spot is, therefore, called the "*oxydizing flame*" of the blow-pipe.

The opposite actions of the different portions of the blow-pipe flame may be illustrated by the effects which they produce upon a piece of flint-glass, which contains oxyd of lead, united with silica. In the reducing flame the silicate of lead is partially decomposed, and the glass at this point becomes black and opaque from the reduction of the oxyd of lead to the metallic state; but by placing the blackened part for a few seconds in the oxydizing flame, oxygen is again absorbed by the metal, and the transparency of the glass is restored.—MILLER.

FIG. 176.

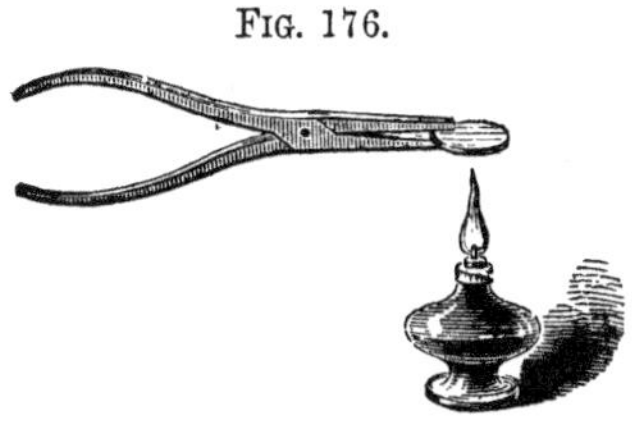

So also if we hold a brightly polished cent over the flame of a spirit-lamp (see Fig. 176) the parts exposed to the exterior of the flame will become covered with an iridescent coating of oxyd, while those over the center of the flame remain bright. By moving the coin, after it has become thoroughly heated, to and fro over the flame, a very beautiful play

QUESTIONS.—What is the constitution of the blow-pipe flame? What is the reducing and what the oxydizing flame? How may their two actions be illustrated?

of colors will be observed, the metal being alternately converted into oxyd, and the oxyd into metal.

486. Carbon, during the act of combustion, as in an ordinary flame, assumes two consecutive phases, viz., while it is evolving heat and light it is a solid, but immediately after it becomes a gas. It is this property which renders carbon, of all combustible bodies, the most suitable for heating and illuminating purposes—questions of cost and convenience being set aside. Phosphorus burns in the air with a more brilliant light than carbon—yet this substance could not be used as an agent for producing light and heat, since the solid products of its combustion remain solid, and being deposited on contiguous objects, soon smother the combustible beneath its own ashes. Zinc, when highly heated, burns in the air with a brilliant flame, but the products of its combustion—white oxyd of zinc—fall about the illuminating center in a miniature shower. The ordinary product of the combustion of carbon, on the contrary, is a gas, carbonic acid, which in virtue of its gaseous qualities escapes into the atmosphere, and combustive action remains unimpeded. Had, however, the results of its combustion been a permanent solid, "the world would have been buried beneath a covering of ashes."*

CHAPTER VIII.

THE METALLIC ELEMENTS.

487. **History.**—Of the whole number of elementary substances included in the class of metals, fully one half are so rare, that they are known only to the chemist and the mineralogist; of the remainder, some fourteen or fifteen only admit of any extensive practical applications. But eight metals were supposed to be known to the ancients.

* There can scarcely be conceived a more beautiful balance of powers designed for the accomplishment of a specific end, than this fixation of carbon in a pure state, and the volatility of its oxygen compounds; yet so familiar has the result become to us—so unnoticed by its very perfection—that an effort of chemical reasoning is required to enable us to appreciate it. The enormous quantity of ponderable, yet invisible carbon removed in the draught of our larger fireplaces is, on its first announcement startling; yet nothing admits of more satisfactory proof. Through an average sized iron blast furnace there rushes hourly no less a quantity of atmospheric air than six tons, carrying off fifty-six hundredths, or more than half a ton of carbon in the form of carbonic acid.—Faraday.

Questions.—What two phases does carbon assume in combustion? Why is it the most suitable of all bodies for combustion? Why could we not use phosphorus as an illuminating agent? What is said of the relative abundance of the metals?

488. **Properties.**—The metals, as a class, are characterized by a peculiar luster, termed metallic; a property exhibited in the highest degree by burnished steel, and the reflecting surfaces of mercury in glass mirrors. They are also possessed of a high degree of opacity, and are good conductors of heat and electricity.

Density.—In density the metals differ greatly; potassium and sodium being lighter than water, while gold and platinum are the most dense of all substances, being respectively nineteen and twenty-two times heavier than an equal bulk of water.

Hardness.—Titanium and manganese are the hardest of the metals, being harder than steel; lead may be scratched by the finger-nail; potassium and sodium are as soft as wax; while mercury, at ordinary temperatures, is a liquid.

Malleability and Ductility.—The most malleable of the metals are gold, silver, copper, tin, cadmium, platinum, lead, zinc, iron, nickel, potassium, sodium, and frozen mercury—in the order given. These may all be hammered out into plates, or even into thin leaves.

The same metals are likewise ductile, or may be drawn into wires, although the ductility of different metals is not always proportional to their malleability. The most ductile of the metals are gold, silver, platinum, and iron.

In the manufacture of gold-thread, by recently improved processes, gold in combination with silver is drawn into wire, by forcing it through smooth conical holes perforated in rubies—so fine, that a single ounce is made to stretch over a length of sixty miles.

Tenacity.—The tenacity of the metals, or the power which they possess of resisting tension without breaking, is determined by ascertaining the weight required to break wires of them having the same diameter. Iron appears to possess this property in the greatest, and lead in the least degree. A wire of iron 7-100ths of an inch in diameter, will sustain a weight of 444 lbs.; a wire of copper of the same diameter, 300 lbs.; of gold, 137; of lead, 24.

The tenacity of metals, however, varies greatly in the same metal, with its purity and the method by which it has been wrought. Recent experiments, made under the direction of the U. S. War Department, have shown that the cohesive strength of iron is greatly increased by fusing it a number of times up to a certain point—its capacity to resist transverse strains being increased thereby sixty per cent. The tenacity of iron is closely dependent on its density. Thus cast-iron, having a density of 6·900 has a tenacity five times less than iron of a density of 7·400. Iron castings of the greatest weight,

---

QUESTIONS.—What are the leading characteristics of the metals? What is said of their density? Of their hardness? What metals are the most malleable? What most ductile? What are illustrations of the ductility of the metals? How is the tenacity of a metal determined? What metals possess this property in the greatest and least degree? How may the cohesive strength of iron be increased? What connection is there between the tenacity of iron and its density?

according to their size, are by far the strongest, and weighing them is a ready means of judging comparatively of their strength.

A corrugated sheet of metal, or one that is doubled into ridges and folds, will resist a far greater crushing force than a flat surface. In the case of copper, the ratio of strength has been proved to be as great as 1 to 9.

Fusibility.—All the metals admit of being fused by the application of heat, but the temperatures at which they liquefy are very various. Mercury, for example, remains fluid at a temperature as low as —39° F., while platinum, iridium, rhodium, and several others, require the intense heat of the voltaic battery or the oxyhydrogen blow-pipe to effect their fusion.

Welding.—Some metals acquire a pasty or adhesive state before undergoing complete fusion, in which, if two clean surfaces be presented to each other, and strong pressure or hammering be employed, they unite or weld together, so as to form one continuous mass. The metals which possess this property are iron, platinum, palladium, and the metals of the alkalies.

Volatility.—At higher temperatures than is required for their fusion, all the metals are probably volatile. Seven of the metals are so volatile as to admit of distillation from the compounds which contain them. They are mercury, arsenic, tellurium, cadmium, zinc, potassium, and sodium.*

489. **Alloys.**—Combinations of the metals with metals are termed Alloys, many of which are most extensively used in the arts, as brass, bronze, bell-metal, type-metal, German silver, etc.

490. **Amalgam.**—When the metals combine with mercury, the resulting product is called an *amalgam*.

It is sometimes questioned whether alloys are true chemical compounds; but the general opinion at the present time is, that they are mixtures of definite compounds, with an excess of one or other metal. The evidence in favor of this view is, that some definite compounds of the metals occur naturally; and when an alloy is formed, the specific gravity of the compound is either above or below that of the mean of the metals employed; the fusing point, also, of an alloy is generally much lower than the mean of the metals which compose it. This is strikingly shown in an alloy called the "fusible metal," which is composed of 8 parts of bismuth, 5 of lead, and 3 of tin, and melts at 203° F.—a temperature more than 200° below the melting point of tin, the most fusible of its constituents, and 400° below that of lead. Its low fusibility may be illustrated by melting a quantity of it in a paper crucible.

---

* Beams of wood suspended over copper smelting furnaces have been observed to be pervaded throughout their entire structure with minute beads of metallic copper—the copper having been raised in vapor, and so deposited within the fibers of the wood. Gold may be seen to undergo volatilization in the focus of an intensely powerful burning-glass; and fine wires of the most refractory metals may be dispersed in vapor by transmitting a powerful electric discharge through them.—MILLER.

QUESTIONS.—What effect has corrugation on the strength of a metal? What is said of the fusibility of the metals? What is welding? What metals can be welded? What is said of the volatility of the metals? What are alloys? What are amalgams?

491. All the metals have the property of assuming the crystalline form but it is not always easy to place them under circumstances favorable to their doing so. Some of them occur in nature, in a crystallized state, particularly gold, silver, copper, bismuth, and platinum.

492. All the metals are capable of uniting with oxygen, but they differ greatly in their affinities for this element. The greater number combine with it at all temperatures, and are reduced (deoxydized) with difficulty. Others on the contrary, like gold and platinum, can not be made to combine with oxygen directly; and their oxyds are decomposed at a slight increase of temperature.

The metallic oxyds differ greatly in their properties. Some of them possess basic characters more or less marked; others will not combine with either acids or alkalies; while a third class have distinctly acid properties. The strong bases are all protoxyds, containing single equivalents of metal and oxygen; the peroxyds are generally neutral, while the metallic acids contain the largest quantities of oxygen.

493. **Classification of the Metals.**—The metals may be arranged in four classes, viz.: 1. The metals of the alkalies; 2. The metals of the alkaline earths; 3. The metals of the earths; 4. The heavy metals, or metals proper.

The latter class may be again subdivided, according to the affinity of the metals contained in it for oxygen, into two groups—the noble and the common metals. The former resist the action of oxygen, like gold, silver, etc.; while the latter, like iron, lead, copper, etc., unite with it readily.

CHAPTER IX.

THE METALS OF THE ALKALIES.

THE metals which by oxydation produce alkalies are Potassium, Sodium, Lithium, and a hypothetical substance, Ammonium, the radical of Ammonia.

SECTION I.

POTASSIUM.

Equivalent, 39·2. *Symbol*, K (Kalium). *Specific gravity*, 0·865.

494. **History.**—Potassium was discovered by Sir Humphrey Davy in 1807, who obtained it by decomposing

QUESTIONS.—Do all the metals crystallize? What is said of the affinities of the metals for oxygen? What are the characteristics of the metallic oxyds? How may the metals be classified? What are the noble metals? When and by whom was potassium discovered?

hydrate of potash (KO, HO) by the action of a powerful galvanic battery.

The discovery of potassium marks an era in the progress of chemistry. The alkalies and the alkaline earths had long been suspected to be compound bodies, but up to this period they had resisted all attempts to decompose them. When once, however, potassium had been separated from its compounds, and potash had been proved to be an oxyd of this metal, the decomposition of the other alkalies and earths, and the discovery, in quick succession, of sodium, barium, strontium, and calcium, followed as a necessary consequence.

495. **Distribution.**—Potassium is widely diffused in nature, but always in combination with other bodies. Many of the minerals which compose the crystalline rocks, such as feldspar, mica, etc., contain potash united with silica—silicate of potash. As these rocks crumble down into soils, potash assumes a soluble form, and is gradually taken up by plants, and accumulated in their structure. When plants are burned, the potash thus absorbed constitutes a part of their ashes, and from these nearly all our supplies of this substance are derived. Potassium also exists in sea-water, as chloride of potassium.

496. **Preparation.**—The original method of preparing potassium through the agency of the galvanic battery is troublesome and expensive, and a new method has been devised, which consists essentially in subjecting a mixture of finely pulverized charcoal and carbonate of potash in an iron retort to an intense heat; decomposition of the alkali ensues, and the potassium distils over in metallic globules which are collected in a vessel of naphtha.

497. **Properties.**—When a globule of potassium is freshly cut open, it appears as a brilliant, silver-white metal; but the exposed surface instantly tarnishes by contact with the air, and in a few minutes becomes covered with a white coating of oxyd (potash). At common temperatures it is soft, and may be molded like wax; at 32° F. it is brittle and crystalline. Its attraction for oxygen is so great, that it can only be preserved in a pure state in exhausted and sealed glass tubes, or under the surface of some liquid, like naphtha, which contains no oxygen. At high temperatures it will remove oxygen from almost all bodies which contain this element, with which it is brought in contact. The powerful attraction of potassium for oxygen may be illustrated by throwing a small piece of the metal upon the surface of water, in which case a part of the water is immediately decomposed—its oxygen combining with the potassium to form potash, whilst the liberated hydrogen, taking fire from the heat evolved, burns in connection with a portion of the volatilized metal, with a beautiful rose-red flame (see Fig. 177);

QUESTIONS.—What consequences followed the discovery of potassium? What is said of its distribution? From whence are the chief supplies of potassium and its compounds obtained? How is potassium practically obtained? What are its properties? What is said of its attraction for oxygen? How may this be illustrated?

the potassium at the same time fusing, assumes the spheroidal state (§ 154), and moves over the surface of the water with great rapidity, finally disappearing with an explosive burst of steam, as the globule of melted potash, which is formed by oxydation, becomes sufficiently cool to come in contact with the water. If this experiment, which is one of the most beautiful in chemistry, be made on a vessel of water reddened with a vegetable color, the alkali produced changes this color to blue or green.

FIG. 177.

498. **Compounds of Potassium.**

Protoxyd of Potassium, Potash, or Potassa, KO.—The only known method of obtaining this oxyd free from water, is by exposing potassium to dry air, when it oxydates to a fine white powder. If once united with water, no degree of heat is sufficient to expel the water.

The potash of commerce and of the laboratory is always a hydrate (KO, HO). It is prepared by dissolving carbonate of potash in ten or twelve times its weight of water, in a clean iron vessel, and adding to the boiling solution a quantity of good quick-lime equal in weight to half the carbonate of potash used. The lime should be previously slacked, made into a cream with water, and added in small portions at a time, so that the liquid may be kept at the boiling point. The lime abstracts the carbonic acid from the potash, and forms carbonate of lime; which, being insoluble, is precipitated, leaving hydrate of potash in solution. The clear solution, if properly prepared, will not effervesce on the addition of hydrochloric acid, thus showing that all the carbonic acid has been transferred from the potash to the lime. The clear liquor, which is known as solution of *caustic potash*, when drawn off by a syphon from the precipitate, and evaporated to dryness, yields a grayish-white solid, with a crystalline fracture—the crude potash of commerce. This, melted and cast into sticks, constitutes the caustic or fused potassa of the shops (*lapis infernalis*), and is used in this state by the surgeons as a cautery.

499. **Properties.**—Hydrate of potash, after fusion, is a hard, grayish-white substance; very deliquescent, and dissolving freely in water and alcohol. Both in the solid state and in solution, it rapidly absorbs carbonic acid from the air, and must therefore be preserved in closely-stopped bottles.

Hydrate of potash possesses in solution, the properties termed alkaline, in the very highest degree. It neutralizes the most powerful acids; restores the blue color to reddened litmus, changes the blue infusion of cabbage into green, but in a short time entirely destroys these colors. It has a peculiar odor, an acrid and disgusting taste, characteristic of the alkalies, and quickly destroys both animal and vegetable matters; for this reason, its solution can not be filtered, except through pounded glass or sand, and is always best clarified by allowing the impurities to subside, and then decanting off the clear

QUESTIONS.—How may potash free from water be obtained? What is the composition of commercial potash? How is it prepared? What is caustic potassa? What are its properties?

liquor. Hydrate of potash, when handled, imparts to the fingers a peculiar, soapy feel, which is occasioned by a gradual solution of the skin (cuticle).

The affinities of potassa when heated are so powerful that but few substances are capable of resisting its action; those which contain silica are decomposed by it, and even platinum itself is oxydized by it. With the fats and fixed oils it forms soaps, which are true salts, composed of a fatty acid and the alkaline base. Its applications also in chemistry and in the arts are almost innumerable.

500. Potassa is the strongest base known in chemistry; consequently, it may be used to effect the decomposition of almost every salt. This may be illustrated by adding a solution of potash to a solution of either the sulphates of iron (green vitriol) or copper (blue vitriol), in water; the potash immediately unites with the acid, and the insoluble metallic oxyd is precipitated

Potash is a fatal corrosive poison.

501. **Carbonate of Potash, KO,CO_2.**—*Pearlash.*— This important salt is obtained almost exclusively from the ashes of land plants; the ashes of marine plants, on the contrary, contain soda, and but comparatively little potash.

In countries where wood is most abundant, as in some parts of the United States, Canada, Russia, etc., it is burned exclusively for the sake of its ashes. These are collected, placed in large tubs (leach tubs), and treated with water; the water soaking through the ashes, dissolves out the potash salts, together with various other soluble mineral substances, and is converted into ley; this when evaporated to dryness, yields an impure carbonate of potash, which is sold in commerce in immense quantities, under the names of *pot* and *pearl-ashes.*

The weight of ashes furnished by different plants varies in different species and soils. Herbaceous plants yield more than woody ones; and the leaves, bark, and young shoots are the parts which furnish the greatest quantity of alkali. Potash does not exist in plants in the form of carbonate, but is accumulated in their substance in combination with certain organic acids. Thus, potash in the vine is combined with tartaric acid, and in the sorrel with oxalic acid. When plants are burned, these acids are destroyed, and the potash, uniting with carbonic acid formed during the combustion, is obtained in the form of a carbonate.

Carbonate of potash has strong alkaline properties, and dissolves in about twice its weight of water.

502. **Bi-Carbonate of Potash, $KO, 2CO_2$** is a compound containing double the quantity of carbonic acid that ordinary potash does; it is

QUESTIONS.—What gives to potash its peculiar feeling? What is said of its affinities and uses? What of its basic properties? From what source is carbonate of potash obtained? What is the process of preparing it? Under what name does it occur in commerce? What is said of the amount of ash yielded by plants? In what state does potash exist in plants? What is bi-carbonate of potash?

very generally known under the name of "saleratus," but this term is often applied to designate any purified carbonate of potash.

503. Nitrate of Potash, KO, NO_5. — *Saltpeter*, *Niter*.—This salt occurs somewhat abundantly as a natural product. The chief sources of its supply are certain districts of the East Indies, where it is found disseminated through the soil, or as an efflorescence upon the surface. It is obtained in a separate state by treating the earth with water, and allowing the solution to crystallize. It is supposed to be produced by the decomposition of organic matters containing nitrogen in soils containing potash and lime.

In Europe saltpeter is formed artificially by mixing animal refuse of all kinds with old mortar, wood-ashes, etc., in heaps, exposed to the air, but sheltered from the rain. These heaps are watered from time to time with putrid urine, and after the lapse of two or three years the mixture is washed, and the salt crystallized out. A cubic foot of refuse may in this way be made to yield as much as 20 ounces of niter.

The earth on the floor of many caverns, as the Mammoth Cave of Kentucky, often becomes strongly impregnated with nitrate of lime, which, when leached with wood ashes, or treated with potash, is decomposed, and yields nitrate of potash. In this way saltpeter was manufactured for the Government during the war of 1812.

504. Properties.—Saltpeter crystallizes in long, six-sided prisms, and is freely soluble in water; its solubility increasing in a remarkable manner with the temperature of the water; thus, 100 parts of water at 32° F. dissolve 7 parts; at 65° F., 29 parts; and at 212° F., 400 parts. The taste of saltpeter is cooling and saline; it is an antiseptic,* and is used in brine for preserving the natural color of salted meats.

Owing to the great quantity of oxygen which saltpeter contains, and the facility with which it parts with it, it is extensively used as an oxydizing agent. When thrown upon burning coals it deflagrates brilliantly. If paper be dipped in a solution of niter, and dried, it forms what is well known as "*touch-paper*," which, when once kindled, steadily smoulders away till consumed, and is hence largely employed in firing trains of powder, fireworks, etc.

The occurrence of fearful explosions, when warehouses containing saltpeter in large quantities have been consumed by fire, has occasioned much speculation as to whether ignited saltpeter will, under any circumstances, explode. The facts in regard to this subject are as follows;—saltpeter, when burned by itself, will not explode; but the oxygen, which is liberated during its ignition, by mingling with the carbonaceous gases evolved during the combustion, at the same time, of other substances, may produce explosive compounds.

* The name antiseptic is given to those substances which resist and retard the decomposition of organic substances, such as saline bodies, acids, etc.

QUESTIONS.—What is saleratus? From whence is saltpeter mainly obtained? What is supposed to be its origin? How may saltpeter be formed artificially? What are the properties of saltpeter? What is "touch-paper?" Will saltpeter explode?

505. **Gunpowder.**—The principal use of saltpeter is for the manufacture of gunpowder, which consists of a mechanical mixture of niter, sulphur, and charcoal, in proportions which very nearly correspond to 1 equivalent of niter, 3 of carbon, and 1 of sulphur; thus:—

			In 100 parts.
Niter,	1 eq.	101	74·8
Sulphur,	1 eq.	16	13·3
Charcoal,	3 eq.	18	11·9
		135	100·0

The great explosive power of gunpowder is due to the sudden conversion of the solid grains into gases (principally nitrogen and carbonic acid); these, at the ordinary temperature of the air, would occupy a space equal to about 300 times the bulk of the powder used; but from the intense heat developed at the moment of the explosion, the expansion amounts to at least 1,500 times the volume of the powder.*

506. **Manufacture of Gunpowder.**—In the manufacture of gunpowder, the three materials, in the state of the greatest purity, are first pulverized separately, and then mixed in the proper proportions. They are then slightly moistened, and further ground and blended together, in charges of 42 lbs. each, by means of large cylinders or wheels of iron, weighing several tons each, which roll round over the powder in a large wooden tub. The mixture is then spread in layers of about an inch in thickness, between copper plates, and subjected to an immense hydraulic pressure. A thin, hard cake is thus obtained, which is broken into small fragments, or *granulated*, by subjecting it to the action of toothed, brass rollers, of different successive gauges. The grains are next sorted by means of sieves of different sizes; after which they are thoroughly dried by steam-heat, and finally polished and glazed by rotating them in wooden revolving cylinders, with a small quantity of "black lead."

The object of granulating the powder is to favor the rapidity of the explosion, by leaving interstices through which the flame is enabled to penetrate, and kindle every grain at the same moment. Powder, in the form of fine dust, burns rapidly, but does not explode. The firing of gunpowder is not absolutely instantaneous, inasmuch as gun-cotton and fulminating mercury explode much more rapidly—which facts prove duration in the explosion

* The expansive force of gunpowder depends almost entirely upon the circumstances under which it is fired. Count Rumford showed, during the last century, that if powder be placed in a closed cavity, and the cavity be two thirds filled, the force will exceed 150,000 lbs. upon the square inch; and he estimated that if the cavity were entirely filled, and restrained to its original dimensions, the force would rise to 750,000 lbs. per square inch. Recent experiments, by Mr. Treadwell of Boston, also tend to confirm these conclusions. On the other hand, if powder be fired in constantly-maintained vacuum, it would not rend walls made of cartridge-paper, if a single end were left open to its escape.

QUESTIONS.—What is gunpowder? To what is the explosive force of gunpowder due? How does its force vary? How is gunpowder manufactured? Why is powder made in grains? Is the explosion of gunpowder instantaneous?

of powder.* Substances which explode more rapidly than gunpowder are not adapted for the movement of projectiles, inasmuch as sufficient time is not given to allow the charge to receive the full advantage of the expansive force of the gases generated; their action, therefore, is not to project the ball, but to burst the gun.

The goodness of gunpowder may be tested by placing two small heaps upon clean writing-paper, three or four inches asunder, and firing one of them with a red-hot wire; if the flame ascends quickly, with a good report, leaving the paper free from white specks, and not burnt into holes; and if no sparks fly off to ignite the contiguous heap, the powder is very good; but if these tests fail, the ingredients are badly mixed or impure.

SECTION II.

SODIUM.

Equivalent, 23. *Symbol*, Na (Natrium). *Specific gravity*, 0·972.

507. **History and Distribution.**—This metal was first obtained by Davy, immediately after the discovery of potassium, by the voltaic decomposition of soda. It is now prepared very cheaply from the carbonate of soda, by a process analagous to that followed in the preparation of potassium.

Sodium, in combination, occurs most abundantly in the mineral kingdom, though it is not so widely diffused as potassium. Its great storehouse is common salt, from which substance most of the soda of commerce is obtained. "As potassium is in some degree characteristic of the vegetable kingdom, so

* While the logical solution of this question adds but little to our knowledge, we are able to infer, from certain experimental results, the course of action which accompanies or causes the amazingly rapid explosion of a quantity of powder confined in a close cavity. "Thus, when the fire reaches the charge from the touch-hole, the nearest grains become kindled, the hot fluid evolved is thrown further into the charge, and the burning succeeds successively until the pressure becomes so great as to condense the air contained between the grains sufficiently to produce the heat required for firing these grains, which are then consumed more or less rapidly as they are fine or coarse. We have then, first, the burning, in succession, of a small part of the charge; then the immensely rapid, though not instantaneous, kindling of every grain composing it; and then the consumption of these grains, which is not accomplished without time. It is a task for the conception to grasp these events, following one another in distinct succession; each having its beginning, middle, and end, and all being compressed in a period not exceeding 1-200th of a second. When we have mastered the imagination of these we may go further, and combine with them the connected and contemporaneous action of the ball, which passes from rest to motion, and through every gradation of velocity up to 1,600 feet per second, and leaves the gun as our historical period of 1-200th of a second expires."—TREADWELL.

QUESTIONS.—Why are compounds more explosive than gunpowder not adapted for moving projectiles? How is the goodness of powder tested? What is said of sodium? What of its occurrence in nature?

sodium is the alkaline metal of the animal kingdom, its salts being found in all animal fluids."

508. **Properties.**—Sodium is a white metal, having the aspect of silver. It resembles potassium in its properties, but does not oxydate so readily as potassium, and when thrown upon water, does not inflame, unless the water has been previously heated. Sodium and all its salts, when ignited, communicate to flame a rich yellow color; this reaction may be illustrated by holding a piece of soda glass, or any mineral containing soda, in the flame of a blow-pipe.

509. The compounds of sodium have mainly the same composition and properties as those of potassium.

510. **Caustic Soda,** or the *Hydrate of Soda*, NaO, HO, is prepared by decomposing carbonate of soda with quick-lime, in the same manner as has been already described for caustic potash. Its properties and appearance are also exactly similar to those of caustic potash.

511. **Chloride of Sodium, NaCl.**—*Common Salt.*—This important and well-known compound is formed when sodium is burned in chlorine gas, and also when soda or its carbonate is neutralized by hydrochloric acid.

The union of these two elements is attended with a most remarkable condensation of volume. Thus 24 parts by measure of common salt contains no less than 25·8 parts by measure of sodium (more than its own bulk), and no less than 30 parts by measure of liquid chlorine; or in other words, 55·8 parts by bulk are compressed by the action of the force of chemical affinity into 24. "No known mechanical force," says Faraday, "could have accomplished this result;* and it is also strange that such an amount of condensation—of squeezing together of atoms—should be co-existent with such perfect transparency, for common salt is even more transparent than glass, allowing a certain kind of radiant matter to pass which stands on the confines of light and heat." (§ 206.)

512. Common salt is found pure or native in the earth in rock-masses (rock-salt), in various countries, and is regularly mined or quarried. The celebrated mine near Cracow, in Poland, is located in a bed of rock-salt which is estimated to be 500 miles in length, 20 broad, and not less than 1200 feet thick.

Salt also exists in solution in all sea-water, in a proportion of about 2·7 per cent., which amounts to nearly 4 oz. per gallon, or to a bushel in from 300 to 350 gallons. Salt manufactured from sea-water by solar evaporation, is termed "bay," or "solar salt." The evaporation is not carried to dryness, but when the greater part of the chloride of sodium is deposited in crystals,

* The student, in this connection, will do well to bear in mind, that physicists are not yet fully agreed as to whether a liquid is capable of any reduction of volume by any application of mechanical pressure.

QUESTIONS.—What are its properties? What is caustic soda? What is common salt? How may it be formed artificially? What singular circumstance attends the union of its elements? What is rock salt? What proportion of salt exists in sea-water? How is salt manufactured from this source?

the mother-liquor is drawn off. This, which from its bitter taste is technically termed the "bittern," retains most of the other salts contained in sea-water, *i. e.*, the compounds of magnesia, lime, bromine, etc.

Salt is also manufactured in large quantities, especially in the United States, by evaporating the water of saline springs. From this source 6,000,000 bushels were manufactured in the State of New York (principally in Onondaga County) and 3,500,000 bushels in the State of Virginia, during the year 1356. The water of the Onondaga salt-springs contain about one seventh part of dry salt. The estimated amount of salt manufactured from all sources in the United States during the year 1856, was upward of twelve millions of bushels.

The appearance of salt varies, according to the rate at which evaporation is conducted. When boiled down rapidly, it forms the fine-grained salt used upon our tables; if evaporated more slowly, the hard, crystallized salt, preferred for the packing of fish and meats, is obtained.

Common salt crystallizes in cubes, which are anhydrous, but crackle or decrepitate, when heated, from the water mechanically confined between their plates. If the evaporation of the solution of salt takes place slowly, the cubical crystals are large; but if it be rapid, they are small, and curiously-arranged in what is called a "hopper-shaped" form. Thus, let us suppose a small cubical crystal has formed on the surface of the solution. From its greater density, the crystal has a tendency to fall to the bottom of the liquid, but capillary attraction keeps it upon the surface. (See Fig. 178.) New crystals soon form, which are joined to the first at the four upper edges, and constitute a frame above the first little cube. (See Fig. 179.) As the whole descends into the fluid, new crystals are grouped around the first frame, constituting a second. (Fig. 180.) Another set, added in the same way, gives the appearance shown in Fig. 181. The consequence of this successive arrangement is, that the crystals are grouped into hollow, four-sided pyramids, the walls of which have the appearance of steps, because the rows of small cubic crystals retreat from each other. (See Fig. 182.)

Fig. 178.

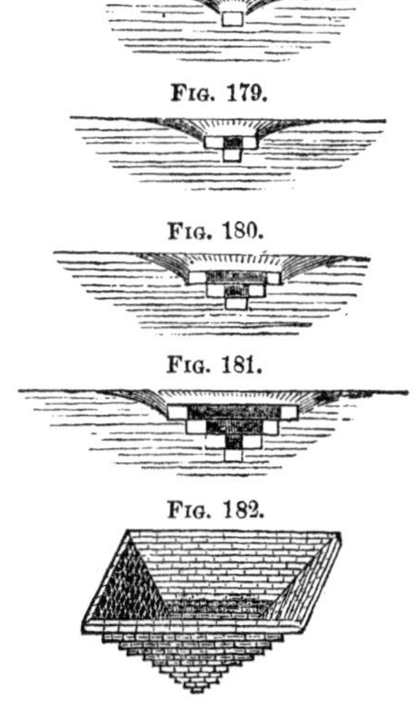

Fig. 179.

Fig. 180.

Fig. 181.

Fig. 182.

Common salt is equally soluble in hot and cold water; 100 parts of water dissolve 37 parts of it; so that a saturated solution, or the strongest possible

QUESTIONS.—From what sources is salt principally manufactured in the United States? What occasions the variations in the appearance of salt? What is said of the crystallization of salt? What of its solubility?

brine, contains 37 per cent. It is an essential constituent of the food of both man and animals, who languish if it be supplied in insufficient quantities.*

513. **Sulphate of Soda, NaO,SO_3+10HO.**—This compound is popularly known as "Glauber salts," from its discoverer, Glauber. It has a saline, bitter taste, and is occasionally used in medicine as a purgative. It is found naturally as a mineral, and occurs also in sea-water, and in many mineral springs; it is generally prepared, however, by decomposing common salt with sulphuric acid, as in the process for preparing hydrochloric acid.

Glauber salts possess the peculiar property of being more readily soluble in water at 90° F. than in water at a higher, or at a boiling temperature. It crystallizes readily from a saturated solution in long four-sided prisms, which contain more than half their weight of water; exposed to air, this water gradually evaporates, and the crystals crumble to a fine powder—effloresce. A very interesting experiment may be performed by closing hermetically a flask containing a boiling saturated solution of this salt; in this condition, the solution may be kept for months without crystallizing, but the moment air is admitted, the whole becomes a semi-solid mass of crystals.

514. **Carbonate of Soda, NaO,CO_2+10HO.**—*Sal-Soda, Soda-Ash.*—The preparation of this salt constitutes one of the most important branches of chemical manufacture; immense quantities of it being consumed in the production of glass, in the fabrication of soap, in the operations of bleaching, and in the preparation of the salts of soda.

The material from which carbonate of soda is now manufactured, is common salt, and the details of the process are essentially as follows: a charge of 600 lbs. of salt is placed upon the hearth of a well-heated reverberatory furnace,† and an equal weight of strong sulphuric acid is poured upon it

* "Salt," says Mungo Park, "is one of the greatest of all luxuries in Central Africa and the continued use of vegetable food creates so painful a longing for it, that no words can describe the sensation." From time immemorial, it has been known that without salt man would miserably perish, and among horrible punishments, entailing certain death, that of feeding culprits on saltless food is said to have prevailed in barbarous times. The explanation of this is, that the blood contains a very large percentage of common salt; and as this is partly discharged every day through the skin and kidneys, the necessity of continued supplies of it to the healthy body becomes apparent. The bile also contains soda as a special and indispensable constituent, and so do all the cartilages of the body. Stint the supply of salt, therefore, and neither will the bile be able properly to assist the digestion, or the cartilages to promptly repair their waste.—JOHNSON.

† A reverberatory furnace (Fig. 183), used extensively in the manufacture of soda-ash, the puddling and refining of iron, and in the smelting of metals, is a furnace so arranged that the heating is effected, not by the fuel itself, but by the flame passing from the fire-place, *f*, under the influence of a powerful draft, over a bridge into a chamber, where the

QUESTIONS.—What of its necessity to man and animals? What are Glauber salts? What is said of them? What of their solubility? What of their crystallization? What is soda-ash? What is said of carbonate of soda? From what is it manufactured? Describe the process. What is a reverberatory furnace?

through an opening in the roof of the furnace. Hydrochloric acid is disengaged, which is usually allowed to escape up the chimney (§ 360), and the salt is converted into sulphate of soda. This operation is completed in about four hours, and requires much care and skill.

The sulphate thus formed is next reduced to powder, and mixed with an equal weight of chalk or limestone (carbonate of lime), and half as much fine coal. The mixture is then heated to fusion, with constant stirring, about 200 lbs. being operated on at once. By this treatment double decomposition is effected, the sulphate of soda being converted into carbonate of soda, and the carbonate of lime into sulphuret of calcium. The mass, when cold, is treated with water, the carbonate of soda dissolved out, and the solution subsequently evaporated to dryness. The product constitutes the soda-ash or British alkali of commerce (anhydrous carbonate of soda), and when of good quality contains from 48 to 52 per cent. of pure soda.*

Fig. 183.

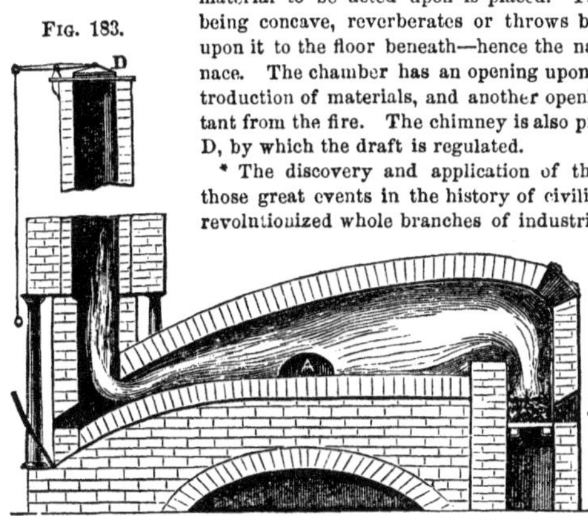

material to be acted upon is placed. The roof of this chamber being concave, reverberates or throws back the flame striking upon it to the floor beneath—hence the name, reverberatory furnace. The chamber has an opening upon the side, A, for the introduction of materials, and another opening at the end most distant from the fire. The chimney is also provided with a damper, D, by which the draft is regulated.

* The discovery and application of this method was one of those great events in the history of civilization which created or revolutionized whole branches of industrial art, and by cheapening the production of great classes of articles of convenience and necessity, materially improved the condition of the human race. The process in question was devised by Leblanc, a French chemist, toward the close of the last century. It remained for a long time unnoticed, and it was not until 1820 that any successful trial was made with it in England. Previous to this, all the soda of commerce was obtained from the ashes of seaweeds, which were sold in the market under the names of Spanish barilla and kelp; the former being produced on the coasts of France and Spain, and the latter chiefly on the coast of Scotland. Only a small quantity of the weight of these substances, however, was an alkali. The barilla contained about 18 per cent., and was sold for about $50 per ton; and the kelp only 5 or 6 per cent., and was worth $20 per ton. It is obvious, therefore, that the soap and glass-maker, in buying these substances, would, in the one case, purchase 95 parts of worthless material, and in the other 82 parts; we say worthless, because of no service in the fabrication of soap or glass. It would seem, therefore, that the introduction of a strong and cheap alkali, would have been hailed by the manufacturers as one of the greatest advantages; but the fact was quite the contrary, and the chemists and manufacturers found it extremely difficult to dissipate the prejudice in favor of kelp and

---

Question.—What is said of the history and introduction of carbonate of soda?

515. **Bi-Carbonate of Soda, NaO, HO, $2CO_2$,** is obtained by passing carbonic acid gas into a solution of carbonate of soda, or by exposing soda-ash to the carbonic acid generated from fermenting grain, as in distilleries, etc. This salt is often sold under the name of "soda saleratus."

516. **Alkalimetry.**—As the purity and value of the commercial carbonates of potash and soda differ greatly, it becomes important to the buyer and the manufacturer to be able to determine rapidly and accurately the quantity of available alkali in a given sample. This operation, termed alkalimetry, consists in ascertaining how much dilute sulphuric acid of a standard strength is required to neutralize exactly a known weight of a particular specimen. A good article will require more acid than a poor one; consequently, the amount of alkali present may be estimated from the quantity of acid consumed. In practical operations, an instrument called an alkalimeter is employed. This consists of a graduated glass cylinder, or tube, divided into degrees (graduated)—Fig. 184)—in which th acid used is measured instead of being weighed. For this purpose a test acid must be prepared, of such a strength that one degree of it will exactly neutralize one grain of pure alkali (potash, or soda). The number of degrees then consumed in neutralizing the alkaline properties of a known weight of a sample, in solution, will indicate at once, in per cents., the quantity of pure alkali in the article tested.

Fig. 184.

0 10 20 30 40 50 60 70 80 90

517. **Nitrate of Soda,** *Soda-Saltpeter, Cubic Niter,* NaO, NO_5, is a native product, occurring in great quantities in Peru and Chili, S. A. It resembles nitrate of potash in its properties, but can not be used in the manufacture of gunpowder, as it freely absorbs moisture from the atmosphere. It is used, however, extensively in the manufacture of nitric acid, and to some extent in agriculture, as a fertilizer.

---

and barilla. When, however, the soda-ash was once introduced, it so reduced the expense of making soap, that the operation of alkalizing the fats, which had before cost $40 per ton, was effected, in one third the time, for $10 per ton. Similar results followed its application to the manufacture of glass; and the business of manufacturing soda-ash increased so fast, that in 1837, seventeen years after the establishment of the first manufactory in England, the quantity produced was 72,000 tons, and at the present time it is upwards of 200,000. The saving to the English nation in the manufacture of soap alone, from the introduction of Leblanc's process, taking as a basis the former price of barilla, and the present consumption and price of soda-ash (1 ton of the latter being equivalent to 8 tons of kelp and 3 of barilla), was estimated in 1847 as equal to twenty millions of dollars per annum; while the benefit to the world at large has been, that the prices of soap and glass have been reduced so low, that the poorest are not debarred from their unrestricted use.

Questions.—What is said of bi-carbonate of soda? What is alkalimetry? What of nitrate of soda?

SECTION III.

LITHIUM.

Equivalent, 6. *Symbol*, L.

518. This rare metal forms the basis of the third alkali, lithia, and resembles sodium in appearance and properties. The alkali, lithia (oxyd of lithium), occurs in small quantities in a few varieties of minerals, and is rarely met with.

SECTION IV.

AMMONIUM (HYPOTHETICAL).

Equivalent, 18. *Symbol*, NH_4.

519. The alkali ammonia so closely resembles potassa and soda in its properties and in its salts, that chemists at the present time generally regard it as the oxyd of a compound metal, as the other alkalies are oxyds of simple metals. The name applied to this hypothetical metal is Ammonium, its composition being 1 atom of nitrogen, and 4 atoms of hydrogen.

All attempts to isolate this substance have failed, from its tendency to separate into ammonia and hydrogen gas. It can be apparently obtained, however, in combination with mercury. This fact may be easily illustrated by the following experiment:—A little mercury is put into a test-tube, with a grain or two of potassium or sodium;* on the application of moderate heat, over a spirit-lamp, combination ensues, with an evolution of heat and light. When cold, the fluid amalgam is put into a little porcelain cup, and covered with a strong solution of sal-ammoniac (chloride of ammonium). A double decomposition immediately ensues: the chlorine and sodium unite to form common salt, while the mercury at the same time commences to increase in bulk, and ultimately swells up until it acquires eight or ten times its original volume, assuming a pasty consistence, without losing its metallic luster. The new substance, exposed to a temperature of 0° F., crystallizes in cubes, but if left to itself, is quickly decomposed, at ordinary temperatures, into fluid mercury, ammonia, and hydrogen. Now it is evident that the mercury has combined with something; but in no case where mercury or any other metal

* The proportions should be about 100 of mercury to 1 of potassium or sodium, by weight.

QUESTIONS.—What is said of lithium? What of ammonium? How may the apparent production of this substance be illustrated?

combines with a non-metallic substance, is there ever a retention of metallic properties after combination, as in this instance; therefore, the inference is, that the substance which has entered into combination with the mercury is a metal—ammonium.

The fact that a compound body—cyanogen—is generated from carbon and nitrogen, which comports itself in every respect like the non-metallic element chlorine, removes every difficulty in the way of our conceiving that a compound may also be formed from nitrogen and hydrogen, which may have the properties of a metal.

According to the ammonium theory, all the salts of ammonia are derived from this radical, and correspond in constitution to the salts of the simple metals.

520. **Chloride of Ammonium, $NH_4\,Cl$.**—*Sal-Ammoniac.*—This substance, which is a compound of ammonium and chlorine, is the most important of all the salts of ammonium, and occurs naturally as a volcanic product. It was formerly imported from Egypt, as a product of distillation from dried camel's dung, and from its having been originally procured from a district in Northern Africa, near the temple of Jupiter Ammon, the name ammonia originated. It is now, however, manufactured in large quantities, from the ammoniacal liquors formed in the manufacture of coal-gas, and from the condensed products of the distillation of bones and other animal refuse, in the preparation of animal charcoal. These are first treated with hydrochloric acid, and the resulting liquors evaporated to dryness. The residue is then subjected to heat in iron vessels, when the chloride of ammonium volatilizes in dense white fumes, which condense, on cooling, into white, semi-transparent, fibrous masses, the sal-ammoniac of commerce.

Sal-ammoniac has a sharp, acrid taste, corrodes metals powerfully, and is readily soluble in water. It does not, however, possess the characteristic odor of ammonia. It constitutes the source from whence most of the salts of ammonia are prepared.

521. **Ammonia, NH_4O.**—*Volatile Alkali, Hartshorn.*—This alkali exists in the atmosphere, in the juices of certain plants, in clayey and peaty soils, and is freely evolved, in combination, from the craters of volcanoes.

522. **Preparation.**—Ammonia can not, under ordinary circumstances, be formed by the direct union of its elements. A series of electric sparks, however, passed through a mixture of hydrogen and nitrogen, will, after a time, generate a limited quantity of it. The production of ammonia, on the contrary, by the indirect combination of hydrogen and nitrogen, is a circumstance of continual occurrence. It especially takes place during the spontaneous decomposition of animal and vegetable substances which contain hydrogen and nitrogen, and in almost every process of oxydation in the

---

QUESTIONS.—Have we any reason to doubt the possibility of the existence of a compound metal? What is sal-ammoniac? What is said of its natural occurrence? What of its manufacture? What is said of the natural occurrence of ammonia? What of its production?

presence of moisture; in the latter case, the hydrogen, at the moment of liberation (in a nascent state) from the water by deoxydation, enters into combination with the nitrogen of the atmosphere.

523. Ammonia is usually obtained by subjecting a mixture of quick-lime and sal-ammoniac to a gentle heat in a flask or retort;—the lime decomposes the chloride of ammonium, forming chloride of calcium, and liberating free ammonia, which latter escapes as a colorless, transparent gas. The same mixture slowly evolves ammonia at ordinary temperatures, and is sometimes used for the filling of smelling-bottles. For experimental purposes, ammoniacal gas is best prepared by heating a strong solution of ammonia in a glass retort, and collecting the evolved gas over mercury, or by displacement, as is represented in Fig. 185. When collected by displacement, the gas must be allowed to pass into the bottle until a piece of reddened litmus paper held to the mouth is immediately turned blue. The tube is then withdrawn, and the stopper, slightly greased, is inserted.

Fig. 185.

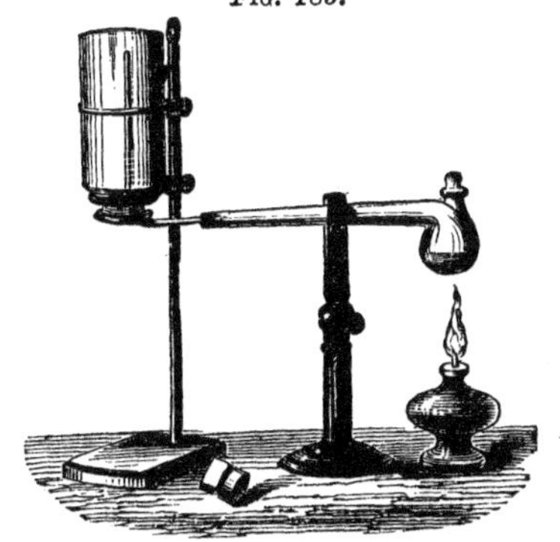

524. **Properties.**—Ammonia thus produced is a gas, which is easily condensed to a liquid by a reduction of temperature (—40° F.) or by pressure. It has an extremely pungent smell, and instantly kills an animal immersed in it; but when largely diluted with air, it is an agreeable stimulant. From the fact that ammonia was formerly prepared by distilling the horns of deers and harts, it is often popularly called "hartshorn."

Ammonia does not support the flame of burning bodies, but is slightly combustible. A jet of gas directed across the stream of hot air issuing from a lighted Argand lamp burns with a pale green flame. It acts strongly as an alkali, turning vegetable-blues green, restoring the blue color of reddened litmus, and neutralizing the most powerful acids. The change, however, of vegetable colors produced by ammonia, owing to its great volatility, is not permanent; but the vegetable substances regain their colors after a time by exposure to the air, which is not the case when the change is effected by the fixed alkalies. Ammonia is, therefore, often called the "*volatile alkali.*"

Any volatile or gaseous acid brought into an atmosphere containing ammonia, produces a white cloud, from the formation of a solid salt. This property is often employed to detect the presence of ammonia in quantities

---

QUESTIONS.—How is ammonia obtained practically? What are the properties of ammonia? Why is ammonia sometimes called hartshorn? How may the presence of ammonia be detected?

too small to be recognized by their odor. The reaction may be illustrated by bringing a rod of glass, or a strip of wood moistened with dilute hydrochloric acid, near to a vessel or substance evolving ammonia;—chloride of ammonium being formed. (See Fig. 186.)

FIG. 186.

Water dissolves ammoniacal gas in large quantities, and with great rapidity;—water at 50° F. absorbing about 670 times its volume. When a piece of ice is introduced into a jar of gas standing over mercury, it instantly liquefies, and by condensing the gas forms a vacuum. The almost instantaneous absorption of this gas by water may be also illustrated by closely fitting a perforated cork and tube into the mouth of a jar containing ammonia, and inverting the jar in a vessel of water. (See Fig. 187.) The first portions of water that enter the jar absorb the gas so rapidly, that a vacuum is created, and a miniature fountain produced.

FIG. 187.

525. **Solution of Ammonia**.—The aqueous solution of ammonia, known as *aqua ammonia*, liquid ammonia, etc., is a reagent much used in pharmacy and chemistry. It is a colorless, transparent liquid, and has all the pungent and alkaline properties of the gas.

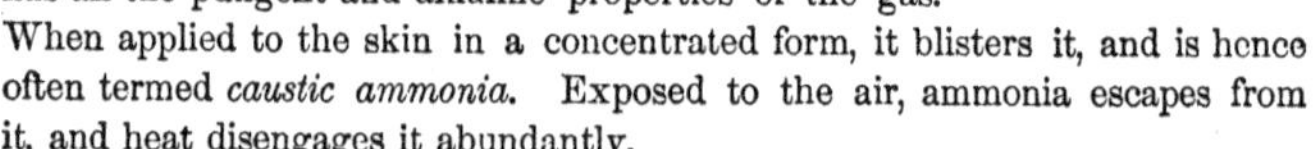

When applied to the skin in a concentrated form, it blisters it, and is hence often termed *caustic ammonia*. Exposed to the air, ammonia escapes from it, and heat disengages it abundantly.

526. There are several *carbonates of ammonia*. The ordinary *sal-volatile* of the shops, which constitutes the basis of the well-known "smelling-salts," is a sesqui carbonate of ammonia, $2NH_4O, 3CO_2$. It is a white solid, highly volatile, and when exposed to the air absorbs carbonic acid, and becomes converted into an inodorous bi-carbonate. This salt is frequently used by bakers in the place of yeast, for raising bread, cake, etc.—heat converting it into gas, which, escaping from the dough, renders it light and porous.

527. **Hydrosulphuret of Ammonia, Sulphide of Ammonium, $NH_4, S + HS$**.—This reagent, which is extensively employed in chemical analysis, is formed by transmitting sulphuretted hydrogen through a solution of ammonia to saturation. The solution thus prepared should be kept cold and in closed glass bottles.

528. **General Properties of the Alkalies**.—The alkalies are the strongest bases known in chemistry. They are all soluble in water, have alkaline properties in the most marked degree, and exert a caustic and decomposing action upon organic substances.

Most of the salts which the alkalies form with acids are soluble in water.

QUESTIONS.—What is said of the absorption of ammonia by water? How may this be illustrated? What is aqua ammonia? What are its properties? What is said of carbonate of ammonia? What is hydrosulphuret of ammonia? What are the general properties of the alkalies? What is said of their salts?

This is especially true of their carbonates, which also exhibit alkaline propit is much used in chemical analysis as a test for the presence of sulphuric acid in solution—which unites with baryta to form a white, insoluble sulphate.

531. **Strontium.**—*Equivalent*, 44; *symbol*, Sr.—Strontium is a white metal, greatly resembling barium.

Its oxyd, strontia, occurs in nature as a carbonate (the mineral, *strontianite*) and more abundantly as a sulphate (*celestine*). The most remarkable characteristic of the strontia salts, is that of communicating a magnificent crimson tint to the flame of burning substances. The red fires of the pyrotechnists are composed of nitrate of strontia, chlorate of potash, sulphur, and antimony. This reaction may be illustrated by inflaming a little alcohol, in which chloride of strontium has been dissolved.

SECTION II.

CALCIUM.

Equivalent, 20. *Symbol*, Ca.

532. Calcium is a light, yellow metal, of the color of gold alloyed with silver. It is very malleable, and can be hammered into leaves as thin as writing-paper. It melts at a red heat, and oxydizes in the air at ordinary temperatures. In combination, *as lime*, it forms one of the most abundant and important constituents of the crust of the globe.

533. **Lime, CaO.**—*Oxyd of Calcium.*—Lime is obtained in a state of purity by heating pure carbonate of lime (calcareous spar) in an open crucible, for some hours, to full redness: the carbonic acid is driven off by the heat, and the lime remains. For commercial purposes, it is prepared by heating common limestone, which is an impure carbonate of lime, in a stone kiln or furnace, the interior of which is somewhat in the form of a hogshead, and is filled with alternate layers of limestone and fuel. The lime, as it is burned, gradually sinks down, and is removed by openings at the base of the furnace, while fresh supplies of fuel and limestone are supplied at the top. In this way the furnace may be kept in action for a great length of time without interruption.

534. **Properties.**—Lime as thus prepared is termed "quicklime," or caustic lime, and in a state of purity has resisted all attempts to fuse it. When water is poured upon quicklime, it swells up, and enters into combination with the water, forming hydrate of lime, or slacked lime. If the proportion of water is about half the weight of the lime employed, a light, dry pow-

---

QUESTIONS.—What is said of strontium? What is a characteristic of its salts? What is calcium? How is lime prepared? What is quicklime? What is slacked lime?

der is formed, accompanied with a powerful evolution of heat—sufficient to occasion the ignition of wood. The hydrate which is thus formed is a definite compound of 1 equivalent of lime with 1 equivalent of water. Lime, also, when exposed to the air, slowly attracts both water and carbonic acid, and crumbles to white powder—"air-slacked lime."

Lime is soluble in about 700 parts of water, forming what is called "lime-water." It is more soluble in cold than in hot water, the latter dissolving only half as much as the former. Lime-water is characterized by a nauseous taste, and decided alkaline properties. It restores the blue of reddened litmus, and changes the blue infusion of cabbage to green. Exposed to the air, it gradually absorbs carbonic acid; a pellicle of carbonate of lime forms upon its surface, which, if broken, is succeeded by another pellicle, until the whole of the lime is separated from the solution, in the form of an insoluble carbonate.

Lime diffused through water forms *milk* or *cream of lime.*

Quicklime exerts a corrosive and destructive action upon the skin, nails, and hair, and upon some vegetable substances. Advantage is taken of this property to remove the hair from hides, preparatory to tanning, by immersing them in milk of lime.*

Lime is also largely employed as a manure, and is particularly valuable upon very rich vegetable soils, such as those formed from reclaimed peat-bogs; its effects in these cases are due to the decomposition of the organic matter, which it renders soluble and capable of assimilation, by plants. Lime formed from limestone, which contains much magnesia, is unsuited for agricultural purposes. Lime should not be mixed with manures in the state of decomposition, since it liberates the ammonia contained in them, and impairs their value as fertilizers.

535. **Mortars and Cements.**—The most important practical application of lime is for the manufacture of mortars and cements. Pure lime, when made into a paste with water, forms a somewhat plastic mass, which sets into a solid as it dries, but gradually cracks and falls to pieces. It does not possess sufficient cohesion to be used alone as mortar. To remedy this defect, and to prevent the shrinkage of the mass, the addition of sand is found to be necessary.

The proportions of lime and sand in good mortar, vary; the amount of

* According to Dr. John Davy, of England, the opinion popularly entertained, that quicklime exercises a corroding and destructive influence upon animal and vegetable matter in general, and that animal bodies exposed to its action rapidly decompose and decay, is wholly erroneous. The results of numerous experiments made by him, seem to show, that with the exception of the cuticle, nails, and hair, lime exerts no destructive action on animal tissues, but that its influence is antiseptic. In the case of vegetable substances, also, the action was similar, and instead of promoting, it arrested fermentation.

QUESTION.—When is lime said to be air-slacked? What is said of the solubility of lime? What are the properties of lime-water? What is cream of lime? What is said of the caustic action of lime? What of its uses in agriculture? What is mortar? What is the necessity of sand in mortar?

sand, however, always exceeding that of lime, and generally in the proportion of 4 to 1. The more sand that can be incorporated with the lime the better, provided the necessary degree of plasticity is preserved. That sand is most suitable for mortar which is wholly silicious, and whose particles are sharp, or not rounded by attrition.

The cause of the hardening of mortar is not thoroughly understood; the explanation generally given is, that the water gradually evaporates, and the lime, by a sort of crystallization, adheres to the particles of sand, and unites them together. A portion of the lime, also, by absorption of carbonic acid from the air, is gradually converted into carbonate of lime. In the course of time, also, a chemical combination takes place between the silica of the sand and the lime, forming a compound of silicate and hydrate of lime, which possesses great hardness. This reaction explains the remarkable hardness often observed in the mortar of old buildings.

It is an advantage to moisten bricks and stones before applying mortar to them, in order that they may not absorb water from the mortar, and thus cause it to set too rapidly. The completeness of the hardening of mortar, depends upon a thorough intermixture of the lime and the sand.

536. **Hydraulic Cements.**—Ordinary mortar, when placed in water, gradually softens and disintegrates, while the lime dissolves away; it can not, therefore, be used for subaqueous constructions. Some limestones, however, which contain about 20 per cent. of clay (silicate of alumina), afford lime which possesses the property of hardening under water. Such limes are known as hydraulic limes, or cements, and may be artificially imitated by mixing with ordinary lime a due proportion of clay not too strongly burnt.*

Concrete is a mixture of hydraulic lime with small pebbles, coarsely broken.

537. **Carbonate of Lime, CaO,CO_2.**—This substance is one of the most abundantly diffused compounds in nature. In its amorphous condition it forms the different varieties of limestone, chalk, and calcareous marl; it is also the principal constituent of corals and shells, and enters, to some extent, into the composition of the bones of animals.

The term limestone is applied to those stones which contain at least half their weight of carbonate of lime; and according to the other prevailing ingredients, a limestone may be argillaceous (clayey), magnesian, ferruginous (containing iron), bituminous, fœtid, etc.

* The rapidity with which different kinds of hydraulic limes set, varies with their composition. If the clay do not exceed 10 per cent. of the mass, the mortar requires several weeks to harden. If the clay amount from 15 to 25 per cent., it sets in two or three days; and if from 25 to 35 per cent. of clay be present, it sets in a few hours. The substance to which the term *Roman cement* is applied, is a lime of this latter composition. In order that hydraulic lime should properly harden, it should not be submerged until it begins to set.—Miller.

Questions.—What is the cause of the hardening of mortar? What advantage is it to moisten bricks, etc., before applying mortar? What are hydraulic cements? What is Roman cement? What is concrete? What is said of the distribution of carbonate of lime? What is a limestone?

The term marble is applied to those varieties of compact limestone which are capable of being worked in all directions, and also of taking a good polish.

Carbonate of lime is found in a greater variety of crystalline forms than any other known substance. Its primary form is a rhombohedron, as seen in double refracting, or Iceland spar (see Fig. 187); but of this figure over 650 modifications are known to mineralogists. Carbonate of lime also crystallizes in another primary form, that of six-sided prisms, as in the mineral aragonite.

Fig. 187.

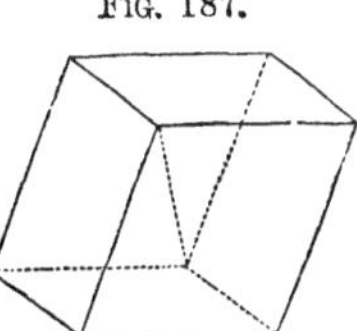

538. Carbonate of lime dissolves in pure water to the extent of about two grains to the gallon, but in water charged with carbonic acid it is taken up freely, and again deposited as the gas escapes—often in anhydrous crystals. It is in this way that the enormous rock masses of crystalline carbonate of lime are supposed to have been formed. This action, which has been before alluded to, (§ 434), is beautifully illustrated in the formation of stalactites and stalagmites in caverns. Water charged with carbonic acid and carbonate of lime, falls in drops from the roof of the cavern; but each drop before falling remains suspended for a time, during which a part of the carbonic acid escapes, and a minute portion of carbonate of lime is left behind. It also deposits another minute portion of calcareous matter on the spot upon which it falls, and as the drops are formed nearly on the same spot for years together, a dependent mass like an icicle is formed from the roof—the stalactite; while another incrustation gradually rises up from the floor beneath it—the stalagmite. In the process of time the two may meet and form a continuous column. (See Fig. 188.)

539. **Building Materials.**—Carbonate of lime is a material much used in architecture and building, but all its varieties are not equally valuable for this purpose. Those varieties of marble which exhibit large crystals, or contain disseminated throughout their mass crystals of sulphuret of iron, have comparatively little strength, and are liable to disintegration. The stone of which the Washington Monument at Washington is constructed, is an example. On the other hand, very fine-grained porous limestones, and also those varieties of porous sandstones which are termed free-stones, are ill-adapted for the external portions of buildings, since they are liable to split into flakes after a few years' exposure to the weather. This generally arises from the absorption of water, and its expansion by freezing in the interior of the stone during winter. A simple and ingenious method of ascertaining whether a stone is liable to this defect, is to thoroughly soak a smoothly-cut block, one or two inches on a side, in a solution of sulphate of soda. On subsequently drying the block in the air, the sulphate of soda crystallizes in

---

Questions.—What is marble? What is said of crystallized carbonate of lime? What is the supposed origin of crystallized carbonate of lime? What are stalactites and stalagmites? Explain their formation? What is said of the adaptability of carbonate of lime to building purposes? Why are porous stones liable to disintegrate? How may the durability of a stone be tested?

the pores of the material, and tends to split off fragments from its surface. The resistance which the stone opposes to this action affords a basis for estimating its durability.*

Fig. 188.

540. **Sulphate of Lime, CaO, SO_3.** — *Gypsum.* — This salt, as commonly met with, is a hydrate—CaO, SO_3+2HO—and occurs abundantly in nature. In transparent plates it is termed "selenite," but in a fibrous, granular, compact, or earthy form it constitutes the different varieties of gypsum and alabaster. When ground to a fine powder, it is known in the arts as "Plaster of Paris," from the circumstance of the mineral being extensively found in the vicinity of the French capital.

Gypsum is extensively used in agriculture as a manure; but its most remarkable property, and the one for which it is the most valued, is the power

* In selecting a stone for architectural purposes, we may be able to form a very good opinion of its durability and permanence, by visiting the locality from whence it was obtained, and observing the condition of the natural surfaces exposed to the weather. For example, if the rock be a granite, and it be very uneven and rough, it may be inferred that it is not very durable: that the feldspar, which forms one of its component parts, is more readily decomposed by the action of moisture and frost than the quartz, another ingredient, and therefore that it is very unsuitable for building purposes. Moreover, if it possess an iron-brown, or rusty appearance, it may be regarded as highly perishable, owing to the attraction which this metal has for oxygen—causing the rock to increase in bulk, and so disintegrate.

The following is the comparative strength of some of our best-known building materials in resisting a crushing force. The best varieties of Quincy granite (sienite) will sustain a pressure of 29,000 lbs. per square inch; good compact red sandstone, 9,000 lbs.; a variety of sandstone called the "Malone," from northern New York, 24,000 lbs.; ordinary marbles, from 7,000 to 10,000; the poorer varieties of sandstone, like that composing the body of the capitol at Washington, 5,000.

Questions.—What is the constitution of gypsum? Under what names is it known? For what is it used?

it possesses, after it has been deprived of water by a heat not exceeding 300° F., of again combining with water and forming a hard, compact mass. When the dried powder, known as "boiled plaster," is made into a thin paste with water, the mixture becomes solid in a few minutes; a chemical combination being formed of 2 equivalents of water and 1 of sulphate of lime, which eventually becomes as hard as the original gypsum. This power of resolidifying renders gypsum applicable for taking copies of objects of every description, and for the construction of molds and models.

If the powdered gypsum is subjected to a heat much exceeding 300° F. it loses its property of solidifying when mixed with water. By mixing gypsum with 1 or 2 per cent. of alum, sulphate of potash, or borax, it forms, when mixed with water, a material much harder than ordinary plaster, and capable of taking a high polish. Artificial colored marbles, called "*Scagliola*," are formed of gypsum, alum, isinglass, and coloring materials, incorporated into a paste. Stucco is a combination of Plaster of Paris with a solution of gelatine, or strong glue.

541. **Hyposulphite of Lime, CaO,S_2O_2** is an abundant constituent of the refuse lime of gas-works, and by exposure to the air gradually passes into sulphate of lime (gypsum). Gas-lime has been used for agricultural purposes, but it probably possesses little or no value as a fertilizer. It has, however, been recommended for mossy land and for composts. All the hyposulphites act as depilatories, or hair-removers, and many of the depilatory powders sold by druggists are compounds of this character.

542. **Chloride of Calcium, $CaCl$,** is formed by dissolving carbonate of lime in hydrochloric acid. The saturated solution evaporated to dryness, and the residue fused, yields a white crystalline solid, which possesses so great an attraction for moisture, that it is used for drying gases, and for depriving alcohol, ether, and other liquids, of water, by distilling them in contact with it. When mixed with snow or ice, it forms a powerful freezing mixture.

SECTION III.

MAGNESIUM.

Equivalent, 12.—*Symbol*, Mg.

543. Magnesium is a malleable metal of the color of silver, and in combination, is an abundant constituent of the crust of the earth. Associated with lime, as a double carbonate of lime and magnesia (oxyd of magnesium), it forms magnesian limestone, or dolomite. United with silica, as a silicate of magnesia, it enters more or less extensively into the formation of many rocks, and a great variety of minerals—such as soapstone or steatite, serpentine, talc,

QUESTIONS.—What are its properties? How may plaster of Paris be hardened? What is scagliola? What is stucco? What is said of hyposulphite of lime? What of the agricultural value of gas-lime? What peculiar property do all the hyposulphites possess? What is said of chloride of calcium? What is said of magnesium and its distribution? What is dolomite? Of what minerals is magnesia a principal constituent?

meerschaum, etc.—all of which are nearly pure silicates of magnesia. The presence of oxyd of magnesium in rocks or minerals in considerable quantity, may be recognized by a peculiar slippery or greasy feeling which it imparts to them—hence the name "soapstone." Magnesium, also, exists abundantly in all sea-water, in combination with chlorine, iodine, and bromine.

544. **Oxyd of Magnesium, MgO.** — *Calcined Magnesia.* — This substance, forming a white, very light, bulky powder, is left when carbonate of magnesia is heated to redness. It is much used in medicine as a mild and gentle aperient.

545. **Sulphate of Magnesia, MgO,SO_3,** constitutes the well-known purgative medicine, *Epsom Salts.* It is manufactured largely from the bittern, or mother-liquor left after the partial evaporation of sea-water, by the addition of sulphuric acid to the solution of chlorides, and also by treating serpentine rock with sulphuric acid. It possesses a bitter, disgusting taste, and readily crystallizes from solution in small prismatic crystals.

546. **Carbonate of Magnesia, MgO,CO_2.**—The common, white magnesia of the shops is formed by precipitating a solution of sulphate of magnesia by a solution of carbonate of soda. It is insoluble in water, but a solution of carbonic acid dissolves it, and forms the popular medicine known as Murray's "fluid magesia." Carbonate of magnesia also occurs as a mineral.

547. **Properties of the Alkaline Earths.** — The alkaline earths are, next to the alkalies, the strongest chemical bases. They have a caustic action, but far less so than the alkalies, and form with fats, soaps which are insoluble in water. The carbonates of the alkaline earths are insoluble in water, and when exposed to a powerful heat, part with their carbonic acid—in this respect, being the opposite to the carbonates of the alkalies.

CHAPTER XI.

METALS OF THE EARTHS.

548. The metals of the earths are, Aluminum, Glucinium, Zirconium, Thorium, Yttrium, Erbium, Terbium, Cerium, Lantanium, and Didymium.

Of these, all but the first, aluminum, are extremely rare, and comparatively unimportant. Glucinium is the metallic base of the earth glucina, which is the characteristic constituent of the emerald and the beryl. Zirconium is the metallic base of the earth zirconia, which is found in the gems, zircon and hyacinth. The others possess few points of general interest.

QUESTIONS.—What is a characteristic of magnesian minerals? What is calcined magnesia? What are Epsom salts? How are they obtained? What is said of carbonate of magnesia? What are the characteristic properties of the alkaline earths? What are the metals of the earth? What is said of their occurrence in nature?

SECTION I.

ALUMINUM.

Equivalent, 13·7. *Symbol*, Al. *Specific gravity*, 2·5.

549. The metal aluminum was first obtained by Wöhler, an eminent German chemist, in 1827. Comparatively little, however, was known of it until within the last few years, but processes have been recently devised by its discoverer and M. Devillé of Paris, by which it is obtained, in considerable quantities, at a cost which (at present) renders it about twice as valuable as silver.

Pure aluminum is a beautiful, white metal, closely resembling silver in color and hardness. Its most striking characteristics are, that, while it closely resembles in appearance the dense, heavy metals, it is in fact lighter than glass; and, also, its power of resisting oxydation—not tarnishing by exposure to air or moisture, or even when heated to a red-heat. It fuses at a temperature below the melting point of silver, is malleable, ductile, and remarkably sonorous. Nitric and sulphuric acids, even when concentrated, scarcely attack it at ordinary temperatures; but it dissolves freely in hydrochloric acid, and even in strong vinegar (acetic acid). Aluminum derives its name from *alum*, into the composition of which it enters.

The properties of aluminum are such as to give it a high industrial value; and it has been applied to some extent for economic purposes.

550. **Oxyd of Aluminum, Alumina, Al_2O_3.**—This is the only known oxyd of aluminum (a sesquioxyd). It occurs in a state of purity, with the exception of a little coloring matter, in the sapphire and the ruby; the first of which is blue, and the latter red. These gems are only inferior in hardness, luster, and value, to the diamond. Emery (corundum), which, from its hardness, is so largely used in grinding and polishing, is also nearly pure alumina. Next to silica, alumina, in combination, is the most abundant mineral constituent of the crust of the earth.

By mixing a solution of alum with an excess of ammonia, we obtain a white, semi-transparent, bulky precipitate—hydrate of alumina, Al_2O_3+3HO. This, washed, dried, and strongly ignited, furnishes a pure alumina, in the form of a white powder, almost insoluble in acids, and infusible, except before the oxyhydrogen blow-pipe.

551. **Alum.**—Common alum is a combination of the sulphate of alumina and the sulphate of potash, with 24 equivalents of water. The constitution of this double salt may be represented as follows: $Al_2O_3, 3SO_3+KO,SO_3+24HO$. When alum is heated, it froths up, loses its water of crystallization, and is converted into a white, porous mass, many times the volume of the salt employed; in this condition it is known as anhydrous, or burnt alum.

Alum is occasionally found as a natural product in the earth, but for indus-

QUESTIONS.—What is said of aluminum? What are its properties? What is the formula of alumina? In what substances is it found pure? What is said of hydrous and anhydrous alumina? What is alum? Give its formula? What is burnt alum?

trial purposes it is manufactured artificially. The sulphate of alumina, which enters into its composition, may be obtained by dissolving alumina from common clay by sulphuric acid, or by exposing certain aluminous (clayey) slates and shales, which contain sulphuret of iron (iron pyrites), to the action of the air, or to a moderate heat; under these circumstances, the sulphuret of iron is decomposed, its sulphur uniting with oxygen to form sulphuric acid, which, subsequently, combines with the alumina of the clay to form sulphate of alumina. This salt, obtained in solution from the clay by washing, is mixed in large casks with sulphate of potash, in proper proportions, and the whole allowed to stand. The formation of alum immediately commences, and after the lapse of a few weeks, the interior of the cask becomes lined with a thick mass of crystals. The staves of the cask are then removed, and an enormous mass of alum crystals, of the shape of the cask, is left standing. (See Fig. 189.) These, when drained and broken up, furnish alum ready for market.

FIG. 189.

Ordinary alum has a sweetish, astringent taste, and crystallizes very readily in regular octohedrons.

552. The constitution and formation of alum affords a good illustration of the principle of isomorphism. For example, we may substitute in its manufacture in the place of sulphate of potash, sulphate of soda, or sulphate of ammonia, and thus obtain soda, or ammonia alums, which crystallize in the same form as the potash alum, and possess similar properties; or we may

QUESTIONS.—How is alum manufactured? What are its properties? How does the constitution and formation of alum illustrate isomorphism?

substitute in the place of the sesqui-oxyd of alumina Al_2O_3, sesquioxyds of iron, chromium, or manganese, without changing the original octohedral, crystalline form. These substitutions will be more clearly understood from an examination of the annexed table:

Potash alum	Al_2O_3, $3SO_3+KO$, SO_3+24HO
Soda alum	Al_2O_3, $3SO_3+NaO$, SO_3+24HO
Ammonia alum	Al_2O_3, $3SO_3+NH_4O$, SO_3+24HO
Iron alum	Fe_2O_3, $3SO_3+KO$, SO_3+24HO
Chrome alum	Cr_2O_3, $3SO_3+KO$, SO_3+24HO

All these compounds are called alums, and are said to be isomorphous, because they possess a similar chemical constitution, and the same crystalline form. They may be easily prepared by dissolving together in water their simple constituent salts in proper proportions, and allowing the solution to crystallize. Potash, soda, and ammonia alums are white, chrome alum a deep purple, and iron alum a pale purple, or red.

Alum, and the compounds of alumina formed from it, are largely used in dyeing, calico printing, and in tanning. Alumina has a very great attraction for certain kinds of organic matter, and especially for coloring substances. To such an extent is this the case, that the hydrate of alumina is extensively employed in the place of animal charcoal for decolorizing animal and vegetable solutions. If cloth is soaked in a solution of alumina, prepared from alum, a portion of the earth attaches itself to the fibers; and if subsequently plunged into a bath of coloring matter, it becomes permanently dyed. Most coloring substances, without this treatment, would be removed by washing; but the presence of alumina seems to serve as a bond of union between the color and the fiber, which renders the adhesion of the dye permanent; a few other substances, such as binoxyd of tin, and the sesquioxyds of chromium and iron, act in the same manner, and are called mordants (from the Latin *mordeo, to bite* in).

When alum is added to a colored vegetable or animal solution, and the alumina precipitated by the addition of an alkali, it carries down with it the greater portion of the coloring substance, and forms a class of pigments called *lakes*. Carmine is a *lake* prepared in this way from a solution of cochineal.

553. **Silicates of Alumina.**—The salts of silicic acid and alumina comprise a great number of important and interesting mineral substances.

554. **Clay.**—All the varieties of clay consist of hydrated silicate of alumina, more or less mixed with other matters derived from the rocks, which by their decomposition have formed clay; such as potash, uncombined silica, oxyd of iron, lime, and magnesia. According as one or the other of these ingredients predominates, the character of the clay and its adaptation to specific purposes will vary.

QUESTIONS.—What are the uses of alum? What property characterizes hydrous alumina? How does alumina act in dyeing? What are lakes? What is carmine? What is clay?

Clays which are nearly free from oxyd of iron or carbonate of lime, are termed *fire-clays*, and are used for the manufacture of fire-bricks and crucibles; such clays are of rare occurrence. *Pipe-clay*, used for the manufacture of tobacco-pipes, is a fine white clay, nearly free from iron. When the proportion of carbonate of lime in a clay is considerable, it constitutes what is known as a *marl;* if the aluminous constituent predominates, it forms an aluminous marl; if the carbonate of lime be in excess, it is a calcareous marl; the latter is highly valued in agriculture as a fertilizer for light, sandy soils. *Loam* is a mixed substance containing much clay, some sand, iron, and a varying proportion of organic matter. Ochres are clays colored red or yellow by oxyd of iron; they are extensively used as paints. *Fuller's earth* is a porous silicate of alumina, which has a strong adhesion to oily matters; if made into a paste with water, and allowed to dry upon a spot of grease on a board or cloth, it removes most of the oil by capillary attraction. It owes its name to the fact that it is employed to remove the grease applied to wool in spinning.

555. Clay emits a peculiar odor when breathed upon, which is known as an argillaceous odor. When mixed with a soil, it gives it firmness and consistency, and retains the moisture, ammonia, carbonic acid, and organic matters which contribute to the support of plants. In this way it indirectly ministers to the wants of vegetation, although alumina itself is not known to enter as a constituent into the structure of either plants or animals.

Among other important minerals of which silicate of alumina is a principal constituent, may be mentioned feldspar, mica, all the varieties of slates, and lavas, trap, basalt, porphyry, etc. The gems, topaz and garnet, are also in great part silicate of alumina.

The beautiful artificial blue pigment known as *ultramarine* consists mainly of silicate of alumina fused with sulphide of sodium.

556. **General Properties of the Earths.**—The earths are entirely insoluble in water, and do not combine with carbonic acid. They possess weak basic properties, and alumina in some instances may even act the part of an acid. The metals of the alkalies, the alkaline earths, and the earths, are all of a low specific gravity, and are sometimes called, on this account, the light metals, to distinguish them from the other metals, which are dense and heavy.

QUESTIONS.—What is fire-clay? What is pipe-clay? What are marls? What is loam? What are ochres? What is "fuller's earth?" What are the properties of clay? What minerals are mainly composed of silicate of alumina? What is ultramarine? What are the general properties of the earths?

CHAPTER XII.

GLASS AND POTTERY.

557. **Glass** is a compound substance produced by fusing together, by a high and long-continued heat, mixtures of the silicates of potash, soda, lime, magnesia, alumina and lead—the nature and proportions of the ingredients varying according to the purpose for which the glass is to be used.

Silica fused with the alkalies, potash, or soda, readily yields a transparent glass of easy fusibility, but not adapted for economic purposes, since it is unable to resist the action of water and acids. If the proportions are 3 of alkali to 1 of silica, the compound is so readily soluble in water as to be designated as "soluble glass." (§ 414.) By increasing the proportion of silica, we can greatly diminish the solubility of the alkaline silicates, but not entirely so. On the other hand, silica fused with lime, magnesia, baryta, or alumina, yields compounds which resemble porcelain rather than glass, are entirely insoluble, and melt at only a high temperature. No single silicate is, therefore, adapted by itself to form glass, but by judicious mixture of the various silicates we can contain compounds which are transparent, free from color, fusible at a moderate heat, and insoluble in water.*

The temperature at which glass fuses depends upon the amount of silica it contains; the greater the proportion, the less the fusibility.

558. The principal varieties of glass are as follows:—

Common, colorless, or white glass, which is used for making tumblers, window-glass, and looking-glasses, is a compound of silicate of potassa or soda, with silicate of lime. The character of the glass, however, varies very much according as one or the other of the alkalies is used. Glass composed of simply the silicates of potash and lime, is exceedingly transparent, very hard, and of difficult fusibility. It is highly prized in the laboratory for its adaptation to certain chemical requirements. The celebrated Bohemian glass—the finest

* In strictness, the best-made glass is to a certain extent soluble. If very finely-powdered window-glass be placed on turmeric paper, and moistened, it will exhibit an alkaline reaction. Windows in old houses often show prismatic colors, owing to the circumstance, that the long-continued action of rain and moisture has washed out the alkali of the glass, and left an irregular condition of surface, which occasions a refraction of light. Specimens of ancient glass which have been dug out of the earth, often exhibit a pearly luster, resulting from pure silica, the alkali having been slowly removed by long exposure to damp.

QUESTIONS.—What is glass? Why is a mixture of silicates necessary for the formation of durable glass? What is said of the fusibility of glass? What is the composition of common white glass? What is the character of potash-glass? What is Bohemian glass?

glass produced, is a silicate of potash and lime, with a little silicate of alumina. By substituting soda in the place of potash, we obtain a more fusible, but a less transparent glass; varieties of glass with this composition, are known as "crown glass," plate-glass, window-glass, etc. The presence of soda in glass imparts to it a blueish-green tinge, which is not observed when potash alone is used.

Green Bottle Glass, and other inferior desciptions of glass used for the manufacture of articles in which color is not regarded, consist of an alkali, silica, lime, and alumina; the cheapest and most ordinary materials being used, such as wood-ashes and common salt, as alkaline products, common sand, clay, gas-lime, and the refuse lime and alkali left after the manufacture of soap. The green color of bottle-glass is due mainly to the presence of oxyds of iron and manganese.

Flint-Glass, so called from the circumstance, that the silica used in its manufacture was formerly derived from pulverized flints, is a mixture of silicate of potash and silicate of the oxyd of lead. It fuses at a lower temperature than the ordinary varieties of glass, has a beautiful transparency, and a comparative softness, which enables it to be cut and polished with ease. Glass which contains lead possesses the property of refracting light in a remarkable manner, and is consequently employed for the construction of lenses for optical instruments, glass prisms, chandelier-drops, etc.; it is, also, the basis of the artificial gems known as *paste*, which are colored by metallic oxyds.

559. The silica used for the manufacture of fine glass is generally in the form of pure white sand, entirely free from oxyd of iron. Such sand is by no means common, the finest in the world being at present found among the Green Mountains of Western Massachusetts, from which localities large quantities are annually exported to Europe. The silica of the Bohemian glass is obtained by pulverizing masses of pure white quartz. The alkali used is a refined carbonate of potash or soda. These two ingredients, with a proper proportion of air-slacked lime, or oxyd of lead, are thoroughly mixed, and fused in large crucibles of refractory fire-clay, in a circular reverberatory furnace. This furnace is usually in the form of a truncated cone, 60 to 80 feet high, and 40 to 50 feet in diameter at the base. The furnace is at the center of the cone, and the glass-pots, to the number of 4 to 10, are arranged around the circumference, and opposite to openings in the walls of the furnace. Fig. 190 represents the exterior of the furnace, and the general appearance of a glass-house.

The fire of a glass furnace is never allowed to slacken, and the melting-pots remain permanently in their situations for several months, being charged from the exterior. A heat of about forty-eight hours is requisite to convert the crude materials into a liquid, homogeneous glass.

Questions.—What is the character of soda-glass? What is the composition of green bottle-glass? What is flint-glass? In what form is the silica used in the manufacture of glass? What its alkali? How is glass formed?

The details of the working and molding of glass are purely mechanical, and a description of them is foreign to the object of this work.

560. **Colored Glass.**—Glass is colored by the addition to it, in a fused state, of small quantities of the metallic oxyds, which dissolve in it without

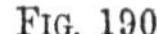

FIG. 190.

affecting its transparency. Thus, oxyd of cobalt imparts a deep blue; oxyd of manganese, a purple or violet; oxyd of copper, a green; oxyds of iron, a dull green or brown; and oxyd of gold, a ruby or rose color.*

* Cut-glass ornamental articles, which exhibit different colors upon the same specimen, and at different depths in the thickness of the glass, are manufactured in the following manner: the object is first formed in white, transparent, and colorless glass; then, being allowed to cool until it acquires solidity and consistency, it is dipped for a moment in a pot of colored glass in a state of fusion, and being suddenly withdrawn, it carries away upon it a thin coating of colored glass, which immediately hardens upon it, and becomes incorporated with it. The article is then shaped by the processes of the glass maker, and if it be afterwards cut, those parts which are cut will disclose the clear, transparent glass, while the parts not cut remain coated with the color. It is by this process that all the effects

QUESTION.—How is glass colored?

561. **Enamel** is a term given to glass which is rendered milk-white opaque by the addition of binoxyd of tin. Examples of such enamels are to be seen in watch-dials, and in the so-called porcelain transparencies. Colored enamels are produced by the addition of metallic oxyds to white enamels.

562. **Annealing**.—If glass be allowed to cool suddenly after fusion, it becomes exceedingly brittle, and articles made from it are liable to break in pieces from the least scratch or jar, or even from a slight but sudden change of temperature, as when transferred from a cold to a warm room.

This property is strikingly illustrated by what are called Prince Rupert's drops, which are little pear-shaped masses of glass, formed by dropping melted glass into cold water. (See Fig. 191.) These may be subjected, without breaking, to considerable pressure, or even to a smart stroke, but if the little end of the drop be nipped off, the whole mass instantly flies in pieces with a sort of explosion, and is converted into powder. This effect appears to be due to the fact, that the particles of which these little masses are composed, are in a state of unequal tension, owing to the formation of a solid coating upon the exterior, while the interior parts are still fluid; the latter being thereby prevented from expanding, as they become solid. The drops will bear a concussion because the mass then vibrates as a whole, but if the end be broken, a vibratory movement is communicated along the surface without reaching the internal parts; this allows them some expansion, which overcomes the cohesion of the outer coating, and the whole at once flies in pieces. To obviate, therefore, this tendency to brittleness, all glass articles, after their manufacture, are subjected to the operation of annealing, which is a very slow and gradual process of cooling, by which the parts are enabled to assume their natural position with regard to each other. In some cases, several days, or even weeks, are required for the cooling of particular articles.

FIG. 191.

563. **Pottery and Porcelain**.—The basis of all earthenware, porcelain, and china, is silicate of alumina (clay).

Pure silicate of alumina, however, contracts greatly and unequally on drying, and, consequently, is unfitted to be used by itself for fictile purposes. This difficulty is, however, overcome by the addition to the clay of a proportion of silica, and to compensate for a loss of tenacity in the clay thereby occasioned, it is also customary to incorporate with the mass some fusible material, as an alkali, silicate of lime, etc., which, at the temperature required for baking the ware, fuses, becomes absorbed by the more infusible portion,

which are seen in ornamental articles, which consist partially of colored, and partially of clear glass, are produced. Additional colors may also be combined on the article in the same manner, and by cutting a surface so coated, to different depths, varieties of effects may be produced, involving a display of two or more colors.

QUESTIONS.—What are enamels? What effect is produced by allowing glass to cool suddenly? How is this illustrated by Prince Rupert's drops? What is annealing? What is the basis of all earthenware? Why can not pure clay be used alone?

and binds the whole, on cooling, into a solid mass. According to the greater or less proportion of these fusible materials, the ware is more or less transparent, or resembles glass in a greater or less degree.

564. **Porcelain** is the name applied to the finest varieties of earthenware. It is composed of a very pure, white clay, called "kaolin" (derived from the decomposition of feldspar), very finely-divided silica, prepared by crushing and grinding calcined flints, and a little lime. The utmost pains are taken to thoroughly incorporate these ingredients, and to avoid the introduction of particles of grit, or other foreign bodies. The mixture, having the consistency and appearance of dough, is then fashioned upon a peculiar kind of lathe—called a "potter's wheel,"—or in molds of plaster of Paris, into ware, —dried, and baked in a kiln or oven for a period of about 40 hours. The porcelain in this condition is technically termed *biscuit*, and is compact and solid, but so porous as to readily imbibe water, and even allow it to filter through its substance. This difficulty is remedied by covering the ware with a glassy coating called a glaze, which generally consists of a more fusible mixture of the same materials as the porcelain itself. These, in a state of fine powder, are made into a cream with water, and into this the ware is dipped for a moment, and then withdrawn; the water sinks into its substance, leaving the powder evenly spread upon the surface, which, when submitted to a moderate heat, fuses, and forms a uniform, vitreous coating. In ornamented porcelain, the designs are printed or painted upon the surface with various metallic oxyds, which develope their colors only after fusion with the ingredients of the glaze.

The material called "Parian," of which statuettes, etc., are manufactured, is a carefully-prepared variety of porcelain.

The details of the manufacture of the ordinary varieties of "stone" and "earthen" ware, are in principle the same as those involved in the manufacture of porcelain, less care, however, being taken in the selection of materials, and less labor being bestowed upon their preparation. The coarser kinds of earthenware are sometimes covered with a yellowish-white glaze, of which oxyd of lead is an important ingredient. The use of such vessels in culinary operations is highly objectionable, inasmuch as the lead is liable to be dissolved off by acids, and act as a poison.

Bricks and common pottery-ware owe their red color to the iron naturally contained in the clay of which they are composed, which, by heating, is converted into red oxyd of iron. Some varieties of clay, like that found near Milwaukie, contains little or no iron; and, consequently, the bricks made from it are all light-colored.

QUESTIONS.—What is the composition of porcelain? Describe its manufacture. How is porcelain ornamented with colored figures? What is "Parian?" How does the manufacture of earthenware differ from porcelain? How is earthenware sometimes glazed? Why is the use of vessels glazed with lead dangerous? Why are bricks and flower-pots red?

CHAPTER XIII.

THE COMMON, OR HEAVY METALS.

SECTION I.

IRON (*Ferrum*).

Equivalent, 28. *Symbol*, Fe. *Specific gravity*, 7·8.

565. **Natural History and Distribution.**—Iron is the most abundant, the most widely diffused, and the most useful of all the metals. It is the only metal which enters into the structure of all the vertebrate animals, as an essential constituent (existing always in the blood), and the only one whose oxyds are not injurious to either animals or plants.

Iron in a metallic and malleable state, alloyed with nickel, cobalt, and small quantities of other metals, is found upon the surface of the earth in large masses of meteoric origin. These masses are so peculiar in their composition and structure, and differ so essentially from all terrestrial substances, that although they may not have been seen to fall, they are easily recognized. Some of these extraordinary bodies are from 15 to 20 tons weight; one observed to fall from the atmosphere in an ignited state in South America in 1844, was upward of a cubic yard in dimensions. A specimen in the cabinet of Yale College weighs 1,635 lbs., and one in the Smithsonian Institution, 252 lbs. The occurrence in nature of metallic iron of a terrestrial origin is exceedingly rare. It is, however, said to be occasionally found associated with ores of platinum, and also in little nodules inclosed in masses of iron ore—the latter being evidently the result of electro-galvanic agency. Recent investigations by Hayes of Boston have also rendered it probable that a deposit of native iron exists on the West Coast of Africa, in the vicinity of Liberia.

Iron in a state of perfect purity is not found also as an article of commerce—the very best artificial irons always containing some carbon, and generally minute quantities of silica, sulphur, and phosphorus. Chemically pure iron may, however, be obtained by reducing the pure peroxyd of iron at a red-heat by a current of hydrogen gas.

566. **Compounds of Iron with Oxygen.**—Iron forms three definite compounds with oxygen: 1. Protoxyd, FeO; 2. Sesquioxyd, commonly called the peroxyd, Fe_2O_3; 3. Ferric acid, FeO_3. Another oxyd, Fe_3O_4, found native in large quantities, and known as the black, or magnetic oxyd of iron, is by some regarded as a distinct oxyd, and by others as a compound of protoxyd and sesquioxyd.

QUESTIONS.—What is said of iron? Is malleable iron found in nature? Is the iron of commerce pure? How may chemically pure iron be obtained? What are the compounds of iron and oxygen?

567. **Protoxyd of Iron, FeO**, does not occur in nature except in combination. It is a powerful base, and unites with the acids to form salts which have a greenish color and a styptic taste—properties which are possessed in a very marked degree by green vitriol, which is a sulphate of the protoxyd of iron. Protoxyd of iron may be easily obtained in the form of a hydrate, by dissolving pure sulphate of iron in water recently boiled and adding an alkali to the solution. The bulky precipitated hydrate is at first nearly white, but absorbing oxygen from the air, it soon becomes brown, and finally red, from its conversion into sesquioxyd. In a moist state, this hydrate constitutes the most effectual antidote in poisoning by arsenic.

568. **Sesquioxyd of Iron, Fe_2O_3**, *Peroxyd*,—is found native in great abundance, and constitutes some of the most valuable of the ores of iron. It is in this state of oxydation that iron is generally found in soils and minerals, assuming oftentimes a deep red color (red oxyd) as in ocher, burnt clay, etc. The substance called *rouge*, *crocus*, or *colcothar*, used for polishing glass or metals, is this oxyd in a state of fine powder, prepared by igniting the sulphate of iron.

569. **Black, or Magnetic Oxyd of Iron, Fe_3O_4**, occurs abundantly in nature, constituting the common magnetic iron ore, and the native *loadstone*, both which acquire magnetic properties from the inductive influence of the earth. It is also the principal constituent of the scales of oxyd which are detached during the forging of wrought-iron.

570. **Ferric Acid, FeO_3**, may be formed by heating 1 part of peroxyd of iron with 4 parts of saltpeter to full redness for an hour, in a covered crucible. A brown mass is thus obtained—ferrate of potash—which digested with water yields a beautiful violet-colored solution.

571. **Ores of Iron**.—The ores of iron are extremely numerous. The following are some of the most valuable: 1. The *magnetic*, or *black oxyd*, which has a black color and a metallic luster. It is found in beds in the primitive rocks, and sometimes constitutes entire mountains, as the iron-mountains of Missouri. It is one of the richest of the ores of iron, and contains about 70 per cent. of pure iron. The superior iron of Sweden and Russia is prepared from it. The *specular iron*, or *red iron ore*, consists mainly of sesquioxyd of iron; under this class are included the ores known as red and brown hematites, and bog-iron ore. Red hematite often occurs in fibrous crystallized nodules, forming beautiful cabinet specimens. (See Fig. 192.) All the ores of this

Fig. 192.

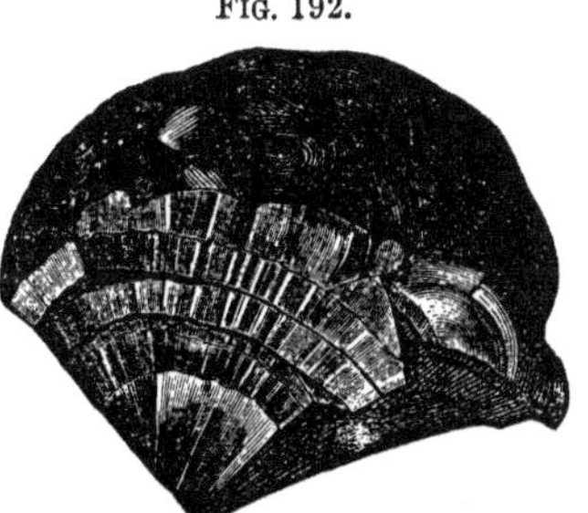

QUESTIONS.—What is said of the protoxyd? How may it be prepared? What is said of the sesquioxyd? What is rouge? What is said of the black oxyd? What of ferric acid? What are the principal ores of iron?

class yield reddish brown powders, and may thus be distinguished from the black oxyd;—they contain about 63 per cent. of iron; 3. *Clay-iron stone* is an impure carbonate of iron, mingled with varying proportions of clay, lime, magnesia, and manganese. This ore occurs extensively associated with coal, and contains about 33 per cent. of metallic iron; it is the chief source of the enormous quantity of iron manufactured in Great Britain. All clays which are capable of yielding 20 per cent. of iron are called ores.

572. **Bi-Sulphuret of Iron,** FeS_2,—*iron pyrites*,—although a very abundant mineral, is not used as a source of metallic iron; it occurs in cubical crystals (see Fig. 193) and fibrous radiated masses; from its bright yellow color and metallic luster it is often mistaken for gold (fool's gold), but its character may be easily determined by the sulphurous odor which it evolves by heating.

Fig. 193.

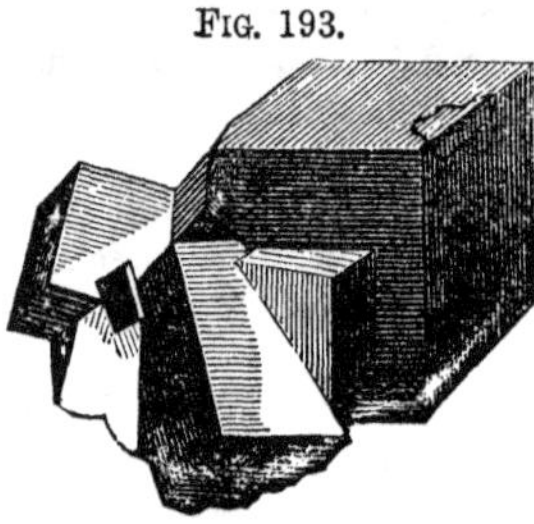

573. **Protosulphate of Iron,** FeO, SO_3+7HO.—*Copperas; Green Vitriol.*—This salt may be readily formed by dissolving metallic iron in sulphuric acid, but for commercial purposes it is prepared on a very large scale by exposing iron pyrites to the action of air and moisture,—the sulphuret of iron, by the absorption of oxygen, yielding sulphuric acid and oxyd of iron. The salt produced is then dissolved out with water, and the solution allowed to crystallize. In this way it is prepared in great quantities at Stafford, Vermont.

Copperas forms beautiful, transparent, bluish-green crystals, which effloresce in dry air, and become covered with brownish-white crust. In combination with certain astringent vegetable matters, as tannin, extract of galls, etc., it forms permanent black dyes, and is hence much used in the arts for dyeing, and for the manufacture of inks.

574. Iron is employed in the arts in three different states, viz., as *crude*, or *cast iron*, as *wrought*, or *malleable iron*, and as *steel*.

575. **Cast Iron,** the metal obtained by smelting the ore with carbon, is a chemical compound of iron and carbon—a carbide, or carburet of iron, containing also, as impurities, small quantities of uncombined carbon and silicon, and generally some phosphorus, sulphur, aluminum, and calcium. It is fusible at a glowing white-heat, is brittle, and can neither be forged or welded. The proportion of carbon in different varieties of cast-iron varies, but in no instance does it exceed 5 per cent. The proportion of silica varies from 3·5 to 0·25 per cent.

In commerce, two varieties of cast-iron are recognized, viz., white and gray metal. The former contains more carbon, and is harder, more brittle,

---

QUESTIONS.—What are iron pyrites? What is copperas? How is it prepared? What are its uses? In what three conditions is iron employed in the arts? What is cast-iron? What two varieties are recognized? What are their respective properties?

and more fusible than the latter. It is also characterized by a silvery whiteness, and a lamelar crystalline fracture. Gray metal, on the contrary, is very soft, dark in color, and of a granular texture; it admits of being filed and drilled with ease, which white metal does not. If white iron be melted and allowed to cool very gradually, a portion of its carbon crystallizes out as graphite, and gray cast-iron is produced. The gray metal is best adapted for castings, and the white for the manufacture of bar iron and steel.

576. **Smelting of Iron.**—The operation of smelting iron, or the reduction of its ores to a metallic state, is effected through the agency of the blast-furnace, which is a tall, chimney-like structure, constructed of stone in a conical form, and lined upon the interior with the most refractory fire-brick. Its internal cavity, represented in section in Fig. 194, resembles in shape a long, narrow funnel, inverted upon the mouth of another shorter funnel, and is divided into the central portion, *b*, called the shaft; the boshes, *e*, or the part of the furnace sloping inward; the crucible, *t*, and the hearth, *h*. The top, or mouth of the furnace serves both for charging it, and for the escape of gases. A steady and intense heat is maintained by means of strong blasts of air driven into the furnace by powerful blowing apparatus through a number of blast-pipes, or tuyeres, *a a*, at its base. The amount of air thus supplied exceeds, in some large furnaces, 12,000 cubic feet per minute. It was formerly the practice to use the air at ordinary temperatures (cold blast), but within a comparatively recent period the production of iron has been very greatly cheapened and increased by heating the air to a temperature of about 500° F. before it enters the furnace (hot-blast).

Fig. 194.

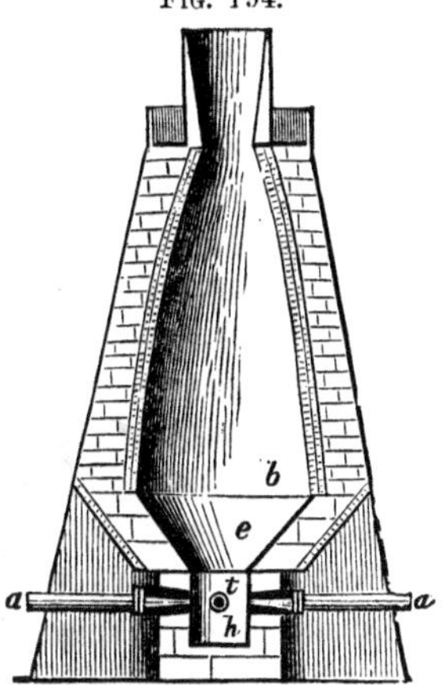

At the commencement of operations, the furnace is first heated with coal only, for about 24 hours, in order to raise it to the proper temperature; but when working regularly, it is charged alternately with coal and a mixture of ore and limestone broken into small pieces, until it is completely filled with successive layers of fuel and of ore. The ore before smelting is generally roasted, or heated separately, in order to expel from it water and carbonic acid, and render it dry and porous. The limestone added serves as flux—that is, it renders the silica, clay, and other foreign matters associated with the ore readily fusible—forming a dark-colored glass termed "slag." As soon as the ore has become thoroughly ignited, its oxygen unites with the carbon of the fuel to form carbonic oxyd, while the metal fuses, and together with the slag flows down to the bottom of the furnace. Here the slag, being

QUESTIONS.—Describe the construction of a blast-furnace? How is iron reduced from the ore? Why is limestone used in the smelting of iron?

the lightest, floats upon the top of the melted metal, and from time to time is raked off through apertures contrived for the purpose—the iron being drawn off by openings at a lower level. As the contents of the furnace are removed from below, or consumed, fresh materials are supplied from above, so that the process of smelting goes on uninterruptedly, day and night, for years, or until the furnace requires repair. The melted iron drawn off from the blast furnace is run into rude molds of sand, and when solidified constitutes crude cast-iron, or the pig-iron of commerce.

577. Malleable, or Bar Iron, is cast-iron deprived of its carbon and other impurities. It is not fusible at a white heat, and may be forged and welded.

The manufacture of bar-iron, or the purification of the crude pig-iron, is effected by exposing cast-metal to the regulated action of oxygen at a high temperature, whereby the carbon, and other oxydizible impurities which it contains, are burnt out of it, and the iron left pure. The details of the process are essentially as follows:—the crude pig-iron is first remelted and suddenly cooled, by which it loses a part of its carbon and silica, and is rendered white,

Fig. 195.

crystalline, and exceedingly hard. In this state it is known as *fine metal.* Broken into fragments, it is next introduced in charges of about 500 lbs. weight, into a kind of reverberatory furnace, called a puddling furnace, and again melted. The workmen then, by means of long iron bars, stir up (puddle) the fused mass, and thoroughly expose it to the influence of the heated air circulating above it. (See Fig. 195.) As the operation proceeds, the metal passes from a liquid to a pasty condition, emits blue flames (carbonic oxyd), gradually grows tough and less plastic, and finally becomes pulverulent. At this point the heat is raised to the highest intensity, and air is carefully excluded

---

Questions.—What is malleable or bar-iron? What is the principle of its preparation? Describe the first step of the process? What is puddling?

by closing the furnace. After a time, the metal softens sufficiently to enable the puddler to collect it in balls (called blooms), upon the end of an iron bar, which are then withdrawn from the furnace, and subjected, while in a state of intense heat, to the action of a massive hammer, moved by machinery. A melted slag (silicate of the oxyd of iron) is thus forcibly squeezed out of the metal, and the particles of iron are brought nearer to each other. The iron is then fashioned into a bar, by passing it between grooved rollers; and the bar thus obtained is cut into lengths, piled up in a reverberatory furnace, reheated and re-rolled. For the best qualities of iron, this process of doubling upon itself, reheating and re-rolling, is repeated several times, in order to render the fibers of the iron parallel to each other—an arrangement which greatly increases the tenacity of the metal. These operations, when properly performed, free the iron from all but mere traces of the impurities contained in the crude metal. The complete separation, however, of phosphorus and sulphur, when present, is a matter of great difficulty; and these two elements, above all others, are the most injurious to iron—rendering it brittle and rotten.*

578. **Malleable Iron Castings.**—Small articles of cast-iron, such as stirrups, bits, door-latches, etc., may be rendered malleable in a degree, by closely packing them in powdered hematite (peroxyd of iron) in tight fire-brick cases, and subjecting them to a red heat, in what is called an annealing furnace, for a period of time varying from six to ten days, finally allowing them to cool slowly. In this case, the character of the iron is changed, by a removal of a part of its carbon, through the agency of the oxygen of the powdered hematite.†

579. **Steel** is a chemical compound of carbon and iron—a carburet or carbide of iron—containing, however, a much less proportion of carbon than cast-iron.

The quantity of carbon in good steel varies between 0·7 and 1·7 per cent; but steel which possesses the greatest tenacity, has been found to contain from 1·3 to 1·5 per cent. of carbon, and about 0·1 of silicon.

What is called *Natural Steel* is produced directly from the best cast-iron,

* The presence in bar-iron of 0·033 per cent. of sulphur, is sufficient to destroy its property of welding, and render it brittle when hot. Such iron is termed "hot short." Iron, on the contrary, which contains phosphorus, may be readily forged and welded when hot, but breaks when cold; it is accordingly known as "cold short." The discovery of a ready method of effectually separating these two elements from iron, is regarded as one of the great problems of chemical science which yet remains unsolved.

† Sheet-iron is bar-iron rolled while hot to the requisite degree of thinness. It is a very popular notion, that the so-called "Russian sheet-iron" is manufactured in Russia by a secret process; but such is not the case. The iron in question is, in the first instance, a very pure article, rendered exceedingly tough and flexible by refining and annealing. Its bright, glossy surface is partially a silicate and partially an oxyd of iron, produced by passing the hot sheet, moistened with a solution of wood-ashes, through polished steel rollers.

QUESTIONS.—What are malleable iron castings? What is steel? What is the percentage of carbon in steel? How is natural steel produced?

by exposing it, in a melted condition on the hearth of a furnace, to the action of a current of air; the oxygen of the air burns off a portion of the carbon from the cast-iron, and steel remains. The preparation of natural steel, therefore, is an intermediate stage in the conversion of cast into wrought-iron. Steel thus obtained is of an inferior quality, and is used for making cheap and coarse instruments. The best qualities of steel are obtained by a process called *cementation*, which is an operation just the reverse of that by which natural steel is formed. It consists in imbedding bars of the best refined malleable iron in powdered charcoal contained in large boxes of fire-brick in such a way that all access of air from without is entirely excluded. The boxes are then subjected, in a furnace, to a most intense heat, for a period varying from five to ten days, during which time the carbon of the charcoal completely penetrates the mass of the iron, and converts it into steel. The steel, when withdrawn, has a peculiar, rough, blistered appearance, and is hence known as blistered steel. Small bars of blistered steel, made into faggots and welded together, at a high temperature, under a tilting, or trip hammer, forms "*tilted steel;*" this, broken up, reheated, and re-welded, forms "*shear steel,*" so called, because it was originally thus prepared for making shears to dress woollen cloth. The quality of the steel is greatly improved by these successive processes of reheating and re-hammering. *Cast steel* is prepared by melting blistered steel, casting it into ingots, and then drawing it into bars under the hammer; it is the most perfect variety of steel, and is employed for all fine cutlery.

Case-hardening.—It is sometimes desirable to convert articles manufactured from soft iron superficially into steel; this is termed case-hardening, and is usually performed by heating them for a short time in contact with powdered charcoal, or sprinkling their surfaces when red-hot with powdered ferrocyanide of potassium.

580. The chemical changes which take place in the conversion of bar-iron into steel are obscure, and it is somewhat doubtful whether we yet fully understand the exact composition of steel. The most recent researches seem to indicate that nitrogen is a constituent of all steel, and that its presence, together with carbon, is essential to its formation. The finest steel known, called Wootz, is produced in a very rude way by the natives of India, and is used for the manufacture of the celebrated sword-blades of the East. The most experienced English manufacturers are unable, with all their resources, to produce steel of an equal quality, and its peculiar excellence has been attributed by Professor Faraday to the presence in its composition of a small quantity of aluminum.*

* Some authorities have supposed that carbon is contained in steel in the form of the diamond, since it seems almost impossible to refer the great differences which exist between cast-iron and steel to merely a minute variation of the proportions of the combined

QUESTIONS.—How is the best steel obtained? What are the different varieties of steel? What is cast-steel? What is case-hardening? What is said of our knowledge of the formation and composition of steel?

581. **Properties of Steel.**—Steel is less fusible than cast-iron, and more so than bar-iron. Its most remarkable property consists in its power of assuming a hardness scarcely inferior to that of the diamond by heating to redness and then suddenly cooling by immersion in cold water; by this treatment it is also rendered extremely brittle and almost perfectly elastic. By reheating the steel and allowing it to cool slowly, it again becomes nearly as soft as ordinary iron, and between these two extremes any required degree of hardness may be attained. In working steel, the articles are first finished in a soft state, and afterward hardened; they are then *tempered*, or raised to such a temperature as is requisite to give them the degree of softness and elasticity required. The workman easily estimates this temperature by observing the color of the thin film of oxyd which appears upon the surface. Thus, a light straw color indicates the degree of heat requisite for tempering razors; a deep yellow, that suitable for scissors, penknives, etc.; while sword-blades, watch-springs, and instruments demanding great elasticity, must be exposed to a much higher degree of heat, or until their surfaces acquire a deep blue color. These various changes in the color of steel may be illustrated by heating a polished steel knitting-needle in the flame of a spirit-lamp.

SECTION II.

MANGANESE AND CHROMIUM.

582. **Manganese.**—*Equivalent*, 27·6; *Symbol*, Mn; *Specific gravity*, 8.—Metallic manganese is a grayish-white metal, resembling some varieties of cast-iron.

It is extremely brittle, and so hard that it is not scratched by a file; a fragment set at a sharp angle may be even substituted in the place of the diamond for cutting glass. It is susceptible of a very high polish, and at ordinary temperatures in the air is not readily oxydized. It dissolves easily in acids. No practical application has ever been made of this metal, and previous to its investigation by Brunner in 1857, very erroneous ideas of its properties were generally entertained. It is now believed to possess a high economic value, especially as an element of certain alloys. Its preparation is, however, difficult.

Manganese is not found in nature as a metal, but as an oxyd it is very widely diffused in the mineral kingdom. Traces of it are very frequently

carbon. In accordance with this view, a theory has been proposed, that the fine cutting properties of a steel blade are due to a minute form of diamond imbedded in the edge; and that the benefit of dipping a razor into hot water before using is owing to the circumstance that the metal is thereby expanded, forcing the sharp edges of the embedded carbon crystals into such positions, that they cut with greater facility.

QUESTIONS.—What are the properties of steel? What is understood by the tempering of steel? What is the appearance of metallic manganese? What are its properties? What is said of its distribution in nature?

found in the ashes of plants, and in river and lake waters. The dark, metallic-like discoloration which may be often noticed on stones and pebbles in the beds of streams flowing over igneous rocks, is due in great part to a coating of oxyd of manganese deposited from the water. The most important and valuable ore of manganese is the black oxyd, also known as the peroxyd, or binoxyd, MnO_2. It is found abundantly at Bennington, Vermont, and in many other localities in the United States.

Seven different oxyds of manganese are described, the two highest of which possess acid properties, and are termed manganic and permanganic acids. Manganic acid is known only in combination with potash, with which it forms a salt—manganate of potash—possessing some very curious properties. It is best prepared by intimately mixing 4 parts of finely-powdered peroxyd of manganese with 3½ parts of chlorate of potash; 5 parts of hydrate of potash dissolved in a small quantity of water, are then added to the mixture, which evaporated to dryness and heated to dull redness for an hour in an earthen crucible, yields a dark green mass. This dissolved in water, gives at first an emerald-green solution, but the color almost immediately and successively changes to dark-green, blue, purple, and finally to crimson. These changes of color are occasioned by a decomposition of manganate of potash, which is hence often called *chameleon mineral;* the final red color retained by the solution is due to the formation of permanganic acid, which is comparatively a stable compound.

The salts of manganese are characterized by a delicate rose-color, which is especially noticeable in crystals of the sulphate. The chief uses of the compounds of manganese are chemical, the black oxyd being extensively employed to decompose muriatic acid, and furnish chlorine (§ 350); it likewise supplies the chemist with his cheapest source of oxygen, and is used as a coloring material in the manufacture of glass and enamels.—MILLER.

583. **Chromium.**—*Equivalent*, 26·4; *Symbol*, Cr.—Chromium is found only as an oxyd in nature, and although abundant in some localities, is very sparingly distributed over the earth. The metal itself, which is obtained with difficulty, is grayish-white, brittle, and is extremely hard; it also resists the action of the strongest acids.

The most abundant ore of chromium, is a compound of protoxyd of iron and sesquioxyd of chromium,—known as "*chrome iron.*" It is found more abundantly in the United States than elsewhere, especially in the vicinity of Baltimore, and at Lancaster, in Pennsylvania.

Almost all the compounds of chromium are characterized by very beautiful colors, and are hence highly valued in the arts as materials for paints, for

QUESTIONS.—What is its principal ore? What is said of its compounds with oxygen? What peculiar properties does the manganate of potash possess? What are the properties of chromium? What is its principal ore? For what are the compounds of chromium remarkable?

dyeing fabrics, and for coloring glass, porcelain, enamels, etc. Oxyd of chromium is the coloring ingredient of the emerald, of the green varieties of serpentine, and probably also of the ruby.

Chromium is prepared for use in the arts by fusing the pulverized ore, chrome iron, with nitrate of potash (saltpeter); by this treatment the chromium absorbs oxygen and becomes converted into an acid—chromic acid—which unites with the potash of the niter to form a yellow salt, the chromate of potash, KO,CrO_3. By adding sulphuric acid to a solution of chromate of potash, we remove one half the base and form a new salt—bi-chromate of potash, $KO, 2CrO_3$—which crystallizes in beautiful red tables, and furnishes the source from whence most of the other compounds of chromium are prepared. It is also in the form of this salt that chromium is best known as an article of commerce.

584. **Chromate of Lead, PbO,CrO_3.**—*Chrome Yellow.*—By adding a solution of bi-chromate of potash to a solution of acetate of lead, we obtain a beautiful yellow precipitate of chromate of lead; this, washed and dried, constitutes the well-known pigment, *chrome yellow.* By mixing chrome yellow with white substances, such as chalk, clay, etc., numerous other shades of yellow are obtained, as Paris yellow, king's yellow, etc.; but by mixing it with Prussian blue, a variety of cheap green pigments are formed, as Naples green, olive green, etc.

585. **Chromic Acid, CrO_3,** is readily prepared by mixing 4 measures of a cold saturated solution of bi-chromate of potash with 5 of oil of vitriol; as the liquid cools, chromic acid is deposited in the form of beautiful crimson needles. The mother liquor is then poured off, and the crystals allowed to dry on a porous brick, closely covered with a bell glass; since they decompose instantly by contact with organic matter. When chromic acid is brought in contact with alcohol or ether, it imparts to these substances a portion of its oxyyen, and the intensity of the chemical action occasioned, produces an immediate ignition. This may be illustrated by projecting a small quantity of chromic acid upon a little alcohol or ether contained in a tumbler.

If we mix in a small mortar as much chromic acid as can be taken upon the point of a knife, with about one quarter as much of powdered camphor (without pressing upon it strongly), and then let fall into the mortar a few drops of alcohol from a considerable height, instantaneous ignition and deflagration ensues—like the burning of gunpowder. The residue in the mortar, after the decomposition, is sesquioxyd of chromium, which presents the appearance of an elegant green, mossy vegetation.—STOCKHART.

QUESTIONS.—How is it prepared for use in the arts? What is the composition of bi-chromate of potash? What is chrome yellow? How is chromic acid prepared? What are its properties?

SECTION III.

COBALT AND NICKEL.

586. **Cobalt.**—*Equivalent,* 29·5, *Symbol,* Co.—Cobalt is a reddish-gray metal, which is never found in nature in a metallic state, except as a constituent, in small proportions, of meteoric iron.

Oxyd of cobalt is remarkable for the magnificent blue color it communicates to glass, and also to porcelain. This may be illustrated by moistening a small bead of borax glass with a solution of nitrate of cobalt, and then fusing it in the flame of a blow-pipe. The substance called *smalt* is a glass, colored blue by oxyd of cobalt, and then finely pulverized; it was formerly much used as a pigment, especially in the manufacture of blue writing-paper; but is now almost entirely superseded by the cheaper *ultramarine.*

587. **Sympathetic Inks.**—The chloride of cobalt, CoCl, is easily obtained by dissolving oxyd of cobalt in hydrochloric acid; the solution, evaporated to small bulk, yields ruby-red crystals, which are freely soluble in water. The liquid resulting from their aqueous solution, is of a deep-blue color when concentrated, but when diluted, is pink. In this latter condition it may be used as a sympathetic ink: characters written on paper with it are invisible, from their paleness of color, until the salt has been rendered anhydrous by exposure to heat, when the letters appear blue. When the paper is laid aside, moisture is again absorbed by it, and the writing once more disappears.

588. **Nickel.**—*Equivalent,* 29·6 ; *Symbol,* Ni.—Nickel is a brilliant, silver-white metal, extremely ductile, and more fusible than iron. It occurs in nature associated chiefly with arsenic, sulphur, and cobalt; but its ores are by no means abundant. It is almost always found native in meteoric iron, sometimes in a proportion of 10 per cent.

The salts of nickel are of a delicate green color, both when in a solid state and when in solution.

The chief use of nickel is in the manufacture of German silver, which is a white, malleable alloy, consisting, in 100 parts, of 51 parts of copper, 30·6 of zinc, and 18·4 of nickel—the latter element imparting to the alloy, when polished, a silver-like appearance.

589. **General Properties of Cobalt and Nickel.**—These two metals are especially remarkable from the circumstance, that they almost

---

QUESTIONS.—What is said of metallic cobalt? What are the characteristics of oxyd of cobalt? What is smalt? What is sympathetic ink? What is said of the properties and distribution of nickel? What of its salts? What of its uses? What is German silver? In what respects do cobalt and nickel resemble each other?

always occur associated together in nature, have nearly the same chemical equivalent, and agree very closely in their chemical properties. They are also the only metals beside iron which are readily susceptible of magnetism.

SECTION IV.

ZINC AND CADMIUM

590. Zinc.—*Equivalent*, 32·5; *symbol*, Zn; *specific gravity*, 6·8 to 7.—Zinc is not found native, but its ores are somewhat abundant.

The most important of its ores are the carbonate of zinc, called *calamine;* the red-oxyd of zinc, found in great purity and quantity at Sterling, New Jersey; and a sulphide of zinc, called *blende*—the latter being usually associated with ores of lead.

591. **Properties.**—Zinc is a hard, bluish-white metal, which exhibits a crystalline fracture when broken. It is brittle at ordinary temperatures, but between 200° and 300° F. acquires considerable malleability and ductility, and may be rolled and wrought with ease; it is a very singular fact that zinc so treated retains its malleability when cold, and it is in this way that the ordinary sheet-zinc of commerce is manufactured. At a temperature of 400° it again becomes so brittle as to admit of being pulverized in a mortar. At 770° it melts, and at a bright red heat it is volatilized. If its vapor be brought in contact with air, it burns with a splendid green light, and is converted into oxyd, which falls in copious white flakes, anciently called "philosopher's wool." Zinc, when exposed to a moist atmosphere, soon tarnishes, and becomes covered with a thin film of oxyd, which protects the metal beneath from any further change. This property renders zinc valuable for a great variety of economic purposes.

By reason of the volatility of zinc at high temperatures, it is reduced from its ores by a process of distillation rather than smelting. This is effected by heating a mixture of pulverized ore and coal in earthen retorts, and condensing the evolved vapors over water, or in receivers from whence free access of air is excluded. The zinc of commerce always contains impurities, generally iron and lead, and sometimes arsenic.

592. **Galvanized Iron** is sheet-iron coated with zinc. It is prepared as follows: the iron is first immersed in dilute sulphuric acid, to remove all scale or oxyd from its surface, and then plunged into a bath of molten zinc, covered with sal-ammoniac. The use of the latter substance is to dissolve off any adhering film of oxyd from the iron, as a complete union of the two metals will not occur unless the surface of the iron is chemically clean. The thin coating of zinc which adheres to the iron renders the latter metal negatively electric, and prevents its oxydation or rusting. (§ 247.)

QUESTIONS.—What is said of the distribution and ores of zinc? What are the properties of zinc? How is zinc reduced from its ores? What is galvanized iron?

593. **Oxyd of Zinc, ZnO.**—*Zinc White.*—Zinc forms only one compound with oxygen, which, when pure, is a white powder. Mixed with drying oils, it is much employed as a white paint, under the name of zinc-white, as a substitute for white lead; it wants, however, the opacity and dead whiteness for which white lead is so much valued; but has the advantage of not blackening by the action of sulphuretted hydrogen, and of not being deleterious to the health of the workmen.*

Sulphate of zinc, ZnO,SO_3, constitutes the white vitriol of commerce; it used medicinally in small doses, and also in the operations of calico printing.

The salts of zinc are colorless, and act powerfully and rapidly as emetics.

594. **Cadmium, Cd,** is a white metal, resembling tin in appearance, but allied to zinc in its properties. It is usually found in very small quantities accompanying the ores of zinc, and has no practical value in the arts.

SECTION V.

LEAD AND TIN.

595. **Lead.**—*Equivalent*, 103·5; *Symbol*, Pb (Plumbum); *Specific gravity*, 11·44.—Lead is rarely found native, but its ores are most abundant. Almost all the lead of commerce is obtained from *galena*, or the sulphide of lead, PbS, an ore which occurs in boundless profusion in the United States, especially in Missouri. The reduction of the metal is easily effected by crushing the ore and subjecting it to a moderate heat in a reverberatory furnace. Galena always contains a proportion of silver, which is sometimes so abundant that the ore is worked for this metal rather than for the lead. When the galena occurs in bold, well-characterized cubes (see Fig. 196), the contained lead is

Fig. 196.

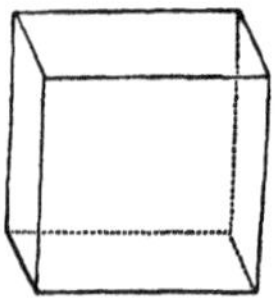

* Zinc-white is at present extensively manufactured from the red zinc ore found at Sterling, New Jersey, by an exceedingly interesting and simple process. The ore, pulverized and mixed with coal, is heated in large oven-shaped retorts of brick, to each of which a wide pipe of sheet-iron is fitted; these extend through the furnace wall and connect with a very large horizontal tube, through which a current of air is kept moving by the revolution of a fan-wheel. The current thus formed flows first through the retorts, and burns the vapor of zinc to oxyd; which is immediately transported by the draft of air through the continuation of the tube to a chamber, where it falls as delicate powder.

Oxyd of zinc, in combination with chloride of zinc, has recently been applied to producing a lustrous hard finish to the walls of rooms, in the place of plaster of Paris. A coat of oxyd of zinc mixed with size, and made up like a wash, is first laid on the wall, ceiling, or wainscot, and over that a coat of chloride of zinc applied, being prepared in the same way as the first wash. The oxyd and chloride effect an immediate combination, and form a kind of cement, smooth and polished as glass.

QUESTIONS.—What is said of oxyd of zinc? How is it prepared? What is said of sulphate of zinc? What of cadmium? What is said of the distribution of lead? What is galena? What is a usual constituent of this ore?

usually nearly pure; but a mineral which will yield 0·36 per cent. of silver, or 120 ounces to the ton, is considered extremely rich.*

596. **Properties.**—Lead is a soft, bluish-gray metal, which may be rolled into tolerably thin sheets, or drawn into wire; its tenacity, however, is very feeble. It fuses at 620° F., and contracts considerably at the moment of its solidification, which circumstance renders it unsuitable for castings. At a temperature above red-heat it is somewhat volatile

The surface of a piece of lead, when freshly cut, presents a high metallic luster, but it soon tarnishes by exposure to the air, owing to the formation of a thin, closely-adherent film of oxyd, which protects the metal from further change. In a perfectly dry atmosphere, lead undergoes no alteration, and even when sealed up in a vessel of pure water, free from air, the metal will retain its brilliancy for an indefinite period; but if it be exposed to the united action of air and *pure* water, it is subject to a powerful corrosion.—MILLER.

597. Lead, when taken into the system in any soluble form, is a dangerous poison; its effects, moreover, do not generally manifest themselves immediately, but the poison accumulates, and produces, often after the lapse of years, a number of different and distressing forms of disease, such as colic, paralysis, etc. The action of water on lead is, therefore, a matter of great importance in a sanitary point of view, since this metal is extensively employed in cisterns and pipes, for the storage and supply of water.

The action of different waters on lead varies considerably. Waters which are very pure and highly aerated—waters which contain nitrates, chlorides, or organic matter, as those flowing from the vicinity of barn-yards, manure heaps, or from swamps and fields, all dissolve lead from the pipes through which they pass; and the constant use of such waters, in the process of time, will introduce sufficient lead into the system to produce disease. Waters, on the other hand, which contain sulphates, carbonates, and phosphates, exert but comparatively little action on lead. Bi-carbonate of lime is especially remarkable for the preservative influence which it exerts; and as this salt is a very common impurity in water, few spring waters act on lead to a dangerous extent. In these cases, a film of insoluble carbonate, sulphate, or phosphate of lead, is formed upon the surface of the pipe, and all further corrosion prevented. Rain-water, as collected from the roofs of houses, is for the most part sufficiently impure, especially in cities, to prevent its action on lead. So general, however, is the action of water upon lead, that it is rare to find any

* So small a quantity of silver as three or four ounces to a ton of lead, may be extracted profitably by a process devised by Mr. Pattinson, of England. It consists in melting the argentiferous lead, and allowing it to cool slowly. Under these circumstances, the lead tends to separate in the form of crystals of pure metal, before the alloy of silver has been cooled sufficiently to solidify. By a careful regulation of temperature, the great mass of the lead may, therefore, be removed mechanically, leaving the silver concentrated in a small bulk of alloy.

QUESTIONS.—What are the properties of lead? What changes does lead undergo in the air? What is said of the poisonous influence of lead? What of the action of water on lead? What salts arrest the action of water on lead? How do they effect this?

water that has been kept in contact with this metal for a considerable period, which does not contain some traces of it. Stone and wooden cisterns, and tin, iron, or wood pipes, are therefore, greatly to be preferred to lead. Where lead service-pipes are used, it is always advisable to allow the water to run for some time before using.

598. **Oxyds of Lead.**—Four distinct oxyds of lead are recognized, the most important of which are the protoxyd, PbO, and the peroxyd, PbO_2.

Protoxyd of Lead, *Litharge, Massicot,* PbO, is a yellow powder, usually obtained on a large scale, by the oxydation of molten lead in a current of air. Its production may be illustrated by fusing a small piece of lead on charcoal, before the exterior flame of a blow-pipe. The melted lead oxydating, is at first converted into a grayish powder—a mixture of oxyd of lead and metallic lead—but by continued blowing, the gray color is changed into a brilliant yellow—litharge. This oxyd of lead is a powerful base, and is extensively used in the arts, especially in the glazing of pottery and the manufacture of flint glass. United with fatty acids it forms insoluble soaps (the well-known diachylon, or lead plaster); and boiled with linseed-oil, it constitutes a varnish much used by cabinet-makers.

Red-Lead, or Minium, $2PbO . PbO_2$, is a compound of protoxyd and peroxyd of lead. It is formed by exposing protoxyd of lead, for a long time, to the action of air, at a temperature below fusing. It possesses a brilliant red color, and is much used in the arts in the manufacture of glass, as a pigment, and for the coloring of red sealing wax and of paper.

599. **Carbonate of Lead, PbO, CO_2.**—*White-Lead.*—This salt occurs in nature, but is prepared artificially, in immense quantities, for use as a paint. Pure carbonate of lead is a soft, white powder, insoluble in water, but easily dissolved by dilute nitric or acetic acids.

Two methods are in use for making white lead. One of these consists in dissolving litharge in acetic acid, and precipitating the lead as carbonate, by a current of carbonic acid gas. The best white lead is, however, made by a process known as the "Dutch method." A great number of small earthen pots are partially filled with weak vinegar, and a thin sheet of lead, coiled in a spiral, placed in each. (See Fig. 197.) The pots are then each covered with a plate of lead, and arranged in rows and tiers, one above the other, to the height of 15 or 20 feet, and the whole finally covered with decomposing stable manure or spent tan. After the lapse of several months, the rolls of lead are found to be mostly or entirely converted into a pure white carbonate, which only requires washing and grinding to fit it for immediate use. The theory of this curious process is as follows: the

FIG. 197.

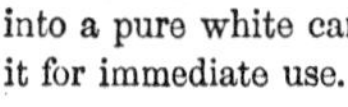

QUESTIONS.—What oxyds of lead are there? By what names is protoxyd of lead known? How is it prepared? What are its properties and uses? What is red-lead? What other name is applied to it? What are its uses? What is white-lead? How is it prepared?

heat of the decomposing dung, etc., volatilizes a portion of the vinegar, as acetic acid, and under the influence of air and acid fumes, a crust of acetate of lead is formed upon the surface of the metal. The carbonic acid, generated from the slow decay and decomposition of the materials of the hot-bed, readily converts this salt into carbonate of lead, leaving the acetic acid free to combine with an additional portion of lead, which is, in turn, again decomposed; and this action is repeated until the whole of the lead is converted into a carbonate. White lead is largely adulterated with sulphate of baryta, but the fraud may be easily detected by digesting a sample in nitric or acetic acids, when the baryta will remain undissolved.

600. The most ready solvent for lead is nitric acid; dilute sulphuric and hydrochloric acids not acting upon it to any great extent.

The presence of lead in solution may be easily detected by any of the following tests: with sulphuric acid it forms a white, insoluble precipitate—sulphate of lead; with sulphuretted hydrogen, a black sulphide, visible, when the quantity of lead present is very minute, only after a little time; and with solutions of bi-chromate of potash or iodide of potassium, yellow precipitates.

FIG. 198

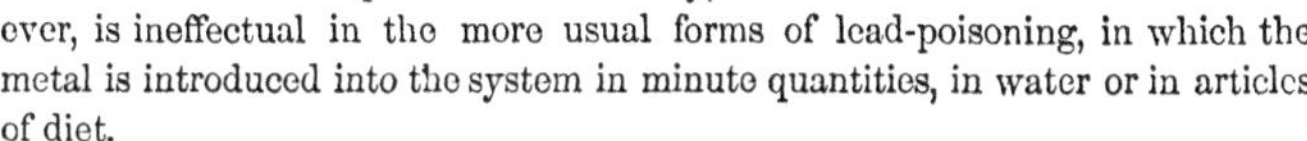

Zinc precipitates lead from its solution by voltaic action, in the form of crystalline metallic particles, forming what is known as the lead-tree. (Fig. 198. § 255.)

In case of poisoning by a dose of soluble lead salts, the best antidote is Epsom salts (sulphate of magnesia), with which the lead forms an insoluble and inert sulphate. This remedy, however, is ineffectual in the more usual forms of lead-poisoning, in which the metal is introduced into the system in minute quantities, in water or in articles of diet.

601. **Alloys of Lead.**—The alloys of lead are numerous and important. Shot is an alloy of lead, with a small proportion of arsenic, which hardens it, and facilitates its separation into globules. In the manufacture of shot, the lead is melted at the top of high towers built for the purpose, and poured into a vessel perforated in the bottom with numerous small holes. The lead, in running through, is separated into drops, which falling through the height of the tower, become spherical, and cool before reaching a reservoir of water placed for their reception at the base of the tower. For the manufacture of the largest sized shot, a tower of at least 150 feet high is required. Shot are proved, and the different sizes separated, by rolling them down an inclined board. Those which are irregular in shape, roll off at the sides, or stop, while the perfectly spherical ones continue in a straight course.

QUESTIONS.—What is the most ready solvent of lead? How may the presence of lead in solution be detected? What are antidotes to lead-poisoning? How are shot manufactured? What is their composition? How are shot proved?

Type-metal is an alloy of 3 parts lead and 1 of antimony. This alloy is sufficiently fusible to allow of its being readily cast, and it expands at the moment of solidification, and copies the mold accurately. *Solder* is an alloy of lead and tin.

602. **Tin.**—*Equivalent*, 59; *Symbol*, Sn (Stannum); *Specific gravity*, 7·29.—Tin occurs in nature usually as an oxyd, but sometimes as a sulphide.

Its ores are very sparingly distributed over the earth—the most important tin-mines being those of Cornwall, England, and Malacca, in Southern Asia. It is also mined to a limited extent in Germany, and in a few localities in Mexico and South America. It has hitherto only been discovered in one locality in the United States, at Jackson, N. H., in very small quantities.

Tin of two qualities, as regards purity, is recognized in commerce, viz., "block" or "bar" tin, and "grain" tin; the latter being a refined metal.

603. **Properties.**—The properties which characterize tin and render it valuable in the arts, are its malleability, fusibility, softness, silver-like color and luster, and especially its slight affinity for oxygen, which enables it to retain its brilliancy at ordinary temperatures, in the presence of air and moisture. Tin melts at 442° F. If heated above this point it is not sensibly volatilized, but becomes rapidly oxydized, and burns with a brilliant white light. When a bar of metallic tin is bent backward and forward, it has a peculiar creaking or crackling sound, which is termed the "tin cry;" this is due to the crystalline texture of the tin, the crystals moving upon each other. Tin is almost insoluble in dilute sulphuric acid, and dissolves slowly in hydrochloric acid. Nitric acid acts upon it with great violence, not dissolving the metal, but converting it into a white powder, the binoxyd of tin. This reaction may be easily illustrated by pouring dilute nitric acid upon a little tin-foil in a porcelain dish. The binoxyd of tin thus formed, when rendered anhydrous by ignition, constitutes the *putty powder* used for polishing glass, and for the manufacture of enamels.

604. With oxygen tin unites to form several oxyds, the principal of which are the protoxyd, SnO, and the peroxyd or binoxyd, SnO_2. This last oxyd in its hydrated condition, has the character of an acid, and is then known as stannic acid, SnO_2,HO. Tin also unites with chlorine to form two chlorides, the protochloride, SnCl, and the perchloride, $SnCl_2$. The last-named chloride is an old and very curious compound, which was formerly called the "*fuming liquor of Libavius.*" It is a dense, fuming liquid, prepared by exposing melted tin to the action of dry chlorine. A preparation of bi-chloride of tin is extensively used in dyeing as a mordant. A *bi-sulphuret of tin*, SnS^2, formed by exposing to a low red heat in a glass flask a mixture of 12 parts

---

QUESTIONS.—What is type-metal? What is solder? What is said of the occurrence and distribution of tin? What two qualities of tin are known in commerce? What are the characteristic properties of tin? What is "tin-cry?" What is the action of acids upon tin? What is putty powder? What are the principal oxyds of tin? What is said of the chlorides of tin? What is the composition of bronze powders?

tin, 6 mercury, 6 sal-ammoniac, and 7 flowers of sulphur, is of a brilliant gold color, and is known as *mosaic gold.* It constitutes the well-known bronze powders used in printing and in the manufacture of paper-hangings.

605. The useful applications of metallic tin are very numerous. *Tin-plate*, of which ordinary articles of tin-ware are made, is sheet-iron superficially coated with tin. It is prepared by first rendering the surface of the iron chemically clean by the action of acids, and then immersing the sheet-metal for a considerable length of time in a bath of molten tin; the union of the two metals, thus effected, is not a case of simple adhesion, but the tin forms with the iron an alloy. *Britannia-metal*, which is much used for the manufacture of tea-pots and cheap spoons, consists of equal parts of tin, brass, antimony, and bismuth. *Pewter* of the best quality, consists of 4 parts tin and 1 of lead. Ordinary brass pins are tinned, or whitened, by boiling them in a vessel containing tin in a finely-divided state, and a solution of cream of tartar in water. The acid of the cream of tartar effects a solution of some of the tin in the first instance, which on longer boiling separates as a metal upon the more electro-positive brass.

SECTION VI.

COPPER AND BISMUTH.

606. **Copper.**—*Equivalent*, 31·7; *Symbol*, Cu (Cuprum); *Specific gravity*, 8·9.—Copper is frequently found native, generally in small crystals, but sometimes in immense masses, as in the mines on Lake Superior. Its ores, also, are numerous and abundant—the most important being the sulphurets of copper, and the red oxyd. Carbonate of copper, constituting the beautiful green mineral *malachite*, is also found in sufficient abundance in some localities to admit of working—especially in Siberia, Eastern Africa, etc.

607. **Properties.**—Copper is one of the metals which has been longest known to man, and was extensively used long before the working of iron was understood. It is moderately hard, tenacious, ductile, and malleable, and is the only metal, with the exception of titanium, which has a red color. Copper requires a high degree of temperature for fusion (1990° F.), and when exposed to an intense heat is somewhat volatile—its vapor burning with a green flame. It is one of the best conductors of heat and electricity.

At ordinary temperatures, in a dry atmosphere, copper remains unchanged, but when exposed to a moist air it quickly tarnishes, and ultimately becomes covered with a strongly-adherent green crust, consisting chiefly of carbonate. Pure water has little or no action upon copper, but in sea-water, or solutions of the chlorides, it is gradually corroded. The corrosion and waste of the copper sheathing of ships is due chiefly to the oxygen contained in sea-water,

---

QUESTIONS.—What is tin-plate? What is Britannia-metal? What is pewter? How are pins whitened? What is said of the occurrence of copper in nature? What are its chief properties? What is the durability of copper in various situations? What occasions the corrosion of copper sheathing?

and to the friction of the water; the corrosion being greatest at those points where the bubbles of air inclosed in the water, by the surging at the bow, rise to the surface and break against the bottom of the vessel. Corrosion of a ship's sheathing is also slow in mid-ocean compared to what it is at the mouths of tropical rivers, or in harbors, where the water is filled with particles of organic matter in a state of decomposition.

608. The most ready solvent of copper is nitric acid. (§ 344) Many of the weak vegetable acids also attack metallic copper, but dilute sulphuric and hydrochloric acids have scarcely any action upon it.

609. The two principal oxyds of copper are the protoxyd, or black oxyd, CuO, and the suboxyd, or red oxyd, Cu_2O.

610. **Protoxyd of Copper** is the basis of all the ordinary salts of copper. It is prepared by heating metallic copper to redness in a current of air, and then suddenly quenching it in cold water; or more conveniently by decomposing the nitrate of copper by heating it to redness—the product in the first instance being black scales, and in the last a dense black powder. It may also be obtained as a hydrate of light blue color by the addition of caustic potash to a solution of any of its salts, (as blue vitriol). Protoxyd of copper is a compound of considerable importance in chemistry and the arts; when mixed with organic substances, and heated, it gives up all its oxygen, and is hence much used to effect the complete combustion of these bodies in a process by which their elementary composition is determined; it is also used for imparting to glass and porcelain a beautiful green color.

FIG. 199.

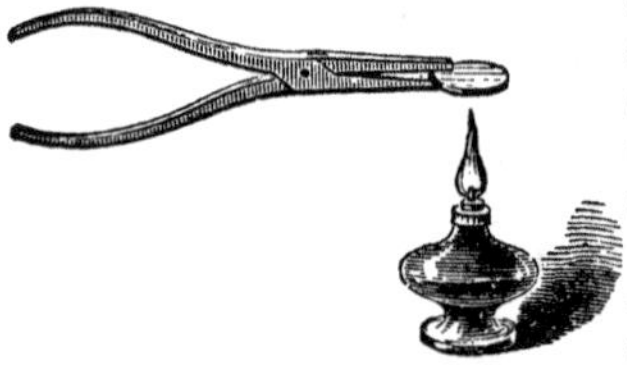

Suboxyd of copper is found native, and may be prepared by heating a mixture of 5 parts of black oxyd and 4 parts of fine copper filings in a covered crucible; the red coating which is formed when metallic copper (as a cent, for example, see Fig. 199) is slightly heated and suddenly cooled, is also suboxyd of copper. Its principal industrial value is for imparting to glass a beautiful ruby or purple color.

611. **Sulphate of Copper**, *Blue vitriol*, CuO,SO_3, is one of the most important of the salts of copper, and is formed by heating metallic copper with concentrated sulphuric acid. It crystallizes in beautiful blue crystals, and is soluble in 4 parts of cold and 2 of boiling water. Large quantities of this salt are used in calico-printing, in the preparation of most of the other salts of copper, and as an agent for exciting galvanic batteries. Wood steeped in a solution of sulphate of copper is protected against dry-rot, and a wash of it on the wood-work of cellars is highly serviceable in preventing the formation of mold. Animal substances thoroughly imbued with it and then dried, remain unaltered.

QUESTIONS.—What is the most ready solvent of copper? What are the two principal oxyds of copper? What is said of protoxyd of copper? What of suboxyd of copper? What is the composition of blue vitriol? What are its uses and properties?

612. **Nitrate of Copper, CuO, NO_5**, is a beautiful blue, efflorescent salt, formed by dissolving metallic copper in nitric acid. It is highly corrosive, and easily decomposed. Its tendency to decomposition may be illustrated by closely enveloping a few moist crystals of nitrate of copper in tin-foil, and placing the parcel upon a porcelain dish; the affinity of the tin for the nitric acid in a short time occasions intense chemical action, accompanied by the phenomenon of ignition; a paper also, moistened with a strong solution of this salt, cannot be rapidly dried without taking fire from the decomposition of the nitric acid.

613. **Verdigris.**—*Sub-Acetate of Copper.*—Verdigris is a salt of acetic acid (the acid of vinegar) and oxyd of copper. It may be formed experimentally by moistening from time to time a copper coin with vinegar, which occasions the production of a green coating. It is prepared on a large scale, either directly from copper and vinegar (green verdigris), or indirectly by spreading the refuse of pressed grapes upon plates of copper exposed to the air; in this latter case the juice adhering to the mash gradually changes to vinegar, and forms blue, or French verdigris. This salt is much used in the arts as a pigment.

614. **Characteristics of the Salts of Copper.**—Most of the salts of copper have a blue or green color, and nearly all are soluble in water. They are distinguished by a nauseating metallic taste, and when taken internally act as deadly poisons, producing violent vomiting, followed by exhaustion and death. Black oxyd of copper is soluble in oils and fats, so that greasy matters boiled in an copper saucepan which is not kept bright, are liable to become impregnated with the metal; verdigris may also be introduced into food from the cooking of acid vegetables or fruits in copper vessels; the use of copper in domestic economy ought, therefore, to be dispensed with as far as practicable. The best antidote against copper poisoning is raw white of eggs, the albumen of which, by forming an insoluble compound with the metal, renders it inert. Milk, or sugar mixed with iron filings are also efficacious.

615. **Alloys of Copper.**—The alloys of copper are extensively used in the arts. *Brass* is an alloy of copper and zinc; the usual proportions being 66 parts of copper and 34 zinc. By varying the proportions, however, the varieties of brass known as "red metal," "pinchbeck," "Muntz," or sheathing metal, etc., may be obtained. *Gun-metal*, used in the fabrication of brass ordnance, is an alloy of 90 parts of copper and 10 of tin. *Bell-metal* and *speculum-metal* contain a larger proportion of tin. *Bronze* for statuary consist of 91 parts copper, 2 of tin, 6 of zinc, and 1 of lead. The brass of the ancients was an alloy of copper and tin.

QUESTIONS.—What is said of nitrate of copper? What is verdigris? How is it prepared? What are the characteristics of the salts of copper? Why is the use of copper vessels in culinary operations unadvisable? What is the best antidote against copper poisoning? What is brass? What is gun-metal—bell-metal—bronze?

616. **Bismuth** is a reddish-white, hard, brittle metal, which is generally found native in small quantities.

It crystallizes from fusion in cubic crystals of great brilliancy. Its principal employment is in the preparation of alloys, a slight admixture of it increasing the fusibility of several metals to a remarkable extent. Oxyd of bismuth is used to some extent in medicine, and also as a cosmetic (pearl powder).

SECTION VII.

URANIUM, VANADIUM, TUNGSTEN, COLUMBIUM, TITANIUM, MOLYBDENUM, NIOBIUM, PELOPIUM, ILMENIUM, ETC.

617. All these metals are very sparingly distributed over the surface of the earth, and some of them are so rare, that they have been seen by only a few chemists. Uranium and titanium are used to some extent for the coloration of porcelain and enamels; and molybdenum, in combination, as molybdate of ammonia, constitutes the most delicate known test of the presence of phosphoric acid in solution.

SECTION VIII.

ANTIMONY AND ARSENIC.

618. **Antimony.**—*Equivalent,* 12·9; *Symbol,* Sb. (Stibium).—Antimony is a blueish-white metal, with a highly crystalline texture, so brittle that it may be easily reduced in a mortar to a fine powder.*

When exposed to air and moisture, at ordinary temperatures, it undergoes no change; but if heated, it burns brilliantly, emitting copious white fumes, which consist chiefly of a teroxyd of antimony. A very interesting experiment consists in fusing a little of the metal on charcoal before the blowpipe, and projecting the melted globule upon the floor or an inclined board; the moment it strikes the hard surface, it bursts into a multitude of little spheroids, which radiate in all directions, leaving a trail of white smoke (oxyd) behind them. Antimony is not used by itself in the arts, but it enters into the composition of several important alloys, as type metal, Britannia metal, etc. Finely-powdered antimony, sprinkled into a jar of chlorine gas, ignites, and occasions a miniature shower of fire.

* Antimony was first made known by Basil Valentine, an alchemist and monk, of Erfurth, Germany. The etymology of its name is said to be due to the following circumstance: its discoverer having observed that its effects, when administered to hogs, were beneficial, tried it upon his fellow-monks. The result of the experiment, however, was that the monks sickened and died—hence the name *antimoine*, anti-monk, *antimony.*

QUESTIONS.—What is said of bismuth? What are its uses? What is said of uranium, titanium, and molybdenum? What of antimony? What are the properties of antimony? What its industrial uses?

619. Antimony forms three oxyds, the most important of which are, the teroxyd of antimony, SbO_3, and antimonic acid, SbO_5.

620. The compounds of antimony are remarkable for their medicinal effects, and are classed in pharmacy among the important remedial agents. When taken, however, in inordinate doses, they act as poisons. *Tartar emetic* is a double salt of tartrate of potash and tartrate of antimony. It is formed by boiling oxyd of antimony with cream of tartar, which last is a salt of potassa and tartaric acid, containing free acid; this free acid combines with the oxyd of antimony, and thus forms a double salt. Tartar emetic, dissolved in sherry wine, in the proportion of two grains of the former to a fluid ounce of the latter, forms the well-known "wine of antimony."

Sulphuretted hydrogen, added to solutions of antimony (as a solution of tartar emetic in water), precipitates the metal in the form of a peculiar and highly characteristic, orange-colored sulphuret.

621. **Arsenic.**—*Equivalent*, 75; *Symbol*, As.—Arsenic is sometimes found native, but generally occurs in the form of an alloy with some other metal, especially with iron, cobalt, nickel, copper, or tin.

The greater part of the arsenic of commerce is obtained in Silesia, in Germany, by roasting, in furnaces, a double sulphuret of iron and arsenic,—called mispickel, or the arsenides of nickel and cobalt. The arsenic, volatilized by heat in the form of an oxyd—arsenious acid—is condensed and collected in the form of a white powder in large chambers, into which the flues from the furnace open.*

Metallic arsenic may be obtained by heating arsenious acid with powdered charcoal in a tightly-closed crucible. It is a dark, steel-gray colored metal, extremely brittle, and may be easily reduced to powder. It is sold by druggists under the very objectionable names of *fly-powder*, *fly-poison*, *cobalt*, *etc.* When heated, it volatilizes without fusion; and if air be present, it oxydizes to arsenious acid. Its vapor has a remarkable odor of garlic, which is so peculiar and noticeable, that it is regarded as one of the characteristic tests of the presence of this element; this odor is easily recognized by heating a fragment of arsenic, or arsenious acid on charcoal before the blow-pipe.

622. The oxyds of arsenic are two:—Arsenious acid, AsO_3, and arsenic acid, AsO_5.

---

* The opening of these chambers, and the removal of arsenic, is a task of great danger, and is performed about once in six weeks. The workmen engaged in the operation, as protection against the poison, are completely encased in leather, with glazed apertures for the eyes. They also wear, in addition, damp cloths over their mouths and nostrils, in order to prevent the inhalation of minutely-divided particles.

QUESTIONS.—What are the chief oxyds of antimony? What is tartar-emetic? What is wine of antimony? What is a characteristic test of antimony in solution? In what form does arsenic occur naturally? How is the arsenic of commerce prepared? What is said of metallic arsenic? What of its oxyds?

Arsenious Acid, AsO_3.—*White Arsenic, Rat's-bane.*—This oxyd is the substance to which the name arsenic is popularly applied, and is the well-known poison. It occurs in commerce as a white powder, but when freshly sublimed it assumes the appearance of a semi-transparent solid, which gradually becomes opaque and white, like porcelain. It is soluble in about 11 parts of boiling water, but to a very much less extent in cold water. Its solution is colorless, and almost tasteless, which circumstances greatly facilitate its employment for criminal purposes. It dissolves freely in hot hydrochloric acid, and in solutions of the alkalies.

Arsenious acid combines with bases to form *arsenites:* arsenite of potash is used in medicine under the name of *Fowler's solution;* and arsenite of copper constitutes the delicate and beautiful green pigment known in commerce as *Scheele's green.* Its poisonous properties have also been taken advantage of for the destruction of vermin. To destroy rats and mice, the poison should be mixed with flour or lard, but not in too large a quantity, or these animals will not touch it.* Arsenious acid, when placed in contact with organic substances, prevents their decay, and may be hence used with advantage for the preservation of stuffed and dried objects of natural history.†

623. **Arsenic Acid,** AsO_5, is formed by treating arsenious acid with nitric acid, and evaporating the solution to dryness. It unites with metallic oxyds to form *arseniates:* the arseniate of potash being used to a very great extent in calico printing, not so much to produce colors as to prevent their adherence to certain portions of the fabric.

Arsenic combines with hydrogen to form a volatile and highly poisonous gas—*arseniuretted hydrogen.* There are also several compounds of arsenic and sulphur, which are used as pigments and in pyrotechny: *realgar,* AsS_3, is a beautiful red pigment, and is a principal constituent of the so-called *white Indian fire,* often used as a signal-light; *orpiment,* AsS_5, is a golden yellow pigment;—both of these substances are found native.

624. Arsenic forms alloys with most of the metals, which are generally brittle and easily fusible. Its presence in iron is highly injurious.

---

* If the poison is put in stables, the receptacles of meal and fodder should be carefully covered over, that the poisoned rats may not vomit the poison into them.

† It is best used for this purpose in the form of an arsenical soap, which may be prepared by mixing 100 parts of white soap, 100 of arsenious acid, 36 carbonate of potash, 15 camphor, and 12 quicklime. The soap is to be scraped and melted with a little water at a gentle heat; then add the potassa and the lime, and mix them well together—the arsenious acid is afterward added gradually, and well incorporated. The camphor is reduced to powder by rubbing it in a mortar, with the addition of a few drops of strong alcohol, and when the soap is cold this is well mixed in. A portion of the soap dissolved in water is applied to the preparations by means of a camel's hair pencil. It constantly exhales the odor of arseniuretted hydrogen, and effectually destroys insects and their eggs.—DUMAS.

QUESTIONS.—What is said of the arsenic of the shops? What are the properties of arsenious acid? What are its salts termed? What is Fowler's solution? What is Scheele' green? What are the uses of arsenic? What is arsenic acid? What are its salts called? What are its uses? What is said of the other compounds of arsenic?

625. **Characteristics and Tests for Arsenic.**—The compounds of this metal are all highly poisonous, either when taken into the stomach, when applied to wounds, or when inhaled as vapor. The most effectual antidotes, in cases of ordinary poisoning by it, are, first, a powerful emetic, and then the free administration of the hydrated oxyd of iron suspended in water (§ 567); if this is not at hand, calcined magnesia may be used. In the absence of either of these substances, the white of eggs, milk, sugar, and soapsuds are beneficial, (this latter observation applying also to almost all other cases of poisoning). Prompt action is, however, necessary, as arsenic is almost always fatal when time is allowed for its absorption into the system in sufficient quantity.

The frequent employment of arsenic as an agent in poisoning, has induced chemists to study its nature and compounds so carefully, that its detection when present in the body, in the materials which have passed from the body, in food or in liquids, is a matter of certainty. Even though the quantity be too minute to be weighed, its existence in a substance may be absolutely demonstrated and made visible to the eye. Lapse of time can not wholly destroy this chemical evidence;—the body with which the arsenic has become incorporated may decay, but the poison remains unchanged, and may be recognized even after the lapse of years.*

626. An investigation for the detection of arsenic, in cases where a criminal prosecution involving reputation, and perhaps life, depends upon the issue, should be intrusted only to a professional chemist, but a description of the tests employed, and of the methods by which evidence can be accumulated, are matters of general interest.

An exceedingly delicate test known as "Marsh's test," depends upon the property which arsenic possesses of forming a gas with hydrogen, and depositing itself, in the metallic state, upon the surface of a cold plate held over the flame of the burning gas. The experiment is made by generating hydrogen in the usual manner from zinc, water, and sulphuric acid, in a glass flask, and allowing it to escape through a perforated cork and tube of glass drawn down to a fine point. (See Fig. 200.) The hydrogen evolved should first be tested by burning it against a porcelain plate to prove that it is free from arsenic, and then the suspected liquor is to be introduced into the apparatus. (For the purpose of experiment a few drops of a solution of arsenious acid in water, or hydrochloric acid, may be used). If arsenic is present, the flame of hydrogen, when brought in contact with the surface of a cool white

* In cases of arsenical poisoning, putrefaction of the body after death is retarded in a remarkable degree; and in many cases where the body has been disinterred several months after death, it has been found sufficiently preserved from decay to allow many of the principal internal organs to be distinguished. In one instance, in France, conviction of poisoning by arsenic was obtained on evidence procured by the celebrated chemist Orfilá, from the remains of a person who had been dead for a lengthy period of years.

QUESTIONS.—What is said of the poisonous effects of arsenic? What of its antidotes? What is said of its detection in the body, or in other substances? What is Marsh's test?

FIG. 200.

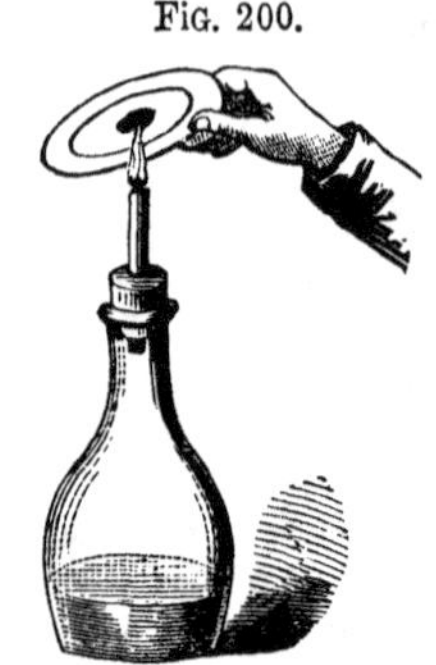

plate, or saucer, will deposit a smooth black or brown spot (*a little metallic mirror*). One other metal—antimony—will give the same reaction, but a drop of an aqueous solution of chloride of lime instantly dissolves the arsenic spot, but leaves the antimony unaltered. If the arseniuretted hydrogen gas be conducted through a glass tube heated at one point over a spirit-lamp, metallic arsenic will be deposited in the colder part of it, forming a beautiful incrustation.

Sulphuretted hydrogen precipitates arsenic from its solutions in the form of a sulphuret of arsenic, of a rich lemon color. This is a very accurate test, and so delicate that the yellow tint is apparent when only a ten-thousandth part of arsenious acid is present, and a precipitate when the proportion is as 1 part of arsenious acid to 80,000 of water.

Reduction of the metal from its oxyds or sulphurets is a test much relied on in judicial investigation. This may be effected by introducing a little arsenious acid, or the sulphuret obtained in the last experiment, mixed with finely-powdered charcoal and carbonate of soda, into a glass tube of the diameter of a common quill, care being taken not to soil the sides of the tube. The mixture is then gently heated by the flame of a spirit-lamp, when the metallic arsenic sublimes, and is condensed as a black, lustrous mirror, *c*, in the upper and cool part of the tube. (See Fig. 201.) A slip of bright metallic copper, placed in a hot solution of arsenious, or arsenic acid, acidulated by hydrochloric acid, is soon coated by a gray film of metallic arsenic. This is known as Reinsch's test, and is affirmed to show the presence of a 250,000th part of arsenic in solution. It is a test readily applicable even when the solution is contaminated by the presence of so much organic matter as to impair the accuracy of other reactions.

FIG. 201.

A dose of from 2 to 3 grains of arsenic is generally regarded as fatal, though larger doses are sometimes rejected from the stomach by vomiting. A dose of from 1-15th to 1-30th of a grain is said to warm and exhilarate the system, and increase its vigor, and the peasants of Hungary are reported to be in the habit of using it for this purpose.

QUESTIONS.—What other metal gives the same reaction? How may antimony be distinguished from arsenic in this instance? What is the test by sulphuretted hydrogen? What is the test by reduction? What is Reinsch's test? What amount of arsenic is fatal?

627. Arsenic and antimony are the only metals which are capable of combining with hydrogen. In this and several other respects, they comport themselves like metalloids, and by some chemical authorities arsenic is regarded as a non-metallic element.

CHAPTER XIV.

THE NOBLE METALS.

THE metals included in this class are nine in number, viz., Mercury, Silver, Gold, Platinum, Palladium, Rhodium, Ruthenium, Osmium, and Iridium.

The principal characteristic of these metals is their slight affinity for oxygen, by reason of which their oxyds are easily decomposed by the action of heat alone, the metal remaining in an uncombined state. The temperature required to effect this decomposition is less than red heat, with the single exception of the oxyd of osmium. Mercury and silver are generally found in nature as sulphides; the others usually occur native, and are often associated with each other.

SECTION I.

MERCURY (*Hydrargyrum, liquid silver*).

Equivalent, 100. *Symbol*, H. *Specific gravity*, 13·59.

628. **Natural History and Distribution.**—Mercury is sometimes found native, as fluid quicksilver, but most generally occurs as a sulphide, forming a brilliant red mineral termed *cinnabar*. Its most productive mines are those of Almaden in Spain, Idria in Austria, and New Almaden in Upper California. Considerable quantities are also obtained from localities in Mexico, Peru, China, and Japan. It is reduced from its ores by a process of distillation.

629. **Properties.**—Mercury is a brilliant, silver-white metal, possessing great density, and also the remarkable property of remaining fluid at common temperatures. It solidifies (freezes) at —39° F., in which state it is soft and malleable. When heated to 662° F. it boils, and yields an invisible vapor. The metal also, at all temperatures above 40° F., undergoes a slight spontaneous evaporation—a fact easily proved by the action exerted

---

QUESTIONS.—What are distinguishing characteristics of antimony and arsenic? What are the noble metals? What are their characteristics? Under what circumstances does mercury occur naturally? Where are its principal mines? What are its properties? At what temperature does it solidify? When boil? What is said of its volatility?

upon a sensitive daguerreotype plate suspended a few inches above a vessel containing mercury.

Mercury, when pure, is not tarnished by exposure to air and moisture at ordinary temperatures, but when heated to near its boiling point it slowly absorbs oxygen, and becomes converted into a crystalline, dark-red powder, the red oxyd of mercury. This oxyd, when subjected to a dull red heat, evolves oxygen, and is decomposed into its constituents. It was by means of this substance that Priestley first discovered the existence of oxygen, and Lavoisier determined the composition of atmospheric air.

630. The most ready solvent of mercury is nitric acid, which dissolves it with great rapidity. Hydrochloric acid has no action upon it, and the same is true also of dilute sulphuric acid.

631. When pure mercury is agitated with ether, or oil of turpentine, or rubbed with sulphur, sugar, chalk, lard, etc., it is reduced to so fine a state of division that it loses its metallic appearance entirely, and becomes thoroughly incorporated with the foreign body. In its ordinary state, mercury is inactive as a medicine, but in this state of mechanical division it is readily absorbed by the system, and becomes efficacious. The well-known *blue-pill* is mercury rubbed into a gummy compound, called "confection of roses;" and *mercurial ointment* is mercury incorporated with lard.

632. Mercury combines with oxygen in two proportions, forming a gray, or suboxyd, Hy_2O, and the protoxyd, or red oxyd, HyO. This last oxyd is a red powder, and was called by the old chemists *red precipitate.*

633. Mercury forms two compounds with chlorine, which correspond in constitution to the two oxyds, and are of great importance in medicine and the arts; they are the subchloride and the chloride.

634. **Subchloride of Mercury, Hy_2Cl,** is the well-known medicine, *calomel.* It may be obtained by precipitating a solution of sub-nitrate of mercury with common salt. When pure, it is a heavy, white, insoluble, and tasteless powder.

635. **Chloride of Mercury, $HyCl$,** is known in commerce under the name of *corrosive sublimate.* Its formation may be shown experimentally by heating a globule of mercury in a deflagrating spoon, and plunging it into a jar of chlorine; the metal takes fire and produces the chloride. Practically, it is prepared by subliming a mixture of common salt and sulphate of protoxyd of mercury.

Corrosive sublimate is a dense, white crystalline substance, soluble in 16 parts of cold, and 3 of boiling water—its solution possessing a disgusting and burning metallic taste. It is one of the most deadly poisons known in chemistry. With albumen it unites to form compounds which are nearly insoluble; hence substances which contain albumen, such as white of eggs, milk, etc., are the most effectual antidotes in cases of poisoning by it. Timber,

---

QUESTIONS.—What of its power to resist oxydation? What is its most ready solvent? What of its susceptibility to mechanical division? What is blue-pill? What mercurial ointment? What are its oxyds? What is said of its chlorides? What is calomel? What is corrosive sublimate? What are its properties? What are antidotes to it?

and animal and vegetable substances in general, are effectually protected against decay and the action of insects, by steeping in a solution of corrosive sublimate. This process is known in the arts as *kyanizing*, from its inventor, Mr. Kyan, who first applied it with great success for the protection of ship-timber against the effects of "*dry rot.*" The preservative action appears to be due to the circumstance that the corrosive sublimate unites with the organic substances to produce insoluble and poisonous compounds. A solution of corrosive sublimate in alcohol is much used as a preservative wash for plants in herbariums, and for other objects of natural history.

636. Oxyd of mercury forms several salts with nitric acid, the principal of which are the *subnitrate*, Hy_2O,NO_5, and the nitrate, HyO,NO^5. The last-named salt is used in the arts as a wash for rabbit and hare skins, as it imparts to these furs a property of felting which does not naturally belong to them.

637. **Sulphide of Mercury, HyS.**—This compound is the most abundant of the ores of mercury, and as a mineral product is termed *cinnabar;* but when prepared artificially, it constitutes the beautiful red pigment known as vermilion. Vermilion is prepared by subliming 1 part of flowers of sulphur with 6 of mercury. The product is a blackish-red crystalline mass, which by friction and pulverization assumes a magnificent scarlet color.

638. **Uses.**—Mercury is used extensively in the arts in the construction of philosophical instruments (barometers, thermometers, etc.), in the extraction of gold and silver from their ores, in gilding, and in medicine.

639. **Alloys of Mercury** with other metals are termed *amalgams.* An amalgam of 4 parts of tin to 1 of mercury constitutes the material employed for the silvering of mirrors. A strip of copper becomes amalgamated if rubbed with a solution containing mercury. If we make a stroke across a brass plate with a stick or brush dipped in a solution of mercury, and afterward bend the plate at this place, it will break as though it had been cut; the explanation of this is, that the mercury of the solution at once penetrates and combines with the brass, and renders it extremely brittle. Mercury, when brought in contact with bars of lead, tin, and zinc, readily permeates them by a species of capillary attraction; and by employing a bar of lead in the form of a syphon, we may gradually raise and draw off mercury from its containing vessel.

Tin, lead, silver, gold, and several other metals, are dissolved by mercury to a considerable extent, without much loss of fluidity. It has, on the contrary, but little attraction for iron, and on this account it is generally preserved in iron bottles.

The presence of mercury, when in solution, may be detected by placing a

---

QUESTIONS.—What is kyanizing? How does corrosive sublimate act as a preservative agent? What is said of the nitrates of mercury? What is vermilion? How is it prepared? What are the principal uses of mercury? What are alloys of mercury termed? What forms the lustrous coating of mirrors? How does mercury comport itself in contact with the other metals? How may the presence of mercury in solution be detected?

drop of the suspected liquid on a piece of polished gold, as a half-eagle, and touching the metal, through the liquid, with a scrap of zinc, or with the point of a penknife. The part touched will instantly appear white, owing to the deposition of mercury by voltaic action.

SECTION II.

SILVER.

Equivalent, 108. *Symbol*, Ag. (Argentum). *Specific gravity*, 10·5.

640. **Natural History and Distribution.** — Silver is frequently met with in the native state, but most generally it is found in combination with sulphur, mixed with sulphides of lead, antimony, copper, and iron. The mines of Mexico and Peru are the most productive sources of silver; but it occurs in quantities sufficient to pay for working, in Norway, Saxony, Spain, and the Hartz mountains.

641. **Amalgamation.**—Silver is obtained from ores free from lead, as those of South America and Mexico, by a process termed *Amalgamation*, which is founded upon the ready solubility of silver and other metals in metallic mercury. The ore is first crushed to a fine powder, mixed with common salt, and roasted at a low red-heat in a furnace. By this treatment the silver obtains chlorine from the salt, and is changed from a sulphide into a chloride. The resulting products of the furnace, consisting of chloride of silver, oxyds of copper, iron, and earthy matters, are then placed, with water and a portion of scrap iron, in barrels which revolve upon their axes, and the whole agitated together for some time, during which the iron reduces the chloride of silver to a state of metal, and forms chloride of iron; a certain portion of mercury is then added, and the agitation continued. The mercury dissolves out the silver, the copper, and the gold, if there be any, and combines with them to form an amalgam; which, by reason of its great weight and fluidity, is easily separated from the other materials by washing and subsidence. This amalgam is then pressed in woolen bags, to squeeze out the uncombined mercury, and the solid portion heated in a kind of retort, when the last trace of mercury volatilizes, and leaves the silver alloyed with copper or gold behind. In this state it is exported in ingots.*

* This process, as conducted in Mexico and South America by the rude miners, is exceedingly imperfect, and is attended with an enormous loss of quicksilver, by volatilization and the formation of calomel, Hy_2Cl; so much so, that it has been calculated that upwards of six million cwt. of mercury had been wasted in the American mines up to the close of the last century. It must be, therefore, apparent, that the great employment of mercury is in the mining of silver; and previous to the discovery, a few years since, of the rich cinnabar mines of California, the price of mercury (owing to a diminished supply from the mines in Spain and Austria) had risen so high, that many of the richest silver-mines of Mexico and Peru were of necessity abandoned.

QUESTIONS.—What is said of the natural condition of silver? Where are its principal mines? How is silver obtained from its ores by amalgamation?

642. **Liquation.**—Silver containing a large percentage of copper is separated from this metal by what is called the process of *Liquation:* this consists in melting the two metals with a large proportion of lead, and cooling the mixture suddenly in the form of cakes; these are then exposed, on an inclined hearth, to a temperature sufficient to melt the lead, but not the copper, when the former metal runs off, and carries all the silver with it, leaving the solid copper behind.

643. **Cupellation.**—Silver is parted from lead by a process termed *Cupellation.* It consists in exposing the mass, in the first instance, to a red-heat, upon the hearth of a shallow furnace, while a current of air is caused to play upon its surface; the lead rapidly oxydizes, and is converted into litharge, which, in turn, melts and runs off, leaving the metallic silver unoxydized, and in a nearly pure state (*refined silver*). The hearth upon which this operation is conducted, is called a *cupel*, and is formed by molding pulverized bone ashes into the shape of an oval, shallow basin. In order to render the silver thus obtained still purer (*fine silver*), it is again fused under the same circumstances in small cupels (Fig. 202); by which, the last remaining traces of lead, and all other metallic impurities, except gold, are converted into oxyds, and absorbed by the porous bone-ash.

FIG. 202.

644. **Properties.**—Silver has the clearest white color of all the metals. It is malleable and ductile in a high degree, and in hardness is intermediate between gold and copper. It melts at a bright red-heat, 1873° F., expanding forcibly at the moment of solidification; and is not oxydized by exposure, at any temperature, to either a dry or moist atmosphere. Pure silver, however, possesses the remarkable property of mechanically absorbing oxygen, when melted, to the extent of many times its volume. This oxygen is again disengaged at the moment of solidification, and gives rise to the peculiar arborescent appearance often noticed on the surface of masses of silver. Silver has a powerful affinity for sulphur; and when exposed to air containing very minute quantities of sulphurous acid, or sulphuretted hydrogen, it soon becomes superficially blackened or tarnished, from the formation of a thin film of sulphide upon its surface.*

The best solvent of silver is nitric acid, which acts upon the metal with great rapidity; if the silver contains any gold, it will be left undissolved as a dark powder. Solution of silver coin in nitric acid appears of a bluish-green color, from the copper it contains. Hydrochloric acid scarcely acts upon silver, and sulphuric acid only when hot.

* The air of large towns or cities, and the air of rooms in which mineral coal or coal-gas is burnt, always contains sufficient of the gaseous compounds of sulphur to gradually tarnish silver.

QUESTIONS.—How is silver obtained by amalgamation freed from copper? What is this process termed? How is silver freed from lead? What is a cupel? What are the properties of silver? What are the relations of fused silver and oxygen? What of silver and sulphur? What are the solvents of silver?

645. **Oxyds of Silver.**—Silver forms three oxyds—the suboxyd, Ag_2O; the protoxyd, AgO; and a peroxyd, AgO_2.

646. **Protoxyd of Silver** is the only oxyd which forms permanent salts, and may be procured by adding potash or soda to a solution of the nitrate, or any soluble salt of silver. It is a dark-brown or black powder, soluble in ammonia, and to a slight extent in pure water. Its solution in cyanide of potassium constitutes the silver solution used in electro-plating. Oxyd of silver is decomposed at a temperature below red heat, and to some extent also by the action of solar light.

647. **Nitrate of Silver, AgO,NO_5.**—This is the most important of the salts of silver, and may be obtained in the form of colorless, transparent, tabular crystals, by dissolving silver in nitric acid, and evaporating the solution to dryness. The crystals thus obtained are readily soluble in water, and when fused and cast into slender sticks, they constitute the *lunar caustic* of the surgeon.*

Nitrate of silver, when perfectly pure, undergoes no change by the action of light, but when exposed to light in contact with organic matter, it blackens rapidly. Stains thus produced by it can not be removed by washing with soap and water; hence nitrate of silver constitutes an essential ingredient in the composition of hair-dyes, and the indelible inks used for marking linen. Ivory, marble, and other bodies, may be stained a permanent black by soaking in a solution of this salt, and then exposing to the direct action of the sun's rays. The black coloring matter is by some supposed to be silver in a state of fine division, and by others to be a suboxyd of silver. It may be removed from the hands, or from linen, by the employment of a strong solution of iodide of potassium, or more easily by cyanide of potassium. Nitrate of silver is sometimes given as a medicine; if the administration of it is long continued, a portion of the silver in combination tends to find its way out of the system at the surface of the body; but becoming decomposed by the action of light before it reaches the outer surface of the skin, it imparts to all those portions of the body exposed to light a singular blue or lead-gray color. This color, from the circumstance that it is formed below the outer skin (or cuticle), is perfectly indelible.†

* The corrosive power of lunar caustic is not the result of any specific action of the nitrate of silver but of the nitric acid, which is liberated by the decomposition of the salt when in contact with organic matter.

† A most singular case of this discoloration was to be seen a few years since in the city of New York, in the person of an itinerant book-agent, who was familiarly called the "blue man." The color of this person, owing to an injudicious use of nitrate of silver as a remedy for epilepsy, was generally of a dark, dull blue, varying to brown with shades of green.

QUESTIONS.—What oxyds of silver are there? What is the principal oxyd? How is it prepared? What are its properties? How is nitrate of silver prepared? What is lunar caustic? What action has light upon this salt? Into what articles does it enter as an essential ingredient? How may nitrate of silver stains be removed? What sometimes happens when nitrate of silver is taken into the system?

When a stick of phosphorus is introduced into a solution of nitrate of silver, it soon becomes incrusted with arborescent crystals of the metal. The introduction of mercury into a solution of nitrate of silver also precipitates the metal in beautiful tree-like forms which are called *arbor Dianæ*. Metallic copper at once throws down metallic silver from solutions of the nitrate, and forms nitrate of copper.

648. **Chloride of Silver, AgCl.**—This compound appears as a white, curdy precipitate when hydrochloric acid, or the solution of any chloride (as common salt) is added to a solution of silver. Its formation, under these circumstances, constitutes a most delicate test for the presence of silver in solution, as the chloride of silver is so entirely insoluble in water, that a millionth part of it will occasion a cloudiness of the solution. It is, however, readily soluble in ammonia, and when exposed to the light, quickly assumes a violet color. Chloride of silver, kept in solution by the alkaline chlorides, probably exists in minute quantities in all sea-water. MM. Malagutti and Durocher, eminent French chemists, have estimated, on the basis of recent experiments, that each cubic mile of sea-water contains 10¾ lbs. of silver in the form of chloride.

649. **Uses.**—Pure silver, by reason of its softness, is not used to any extent in the arts; but for coin, plate, etc., it is always alloyed with a proportion of copper, which greatly increases its hardness without materially diminishing its whiteness, and thus renders it less liable to be worn by use. The amount of copper that may be alloyed with silver for the manufacture of coin is always regulated by government. In Great Britain, standard silver is composed of 11 parts of silver and 1 of copper; in the United States, all gold and silver coin consists of nine tenths pure metal and one tenth alloy. In England and France, the government also regulates the purity of silver used for the manufacture of plate; in the United States the manufacturer alloys his silver at discretion.

Silver is frequently employed to give a coating to less expensive metals.

Plating on copper is effected by laying a strip of silver upon a bar of copper, and uniting the two metals (without solder) by hammering and rolling at a temperature just below the fusing point of silver. The compound ingot is then rolled to the required degree of tenuity. *Silvering*, or covering the surface of brass or copper with a thin coating of silver, may be effected by first thoroughly cleaning the surface to be silvered by momentary immersion in nitric acid, and then rubbing, with a mixture of cream of tartar (100 parts), chloride of silver (10 parts), and corrosive sublimate (1 part); the surface is afterwards polished. It is in this way that thermometer scales are silvered. A peculiar *blanched* or "*dead*" appearance may be given to articles manufactured from an alloy of silver and copper, by boiling them in a solution of bi-

---

QUESTIONS.—What is said of chloride of silver? What is a test of the presence of silver in solution? Does silver exist in sea-water? In what state is silver used in the arts? What is standard silver in Great Britain and the United States? How is plating effected? How may articles be silvered? What is dead silver?

sulphate of potash; the acid of which dissolves out the copper from the surface, and leaves the particles of silver isolated.

650. **Silvering of Glass.**—Certain organic substances, such as oil of cassia, oil of cloves, or solution of grape-sugar, possess the property, when added to certain salts of silver in solution, of precipitating the silver in the state of bright, lustrous metal. This principle has been recently applied to the silvering of glass; and many articles of great beauty, such as mirrors, glass-globes, vases, door-knobs, etc., are now coated in this manner.*

SECTION III.

GOLD.

Equivalent, 98·7. *Symbol*, Au. (Aurum). *Specific gravity*, 19.2.

651. **Natural History and Distribution.**—Gold is always found native or in the metallic state; generally in the form of thin scales or grains, sometimes as large, nodular masses,† and sometimes in crystals; the last being always modifications of the cube, or octohedron. (See Figs. 203 and 204.)

FIG. 203.

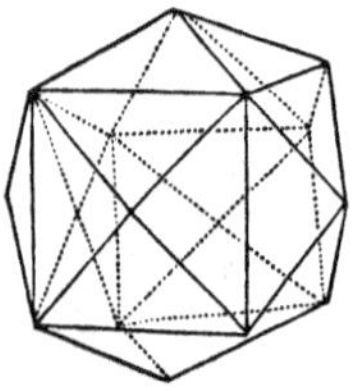

FIG. 204.

Native gold is always alloyed with silver, and is often associated with small quantities of osmium, iridium, copper, antimony, sulphuret of iron, and rarely with tellurium. No regular veins of gold are met with (what are called veins of gold being merely veins of quartz containing disseminated metallic particles). It commonly occurs in the most ancient rocks, or in the alluvial deposits of rivers.

As gold is found naturally in a metallic state, the operations for obtaining it are almost purely mechanical, as washing, etc. When the gold is very finely divided and mixed with earthy matters or other metals, it is separated by a process of amalgamation similar to that already described for obtaining silver. (See § 641.)

* A composition for silvering glass may be prepared as follows:—Mix 30 grains aqua ammonia, 60 nitrate of silver (crystals), 90 spirits of wine, and 90 of water. When the nitrate of silver is completely dissolved, filter the liquid and add 15 grains of grape sugar dissolved in a mixture of $1\frac{1}{4}$ ounces of water and $1\frac{1}{2}$ ounces spirits of wine. For silvering a glass, it is sufficient to leave this solution in contact with the glass for a space of two or three days; but by heating the mixture, the effect may be produced rapidly.

† A mass of gold once found in North Carolina weighed 28 pounds; a mass found in the Ural Mountains, and now in the Imperial Cabinet of St. Petersburgh, has a weight of nearly 80 pounds. Masses, however, of larger size, mingled with quartz, have been since found in both California and Australia.

QUESTIONS.—How may glass be silvered? What is said of the natural occurrence of gold? What metals usually occur associated with it? How is gold obtained from the earth?

652. **Properties.**—Gold possesses a characteristic yellow color and a high metallic luster. It is the most malleable of all the metals, and may be beaten into leaves which do not exceed 1-200,000th of an inch in thickness. It also possesses a high degree of tenacity. When pure, gold is nearly as soft as lead. It fuses at a temperature of 2016° F., and can not be advantageously employed for castings, as it shrinks greatly in solidifying. Gold does not combine directly with oxygen at any temperature; none of the simple acids, with the exception of the selenic, have any effect upon it; neither has sulphur or sulphuretted hydrogen. Chlorine and bromine attack it easily, and it is dissolved by any solution that liberates chlorine. The most usual solvent of gold is *aqua regia*. (See § 361.)

653. **Compounds of Gold.**—There are two oxyds of gold, a protoxyd, AuO, and a peroxyd, or auric acid, AuO_3. Both are unstable compounds, and are decomposed by the action of light. With chlorine, also, gold unites in two proportions to form a protochloride, AuCl, and a perchloride, $AuCl^3$. The last is the most important compound of gold, and is always produced when gold is dissolved in nitromuriatic acid.

By cautiously evaporating the solution of gold in aqua regia, the perchloride may be obtained in the form of yellow crystals, soluble in water, alcohol, and ether. If a solution of crystallized chloride of gold be applied to the skin, or any other organic substance, it imparts to it, on drying, a purple-colored stain. If a few drops be added to a dilute solution of protochloride of tin, a most beautiful purple precipitate is formed, which is known as the *purple of Cassius*. This compound of gold and tin is extensively used in enamel and porcelain painting, and also for imparting to glass a rich rose or purple color.

Polished steel dipped into an ethereal solution of perchloride of gold, decomposes it, and becomes covered with a coat of metallic gold: delicate cutting instruments are gilt in this way. Silk ribbons may be also gilt by moistening them with this solution, and exposing them to a current of hydrogen, or phosphuretted hydrogen gas.

Ammonia added to a solution of chloride of gold, produces a yellowish-brown precipitate, aurate of ammonia, or *fulminating gold;* this compound explodes at a moderate heat, or by friction.

654. **Industrial Uses of Gold.**—Gold intended for coin and most other purposes, is always alloyed with a certain proportion of silver or copper, in order to increase its hardness and durability. Gold for coinage is usually alloyed with copper to the amount of about 10 per cent.

The quantity of pure gold contained in a given mass is expressed by the word *carat*, used in reference to a certain standard number; which number in the United States is 24. Thus, perfectly pure gold is said to be 24 carats

---

QUESTIONS.—What are the characteristic properties of gold? What is said of its relations to oxygen? What of its solubility? What are the principal compounds of gold? How is perchloride of gold prepared? What are its properties? What is the "purple of Cassius?" How is steel gilded? What is fulminating gold? In what condition is gold used in the arts? How is the purity of gold expressed?

fine; when, on the other hand, gold is spoken of as 18 carats fine, it is understood that the mass consists of 18 parts (three fourths) gold, and 6 parts (one fourth) alloy.

The determination of the amount of pure gold or silver in a given mass of metal, is called *assaying;* and as the value of all the various gold and silver coins in the world is regulated by the operation, the various processes contained in this department of chemistry have been carried to a high degree of perfection.

655. **Preparation of Fine Gold.**—The process of obtaining fine gold, or of separating gold from its alloys of silver and copper, depends upon the solubility of silver and copper in nitric acid, and the perfect insolubility of gold in the same liquid. In order to effectually carry out the operation, it is necessary that the silver should amount to at least three times the weight of gold, or otherwise portions of silver will be mechanically protected from the action of the acid, and the separation be incomplete. If, therefore, the alloy be found to contain more than one fourth of its weight of gold, sufficient silver is added to reduce it to this proportion; and hence this method of parting the metals is known in assaying as *quartation.* The gold remaining undissolved in the acid is collected and melted into ingots, while the silver is separated from the copper in solution by precipitation with common salt as a chloride, and subsequently reduced by contact with metallic zinc. The separation of gold from its alloys may also be effected by boiling the gold in sulphuric acid, which dissolves the silver and the copper, but leaves the gold unchanged.

When a solution of protosulphate of iron is added to a solution of perchloride of gold, metallic gold is precipitated in the form of a fine brown powder, which, when diffused in water and viewed by transmitted light, appears green; the gold thus obtained is perfectly pure, and appears dark, by reason of its extreme subdivision. When rubbed and pressed, it regains its characteristic color.

656. **Gold Leaf** is manufactured by first forging the gold into plates, and rolling them into thin ribbons by means of steel rollers. The ribbon is then divided into small squares, which are placed between leaves or sheets of gold-beaters' skin (a thin membraneous substance obtained from the intestines of animals), and the whole beaten with a heavy hammer. As the gold expands, it is divided and subdivided until the required thinness of leaf is obtained.

The commercial value of pure silver is about $16 per pound; a dollar coin weighs an ounce troy. The value of fine gold is about fifteen times greater than that of silver, an ounce being worth from sixteen to eighteen dollars.

Bullion is the term applied to gold and silver reduced from the ore, but not yet manufactured; at the mint it means all gold and silver suitable for coinage.

QUESTIONS.—What is meant by gold 18 carats fine? What is assaying? How is gold parted from its alloys? What is understood by quartation? How may brown metallic gold be obtained? How is gold leaf manufactured? What is the comparative value of silver and gold? What is bullion?

SECTION IV.

PLATINUM, PALLADIUM, RHODIUM, RUTHENIUM, OSMIUM, IRIDIUM.

657. **Platinum.**—*Equivalent*, 98·7.; *Symbol*, Pt.; *Specific gravity*, 21·5. —Platinum (little silver) is not an abundant metal, and is always found native, usually in the form of small flattened grains, but sometimes in nodular masses of considerable size. It is very rarely met with imbedded in rock, but is always obtained from alluvial deposits (sand, etc.) by washing. The principal localities which furnish platinum are situated upon the western slope of the Ural mountains in Russia, in Brazil, and Borneo. It was first recognized as a distinct metal about the middle of the last century (1749).

658. **Properties.**—Platinum is a grayish-white metal, intermediate in hardness between copper and iron; it exceeds in tenacity all the metals except iron and copper, and is only inferior in ductility to gold and silver. It may also be beaten into thin laminæ like gold leaf, and at a white-heat may be welded like iron. The most valuable property, however, of platinum, is its infusibility, which is so great that it resists the most intense heat of a wind furnace, and only yields to the heat generated by the oxyhydrogen blow-pipe, or the voltaic battery. It alloys readily with lead, iron, and many other metals; and the compounds thus formed are much more fusible than pure platinum. Care, therefore, must be taken in using platinum crucibles, not to heat in them oxyds of fusible and easily-reduced metals, as lead, tin, bismuth, etc.; since, in the event of the reduction of the oxyd, the crucible would be destroyed by the formation of a fusible alloy.

Platinum does not oxydize in the air at any temperature, and none of the simple acids have an effect upon it. Aqua regia dissolves it, but less readily than gold; and it is also corroded by heating to redness in contact with the caustic alkalies, or with phosphoric acid in the presence of carbon.

The great infusibility of platinum, and its power of resisting chemical agents, give it a high value as a material for the construction of apparatus to be used in the manufacture of powerful acids, and in chemical analysis. It is also extensively employed by dentists for the setting of artificial teeth,* and to some extent for the bushing of the touch-holes of guns. An attempt was made in Russia some years since to employ platinum for coinage, but it was found to be inconvenient, and the experiment has now been abandoned. The value of crude platinum is about half that of gold; but in its manufactured state it sells for from $18 to $20 per ounce.

The process employed for working it depends upon its property of welding

* The value of the platinum annually required for this purpose at the present time in this country, is estimated at $150,000.

QUESTIONS.—How is platinum found in nature? What are its principal localities? When was it discovered? What are the general properties of platinum? What is said of its infusibility? What of its alloys? What of its solubility? What are its industrial uses? How is it manufactured?

at high temperatures. The crude grains are first purified by dissolving in aqua regia and precipitating as chloride of platinum, which is subsequently reduced to a metallic state by heat. It is then, in connection with scrap platinum, fused into little ingots by the oxyhydrogen blow-pipe, and these are subsequently welded and rolled into bars or sheets. The working of it was formerly confined wholly to France, but within a few years past it has been introduced somewhat extensively as a business in this country.

Platinum exists in two states of minute subdivision, viz., as *spongy platinum*, and *platinum black*. The properties and preparation of spongy platinum have been already described (§§ 48, 296). Platinum black is the metal in a state of such fine subdivision, that it has the appearance of soot. It is easily prepared by slowly heating to 212° F., with frequent agitation, a solution of chloride of platinum, to which an excess of carbonate of soda and a quantity of sugar have been added. The precipitated black powder is collected on a filter, washed and dried. Platinum black possesses the power, in a much higher degree than spongy platinum, of condensing gases, and oxydizing alcohol and ether (§ 469).

659. Platinum forms two oxyds, PtO and PtO_2, and two chlorides, $PtCl$ and $PtCl_2$. The last named compound, the bi-chloride of platinum, is the most important soluble salt of platinum, and is always formed when platinum is digested in aqua regia. Its crystals, obtained by evaporating its acid solution, form with water a rich orange-colored liquid, which is much valued in chemistry as the only reagent which precipitates potassa from its solutions.

660. **Palladium, Rhodium, Ruthenium, Osmium, and Iridium.**—These metals are found only in exceedingly small quantities, and usually occur associated with platinum, which metal they resemble generally in their properties.

Palladium is a white metal, more brilliant than platinum, very infusible, malleable, and ductile. Its hardness, whiteness, and inalterability would render it exceedingly valuable in the arts if it could be obtained in sufficient quantities. The Royal Geological Society of Great Britain award a medal of palladium for eminent discoveries in this department of science. *Iridium* is found alloyed with osmium, very often in California gold, forming the mineral *iridosmine*, which is the hardest of all known alloys. Iridium is a white, hard, brittle metal, more infusible than platinum, and one of the heaviest of the metals, being nearly 22 times heavier than an equal bulk of water. It has been used to some extent for forming the tips of gold pens.

QUESTIONS.—In what two states of subdivision does metallic platinum exist? Give the properties of spongy platinum. How is platinum black prepared? What compounds does platinum form? What is its most soluble salt? For what reaction is bi-chloride of platinum distinguished? What is said of the other metals included in the group of noble metals? What of palladium? What of iridium?

CHAPTER XV.

PHOTOGRAPHY.

661. **Photography** (*light-drawing*) is the art of drawing, or producing pictures, or copies of objects, by the action of light upon certain chemical preparations.

The whole art is based upon the circumstance, that the chemical element of the solar ray is capable of blackening or discoloring certain compound substances exposed to its influence, the principal of which are various salts of silver.* This fact has been long known and recognized, and as far back as 1802, Sir Humphrey Davy succeeded in obtaining images upon paper or white leather prepared with nitrate of silver, by exposure in a camera obscura;—the parts of the surface subjected to a strong light being blackened, while those in the shadow, which were unacted upon, remained white. It was found, however, impossible to arrest the action thus generated, and the image formed soon disappeared by a continuous darkening of the whole surface. The subject appears to have been next taken up by M. Niepcé, a French gentleman of Chalons, who ascertained, in 1823, that when a surface of a peculiar kind of bitumen, known as "Jew's pitch," was exposed in a camera, that the parts strongly acted upon by light became insoluble in oil of lavender, while those unacted upon, or influenced by weaker rays, retained their solubility in a greater or less degree, and could consequently be dissolved off,—thus forming an imperfect picture. This, and other interesting facts, Niepcé laid before the Royal Society of Great Britain in 1827, but they attracted little attention, and in 1829 he formed a partnership with a French artist by the name of Daguerre (who was engaged in experimenting on the same subject), for the future joint prosecution of their investigations. Niepcé died in 1833, but Daguerre continued his experiments, and in 1839 first exhibited, as the result of his labors, the pictures since called in his honor Daguerreotypes. His process was at first kept secret, but was soon purchased by the French Government and made known to the world—a pension of 6,000 francs being awarded to Daguerre, and one of 4,000 to the son of Niepcé. It is also a very singular fact, that substantially the same results made known by Daguerre, were also discovered at about the same time by Mr. Talbot, an

* The influence of light in producing the coloration of fruit may be very prettily illustrated by pasting letters cut in paper upon the surface of a ripening peach exposed to the sun. After the lapse of a few days the characters will be found, on removing the paper, to be distinctly marked in white, on a red, or yellow ground.

QUESTIONS.—What is photography? Upon what does the art depend? What were some of the earliest photographic experiments? What were Niepcé's experiments? Under what circumstances was the daguerreotype process discovered and made known?

English gentleman, who had been engaged in investigating the chemical relations of light for a number of years previous.

662. **Daguerreotype Process.**—The essential features of the daguerreotype process, as discovered by Daguerre and now practised, are as follows: a highly-polished tablet of silver (copper-plated) is selected as the basis of the picture, and exposed to the vapor of iodine. The iodine rapidly attacks the silver, and forms over its surface a thin yellow film of iodide of silver, which is so exceedingly sensitive to the action of light, that it is almost instantly decomposed by it.* The plate thus prepared, and carefully protected from the light, is then exposed to the image formed by the lens of a camera obscura. Relatively the quantity of the light-producing principle, and the quantity of the chemical principle reflected from any object are the same; therefore, as the light, and shadows of the luminous image vary, so will the power of producing change upon the plate vary, and the result will be the production of a picture which will be a faithful copy of nature, with reversed lights and shadows; the lights darkening the plate, while the shadows preserve it white, or unaltered. The time required for producing the impression may vary from 1 to 60 seconds, according to the brightness or clearness of the atmosphere, and the time of day.

If the picture thus formed were left without further care, it would soon fade away, and no trace of it would appear on the surface of the plate. In practice, the plate is not exposed to the influence of light sufficiently long to form upon its surface an image visible to the eye, but the picture is developed, or brought out and rendered permanent, by exposure to the vapor of mercury. This metal, in a state of very fine division, is condensed upon and adheres to those portions of the surface of the plate which have been affected by the light. Where the shadows are deep, there is scarcely a trace of mercury; but where the lights are strong, the metallic dust is deposited of considerable thickness. This deposition of mercury essentially completes and fixes the picture.

The reason why the vapor of mercury attaches itself only to those portions of the plate which have been affected by the chemical influence of light is not definitely known: in all probability, we have involved the action of several forces. It is not, however, necessary that a surface should be chemically prepared to exhibit these results. A polished plate of metal, a piece of marble, of glass, or even wood, when partially exposed to the action of light, will, when breathed upon, or presented to the action of mercurial vapor, show that a disturbance has been produced upon the portions which were illuminated; whereas no change can be detected upon the parts kept in the dark.

The next step of the process is to remove from the plate any iodide of silver which may remain unacted upon, and which would be liable to change

---

* Bromine forms a coating even more sensitive than iodine, and is now extensively used in its place.

QUESTIONS.—What is the first step of this process? What the second? Why does the vapor of mercury develop the picture? What is the concluding part of the process?

on exposing the plate to light. This is effected by dipping the plate into a solution of hyposulphite of soda, which dissolves off all the remaining sensitive coating. The plate is protected to some extent from mechanical injury, and a richer and warmer effect given to the picture, by covering it with a very delicate film of reduced gold. This is accomplished by dipping the plate into a solution of chloride of gold, and heating it over the flame of a spirit-lamp.

The surface of the plate is rendered uneven by the operation of light upon it, so that it admits of being copied by the process of electrotyping.

663. **Paper Photographs.**—The plan of obtaining permanent photographic images upon paper was originally devised by Mr. Talbot of England in 1839. The process first followed consisted in soaking ordinary writing-paper in a weak solution of common salt, and when dry washing it over on one side with a solution of nitrate of silver. This operation was performed by candle-light, and the paper dried by a fire. The sheet thus prepared, when laid under an engraving or leaf, and exposed to diffused daylight for a period of about half an hour, receives a fair impression, with the lights and shadows reversed. The picture thus formed is preserved from further change by immersing it in a solution of salt.

664. **Talbotype.**—In 1841, Mr. Talbot invented the process known as the Talbotype, or Calotype, which is essentially the plan at present followed in obtaining photographs on paper by the camera. The paper (smooth writing-paper) is first brushed over with a solution of nitrate of silver, and then immersed in a bath of iodide of potassium. In this way a surface of iodide of silver upon paper is prepared, which is not of itself sensitive to the action of light. These operations may be conducted in diffused daylight, and a stock of paper may be prepared at once and kept for use. In order to render the paper sensitive to the action of light, it is washed over with a mixture of nitrate of silver with gallic and acetic acids, and then exposed in the camera. Unless the light is very strong, the paper when withdrawn exhibits no image, or a mere outline, but the compound has undergone a very remarkable change; for if the blank sheet be washed over with the mixture of nitrate of silver with gallic and acetic acids, and then gently warmed, an image appears with wonderful distinctness and fidelity, the portions exposed to the strongest lights assuming the darkest tints. The development of the image in this process appears to be due to the reducing agency of the gallic acid, which acts more rapidly upon those portions of the surface which have been most freely exposed to the action of light. The dormant picture may be developed many hours, or even days after it has been produced, provided the paper be kept in the dark. It seems as though the light, without actually producing a decomposition of the particles of the silver salt upon which it falls, gives to them a peculiar condition of unstable equilibrium, which predisposes to decomposition when acted upon by a re-

QUESTIONS.—What was the original process for obtaining paper photographs? Describe the Talbotype.

ducing agent like gallic acid. The picture is preserved in this instance, as in most others, from future change, by dissolving off the exciting agents by solutions of the hyposulphites.—MILLER.

As silver tablets are expensive, and paper somewhat unreliable, glass coated with a sensitive substance has been extensively introduced as a material for receiving the photographic images. Glass is chiefly prepared for this purpose in two ways; by coating it with a thin film of albumen containing iodide of potassium (the albumen process); or by coating it with collodion, containing iodide of potassium (the collodion process).* The surfaces thus formed, when dried and washed with a compound of silver, are ready for exposure in the camera. The collodion film can be rendered so sensitive to light, that a perfect picture can be formed upon it by an exposure continuing for less than one second of time. In what are called *ambrotypes*, the picture is first formed upon a film of collodion and then varnished with a solution of balsam, which is thought to render the image more distinct.

Although the agents indicated are the ones chiefly employed in photography, recent researches have shown that nature abounds in materials susceptible of photographic action. Preparations of gold, platinum, mercury, iron, copper, tin, nickel, manganese, lead, potash, etc., have been found more or less sensitive, and capable of producing pictures of beauty and distinctive character. The juices of many plants and flowers have also been put into requisition, and papers impregnated with them have been made to receive delicate, though in most cases, fugitive images.† Attempts have also been made, with a considerable degree of success, to cause the light not only to draw, but also to engrave the image upon a prepared basis, in such a way that the surface may be used for printing.

665. **Photographs in Colors.**—All attempts to produce photographs in their natural colors have as yet been, on the whole, unsuccessful, although a partial success has, in some instances, been attained to. The circumstance that the rays by which photographic effects are produced are *dark* rays, entirely distinct from the rays constituting color, would appear, *a priori*, unfavorable to a successful result.

* Albumen is prepared for this purpose by beating up the white of eggs with iodide of potassium. Collodion mixture is formed by dissolving gun-cotton in ether, and adding iodide of potassium.

† The terms which have been applied to designate the effects resulting from the use of various materials are very numerous. Thus we have the *Chrysotype*, in which salts of iron and gold are used; *Cyanotype*, in which impressions are produced by salts of iron, in conjunction with those of cyanogen; *Anthotype*, in which juices of the poppy, rose, etc., are employed, and many others.

QUESTIONS.—What materials have been substituted as a basis for photographic action in place of silver and glass? What are the albumen and collodion processes? What is an ambrotype? Is photographic action restricted to a few substances? Illustrate this fact. What is said of photographs in colors?

ORGANIC CHEMISTRY.

ORGANIC Chemistry is that department of science which treats of the chemical nature and relations of those substances which are derived, either directly or indirectly, from organized beings,—animal or vegetable.

CHAPTER XVI.

NATURE OF ORGANIC BODIES.

666. **Composition of Organic Substances.**—The number of elements which enter into the composition of organic substances is extremely limited, the great bulk of all of them being made up of carbon, hydrogen, oxygen, and nitrogen, with which are generally associ ted extremely small quantities of sulphur, phosphorus, iron, and a few other elements. The infinite differences of appearance and properties which organic substances manifest, is due either to a variation in the number of the combining atoms of their constituent elements, or to a variation in the grouping or arrangement of the constituent atoms as respects each other.

Thus, for example, vinegar differs from alcohol only in containing a little more oxygen and a little less hydrogen, while the proportion of carbon is the same in both; the change of properties, which is occasioned by this slight change in compostion, is, however, exceedingly great; on the other hand, the most careful chemical analysis reveals no difference in the composition of woody-fiber, starch, and gum, each consisting of precisely the same elements united in the same proportions. The difference in properties in this case, is supposed to be due to a difference in the grouping of the atoms, somewhat as is represented in Figs. 205, 206, 207.

QUESTIONS.—What is organic chemistry? What is said of the composition of organic compounds? How are so many different organic compounds produced from so few elements? Illustrate this.

The number of such isomeric bodies in organic chemistry is very large, while their occurrence in inorganic chemistry is extremely rare.

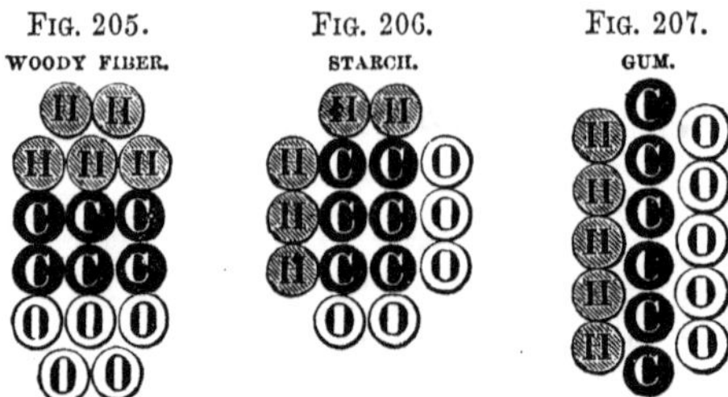

FIG. 205. WOODY FIBER. FIG. 206. STARCH. FIG. 207. GUM.

By far the largest proportion of the substances which make up the structure of plants are composed of but three elements—carbon, hydrogen, and oxygen. Animal substances, on the contrary, are generally characterized by the presence of nitrogen. Bodies which contain nitrogen are designated as *azotized* compounds; and those which are wanting in it, as *non-azotized* compounds.

667. The elements of organic bodies, in uniting with each other, are governed by the same laws of combination which regulate the composition of mineral or inorganic substances. The manner, however, in which the atoms of the constituent elements are associated in the one class of compounds is, in general, altogether different from what it is in the other—inorganic compounds being characterized, for the most part, by a great simplicity of composition, while those of organic origin are remarkable for their very great complexity. Thus water, HO, is composed of 1 atom or equivalent of hydrogen, and 1 of oxygen; Sulphuric acid, SO_3, of 1 of sulphur and 3 of oxygen; hydrochloric acid, HCl, of 1 of hydrogen and 1 of chlorine, etc. On the other hand, alcohol consists of 4 atoms, or equivalents, of carbon, 6 of hydrogen, and 2 of oxygen, its composition being represented by the formula $C_4H_6O_2$; and ordinary sugar, of 12 atoms of carbon, 11 of hydrogen, and 11 of oxygen, or $C_{12}H_{11}O_{11}$. The composition of stearic acid, the basis of stearine, is also represented by the formula $C_{68}H_{66}O_5$, and that of fibrine, the principal constituent of muscular fiber, by $C_{400}H_{310}N_{50}O_{120}PS$.

As a consequence of this complexity of composition, organic substances are, as a class, far more unstable and more liable to decomposition from slight causes than inorganic substances;—the power to resist the action of disturbing forces decreasing, as a general rule, as the number of combined atoms or equivalents increases. It is also a noticeable fact that all those organic

QUESTIONS.—What organic bodies, as a class, are generally wanting in nitrogen? What generally contain it? In what manner do the elements of compound bodies unite with each other? What are characteristics of the composition of organic and inorganic bodies? Illustrate this. What is the consequence of the complexity of the composition of organic bodies? What is a noticeable fact in relation to organic compounds of a high order?

bodies which discharge high organic functions, as the substance of the brain, the nerves, and the blood, have a most wonderfully complex constitution, and are susceptible of disorganization from the slightest causes.*

When organic substances are decomposed by the action of heat, light, electricity, chemical affinity, and even by mechanical action, they do not tend to divide into separate and isolated elements, but to form more simple compounds. Thus 1 (compound) atom of grape sugar, $C_{12}H_{14}O_{14}$, easily divides in 2 atoms of alcohol, $2(C_4H_6O_2)$, 4 of carbonic acid, and 2 of water. If an organic body be exposed to an intense degree of heat, with access of air, its constituents all unite with oxygen to form gaseous compounds, and it is completely consumed—generally after it has been converted into a black, carbonaceous mass. The property of blackening when a body is exposed to heat, which is due to the presence of carbon, is a sure characteristic of its organic derivation.

668. Origin of Organic Substances.—Organic substances have their origin entirely in plants.

The chemist, when he exerts his skill on materials of an organic origin, extracts a series of substances, each proceeding from the other, whose composition becomes more and more simple, until it reaches some species known to mineral chemistry. Thus, from sugar we may extract alcohol and carbonic acid, and from alcohol water and bi-carbureted hydrogen. In the vegetable organization, on the other hand, an operation exactly the reverse takes place. The living structure takes in air, water, and mineral elements, assimilates them, and in virtue of a certain peculiar force, builds them up and disposes them into groups of a certain stability—or into organic products.

* "There is a physical character which will *sometimes* enable us to give a good guess as to the simple or complex constitution of an organic substance—the faculty of crystallization. The power of assuming, on solidification, a distinct and often very characteristic geometrical form, appears to be possessed by all chemical compounds of a definite and constant composition, with the exception of a certain number, principally to be found in a class of organic substances of the most complicated and unstable nature. We know nothing, and apparently at present can know nothing, of the ultimate structure of any substance whatever; but it is not difficult to figure to one's self some idea of the gradual weakening of the molecular forces upon which crystallization depends, whatever the nature of those forces may be, by an increase in their number, and in the multiplicity of directions in which the forces themselves are exerted. It very often happens that in those cases where crystalline texture is altogether absent, we observe in its place an appearance of a very different kind;—we notice that the smallest particles of matter which can be traced by the microscope exhibit a rounded or globular figure instead of the straight lines and angles of the crystallizable compounds. These very frequently appear to aggregate together in strings, or rows, not altogether unlike some of the very lowest structures of the vegetable world, where a commencement of organization is, as it were, just visible. The substances forming the chief constituents of the animal body are in this condition." —*Actonian Prize Essay*, *Fownes*.

QUESTIONS.—What circumstances attend the decomposition of organic bodies? What property indicates the derivation of an organic substance? What is the primal origin of all organic substances? Illustrate this. Do animal structures create organic products?

The force by which this result is brought about is called the *vital* or *life* force; but we know nothing of its nature, and recognize it simply by its effects.

Organic substances thus originated pass into the systems of animals, which possess no power of *creating* or *forming* the materials which compose their structures, and can only *consume* and *transform* that which is supplied to them by plants.

Man has never yet been able to artificially make an organic body; by which assertion we mean to be understood, that he has never been able to take the single or dead elements, and cause them to unite at will so as to form compounds like those produced through the agency of animal or vegetable life. Chemists are, however, able to transform one organic body into another, or to unite materials derived from substances already organized into compounds possessing characters entirely different from those of their constituents. Thus, starch may be transformed into sugar, and sugar into the acid of ants (formic acid); some of the essential oils have also been produced artificially, and within the last few years (1855), Bertholet, an eminent French chemist, has succeeded in making alcohol from sulphuric acid, water, and bicarburetted hydrogen.*

669. **Compound Radicals.**—It has been already shown that cyanogen and ammonium, compound bodies, comport themselves in every respect like radicals, or elements. In organic chemistry many such compound radicals are recognized, some consisting of two elements, carbon and hydrogen, and some of three or four, carbon, hydrogen, oxygen, and nitrogen. Some, like cyanogen, correspond in properties to the metalloids; others, like ammonium, resemble the metals, and both by uniting with oxygen, chlorine, and acids, form oxyds, chlorides, and salts. Each, also, by the addition or grouping round it of other molecules, constitutes the root or basis of a whole class or series of compounds.

Thus, for example, carbon unites to hydrogen in the proportion of 4 atoms of the former to 5 of the latter, C_4H_5, to form a radical called *Ethyle.* Ethyle oxydated, forms oxyd of ethyle, or ether, C_4H_5+O; oxyd of ethyle plus an atom of water, forms hydrated oxyd of ethyle, C_4H_5,O,HO, or alcohol, the formula of which is generally written $C_4H_6O_2$; ethyle, plus an atom of chlorine, forms chloride of ethyle, C_4H_5,Cl, and if sulphur be substituted in the place of chlorine, we have sulphide of ethyle, C_4H_5,S; and in this way, by

* The muscles of animals and the fiber of wood consist of distinct chemical compounds, which the chemist has been able to isolate and study, but not to imitate. It is hoped, and expected by some, that the power will ultimately be attained to of artificially forming those products which, in the form of meat, cotton, flax, etc., are so essential to the welfare of man. The advocates of the possibility of such a result find some support for their views in the fact that two organic bodies, cyanogen and ammonia, are undoubtedly formed artificially in the workings of blast-furnaces, but in what manner it is impossible at present to explain.

QUESTIONS.—Can we artificially accomplish this? What power do we possess? What are compound radicals? What is the character of the radicals recognized in organic chemistry? Illustrate how classes of compounds are formed from such a basis?

the continued addition or subtraction of elements, a great variety of compound bodies may be formed, all referable to one central radical. Ethyle itself may be also obtained from its oxyd, as potassium is derivable from oxyd of potassium, or potassa, although by a different process.

The discovery and recognition of these compound radicals has greatly facilitated the progress of organic chemistry, and has rendered it possible to classify and arrange in groups a great number of bodies, which from their diverse properties would seem to have no connection with each other. Thus, the fats, the oils, the resins, the alcohols, the ethers, with many coloring, odoriferous, and medicinal substances, are now grouped and studied as derivatives from various central radicals, and not as independent principles. There are, however, many organic substances of great importance, the radicals of which have not yet been discovered.

CHAPTER XVII.

ESSENTIAL IMMEDIATE PRINCIPLES OF PLANTS.

670. By the essential immediate principles of plants, we understand those substances which the plant appears to form, through the agency of the vital force, directly from the inorganic elements obtained from without; or those principles which mainly constitute the structure, in a greater or less degree, of all plants, and are essential to their existence.

These substances are also often spoken of as the *proximate* principles of plants, and are conveniently divided into two classes, viz., those which contain nitrogen, as albumen, gluten, vegetable casein, etc., and those which are destitute of this element, as vegetable tissue (woody-fiber), starch, gum, sugar, etc. The separation of an organized substance into its proximate substances, or principles, is called its *proximate analysis;* and its separation into its final or simple elements, its *ultimate analysis.*

In the consideration of the two classes of the proximate principles of plants, it is most convenient to commence with those which do not contain nitrogen as a constituent element.

SECTION I.

VEGETABLE TISSUE, STARCH, GUM, SUGAR, ETC.

671. **Organic Structure.**—Since the discovery of the microscope, unwearied efforts have been made to ascertain the manner in which dead and inert inorganic elements unite to form organized and living structures.

QUESTIONS.—What do we understand by the essential immediate principles of plants? Into what two classes are the proximate principles of plants divided? What is understood by a proximate and an ultimate analysis?

The result of all inquiries have terminated in the establishment of a single fact, viz., that the lowest primary form of organization we can detect, whether of the individual (animal or vegetable) or of its parts, is a *cell*—a little globular or oval body, membranous or solid externally, fluid within. (See Figs. 208, 209, 210.) Beyond this we can not go, or say how it is that the elementary particles of matter are led to assume this form; but the appearance of cells always precedes the formation of circulating vessels, or any of the more complex forms of organic structure.

FIG. 208.

FIG. 209.

FIG. 210

Cells once formed, multiply in number by division (see Figs. 209, 210), and by the introduction of new matter from without, and thus it is that all growth, or increase in volume and weight, in all animals and vegetables, takes place; and an animal or plant is a structure "built up of individual cells, somewhat as a house is built of bricks." Fig. 211 represents a magnified view of the cellular tissue of a rootlet.

FIG. 211.

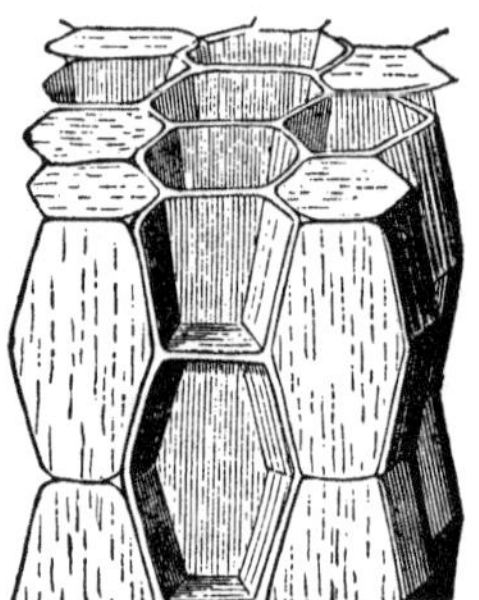

672. The natural figure of a cell is globular, but under varying circumstances it may assume a great variety of forms. The hairs on the surface of plants are cells drawn out into tubes, or are composed of continuous rows of cells. Cotton consists of simple long hairs on the coat of the seed; and each of these hairs is a single cell. Fig. 212 is a microscopic appearance of a section of the stalk of the *calla*, showing the arrangement of the cells, with passages between them.*

* The size of the common cells of plants varies from about the thirtieth to the thousandth of an inch in diameter. An ordinary size is from 1-300th to 1-500th of an inch in diameter; so that there may be generally from 27 to 125 millions of cells in the compass of a cubic inch. Now when it is remembered that many stems of plants shoot up at the rate of an inch or two a day, and sometimes of three or four inches, we may form some conception of the rapidity of their formation. When a portion of any young and tender vegetable tissue, such as an asparagus root, is boiled, the elementary cells separate, or may be readily separated by the aid of fine needles, and examined by the microscope.—GRAY.

QUESTIONS.—In what form does organization first manifest itself? What is a cell? How do plants and animals grow and increase? What is the natural figure of cells?

673. A living cell possesses a wonderful power of influencing chemical action; and what is called "*secretion*" in animals and plants is the result of the exercise of this function. By means of it, the cell first draws in or secretes inorganic matter, and then organizes it, or fits it into its own structure. Different cells manifest very different powers; for example, one kind of cell will decompose carbonic acid, reject the oxygen, and preserve the carbon within its walls or tissues; another will produce out of the inorganic constituents of the air the odoriferous principle of the rose; a third will convert a portion of blood into milk; and yet to the eye they are all alike, "a collection of little wet bladders."

FIG. 212.

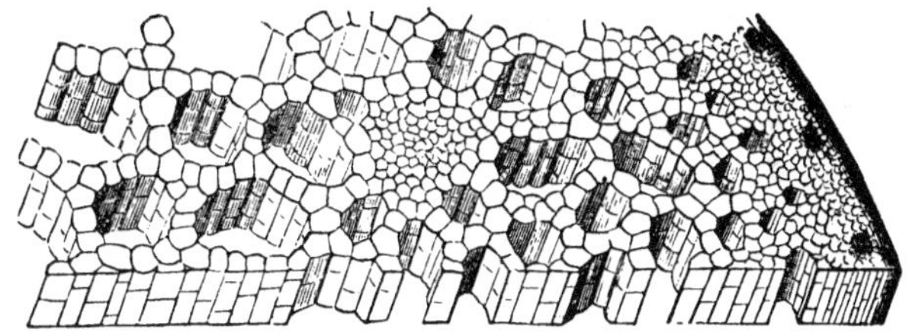

674. **Cellulose, or Cellular Tissue,** $C_{12}H_{10}O_{10}$.—The materials of which the walls of the cells of plants is composed is termed in chemistry *cellulose*, or *cellular tissue*. It consists of three elements, carbon, hydrogen, and oxygen, and has the same composition, when pure, in all plants. It is distinguished among all the substances which enter into the composition of plants by its great resistance to chemical agents—a resistance which allows its separation in a state of purity.

Cellulose is nearly pure in cotton, and in the fibers of the flax and hemp-plants, also in paper and old linen and cotton cloth. The difference between cotton and flax is due simply to a difference in the mechanical construction of their fibers; the fiber of cotton being a flattened tube or hollow ribbon without joints, and with pointed or rounded ends; while the fibers of flax and hemp consist of rounded tubes (cells) bundled or jointed together in parallel directions, and easily separable into shorter and more minute filaments. Cotton fibers have what is called a staple; that is, they are all of the same length, and are, therefore, easily spun by machinery; flax and hemp fibers are, on the contrary, irregular in length, and are more rigid than cotton, and can not be so easily twisted into fine, regular threads. Fig. 213 represents the microscopic appearance of cotton, and Fig. 214 that of flax.

Cellulose is insoluble in water, alcohol, ether, and dilute acids. By treat-

---

QUESTIONS.—What property do cells possess? Illustrate this. What is cellulose? By what other name is it known? In what substances is it nearly pure? What constitutes the difference between flax and cotton? What is meant by the staple of cotton? What are the properties of cellulose?

ing sawdust successively with warm water, alcohol, ether, alkalies and acids, we may remove from the wood all its soluble constituents, and obtain cellulose in a pure condition. By continued contact with chlorine, acids, and alkalies, cellulose is, however, gradually decomposed and destroyed.

FIG. 213.

FIG. 214.

675. **Gun-cotton,** *Pyroxyline.* — When cellulose is subjected to the action of nitric acid, or to a mixture of nitric and sulphuric acids, it gives up a portion of its hydrogen and oxygen (as water), and receives nitric acid in place—becoming transformed thereby, without change of physical appearance, into an explosive substance which is known as *gun-cotton*, or *pyroxyline.*

The process by which gun-cotton is formed is essentially as follows: perfectly clean cotton is soaked for about five minutes in a mixture composed of 1 part concentrated nitric acid, with 2 parts concentrated sulphuric acid; it is then removed, carefully washed with water from every trace of acid, and dried by exposure to the air. As thus prepared, it retains the appearance of cotton, but inflames instantaneously when touched with a hot wire or lighted match, and when struck with a hammer upon an anvil, explodes with great violence. When used in fire-arms, it acts like gunpowder, but its explosive force is at least four times greater than that of powder, and it does not, moreover, foul the gun to the same extent as the latter substance. Its liability to burst the gun and to accidental explosions has, however, caused its rejection for most practical purposes, and in experimenting with it too great caution can not be exercised. By subjecting starch and sugar to treatment with nitric acid, other explosive substances analogous to gun-cotton may be formed.

676. **Collodion.**—Gun-cotton is insoluble in both water and alcohol; it dissolves sparingly in pure ether, but readily in ether containing a small percentage of alcohol. Its ethereal solution constitutes a syrupy liquid which yields by evaporation a thin, transparent, powerfully adhesive substance, insoluble in water. This product, which has received the name of *collodion*, is advantageously used as a substitute for *court-plaster* for the covering of wounds, and also as a sensitive basis for the reception of photographic pictures.

677. **Parchment Paper.**—When paper is exposed to a mixture of

---

QUESTIONS.—What is the action of nitric acid upon cellulose? What is the process of making gun-cotton? What are its properties? What is collodion? What action does sulphuric acid have upon paper?

2 parts concentrated sulphuric acid (*s.g.*, 1·854, or thereabouts) with 1 of water, for no longer time than is taken in drawing it through the acid, it is immediately converted into a strong, tough, skin-like material, to which the name "*parchment paper*" has been applied. All traces of the sulphuric acid must instantly be removed by washing. If the strength of the acid much exceeds or falls short of the limits named, the paper is either charred or transformed into other compounds. By this treatment, in a little more than a second of time a piece of feeble, porous, unsized paper is rendered so strong, that a ring seven eighths of an inch in width is said to be capable of sustaining a weight of 90 lbs. The nature of the change thus effected is not understood, the chemical composition and weight of the paper remaining unaltered. It is, however, somewhat contracted in dimensions, is not affected by water like common paper, and is not decomposed by heat and moisture like common parchment.

678. **Lignine.**—As the growth of the plant continues, the walls of the cells constituting the cellular tissue generally become incrusted on their interior surfaces with a substance formed from the organic matters dissolved in the sap. This substance constitutes the principal part of the weight of wood (*lignum*), and is chemically known as *lignine.* It grows thicker with the age of the plant, and finally fills up the cells, leaving, however, minute pores or conduits for the circulation of the sap. Fig. 215 represents a microscopic section of wood-cells of the birch, nearly filled up by regular depositions of lignine. The difference between the heart-wood and sap-wood, or external wood, of a tree, is due simply to the fact, that the cells of the center are the oldest, and consequently are more densely and compactly filled with ligneous matter than those which have been formed later, and constitute the exterior of the tree. It is by this thickening of the cells that the skins of fruits and the shells of nuts acquire their hardness, and it is simply through variations in the continuance of this process, and in the nature of the materials deposited, that all the different appearances of wood originate; the coloring and resinous matters of wood being deposited in connection with the lignine.

FIG. 215.

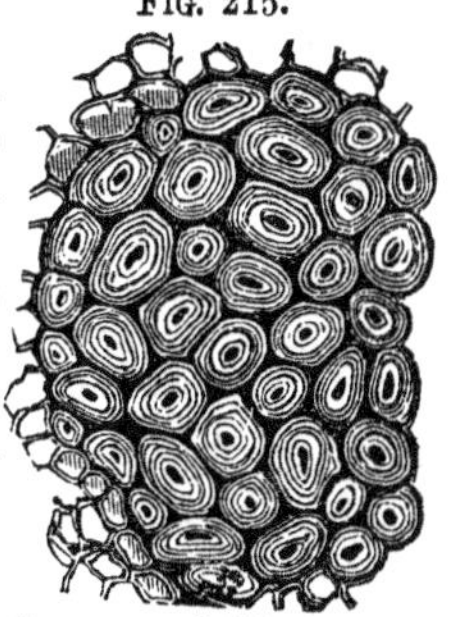

Lignine can not be isolated in a state of purity; it is supposed to differ somewhat from cellulose, or the original cell membrane, in containing a little more hydrogen and carbon; it is, therefore, richer in combustible matter.

679. **Destructive Distillation of Wood.**—When wood is subjected to heat in close vessels (distillation), or with a partial access of air, a great

QUESTIONS.—Describe the process for preparing parchment paper. What is lignine? What constitutes the difference between the heart-wood and sap-wood of a tree? What are other illustrations of the formation of lignine? What is said of the chemical composition of lignine?

variety of products are obtained, which are characterized by singularly different properties. The principal of these are charcoal, which is not volatile, and remains behind, illuminating gas (carburets of hydrogen), carbonic acid, water, pyroligneous acid, and a resinous substance known as "wood-tar." Of these several substances, the two last mentioned only remain unconsidered; they are extensively used in the arts, and are obtained upon a large scale by distilling wood in iron cylinders.

680. **Pyroligneous Acid**, sometimes called *wood vinegar*, is a brown acid liquid, having a strong smoky taste and flavor. It is obtained most abundantly by the distillation of dry beach-wood—a pound of wood yielding nearly one half pound of acid. Its uses and chemical composition will be hereafter noticed.

681. **Creosote** is a colorless, oily fluid, obtained from pyroligneous acid and wood-tar. It possesses a peculiar penetrating odor of smoke, and when applied to the skin of the mouth or tongue, acts as a cautery. Creosote is one of the most powerful antiseptic agents known in chemistry. Hence the etymology of its name, from the Greek κρεως, *flesh*, and σωζω, *I preserve*. Meat steeped for about 24 hours in a solution of 1 part of creosote to 100 of water, is rendered incapable of putrefaction, and acquires a delicate flavor of smoke. It is indeed the presence of this principle in wood-smoke which gives to the latter its characteristic smell, its property of causing lachrymation, and its power of curing meats and fish. Creosote diluted with alcohol is often employed for relieving toothache arising from putrefactive decay in the substance of the tooth, and as a styptic for checking hemorrhage. When taken internally in any quantity it is a corrosive poison, but a very dilute solution is sometimes given in medicine. It is also extensively employed by liquor manufacturers for imparting the peculiar smoky flavor to what is called "Irish whiskey."

682. **Tar.**—There are several varieties of tar. The kind so largely employed in the arts, as in ship-building, is obtained by subjecting to a rude process of distillation the roots and wood of the resinous pine; another variety of tar is obtained from the destructive distillation of hard wood; and a third from the destructive distillation of coal (*coal-tar*).

Wood-tar is insoluble in water, but soluble in alcohol, and is extremely rich in carbon, which gives it in part its black color. When applied to wood it exerts a preservative action by reason of the creosote it contains, and also by preventing the penetration of moisture. On distillation, it separates into a volatile oil (oil of tar) and a non-volatile substance, *pitch*.

From oil of tar a great number of products may be extracted, all of which are compounds of carbon and hydrogen. One of these, called *eupion*, an

---

QUESTIONS.—What are the products which result from the distillation of wood? What is said of pyroligneous acid? What of creosote? To what are the peculiar properties of smoke due? What is said of the antiseptic influence of creosote? What are the uses of creosote? What is common tar the product of? What are the three varieties of tar? What are the properties of wood-tar? What are its products of distillation? What is said of the products of the distillation of oil of tar?

oily, fragrant substance, is the lightest of all known liquids. Another, *parafine*, is a white, crystallizable substance, closely resembling spermaceti in appearance.

Coal-tar is a mixture of solid and liquid hydrocarbons, and is formed abundantly in the production of illuminating gas from coal, which is a vegetable substance. It was formerly regarded as a useless product, but within the past few years it has been rendered valuable by the discovery of economical methods of separating it into its constituents. This is principally effected by distilling at different and carefully regulated temperatures, and condensing the distillates in the order of their volatility.

The first product of distillation is a limpid, oily liquid, called *benzole*. It closely resembles oil of turpentine in appearance and odor, and is highly volatile and inflammable. A current of moist air passed through benzole becomes so thoroughly and permanently impregnated with its vapor, that it may be conveyed away in pipes and burned as an illuminating gas. The application of this property of benzole constitutes the essential feature of the so-called "portable gas generators." Benzole is also used to a considerable extent as a most ready and cheap solvent for various resins, camphor, the essential oils, grease, wax, India rubber, and gutta-percha.

The second important product of the distillation of coal-tar is a heavy oil, not readily volatile at ordinary temperatures. It is known as "*coup*" oil, or heavy oil of coal-tar, and is extensively used for the lucubration of machinery, and for illuminatng purposes.

Similar oils may also be obtained in much larger quantity and more cheaply, by directly distilling the richer varieties of bituminous coal: the products known as "Breckenridge coal oils" being produced in this manner. In addition to these oils, both coal and coal-tar also furnish by distillation a great variety of other products; among which are a white volatile solid called *napthaline*, somewhat resembling camphor in appearance, and exhaling a faint, but agreeable odor, and several less volatile wax-like substances, which have been employed to some extent for the manufacture of candles.

Coal-tar, mixed with gypsum, gum shellac, and other substances, forms a water-proof and durable material for the covering of roofs. By subjecting the products of coal and coal-tar to the action of chlorine and the acids, an almost endless variety of curious compounds may be generated, some of which have important industrial applications. Benzole distilled with nitric acid, yields a highly fragrant substance (*nitro-benzole*), so closely resembling the oil of bitter almonds, that it has almost entirely superseded the latter in the preparation of perfumery and the scenting of soaps. The heavy oil may also be converted by treatment with the same acid into a beautiful lemon-yellow,

---

QUESTIONS.—What is coal-tar? By what process are the constituents of coal-tar separated? What is the first product of its distillation? What are the properties of benzole? What are its uses? What is coup oil? What are some of the other distillates of coal? From what other source beside coal-tar may these products be obtained? What is said of the compounds artificially formed from the distillates of coal?

crystalline substance (*carbazotic acid*), which is capable of imparting to silk and wool a brilliant yellow color.

683. **Mineral Oils**, *Petroleum, Naphtha, etc.*—Oils similar in composition and properties to those obtained from the distillation of coal, are observed to issue from the earth in many localities, and often in considerable abundance. They are supposed to be generated by the action of internal heat upon beds of coal, or upon rocks rich in bituminous matter. The nature of these oils differs greatly—the thinner and purer varieties being generally called *naphtha*, and the more viscid liquids *petroleum.* The most abundant localities of these substances are in Persia, in the vicinity of the Caspian sea, in Italy, and in Birmah. They are also found in many places in the United States—the well-known "*Seneca oil,*" found in the vicinity of Seneca Lake, N. Y., being a product of this character. At Baku, in Persia, extensive beds of marl exist which are saturated with naphtha, and in some parts of this district so much combustible gas or vapor issues from the ground, that it is used by the inhabitants for cooking, and by certain religious sects for the maintenance of a perpetual fire. Naphtha is somewhat used in the arts in the preparation of varnish, as a solvent for certain resins and India-rubber, and by the chemist as a means of preserving the metallic bases of the alkalies—potassium and sodium—from oxydation.

684. **Asphaltum**—*Mineral Pitch, Bitumen*—is another natural product undoubtedly derived from the decomposition of organic matter. It is a black solid, closely resembling petroleum, and melts at about 212° F. It is found abundantly in many localities, especially in the vicinity of the Dead Sea and in the island of Trinidad, W. I., in which latter place it constitutes a lake three miles in circumference and of an unknown depth;—the pitch lake of Trinidad.

685. **Contents of the Cells of Plants**.—The contents of the cells of plants comprise all the immediate products which plants produce.

Growing and vitally active cells are filled with liquid; completed cells may still be filled with liquid or with air, or with solid matter only. The liquid contents of the vegetable tissues, are generally spoken of as *sap;* but this term does not specially refer to any particular substance. Sap, in the first instance, is water impregnated with certain gaseous matters (carbonic acid, ammonia, etc.) and a minute quantity of mineral salts, which are imbibed by the roots of the plant from the soil, and carried upward through the stem. It is, therefore, in the first instance, inorganic in its nature. As, however, it traverses the cells of the plant, it mingles with the soluble assimilated matters which these contain, and becomes changed in character, so that unmixed, crude sap, is never met with in the plant. On reaching the leaves, it becomes further transformed, under the influence of light, into organizable matter, or into matter capable of being assimilated by the cells and converted

Questions.—What are mineral oils? What names are generally applied to these products? What is said of their natural occurrence? What are the uses of naptha? What is asphaltum? Where is it found? What are the contents of the cells of plants? What is sap? Describe the successive transformations of sap?

into organic products. From the sap thus elaborated, the cell manufactures, or secretes, all the constituents of the plant.

Of the non-azotized substances secreted by the plant cells, the most abundant and widely distributed is lignine, which has been already described. In addition to this, three other substances, starch, gum, and sugar, closely allied to lignine in composition, are secreted in greater or less abundance by almost all plants.

686. **Starch**, $C_{12}H_{10}O_{10}$.—This substance presents to the naked eye the appearance of a white powder, but when viewed under a microscope is seen to consist of transparent oval or rounded grains, each of which has a dark spot at one extremity, with fine concentric rings drawn round it. These characteristic appearances are best seen in starch from the potato, with a magnifying power of from 250 to 500 diameters. (See Fig. 216.) The mag-

FIG. 216.

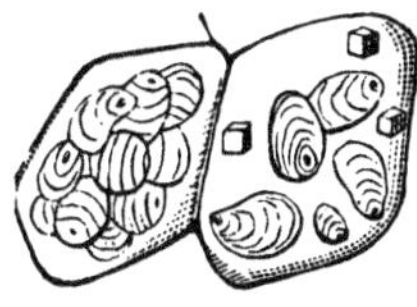

FIG. 217.

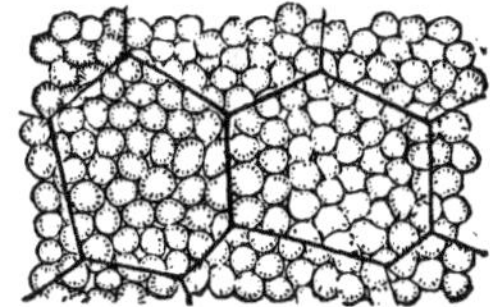

nitude of the starch grains varies extremely in the different plants, and even in the same cell. Thus in the potato the largest grains measure from 1-300th to 1-500th of an inch in their larger diameter, but in the smallest only 1-4400th of an inch. In wheat flour the larger grains are from 1-800th to 1-900th of an inch in diameter. The shape of the grains in the same plant or organ is very nearly uniform, but differs greatly in different plants; so much so, that mixtures of various starches may be easily detected by the microscope.

Starch, while an almost universal product of all vegetable cells, is accumulated more abundantly in some species of plants than in others. In the common potato, each individual cell is so completely filled and distended with an accumulation of starch mingled with water, that the whole root has an appearance of deformity. Fig. 217 represents the manner also in which the starch grains fill up the cells of the maize (*i. e.*, in Indian meal). Starch is particularly abundant in all cereal grains, in all seeds, in the pith and bark of many trees, and in many roots and tubers (as the potato, turnip, carrot, etc.).*

* Wheat flour contains from 39 to 77 per cent. of starch; rice flour about 86 per cent.; Indian meal from 70 to 80 per cent.; rye flour, 50 to 60; buckwheat, 50; pea and bean meal, 42; and potatoes from 13 to 15 per cent of starch, mingled with about 70 parts of water.

QUESTIONS.—Beside lignine, what other non-azotized substances are generally contained in the cells of plants? What is the appearance of starch? What is said of the size and appearance of its granules? In what vegetable substances is starch most abundant?

The starch of commerce is usually obtained from potatoes or wheat. The essential features of its process of manufacture consist in bruising or grinding the vegetable structure to a pulp, and then washing the mass with cold water upon a sieve. In this operation the torn cellular tissue and some other constituents are retained upon the sieve, while the starch granules pass through its interstices with the water. From this liquid the starch separates, on standing, as a fine white powder.

Starch is insoluble in water, as its mode of preparation necessarily implies. When, however, a mixture of starch and water is heated to near its boiling point, the granules swell, burst, and allow their contents to become mingled with the water, producing thereby a nearly transparent, glutinous mass, in which the minute shreds of membraneous matter, comprising the cell-walls, float. The rounded and swollen appearance which potatoes, peas, rice, and most other vegetables assume when boiled, is due to a distension of their starch granules through an absorption of water at the boiling temperature.

The chemical test of starch is iodine, which forms with it a beautiful blue compound, insoluble in water. This reaction may be strikingly illustrated by adding to a tumbler of pure water a single drop of gelatinous starch, and then stirring the mixture with a glass rod moistened with an alcoholic solution of iodine.

The chemical composition of starch is exactly the same as that of cellulose, and it appears to be especially the ready prepared material of vegetable fabric, which the plant accumulates in cells as a provision for future growth.

The substances known as *sago*, *tapioca*, and *arrow-root*, are only varieties of starch; the former being obtained from the pith of various species of the palm, and the two latter from the roots of certain tropical plants.

687. **Dextrine.**—When thick gelatinous starch is boiled for a few minutes with a small quantity of dilute acid, as sulphuric acid, for example, it speedily loses its viscidity, and becomes changed into a fluid as thin and limpid as water. If the acid be now withdrawn, by saturation with chalk (which combines with it to form insoluble sulphate of lime), and the liquid be gently evaporated to dryness, it furnishes a substance resembling gum, which is termed dextrine.* This new body is freely soluble in cold water, and has exactly the same composition as gelatinous starch, but is not colored by iodine.

If, instead of interrupting the action of the acid upon the starch as soon as the mixture has become clear and thin, we continue the ebullition for several hours, adding from time to time small quantities of water to supply the

* Dextrine, so called from the circumstance that when a beam of polarized light is passed through its solution, it causes the plane of polarization to turn to the right.

QUESTIONS.—From what sources is it generally obtained? How is starch manufactured? When starch is boiled in water, what takes place? What is the chemical test of starch? What is its chemical composition? What is sago, tapioca, and arrow-root? What is dextrine? How is starch converted into dextrine? How is dextrine converted into sugar?

place of that lost by evaporation, and finally, neutralizing the acid by chalk, filter and boil down the clear solution to a small bulk, we obtain a syrupy liquid, which on standing for a few days, entirely solidifies to a mass of grape-sugar, exceeding in weight the starch from which it was produced.

How this transformation of starch into dextrine, and dextrine into sugar, is effected is not fully understood. The acid employed undergoes neither change nor diminution, and if not volatile may be recovered, without loss, after the conclusion of the experiment; nothing, moreover, is withdrawn from the air, and no other substances but dextrine and grape-sugar are generated. Chemists, therefore, have very generally adopted the conclusion that the acid occasions the transformation and change in question by its mere presence, and the phenomenon is cited as an example of catalysis. (§ 255, p. 161.) Spongy platinum, as has been already pointed out (§ 296), apparently acts in a similar manner, and is capable of exciting chemical activity in contiguous substances without being itself affected. In the case of the dextrine, as its chemical composition is precisely the same as that of starch, the difference in properties is referred to a change in the arrangement of its constituent atoms of carbon, hydrogen, and oxygen. In the conversion of the dextrine into sugar, the change seems to be effected by a fixation or incorporation into the former substance of an additional quantity of the elements of water, hydrogen and oxygen—as the sugar thus produced sensibly exceeds in weight the starch employed.

688. **Diastase.**—There are, however, several other methods by which these same changes in starch may be effected, in addition to the one noticed. Thus, all seeds in the act of germinating, and all buds in developing, produce from the nitrogen compounds which they contain a very peculiar substance called *diastase.* This body, which the chemist has never yet been able to fully isolate, possesses the same power as the dilute acid of converting a large quantity of starch, first into dextrine and then into sugar. Its action, however, takes place at a much lower temperature than that of ebullition.

This fact may be experimentally shown by mixing a little infusion of malt (germinated barley) with a considerable quantity of thick starch paste, and subjecting the whole to a gentle heat not exceeding 160° F. In a few minutes the mixture, from the production of dextrine, becomes thin like water, and if the temperature be kept up during three or four hours, the liquid will be found to have acquired a sweet taste, and to be rich in sugar. The quantity of dextrine necessary to effect this change is very small—one part in two thousand parts of starch being sufficient to entirely convert the latter into sugar. A boiling heat coagulates the diastase, and by rendering it insoluble destroys its power.

The well-known sweet taste which fruits and vegetables acquire by freez-

QUESTIONS.—What is said of these transformations? How do germinative seeds and buds act upon starch? What is disastase? How may its properties be illustrated? Why are frozen thawed fruits and vegetables sweet?

ing and thawing, is due to the fact that the starch which they contain is converted, in part, by the action of the frost, into sugar.

689. Dextrine is used extensively in the arts as a substitute for gum; *i. e.*, for the stiffening and glazing of muslins, in calico-printing, and in the printing of wall-papers. It is manufactured for industrial purposes by simply roasting dry potato-starch, or subjecting it to a heat of about 400° F. By this treatment the starch acquires a yellowish tint and is rendered soluble in water. Dextrine occurs in commerce under the name of "*British gum.*"

690. **Gum.**—In addition to dextrine, which is found in greater or less quantity in the juices of every plant, the term gum is generally applied to designate certain vegetable substances which possess the same elementary composition as starch, and which are soluble in water, but not in alcohol. In some plants they exist so abundantly, that they exude from the bark as viscid liquids, which subsequently harden into transparent, globular masses. Familiar illustrations of this may be noticed on peach and cherry trees. The term *resin* is rightly applied to those hardened vegetable juices only which do not soften or dissolve in water, but are soluble in alcohol.

The most important gums of commerce are gum arabic, gum senegal, and gum tragacanth.

Gum arabic is the product of a species of *acacia* which grows abundantly in Africa and Arabia; gum senegal, the product of a similar tree, derives its name from Senegal, in Africa, the district from which it was originally exported. Both of these gums are freely soluble in water, and form with it a mucilage much used for paste; the mucilage yielded by gum senegal being somewhat thicker than that formed by gum arabic. The pure gummy substance contained in them may be precipitated from its solution in water by alcohol, and is termed *arabine.*

Gum tragacanth is the product of a shrub found extensively in Asia Minor and Persia, and is composed mainly of a substance termed *basorine.* It swells very much in water, and forms a thick adhesive paste, but can hardly be said to dissolve in it. It is, however, soluble in caustic alkalies.

Gum is an essential constituent of the cereals, and of most seeds, and is abundant in many vegetables. Wheat flour contains about 3 per cent.; rye flour, 11; Indian corn, 2·2; peas, 6·3; kidney beans, 19; potatoes, 3·3; cabbage, 2·8.

691. **Mucilage.**—Many seeds, as flax seed, and many roots, barks, and leaves of plants, as slippery-elm bark, marsh mallow, etc., yield, when digested with water, gummy and stringy liquids. To such products the general name of vegetable mucilage has been applied, and their chemical composition is believed to be the same as that of starch and gum.

692. **Pectine**, or pectic acid, is a gelatinous substance found in the

QUESTIONS.—What are the uses of dextrine? What is "British gum?" To what other substances is the term gum applied? How does a gum differ from a resin? What are the principal gums of commerce? From whence are they derived? What are their general properties? What is said of the general occurrence of gum as a vegetable product? What is mucilage? What is pectine?

juices of all ripe fleshy fruits, and allied in composition to starch, gum, and mucilage. It is the agent which communicates to the juices of fruits the property, when boiled (especially in connection with sugar) and cooled, of hardening into jelly, and is hence sometimes called "vegetable jelly."

693. **Sugar.**—The term sugar is ordinarily used to designate the sweet principle of plants. The chemist, however, at the present day applies it to a large number of bodies, which differ greatly from one another in their properties. Thus we have sugars which are derived from both vegetable and animal organisms—sugars which are sweet, sugars which are slightly sweet, and some which are destitute of sweetness; some sugars, also, are capable of fermentation, others do not undergo this change; some are fluid, but most are solid. All sugars, however, agree, with perhaps a single exception, in one respect—they consist of carbon, hydrogen, and oxygen, with the two latter elements united to the former in exactly the proportions which form water.

Sugar exists in greater or less abundance in all plants, and it is from this source only that we obtain our supplies. It abounds most in the growing parts, in the stems just before flowering, as those of the sugar-cane, maize, maple, etc, in pulpy fruits, and in seeds when they germinate. Like starch, it appears to be a material especially intended to subserve the growth and nourishment of the plant; but unlike it, it exists in the plant only in solution.

All the numerous varieties of sugar may be conveniently arranged in four classes, viz., the cane sugars, the grape sugars, the manna sugars, and the sugar of milk.

694. **Cane Sugar**, $C_{12}H_{11}O_{11}$.—This variety of sugar includes the sugar of the sugar-cane, beet sugar, palm or date sugar, maple sugar, and the sugar of the maize and of the fully ripe sorghum. It is also found in many of our common meadow grasses, and in the juices of melons, carrots, and turnips. Plants which have but little acid in their sap contain for the most part cane sugar; the chemical reason of this is, that cane sugar, by the action of acid substances, is gradually converted into grape sugar, even in the interior of the growing plant.

About eleven twelfths of all the sugar extracted for use is obtained from the sugar-cane, and the yearly production from this source, over the whole globe, has been estimated at 4,500,000,000 lbs. Of this enormous quantity, the population of Great Britain are certainly known to consume at least two elevenths. The method of manufacturing sugar from the cane (and also from the beet) is essentially as follows: the juice extracted from the vegetable structure by pressure, is mixed with a small quantity of hydrate of lime (slacked lime), and rapidly heated to near the boiling point. The action of the lime is twofold: it removes or neutralizes the acid which rapidly forms in the fresh juice, and at the same time unites with and precipitates the glutin-

---

QUESTIONS.— What is its most noticeable property? What is the ordinary signification of the term sugar? Is the term restricted in a chemical sense to any particular substance? In what respect do all sugars agree? What is said of the natural occurrence of sugar? Into what four classes may all sugars be divided? What sugars are included under the name of cane sugars? How is cane sugar manufactured? Why is lime used?

ous matters contained in the juice. The removal of these latter substances is an essential part of the process, as a short exposure to the atmosphere occasions their fermentation, which in turn converts the sweet juice into a sour and spirituous liquid, totally unfit for the manufacture of sugar. The juice, after clarification, is rapidly evaporated in open pans to a thick syrup, and then run into wooden vessels to cool and crystallize, and finally, when crystallized, is allowed to drain in perforated casks. The product remaining after drainage, is the common raw or brown sugar, while the drainings constitute molasses.

695. **Molasses** is uncrystallizable sugar. It does not pre-exist in the juice of the cane, but is produced at the expense of the crystallizable sugar, mainly by the high temperature used in the concentration of the saccharine solution. In improved processes for the manufacture of raw sugar, and always in the refining of sugars, the boiling of the syrups is conducted in what are called "vacuum pans," which are large metallic boilers so constructed that they can be exhausted of air. The boiling point of the syrup, owing to the absence of atmospheric pressure, is thus reduced to about 150° F., and the formation of molasses almost entirely prevented.

The process of manufacturing raw sugar, although apparently most simple, is attended with many difficulties in practice; so much so, that of the 18 per cent. of sugar contained in the cane juice of the West India Islands, not more than 6 per cent., or one third of the whole, is usually sent to market in the state of crystallizable sugar.

696. **The Refining of Sugar** is not generally carried on in connection with manufacture of the crude product. It is effected by dissolving the brown sugars in water, adding albumen (whites of eggs, or bullocks' blood), and sometimes a little lime-water, and heating the whole to the boiling point. The albumen, under the influence of heat, coagulates, and forms a kind of network of fibers, which inclose and separate from the liquid all the mechanically suspended impurities. The solution is then decolorized by filtering through animal charcoal, concentrated by evaporation in vacuum pans, and allowed to crystallize in conical iron molds. The molasses, or drainings which escape from refined sugar, by means of orifices opened in the bottom of these molds, is sold under the name of Sugar-house Syrup, Stuart's Syrup, etc. The time required for the perfect crystallization and separation of the white sugar in the molds is from 18 to 20 days, during which period the syrup is frequently stirred in order to prevent the formation of crystals of a large size.

Fig. 218.

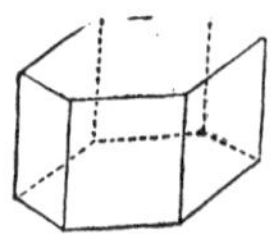

697. **Sugar Candy**.—When a strong solution of refined sugar is allowed to evaporate slowly and uninterruptedly, the sugar separates in the form of large, transparent, colorless crystals, having the form of an oblique, six-sided prisms. See Fig. 218. In this state it is known as "Sugar," or "Rock Candy."

Questions.—What is molasses? How is it formed? How is sugar refined? What is sugar candy?

In many parts of Europe, especially in France, sugar is extensively manufactured from the beet root, the juice of which contains about 8 per cent. of cane sugar. At the present time, about 360 millions of pounds of sugar are annually obtained from this source on the continent of Europe, or about 7 per cent. of all the sugar consumed in the world.

The amount of sugar annually extracted from the date palm (principally in India and the South Pacific), is estimated at 220 millions of pounds; while the quantity annually obtained from the sugar maple of North America is about 45 million pounds.

698. When cane sugar is heated to about 400° F., it gives up two equivalents of hydrogen and oxygen (water), and is converted into a dark-brown substance, termed *caramel.* This body is freely soluble in alcohol and water, and is extensively used for the coloration of spirits—the color of all dark brandies being due to it.

699. Grape Sugar.—*Glucose; Sugar of Fruits,* $C_{12}H_{14}O_{14}$.—This variety of sugar includes the sugar of grapes, of ripe fruits, of honey, and of seeds; together with the sugars artificially produced from starch and woody fiber. It is more generally diffused in nature than cane sugar, and is the product of most plants which contain acids or sour juices.

The white coating upon dried grapes (raisins), figs, etc.; and the white, brittle granules found in the interior of these fruits, is grape sugar;—hence the origin of the name.

Grape sugar may be abundantly obtained from the juice of ripe grapes and pure honey, by washing with cold alcohol, which dissolves the fluid syrup. It may also be prepared by treating starch with sulphuric acid in the manner already described · sugar from this source has received the distinctive name of *glucose,* and is very largely employed in Europe for ordinary sweetening purposes, for confectionary, for adulterating cane sugar, and for the manufacture of spirituous liquors by fermenting and distilling. In the United States, the low price of cane sugar renders its manufacture unprofitable.

In addition to starch, woody fiber of all kinds, paper, cotton, flax, cotton and linen rags, and even saw-dust, may be converted into grape sugar by heating in connection with dilute sulphuric acid. The operation is somewhat slower than when starch alone is employed, which is partially explained by the fact, that the acid first changes the woody fiber into starch, and then the starch into dextrine and sugar.

Almost all the acids, even when very dilute, convert cane sugar into grape sugar.

Grape sugar is sometimes produced in the animal system, and its appearance in the urine in great quantities is a characteristic feature of a very fatal disease termed *diabetes.*

QUESTIONS.—What is said of the production of beet, date, and maple sugars? What is caramel? What are grape sugars? What are familiar examples of grape sugar? How may it be prepared? What other substances besides starch may yield grape sugar? What is the action of acids upon cane sugar? Does grape sugar ever occur in the animal system?

700. **Essential Differences of Cane and Grape Sugars.**—Cane sugars are popularly distinguished from every other variety of sugars by their greater sweetness or sweetening power; three parts being equivalent in this respect to five of grape sugar. Cane sugar dissolves more readily in water than grape sugar; one pound of cold water dissolving three pounds of the former and but one of the latter. Cane sugar, when pure, remains dry and unchanged in the air, crystallizes readily, and when acted upon by sulphuric acid, is blackened; grape sugar, on the contrary, absorbs moisture from the atmosphere, and becomes damp; is not easily crystallized, and when digested in sulphuric acid, dissolves freely without blackening.

As respects chemical composition, cane sugar differs from starch and woody fiber in simply containing an additional equivalent of the elements of water—the formula of the latter being $C_{12}H_{10}O_{10}$, while that of cane sugar is $C_{12}H_{11}O_{11}$. Grape sugar contains relatively less carbon than either starch or cane sugar, its formula being $C_{12}H_{14}O_{14}$.

Grape sugar may be prepared artificially from various substances, but cane sugar can not be so obtained.

These two varieties of sugar may be easily distinguished from each other by their reactions with oxyd of copper: thus, if we add to separate solutions of cane and grape sugars a few drops of sulphate of copper (blue vitriol), and afterward caustic potash in excess, we obtain deep-blue liquids, which exhibit very different characters when heated; the solution of cane sugar retains its blue color, while that of grape sugar throws down a copious reddish precipitate of suboxyd of copper.

Sugar often acts the part of an acid, and is capable of uniting with bases—potash, baryta, lime, etc.—to form salts called *saccharates.* Most of the sugars, when left in contact with certain nitrogenized substances, called yeasts or ferments, become decomposed, and pass into alcohol and carbonic acid. Grape sugar is especially susceptible of this change; and cane sugar, before it undergoes fermentation, always passes into grape sugar.

701. **Manna Sugars,** $C_6H_7O_6$, are distinguished from other sugars in three particulars: they do not contain hydrogen and oxygen united to carbon in the proportions which form water; they are inferior in sweetness to other sugars; and they do not ferment under the influence of yeast. Manna sugars are somewhat extensively distributed in the vegetable organization, and exist most abundantly in *manna,* which is a dried juice of certain species of ash-trees growing in southern Europe. They are also found in the juices of the onion, asparagus, celery, mushrooms, and in several sea-weeds; and may be artificially prepared from ordinary sugar by a peculiar kind of fermentation.

702. **Sugar of Milk.**—*Lactine,* $C_{24}H_{24}O_{24}$.—This peculiar substance is the sweet principle of milk. When the curd is separated in the making of

---

Questions.—What are the essential differences of cane and grape sugar? By what chemical test may the two be distinguished? How does sugar comport itself as respects the bases? What is a property of most sugars? What are the characteristics of manna sugars? What is said of their occurrence? What is said of the sugar of milk?

cheese, the sugar remains in the whey, and may be obtained, in the form of white prismatic crystals, by evaporating the whey to a small bulk, and allowing it to cool. It is much less soluble and less sweet than cane sugar, and in a solid state feels gritty between the teeth. It is principally manufactured in Switzerland, and is used extensively in homœopathic medicine, as envelope for remedial substances. It has hitherto been detected in only one vegetable production—the acorn.

703. The conversion of starch into gummy matter and sugar, and that of the latter into starch, is a very common result in the vegetable kingdom. Unripe fruit, as apples, pears, etc., contain an abundance of starch; this may be proved by applying the tincture of iodine to a freshly-cut surface. When the fruit is completely ripened, this reaction can not, however, be obtained; the starch, therefore, has disappeared, and has been replaced by sugar, as is made evident by the sweet taste which the fruits have acquired.*

SECTION II.

ALBUMEN, CASEINE, GLUTEN.

704. Associated with the non-azotized substances in all plants, is another class of compounds, equally important, but much less abundant, than the former. These are the nitrogenized or albuminous compounds, the principal of which are known as Albumen, Caseine, and Gluten.

705. **Albumen** is widely disseminated through vegetable structures, and also exists abundantly in the animal economy; the white of eggs, and the serum, or thin, transparent part of the blood, being essentially composed of albumen dissolved in water.

Albumen dissolves freely in cold water, and forms a tasteless, glairy, transparent fluid; if heated, however, to about 158° F., it coagulates, or becomes insoluble in either hot or cold water. This change may be especially noticed in the cooking of eggs. Alcohol, creasote, corrosive sublimate, and many other substances, are also capable of transforming albumen from a soluble into an insoluble condition.

* "A similar metamorphosis is also noticed in the potato. The quantity of starch contained in 100 lbs. of the same kind of potatoes has been found to be in August, 10 pounds; in September, 14; in October, 15; in November, 16; in December, 17; in January, 17; in February, 16; in March, 15; in April, 13; in May, 10. Accordingly, the quantity of starch in potatoes increases during the autumn, remains stationary during the winter, and in the spring, after the germinating principle is excited, it diminishes. It is a well-known fact, that in germination, potatoes become soft, mucilaginous, and afterward sweet; the dextrine formed from the starch rendering them mucilaginous, and the sugar formed from the dextrine rendering them sweet. The process of transformation advances still further in the earth; the potatoes becoming softer and more watery, and when the starch is completely consumed in the growth of the young plant, the process of decay commences."—Stockhardt.

Questions.—What fact illustrates the conversion of starch into sugar in nature? What are the principal nitrogenized compounds of plants? What is said of albumen? What are its characteristic properties? Why do eggs harden in boiling?

When water containing a small portion of albumen is heated, the albumen is coagulated, and rises as a scum to the surface, carrying with it any small particles of impurity mechanically suspended in the liquid. It is in this way used for clarifying solutions of sugar and other liquids.

Albumen is found in a soluble state in the sap of plants, in the humors of the eye, in the white of eggs, and in the serum of the blood; and in an insoluble state in the seeds, leaves, and stalks of plants, and in the substance of which the brain and nerves of animals are composed.

706. **Caseine** is a substance of both vegetable and animal origin, and is allied to albumen in its composition and properties. It differs from it, however, in the circumstance that it is not coagulated by heat, although it readily experiences this change under the influence of acids. It is found abundantly in the seeds of leguminous plants; peas and beans containing from 20 to 25 per cent. of their weight of it. It also exists in animal substances, especially in the curd of milk, which is known as animal caseine, and is the chief ingredient in cheese. Vegetable caseine, to distinguish it from animal caseine, is often called *leguminé*, but the identity of the two is well illustrated by the fact that the Chinese make a real cheese from peas. Vegetable caseine may be obtained by macerating peas or beans in tepid water for several hours, and straining through a seive. The liquid which passes through contains caseine in solution, together with some starch, which separates by standing. From the supernatant liquor, which resembles skimmed milk in appearance, caseine may be precipitated by the addition of acetic acid, and when washed and dried, forms a brilliant, transparent mass.

707. **Gluten.**—If flour be made into dough, and worked with the hand upon a seive, or piece of muslin, under a stream of water (Fig. 219), its starch gradually washes away, and there remains upon the seive a white, soft sticky substance, which has received the name of *gluten.* This substance exists in all the cereal grains, and constitutes about 10 per cent. of the weight of pure flour, and from 14 to 15 per cent. of the weight of bran. It is this principle which imparts to flour its plastic and adhesive properties.

The lean part of the muscles of all animals, termed fibrine, resembles the gluten of plants so closely in composition and properties, that it may be regarded as essentially the same substance, and hence gluten is very often called *vegetable fibrine.*

708 **Chemical Composition of Proteine.**—Albumen and gluten are composed of carbon, hydrogen, oxygen, nitrogen, phosphorus, and sulphur; caseine contains the same elements, in nearly the same proportions, with the exception of phosphorus, which does not enter into its composition.

According to the generally received opinion at the present day, all album-

---

QUESTIONS.—How is albumen employed for the clarifying of liquids? What is said of caseine? In what respect does it chiefly differ from albumen? In what vegetable substances does it especially occur? In what animal substance? By what other name is vegetable caseine known? How may caseine be obtained? What is gluten? In what vegetable products is gluten especially found? With what substance of animal origin does it correspond? What is the chemical composition of albumen, caseine, and gluten?

inous matter (and by this term we mean to include albumen, caseine, gluten, and all similar substances, originating either in vegetable or animal structures) are compounds of a peculiar and distinct principle, called *proteine.* The composition of this organic radical is indicated by the formula $C_{36}H_{25}N_4O_{10}$, or by the symbol, Pr. Hence, albuminous substances, as a class, are very generally termed *proteine compounds*—the formula of albumen being 10Pr+P+S; of gluten, 10Pr+P+2S; and of caseine, 10Pr+S.*

FIG. 219.

By dissolving any albuminous substance in caustic alkali, and adding acetic acid to the solution, proteine may be precipitated in the form of a grayish-white, inodorous solid, soluble in water and alcohol, and capable of uniting to form compounds with many acids and bases.

709. **Characteristics of the Albuminous Substances.**—All the albuminous substances, when subjected to heat, exhale an odor similar to that of burnt feathers, and leave, as an ultimate residue, a black, brilliant, spongy coal. When perfectly dried, they are capable of indefinite preservation; but when exposed to the joint influence of air and moisture, they are more susceptible of decomposition than any other class of organic substances—putrefying and calling into existence a multitude of microscopic animalculæ. The decomposition of the albumen contained in wood, especially in what is called the sap-wood, is regarded as the most active cause of its decay. Hence those substances like creosote, corrosive sublimate and the like, which form insoluble compounds with albuminous matter, existing either in animal or vegetable tissues, are the most effectual antiseptic agents; the processes of kyanizing wood, and of smoking fish and meat, being familiar examples of their action. The complete dessication of organic substances, or the extraction of their albuminous constituents by steeping in water, or steam, accomplish the same result.

---

* It is proper to state, in this connection, that the theory which assumes the radical nature of proteine is very strenuously opposed by many chemists, and especially by Dumas, who regards it as a product not pre-existing in albuminous compounds, but as generated by the action of the alkalies on these bodies.

QUESTIONS.—Of what radical are they supposed to be derivatives? What are the characteristics of proteine? How is it obtained? What are the general properties of the albuminous substances?

When albuminous substances are dissolved in caustic alkali, the sulphur which they contain unites with the alkali to form a soluble sulphuret, and the solution blackens paper moistened with sugar of lead. In this way the presence of sulphur in these compounds may be readily demonstrated. When an egg is boiled, the sulphur present in its albumen unites with a little free soda, which is also a constituent of the egg, to form sulphuret of sodium, and it is by the decomposition of this compound that the blackening of silver spoons used in contact with boiled eggs is occasioned.

710. **Nutritive Value of Vegetable Albuminous Constituents.**—As the chief proximate constituents of animal structures, albumen, caseine, and fibrine have the same chemical composition as the albuminous substances produced in the vegetable kingdom, the latter are regarded as the special products provided by nature for the nutriment and support of animals; or in other words, they are the vegetable principles out of which animal fibers and tissues are constructed. All experiments tend to confirm this conclusion, and prove that the value of a vegetable product as an article of food is very nearly in proportion to the quantity of albuminous or nitrogenous compounds which it contains. This subject will be further discussed hereafter.

CHAPTER XVIII.

NATURAL DECOMPOSITION OF ORGANIC COMPOUNDS.

711. So long as organic bodies are pervaded by what is termed the vital principle, so long do they tend to maintain their form and properties essentially unchanged; but when deprived of this influence, they obey the ordinary laws of chemical attraction, and readily undergo decomposition, the products of such decomposition being mainly the result of a separation or falling apart of the complex substances which characterize the living structure, and a re-arrangement of their particles in simpler combinations. The nature of these changes, which vary greatly with the composition of the bodies concerned, and with the conditions to which they are subjected, may be generally considered under three separate heads, viz., as *decay*, *fermentation*, and *putrefaction*.

712. **Decay.**—When vegetable tissue (wood, leaves, straw, etc.) is exposed to the action of atmospheric air and moisture, it absorbs oxygen and

---

QUESTIONS.—How may the presence of sulphur in these bodies be demonstrated? What is supposed to be their special office? What is the proportionate value of a vegetable product as an article of food? What especially distinguishes living from dead organized matter? What change does vegetable tissue undergo when exposed to air and moisture?

undergoes a slow decay, which has been termed by Liebig *eremacausis* (slow combustion). The changes which take place in this process are very nearly the same as in the ordinary combustion of wood, except that they occur much more slowly. In both cases the constituents of the wood, by the addition of oxygen from the air, are converted into carbonic acid and water, and in both cases also the hydrogen is oxydized more rapidly than the carbon, as is shown by the darker color which wood assumes both in combustion and decay. Eremacausis further agrees with ordinary combustion, inasmuch as it can not take place without the access of air, and is uniformly attended with the evolution of heat, and sometimes with light—the total amount of heat evolved being undoubtedly the same in both cases. (§ 469.)

The brown or black matter into which vegetable tissue is converted by decay, has received the general name of humus, or vegetable mold, and is the substance which gives to fertile soils their rich black or brown appearance. Humus is not, however, regarded as a distinct compound, but rather as a mixture of several brown substances, which represent various degrees of decomposition of the original vegetable matter. These substances have received the names of humine, ulmine, humic acid, ulmic acid, geic acid, crenic and apocrenic acids. The two latter are soluble in water, and are mainly the cause of the deep yellow or brown colors which characterize the waters of bogs and swamps. The others are either entirely insoluble, or soluble only in alkaline solutions. The relation which these substances sustain to plants is an important one, and their presence in certain quantity in every soil is essential to its fertility. From the products of their decomposition—carbonic acid and water—plants derive, through their roots, from the soil, their chief supplies of nutriment. They also absorb and retain ammonia, another important element of vegetable nutrition, and to some extent have undoubtedly the power of producing it from the nitrogen of the atmosphere. The humus consumed in vegetation and removed from the soil in the substance of the crop, may be again restored to the land by plowing in straw and animal manures, or green crops (clover, etc.), or by the alternation of plants which leave abundant roots in the soil (fallow plants), with such as have few roots (grains, etc.).

Eremacausis is greatly promoted by heat and moisture, or the presence of the alkalies; it is, on the contrary, arrested or retarded by cold and dryness. Wood, cordage, etc., exposed to the cold of the Arctic regions or to the dry atmosphere of Egypt will remain alike for years unaltered.

713. **Putrefaction.**—The decomposition of vegetable tissue when air is wholly or partially excluded from it, as, for example, when buried in the

---

QUESTIONS.—What is this change called? What is the nature of the change? What is the immediate product of the decay of vegetable tissue called? What is the composition of humus? What produces the discoloration of the water of bogs and marshes? What relation do these substances sustain to vegetation? How may humus consumed by vegetation be restored to the soil? What is the nature of the decomposition which takes place in vegetable tissue when air is excluded?

ground, is essentially different from that of eremacausis. In this case the constituent elements rearrange themselves mutually into new products, either with or without the coöperation of the elements of water; the oxygen gradually uniting with the carbon to form carbonic acid, which separates and leaves as a residue substances rich in carbon and hydrogen—hydrocarbons. It is in this way that bituminous coal, peat, and brown-coal (lignite) have been formed from vegetable matter,* and also the natural gaseous carburets of hydrogen, viz., "marsh gas," obtained by stirring the mud at the bottom of pools (see Fig. 220), and "fire-damp," evolved from rock-strata in mines. (§ 452.) Moist hay, leaves, manure, etc., when piled together in compact heaps undergo similar changes, and are converted into black, carbonaceous products.

Fig. 220.

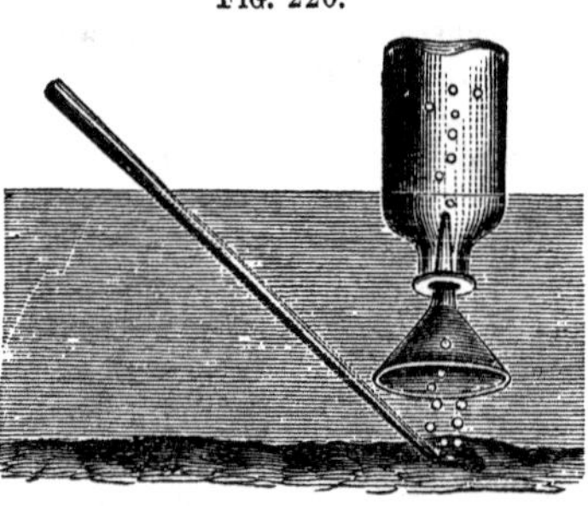

Decomposition of this character is termed putrefaction, and is somewhat analogous to the change which wood undergoes when subjected to dry distillation or incomplete combustion. It differs from eremacausis (or decay), inasmuch as the latter can not take place without the free access of air, the oxygen of which is absorbed by the decaying bodies. The two methods of decomposition may, however, mutually replace each other, since all putrifying bodies pass into the state of decay when exposed freely to the air; and all decaying matters into that of putrefaction when air is excluded.

Nitrogenized animal and vegetable substances, on account of their complex constitution, undergo decay and putrefaction much more readily than non-azotized compounds, and the products of their decomposition are essentially different. Thus the oxygen of the substance unites with the carbon to form

* Peat is mainly the product of the slow decay of certain species of marsh plants under water. Every peat-bog was undoubtedly, in the first instance, a marsh or swamp, which has been filled up and converted into a morass by the annual growth and decay of its surface vegetation. The quantity of vegetable mold which thus accumulates in the course of years is very great, and as the process of decomposition is slow and gradual, the aspect and constitution of the different successive layers of peat vary greatly—those near the surface consisting of the half decayed stems of mosses and of roots, while those of older formation scarcely exhibit any traces of their vegetable origin, and in some instances are converted into a true bituminous coal. In many countries peat is extensively used as fuel, and furnishes by distillation oily products analogous to those obtained by the distillation of coal.

Questions.—What are the products of such decomposition? Illustrate this. What is decomposition of this character termed? How does putrefaction differ from eremacausis? What are the products of the putrefaction of nitrogenized substances?

carbonic acid, while the hydrogen divides itself between the nitrogen, the sulphur, and the phosphorus, and forms ammonia with sulphuretted and phosphuretted hydrogen. It is to the presence of these last-named gaseous substances that the very offensive odors given off during the putrefaction of azotized bodies are to be mainly ascribed.

714. **Fermentation.**—When a nitrogenous substance undergoing putrefaction is brought in contact, under favorable circumstances of temperature and moisture, with a complex organic body of small stability, it is capable of inducing in this latter substance, by the mere agency of its presence, a state of putrefaction or decomposition. In such cases the substance inducing decomposition is termed a "*ferment*," and the decomposition induced, "*fermentation*." For example, a solution of pure sugar may be preserved unaltered for any length of time, but if a minute quantity of putrescent matter containing nitrogen be added to it, fermentation at once takes place, and the elements of the sugar break up into alcohol and carbonic acid. "In the same manner, the most minute portion of milk, paste, juice of beet-root, flesh or blood, in the state of decomposition, causes fresh milk, paste, juice of beet-root, flesh or blood, to pass into the same condition when brought in contact with them."

The method in which ferments act is not well understood, since they do not enter into combination with the fermenting substance or with any of its elements. The theory, however, most usually adopted is, that the molecules of the ferment, or substance already undergoing change, are capable of imparting motion to the molecules of other substances by contact, and that through the impulse thus received, the equilibrium of forces previously existing between the molecules of the body acted on are overcome and destroyed.

715. **Yeast.**—The substance most potent in exciting fermentation in solutions of sugar is a species of microscopic vegetation which is spontaneously developed in the organs of plants, and in a large number of nitrogenous substances, when left to putrify. This organism, which passes into a state of putrefactive decomposition with great readiness, is termed *yeast*, or *ferment.* It is obtained in the greatest abundance when a solution of sugar mixed with albuminous substances of animal or vegetable origin is exposed to the air at ordinary temperatures.

When yeast is added to a solution of sugar, it not only excites fermentation, but if there are albuminous substances present, it occasions the production of an immense additional quantity of yeast. For example, if we add to clear fresh juice of ripe grapes a few particles of yeast, the liquid will in a short time grow thick and give off bubbles of gas, or ferment, and in a few hours a layer of grayish-yellow yeast will collect upon its surface. In the heat of the fermentation the yeast plants are produced in immense numbers, mil-

QUESTIONS.—Explain the meaning of ferment and fermentation. What are examples? What is the theoretical action of ferments? What is yeast? How is it formed? Illustrate this.

FIG. 221.

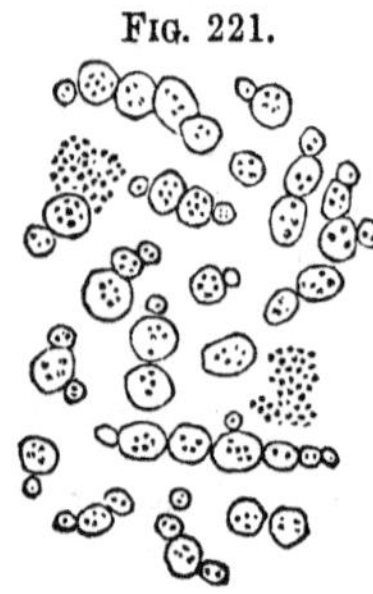

lions of its minute organisms being contained in the space of a single cubic inch. Fig. 221 represents the appearance of the yeast globules under the microscope, and the manner in which they propagate by division. Ordinary brewer's yeast is formed in this manner in the fermentation of infusions of malt. Artificial yeast, or *leaven*, may be prepared by exposing a piece of dough for some days to a moderate temperature, until it acquires a sour, or vinous odor. The fermenting agent in this case is the gluten of the dough in a state of incipient putrefaction. Yeast loses its power of exciting fermentation when perfectly dried, or heated to a temperature of 212° F., or if mixed with alcohol, acids, or alkalies, and finally by the completion of its own decomposition.

716. **Different Kinds of Fermentation.**—The products of fermentation vary under different circumstances. The conversion of saccharine liquids into alcohol and carbonic acid is termed *vinous*, or *alcoholic fermentation.* For the production of this change a temperature of from 50° to 86° F. is necessary. Under 50° F. fermentation does not proceed. All vegetable bodies contain some substances which act as a ferment, and therefore, by the addition of moisture and regulation of the temperature, various kinds of grain containing starch, and ripe fruits containing sugar, will undergo naturally the vinous fermentation. Thus cider is formed from apples, and beer from grain.

A liquid which has already undergone the vinous or alcoholic fermentation, is capable of experiencing another change when exposed to the air in connection with a small quantity of decomposing azotized matter—its alcohol being converted into acetic acid and the liquid into vinegar. This has been called *acetous fermentation.*

There are a variety of substances which, when added to fermentable liquids, even in very minute quantities, have the power of preventing decomposition; such are, for example, the oil of mustard, sulphurous acid, nitrous acid, etc. New cider, it is well known, is kept sweet by the addition of mustard-seed, or by burning sulphur in the barrels previous to filling with liquor.

When azotized matters are beginning to decompose they are at first not able to excite the true alcoholic fermentation in solutions of sugar, but it is necessary for this that their decomposition should be tolerably active and advanced. But even in the early stage of their transformation they are able to effect a very important change in the elements of sugar, and cause it to undergo a peculiar kind of fermentation, the result of which is the production of

QUESTIONS.—How may artificial yeast be prepared? Under what circumstances does yeast lose its power? What is vinous fermentation? What are examples? What is acetous fermentation? What substances are capable of arresting fermentation? Is all decomposing azotized matter capable of inducing alcoholic fermentation? Illustrate this.

an acid called lactic acid, and a viscous substance analogous to sugar. This fermentation, which has been termed *viscous*, or *lactic acid fermentation*, is especially produced when milk or cheese curd is mixed with sugar at a temperature of 86° to 94° F. If, however, the curd of milk is in an advanced stage of decomposition, it produces at the temperature of about 100° F. the vinous fermentation, and the sugar is converted into alcohol and carbonic acid. In this way the Tartars prepare a spirituous liquor from mare's milk, called "*koumiss.*"

717. Lactic acid derives its name from the circumstance that it is the acid which imparts sourness to milk, and is the immediate product of the decomposition of that liquid. Lactic acid, when kept in contact with caseine in the first stage of its decomposition for some time, at a temperature of about 95° F., is itself capable of experiencing a transformation into a sour, pungent smelling liquid termed *butyric acid*, and the change in question is known as *butyric fermentation*. The conversion of starch into sugar by the action of diastase, is also regarded as a species of fermentation, and is termed "*saccharine.*" Several other forms of fermentation in addition to those enumerated, are also recognized, but the most important of them all are the alcoholic and acetous.

718. Organic substances do not possess the power of entering spontaneously into fermentation and putrefaction, but it is necessary that some change in the attraction of their elements should previously take place. This exciting cause is undoubtedly the oxygen of the atmosphere which surrounds all bodies, and we accordingly find that eremacausis always precedes fermentation and putrefaction, and that it is not until after the absorption of a certain quantity of oxygen that the signs of a transformation in the substances show themselves. When the condition of intestine motion is once excited, the presence of oxygen for the continuance of the action is no longer necessary. The smallest particle of an azotized substance in its act of decomposition, also propagates this state of motion to the particles of the substance in contact with it, and although the air be afterward entirely excluded, fermentation or putrefaction will proceed uninterruptedly to its completion. Animal food of every kind, and even the most delicate vegetables, may be preserved unchanged for years, if heated to the temperature of boiling water in vessels from which the air is completely excluded. A fresh exposure to the air at any period will, however, induce fermentation.*—LIEBIG.

* The method of putting up "preserved meats" is essentially as follows; the meat is first placed in a tin cylinder, which is then filled with a properly prepared soup, and a cover, pierced with a minute hole, is soldered on air-tight. The cylinder is next deposited in a bath of chloride of calcium solution (which does not boil under a temperature of 320° F.), where its contents are subjected to heat until sufficiently cooked. When this is effected, and the air in the interior completely expelled by the evolution of steam, the minute orifice in the cover is suddenly and effectually closed with a drop of solder. The

QUESTIONS.—What is the acid of sour milk? What is butyric fermentation? Are there any other kinds of fermentation? Do organic substances possess the power of spontaneous change? What is necessary to effect this? How may animal food be preserved?

719. **Poisons, Contagions, Miasms.**—"When a chemical agent or substance is brought in contact with matter endowed with life (as, for example, if it is introduced into the stomach or any other part of the animal organization), it tends to enter into combination with it, and effect decomposition. This tendency is opposed by the vital principle, and the result will depend upon the strength of their respective actions. If the chemical element is forced to yield to the superior power of the vital action, it is digested, and exercises no chemical influence upon the living organ; when, however, it is able to effect a change in the operation of the vital principle, as in changing its direction, strength, or intensity, without destroying it, it is said to act *medicinally;* but when it obtains an ascendancy over the vital force, and tends to destroy it, it acts as a poison. Food will act as a poison, that is, will produce disease, when it is able to exercise a chemical action by virtue of its quantity; or when either its condition or presence retards, prevents, or arrests the motion of any organ. A medicament administered in excessive quantity may act as a poison, and a poison in small doses, as a medicament. Thus the quantity of a substance and its condition must, obviously, completely change its chemical influence in the system."

Some inorganic poisons, such as arsenic, corrosive-sublimate, etc, exert a destructive action upon animal life, by forming with the component parts of the body compounds which are not susceptible of the changes which it is the office of the vital principle to produce. Other inorganic poisons, like corrosive acids, destroy at once the form and structure of the tissues with which they are brought in contact. In both cases the organs fail to fulfill their offices, and disease or death ensues. "If the quantity of poison is so small that only small portions of the body, which are capable of being regenerated, have entered into combination with it, then eschars (scabs) are produced, and the compounds of the dead tissues with the poison are thrown off by the healthy parts."*

cylinder is then allowed to cool, and form a condensation of its contained vapor, both its ends are pressed inward, and become concave. Thus hermetically sealed, it is exposed in a test chamber, for at least a month, to a temperature above what it is ever likely to encounter; from 90° to 110° F. If the process has failed, putrefaction takes place, and gas is evolved, which will cause the ends of the case to bulge, so as to render them convex instead of concave. But the contents of those cases which stand the test will infallibly keep perfectly sweet and good in any climate, and for any number of years. If there be any taint about the meat when put up, it invariably ferments, and is detected in the proving process.

* "The limit at which substances like arsenic, corrosive sublimate, etc., cease to act as poisons, may be determined with great certainty; for since their combination with organic matters must be regulated by chemical laws, death will inevitably result when the organ in contact with the poison finds sufficient of it to unite with atom for atom, whilst if the poison is present in smaller quantity, a part of the organ will retain its vital func-

QUESTIONS.—When a chemical agent is brought in contact with living matter, what takes place? When will chemical substances act as food, as medicine, and as poison? Illustrate how the quantity and condition of a substance may change its chemical influence on the system? How do inorganic poisons generally produce their destructive effects?

With respect to the action of poisons like Prussic acid, strychnia, etc., no very satisfactory explanation can be given.

In addition to the poisons noticed, "there is a class of substances generated during certain processes of decomposition, which act upon the animal economy as deadly poisons, not by entering into combination with it, or by reason of their containing a poisonous principle, but solely by virtue of their peculiar condition;" in other words, these products being in a state of decomposition themselves, act as ferments, and by their simple presence tend to excite decomposition or disease in the animal substances with which they are brought in contact.

The most striking illustration of this principle is to be found in the case of the wounds which physicians sometimes accidentally inflict upon themselves in the dissection of dead bodies. The knife, in such instances, introduces through the wound a minute portion of matter in the state of decomposition or putrefaction, which acts as a *ferment*, and causes the healthy blood in contact with it to pass into the same decomposed state as itself; the action once commenced, extends with great rapidity, and very often affects the whole body and produces death—injuries to the system of this character being almost beyond the control of medical treatment. The virus of the small-pox, plague, etc., appear to act in like manner, inasmuch as the most careful examination fails to extract from them any poisonous principle. When brought in contact, however, either directly or indirectly, with the blood, they communicate to it their own condition.

Contagion and miasm, or miasmata, are generally included among poisons of this class.

We apply the term contagion to that subtile matter which proceeds from a diseased person, or body, and which communicates disease to another person or body. It is characterized by its ability to reproduce itself. Miasm, on the other hand, is the product of the decay or putrefaction of animal or vegetable substances, and causes disease without being itself reproduced.

The nature of the substances which constitute contagion and miasm is not well understood; according to some authorities, they are merely putrid matters, and according to others, they are microscopical animals or plants, which like yeast, readily undergo decomposition.*

---

tions." The comparative weight of an equivalent, or of an atom of any one of the highly complex substances which make up the animal organism, is, however, so exceedingly great, that a very small amount of poison is sufficient to completely satisfy the combining affinities of a very large quantity of animal substance; the proportion in the case of fibrine and arsenic being as 6361 parts of the former to 1 of the latter.

* Very many curious observations have been made upon these topics. A forest interposed to the passage of a current of moist air charged with pestilential miasmata, sometimes preserves all behind it from its effects, whereas the uncovered portion of a district is exposed to disease. The trees, in such cases, appear to filter the air, and to purify it by

---

QUESTIONS.—What is known respecting the action of poisons, like Prussic acid, etc.? What other class of poisons are mentioned? Explain the manner of their action. What are included under this class? What is contagion? What is miasm?

Mildew is a species of decomposition occasioned by the development and growth of a class of microscopic fungi; (a fungus being a cellular, flowerless plant). The dark spots observed upon awnings, sails, etc., exposed to the weather, are familiar examples of its action. The most effectual agent in preventing mildew is chloride of zinc.

Many of the poisons which act as ferments, and readily excite disease when brought in contact with the blood, such as the contagious matter of small-pox, fevers, etc., are wholly inoperative when introduced into the stomach. The explanation of this is, that they are alkaline or neutral in their properties, and are therefore destroyed or neutralized by the free acid which always exists in the stomach. Poisons of a similar character, however, which have an acid reaction, appear, when placed under the same circumstances, to retain all their frightful properties. The products of the incipient putrefaction of meat and fish are particularly liable to act in this manner. In Germany, especially, the effects of a poison of this character, resulting from a peculiar kind of putrefaction occurring in sausages, and hence termed the "sausage poison," have been very carefully studied. The symptoms which precede death in cases of poisoning by putrefied sausages are very remarkable. "There is a lingering and gradual wasting of muscular fiber, and of all the constituents of the body similarly composed; the patient becomes much emaciated, dries to a complete mummy, and finally dies."

The flesh of animals killed when overdriven or exhausted, is also very liable to produce diseases which, in the rapidity of their action and deadly effect, resemble cholera; the symptoms, however, do not generally manifest themselves until some little time has elapsed after the food has been received into the stomach. The origin of the poison in the meat in these instances is explained as follows: all mental and physical effort is accompanied by and

---

removing the miasmata. Trees also appear to prevent miasmata by absorbing it. The negroes of the South plant the sunflower near their cabins as a preventative against fever and ague. Facts also show that malaria does not prevail in the neighborhood of swamps surrounded with thick forests—the vicinity of the Dismal Swamp, for example, being healthy, while the marshes of the adjacent sea-board are most pestilential." Flint, in his account of the Mississippi Valley, mentions the fact that the wood-cutters on the banks of the streams where the trees had been cut away, were constantly attacked by malarious fevers, while such diseases among the workmen in the forest were comparatively rare, although the ground on which they worked was quite as moist. Every tree which they left to decay on the ground helped to create the poison, while every tree left standing helped to absorb it. Many cases might be cited where the cutting down of woods has had a most unfavorable effect upon the health of the surrounding region. The district around Rome is only a celebrated instance of what is a very common experience. Dampness is not a source of miasmata, but decomposition caused by too rapid drying, whether of vegetable matter or animal infusoria. A ditch which alternates from wet to dry, or a pool which is weekly emptied and replenished as wind and shower follow each other, gives forth a much more deadly poison than ground which is uniformly and steadily saturated."

QUESTIONS.—What is mildew? What are characteristic differences of action in poisons acting like ferments? What are illustrations? What is the character of the flesh of overdriven animals?

requires an expenditure of healthy animal substance. The brain, for example, is undoubtedly used up by thinking, the muscles by exercise, the nerves by excitation. In the healthy state of the system, the waste thus occasioned is at once restored, and the products of decomposition are removed by the organs of secretion, and thrown off from the body. If the functions of the organs of secretion are impeded, the products of decomposition accumulate in the system and occasion disease. In the case of overdriven animals, the products of decomposition consequent upon unusual and excessive physical exertion, remain in the body, because the organs of secretion have not had sufficient opportunity to discharge their office before the animals are slaughtered. The meat, therefore, is full of substances in just that state of decomposition which enables them to act most effectually as ferments, and their presence, therefore, renders the flesh of the most healthy animal unwholesome. It should also be mentioned, that the most severe cases of poisoning of this character seem to occur when the putrefactive fermentation in the meat has only just commenced, and when its presence is hardly discernible by the senses.

720. Every form of disease is occasioned by changes or transformations which take place in organs in a manner different from what occurs in ordinary healthy action. If these transformations are perfected in constituents of the body which are not essential to life, without other parts taking a share in the decomposition, the form of the disease is termed *mild* or *benignant;* but when the changes affect the organs essential to life, the disease is termed *malignant.*—LIEBIG.

CHAPTER XIX.

ALCOHOL AND ITS DERIVATIVES.

721. THE term alcohol is applied by chemists to a series of compounds of a dissimilar but analogous composition, and similar properties. They all consist of carbon, hydrogen, and oxygen, are all liquid at ordinary temperatures, and are characterized by possessing a high degree of volatility and a pungent taste and smell. The most important of the alcohols are wine alcohol, $C_4H_6O_2$, methylic alcohol, $C_2H_4O_2$, and amylic alcohol, $C_{10}H_{12}O_2$. The term alcohol, however, in its ordinary acceptation, refers solely to the spirituous principle resulting from the fermentation of saccharine bodies.

Sugar is the only substance susceptible of vinous fermentation, and the only substance from which alcohol can be derived. Potatoes, the cereal

---

QUESTIONS.—Why is it liable to induce disease? What is the occasion of all disease? When is disease said to be benignant and when malignant? What is the chemical signification of the term alcohol? What is its ordinary meaning? From what substances only can alcohol be produced? How do we produce alcohol from bodies which consist mainly of starch?

grains, and other vegetable products deficient in sugar from which alcohol is obtained, are rendered available for this purpose by first converting their starch into sugar. The various kinds of liquors prepared by means of fermentation, may be conveniently divided into two classes—the *beers*, produced from the nutritive and starch containing grains and roots, and the wines produced from the juices of fruits which contain sugar.

722. **The Beers.**—When a solution of grape sugar is dissolved in water, and a little yeast added, fermentation speedily ensues, and the sugar breaks up into alcohol, water, and carbonic acid; of these several bodies, the two former remain in the liquid, while the latter escapes as bubbles of gas into the air.* When cane sugar is used the results are the same, the yeast, however, in the first instance effecting a transformation of the cane sugar into grape sugar. For the completion of these changes it is not necessary that air should be present.

When the cereal grains, etc., are used for the manufacture of alcohol, the first step, as has been already stated, consists in effecting a change of the starch into this sugar. This transformation may be brought about by the action of dilute sulphuric acid, but in practical operations this agent is rarely used, and the change is effected through the influence of *diastase* (§ 688). In order to arrive at a clear understanding of this phenomenon, it is necessary to first consider the conditions under which diastase originates.

A seed or grain consists essentially, in the first instance, of two substances, starch and gluten, in which is contained a little rudimentary plantlet, called the germ or embryo. It is for the nourishment and support of this embryo, before it has attained sufficient development to be able to derive its own sustenance from the soil or air, that the supplies of starch and gluten contained in the seed are provided. Fig. 222 represents a grain of Indian corn, divided so as to show the embryo embedded in the starch and gluten, which make up the bulk of the seed. Fig. 223 represents, in like manner, a section of an acorn. Under the joint influence of heat and moisture, the embryo of the seed begins to sprout, or germinate,

Fig. 222.

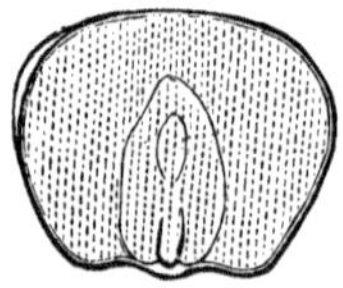

Fig. 223.

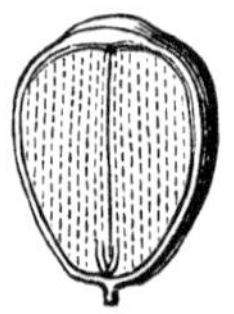

---

*This decomposition may be represented as follows:—

		C	H	O
One atom of grape sugar	=	12	14	14
Two of alcohol	=	8	12	4
Four of carbonic acid	=	4	0	8
Two of water	=	0	2	2
Total,		12	14	14

The yeast, which occasions the decomposition, takes no part in any of the combinations resulting.

QUESTIONS.—When yeast is added to a solution of grape sugar, what takes place? What in the case of cane sugar? How is starch changed into sugar preliminary to the manufacture of alcohol? Of what does a seed consist?

Fig. 224.

and puts forth a tiny stem or axis, bearing upon its summit a pair of small leaves. It has now only to form a root by which to fix itself to the ground, to render it a perfect, though diminutive plant, capable of providing for itself. (Fig. 224 represents a grain of Indian corn in the process of germination.) This root is and can only be formed from the starch and gluten contained in the seed; "but as both these substances are insoluble in water, they can not, in their natural state, pass onwards from the body of the seed to supply the wants of the growing germ. It has been beautifully provided, therefore, that both of them should undergo chemical changes as the sprouting proceeds, and these changes take place at the base of the germ, exactly where and when they are wanted for the formation of the root." The gluten is accordingly first changed into diastase, and this acting upon the starch converts it wholly into grape sugar.

Now the brewer, in the manufacture of spirituous liquors from grains, avails himself of this natural transformation in order to obtain the sugar, which alone is susceptible of vinous fermentation. The grain most usually selected for transformation is barley, which is first moistened in heaps, and spread upon the floor of a dark room to heat and sprout. When the germination has advanced to just the extent sufficient to convert the greater part of the starch into sugar, and the gluten into diastase, the action is arrested by heating the grain in a sort of kiln, which at once destroys the vitality of the germ. The necessity of thus violently arresting the progress of germination, grows out of the fact that the sugar would be wholly consumed by its continuance and converted into vegetable tissue. Barley thus treated is termed *malt.*

The next step of the process consists in bruising the malt, and digesting it with water, gently warmed, in what is called the "mash-tub." The solution obtained contains sugar and diastase, and is termed *wort.* By standing a little time, the diastase acts upon any starch yet remaining in the seed, and converts it into sugar; and it is also capable of changing, in a like manner, any unmalted grain or starch which may be added to the wort at this stage of the process.

The change of all the starch into sugar being effected, the wort is next heated to boiling, which destroys any further action of the diastase. At this point, also, hops are introduced into the wort, which, besides imparting a peculiar bitterness and aroma to the liquid, help to clarify it. The boiled liquor, filtered and clarified, is next run off into shallow vessels, and cooled to a tem-

---

Questions.—What takes place in germination? How does the brewer avail himself of the natural transformation of the starch and gluten of seeds? What is malt? What is the first step of the process of brewing? What is the second? How is fermentation effected? How is fermentation arrested? Is the sugar contained in the wort allowed to entirely decompose?

perature of about 60° F. Yeast is then added, and fermentation allowed to proceed. "In a few hours bubbles of gas will be seen rising from all parts of the liquid, a ring of froth forming at first round its edge, and gradually increasing and spreading until it meets in the center, or until the whole surface becomes covered with a white, creamy foam of yeast. The bubbles of gas then rise and break in such numbers that they emit a low, hissing sound, while the yeast gradually continues to increase in thickness, and at last forms a tough, viscid crust, which the brewer skims off and removes as soon as he judges that the fermentation is complete, (the period of time varying from six to eight days)."

In practice, the fermentation is always checked before the whole of the sugar is converted into alcohol, since, if perfect decomposition were effected, the beer would not keep, but would soon turn sour in the cask. The residue of undecomposed sugar also imparts a sweet, pleasant flavor to the beer.

The liquor is next drawn off into casks, where it undergoes a second fermentation, far more slow and protracted, however, than the first; this effects what is called a ripening of the beer, and is essential to its preservation. At the conclusion of this second fermentation, the liquors must be kept tightly bunged, or corked up, since, as soon as the fermentation ceases, and air gets access to the liquor, oxydation commences, and induces acetous fermentation. The sparkle and foam of bottled liquors is owing to the carbonic acid gas which is generated in this second fermentation, and becomes dissolved in the liquors under pressure.

The varieties of beer depend both upon the difference in their material and the different management in their production. The difference in the colors of ale and porter depends upon the color of the malt employed, which, in turn, is regulated by the length of time the malt is subjected to the heat in the kilns.

723. **Lager Beer.**—Ordinary beers, even after the second fermentation, contain a considerable quantity of albuminous or glutinous matter, which tends to decompose by contact with the air, and convert the alcohol into acid (vinegar). Such liquors, therefore, are with difficulty preserved for a great length of time. In the preparation of lager, or Bavarian beer, the wort is fermented very slowly, and at an extremely low temperature, in large open vessels; by which procedure the yeast produced, instead of rising at the top of the liquor, falls to the bottom, and a separation from the liquor of almost every trace of nitrogenized matters is at the same time effected. The fermentation thus carried on is very complete, and continues for weeks, or even months; the liquor produced being as clear as champagne, and richly charged with carbonic acid. It may also be preserved for years without becoming sour. Lager beer derives its name from the long time it is allowed to lay (*lager*) in vats or casks, in cool cellars, previous to consumption.

Questions.—Does any further fermentation take place? What occasions the sparkle and foam of bottled liquors? What occasions the differences in beer? What is "lager" beer? How is it produced? What is the origin of its name?

724. The intoxicating properties of malt liquors depend entirely upon the alcohol they contain. Of this, there is present in the stronger varieties of ales and beers (English ale, Albany ale, etc.), from 5½ to 10 per cent. by weight; in porter and "brown stout," from 3½ to 6½; in lager beer, from 2 to 3·5 per cent. In addition to alcohol, the malt liquors all contain a certain quantitity of nutritive matters, consisting of undecomposed sugars, nitrogenized substances, oils, the aromatic parts of the hop, and certain mineral salts. In ordinary strong beers, the quantity of these substances varies from 4 to 8 per cent. of the entire weight; in some of the German beers the per centage is much greater; so that beer is, to a considerable extent, food as well as drink.

725. **Wines.**—The expressed juices of ripe fruits containing sugar, contain also a peculiar azotized matter, which causes them to readily undergo fermentation without the addition of yeast. In ordinary summer weather, the clearest juice of the grape will enter into fermentation within a half an hour after its expression, and give off bubbles of gas. The azotized matter which occasions this fermentation will not, however, enter into an active state of decomposition, unless free oxygen has access to it. "Consequently, whole grapes, or those in which the skins remain perfect and entire, may be dried and converted into raisins; but if the skin is once injured, a little air gets in, and fermentation soon commences."

The method of making wine is essentially as follows: the grapes are collected and pressed; the juice, which is called *must*, is poured into vats situated in cellars, where, as the temperature is low, the fermentation proceeds so slowly, that it is not completed until after some months. During the fermentation, the impurities rise to the surface in the form of froth, or yeast, or settle to the bottom of the vats (lees), so that the pure wine is finally drawn off clear, and ready for use. Wines intended to be sparkling or effervescing, are bottled before the fermentation is quite finished, so that the carbonic acid subsequently evolved remains stored up in the liquid.

726. The popular qualities by which wines are known, are their strength, sweetness, acidity, and flavor.

The *strength* of wine depends upon the alcohol it contains, the percentage of which varies greatly in different wines. The weaker hocks and sour wines contain about 9 per cent.; champagne from 5 to 15; claret from 9 to 15; while the stronger madeiras, sherries, and ports, contain from 18 to 24 per cent. The *sweetness* and *fruity* character of wines is due to a portion of grape sugar which has escaped the decomposing action of the fermentation. Of this, there is no sensible quantity present in clarets, Burgundies, hocks,

---

QUESTIONS.—To what are the intoxicating properties of malt liquors due? How much alcohol do they contain on an average? What other substances beside alcohol are contained in malt liquors? Is it necessary to add yeast to the expressed juice of ripe fruits to excite fermentation? Why do not grapes ferment upon the vines? How is wine manufactured? How are sparkling wines prepared? Upon what does the strength of wine depend? State the proportion of alcohol in various wines. Upon what does the sweetness of wines depend?

etc. Sherries contain from 9 to 12 grains of sugar in an ounce; ports from 16 to 30; and the so-called sweet wines (Cyprus, Malmsey, etc.) from 60 to 100 grains. Some wines, like champagne, are artificially sweetened.

All wines, malt liquors, and ciders, contain before undergoing acetous fermentation a variable proportion of free acid, which imparts to them a more or less distinctly sour taste; but in each liquor the characteristic acid is different. Thus, malt liquors contain acetic acid; ciders and the liquors allied to it, lactic acid; while the acidity of wines is due to tartaric acid. In all of them acetic acid is also present in greater or less quantity, as it is always produced when the fermentation of alcoholic liquors is allowed to proceed too far; but lactic acid is not found in malt beer or grape wine in sensible quantity; nor is tartaric acid found in beer or cider. When the fermented juice of the grape is left at rest, the tartaric acid gradually separates from it, and in combination with potash deposits itself as a crust upon the sides of the cask or bottles (cream of tartar). Hence by long keeping good wines become less acid, and every year added to their age increases in proportion their marketable value. Of the common wines, sherry is the least acid, and the Rhine wines of Germany the most so.—JOHNSON.

The agreeable vinous odor of wine is due to the presence of a fragrant ethereal substance called *œnanthic ether.* This body does not exist in the juice of the grape, but is produced during fermentation, and may be isolated in the form of a fetid, highly fluid compound of carbon, hydrogen, and oxygen. In addition, however, to this substance, all wines contain certain fragrant principles which impart to them a peculiar *bouquet*, or flavor, and render wine so different and so preferable to beer, or any artificial mixture of spirit, sugar and water. They exist in wine in very minute quantities, and their chemical composition is not well understood.*

In addition to the substances mentioned, all wines contain small quantities of other vegetable acids, together with various coloring, oily, and albuminous compounds.

727. **Ardent Spirits.**—When fermented liquors are subjected to a moderate heat, the alcohol which they contain, by reason of its greater volatility, separates from the water, and together with a little steam and some odoriferous substances, rises as vapor. When this operation is conducted in

* Some of these peculiar bouquets are only developed by age, a fact which the wine fancier so well appreciates, that he will give many times the original price for a kept wine, and millions of gallons are retained as stock in Europe because of this property. In addition, wines of peculiar localities contain special bouquets which the art of the chemist entirely fails to account for. Thus the celebrated wine of Johannisberg (the most costly of all wines by reason of its flavor) is only produced upon one estate in Germany. The wines of the neighboring valleys, when subjected to analysis, show the same quantities of acid, sugar, and alcohol, but they do not possess the same *bouquet.*

QUESTIONS.—What are the sweetest wines? What is said of the acidity of fermented liquors in general? What is the acid principle of wine? Why do wines acquire sweetness by age? To what is the vinous odor of wine due? What is said of the bouquet of wines? What are ardent spirits?

close vessels (retorts), and the evolved vapors are collected and condensed by cooling (see Fig. 225), liquors containing a large percentage of alcohol are obtained. To such products of distillation only is the term ardent spirits properly applied.

FIG. 225.

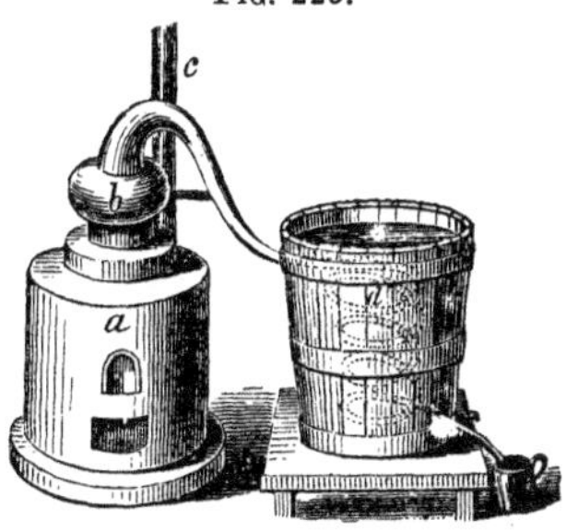

Every different fermented liquor, when distilled, yields an ardent spirit which is characterized by a peculiar flavor, and is distinguished by a name of its own. Thus, *brandy* is the product obtained by the distillation of wine, and *rum* the product of distilling fermented molasses. Whiskey is manufactured from corn, rye, or potatoes in the following manner: the grain or potatoes, boiled or mashed, are mixed with a portion of barley-malt and warm water to form a paste, which is allowed to stand for a time at an elevated temperature. Under these conditions the diastase of the malt converts the starch into sugar, which is then fermented in the usual manner by the addition of yeast. When the fermentation is concluded, the mass is placed in a still, and the spirituous principle distilled over by heat. The condensed product is whiskey, while the residue left in the still, called slops, or swill, is used as food for hogs and cows.* *Gin* is prepared by rectifying (*redistilling*) the spirit obtained from a mixture of fermented rye and barley with juniper berries. By this means it loses the crude flavor it originally had, and acquires the agreeable one of junipers.

The percentage of absolute alcohol contained in ardent spirits intended for consumption (*i. e.*, strong brandy, rum, whiskey, etc.) varies from 50 to 70 per cent. When these are submitted to distillation, a stronger liquor, called *spirits of wine*, is obtained. The product of the redistillation of this last is called *rectified spirits of wine*, or *rectified alcohol*, and contains about 90 per cent. of alcohol and the balance water. It is the strongest alcohol known in commerce. The quantity of water remaining in rectified spirits of wine can not be separated by simple distillation, but is accomplished by mixing the spirits of wine with chloride of calcium, or some other substance which has so strong an affinity for water that it absorbs it, and allows the alcohol to distil over pure. In this condition the alcohol is termed *absolute*, or *anhydrous*. Proof spirit is a mixture of equal parts of water and alcohol.

*** The milk yielded by cows fed on this refuse is considered unhealthy, and is popularly called "swill milk."**

QUESTIONS.—Is the distillate of all fermented liquors the same? What is brandy? What is rum? How and from what is whiskey manufactured? What is gin? What is the percentage of alcohol in ardent spirits? What are spirits of wine? What is rectified alcohol? What is pure alcohol called? How is it prepared? What is proof spirit?

It was formerly the custom to estimate the strength of an alcoholic liquor by igniting a little of it in connection with gunpowder; if the powder was fired, the spirit was considered strong, and called *proof;* if, on the contrary, it contained more than half water, the powder was not ignited, and the spirit was said to be below proof. The quantity of alcohol contained in a solution is now, however, calculated by determining its specific gravity (§ 40), or more conveniently by means of the *alcoholometer* (see Fig. 226), which is so weighted and graduated that it sinks to the topmost point of the scale A, which is marked 100°, in absolute alcohol, and to the lowest degree in pure water, which is marked 1°—intermediate positions indicating proportional mixtures of the two liquids.

Fig. 226.

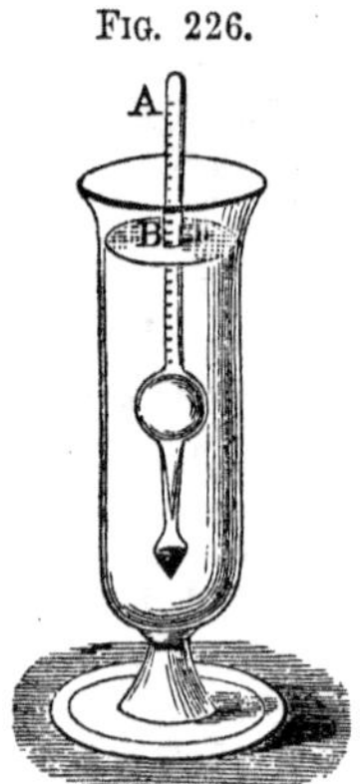

728. **Properties of Alcohol.**—Pure, or strong alcohol, is a highly volatile, mobile liquid, about one fifth lighter than water (sp. g. 0·795) possessing an agreeable, penetrating odor, and a hot, burning taste. It is very combustible, and burns with a pale blue flame without smoke, but with intense heat. It has a strong affinity for water, and absorbs or extracts it from substances with which it is brought in contact. On this account, taken in connection with its property of coagulating or hardening albumen, it acts as a powerful antiseptic, and is much used to preserve organic substances from putrefaction. Strong alcohol has never been frozen.* When taken into the stomach it acts as a deadly poison, but when largely diluted with water it is, as is well known, stimulating and intoxicating. The solvent powers of alcohol are very great; it dissolves a great number of organic substances which are insoluble in water, such as the volatile oils and the resins, together with many acids, salts, the caustic alkalies, and other substances. Alcoholic extracts of medicinal plants, roots, barks, etc., constitute the *tinctures* of pharmacy, and most of the liquid perfumes (*eau de Cologne*, etc.) are solutions of fragrant and volatile oils in alcohol. Many varnishes, also, are formed by dissolving resins in alcohol.

729. **Bread.**—The preparation of bread is properly considered in connection with the subject of vinous fermentation:—

The flour of wheat and other grains which enter into the composition of bread, consists mainly of starch, gluten, and water, together with small pro-

* M. Despretz of Paris, in 1849, succeeded, by the rapid evaporation of liquid protoxyd of nitrogen and solidified carbonic acid, in producing a degree of cold sufficient to deprive alcohol in part of its transparency, and render it thick and viscid.

Questions.—How is the quantity of alcohol in a liquor determined? What are the properties of alcohol? What is said of its solvent powers? What are tinctures? How are liquid perfumes generally prepared? What is the composition of flour?

portions of sugar and gum.* The first step in the process of bread-making, is to mix together, in a suitable vessel, a proper proportion of flour, yeast, warm water, and common salt. This mixture, which is called the *sponge*, is worked up to the consistence of stiff batter, and then left for a few hours in a warm atmosphere, during which time the yeast excites fermentation in the sugar, and occasions its conversion into alcohol and carbonic acid. The gas thus generated does not escape in bubbles, but is retained by the tenacious and viscid dough, which, in consequence, becomes light and porous, and swells up to about twice its original size.

When the fermentation has proceeded sufficiently far, about twice as much flour as was originally taken is added to the sponge, and the two are carefully kneaded together. This is a very laborious part of the operation, but is quite essential to the success of the process, since, if it is not very thoroughly attended to, the half-fermented sponge will not be equally and uniformly distributed throughout the whole of the dough.

If the dough be now put into a hot oven, the fermentation is at first increased and the size and porosity of the loaf are also greatly augmented by the expansion of the carbonic acid gas contained in its cellular spaces. When, however, the whole has been heated to nearly the temperature of boiling water, the fermentation is suddenly arrested; and the alcohol and a large proportion of the water employed in mixing the dough, being at the same time volatilized by the heat, the cellular portions of the baked bread ac-

* Flint-wheat contains, on an average, about 56 parts of starch, 14 of gluten, 8 of sugar, 5 of gum, 2 of bran, and from 10 to 13 parts of water. The manner in which the bran, the gluten, and the starch are respectively distributed throughout the cereal grains, is shown by the following section of a fully-ripe grain of rye, highly magnified. (See Fig. 227.) *a* represents the outer-seed coat, consisting of three rows of thick-walled cells; *b*, the inner seed coat, composed of a single layer of thick-walled cells, having scarcely any cavity; *c*, a layer of cells containing gluten. These three together form the bran. *d* represents the cells containing starch grains in the interior of the seed. The outer coating of the seed contains only 3 or 4 per cent. of gluten, while the inner coating contains from 14 to 20 per cent. All this is separated in the bran. In addition to this, however, gluten is diffused everywhere throughout the mass of grain, among the cells containing starch. As the nutritive quality of any variety of grain depends very much upon the proportion of gluten which it contains, and as the bran embodies a larger proportion of this substance than the white part of the flour, it is obvious, that by sifting out the bran, as is usually done, we render the flour less nutritious. The bran generally constitutes about one fourth part of the whole weight of the grain. When wheat is burned, there is left about 2 per cent. ash, nearly one half of which consists of phosphoric acid: the other constituents being mainly potash, silica, magnesia, soda, oxyd of iron, etc. These mineral ingredients are unequally diffused throughout the seed; fine flour containing the smallest proportion, and the bran the most.

FIG. 227.

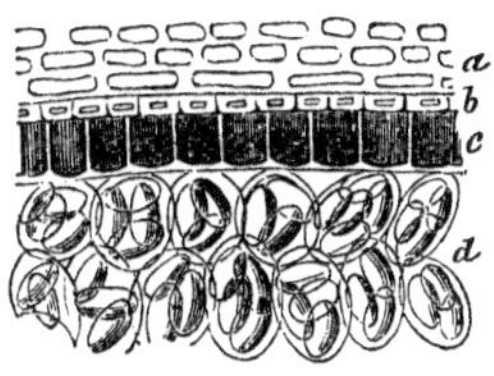

QUESTIONS.—What is the first step in the process of bread-making? What is the necessity of producing fermentation in dough of bread? What is the second stage of the process? What occurs in the baking?

quire so much solidity, that they retain their form and structure permanently. If, however, the heat of the oven is not properly regulated, or if the dough contains too much water, the cellular portions harden too slowly, and on the escape of the carbonic acid, collapse and run together (*slack-baking*). The alcohol which escapes from the bread in baking may, by means of a proper apparatus, be collected and condensed into spirits, and this, in fact, is done in some of the European bakeries.

The yeast, in converting the sugar of the flour into alcohol and carbonic acid, acts also upon the starch, in the manner of diastase, and transforms a portion of it into sugar; so that, although the sugar, which originally existed in the flour, is almost completely decomposed, the amount present in the bread remains very nearly constant. "It is sometimes stated, that, by the ordinary mode of bread-making, a large portion of the most valuable part of flour is destroyed by fermentation. This, however, is not the case. Very little of the azotized matter of the flour is lost during the fermentation of the dough: the chief effect produced is a loss of a portion of the sugar; but as nearly an equal quantity is formed from the starch, the real effect of the fermentation may be said to be principally the loss of about 5 per cent. of starch." —SOLLY.

The addition of common salt to bread renders it more wholesome and digestible, and also assists in its preservation.

The quantity of water in well-baked wheaten bread amounts to about 45 per cent., or, in other words, the bread we eat is about one half water. Bread that has been kept for a few days, loses the characteristic softness which distinguishes it when fresh-baken, and becomes "crumbly," and apparently drier. In this condition it is known as *stale bread.* The change, however, is not due to any lo s of water, but to a change in the internal arrangement of the molecules of the bread.

The solubility of bread, and its consequent ready digestibility, is somewhat increased by toasting, the starch being thereby converted into a modified gum (§ 689).

730. As the process of fermenting bread, in order to render it light and porous, is troublesome, and somewhat uncertain, various attempts have been made to effect the same object by other agencies. The best of the substistituted methods is undoubtedly that in which bi-carbonate of soda and hydrochloric (muriatic) acid are employed. A small, but definite quantity of carbonate of soda is first thoroughly mixed with the flour, and enough pure acid to perfectly neutralize it is then added to the proper quantity of water. The flour and the acid water being then thoroughly incorporated, the acid acts upon the carbonate of soda, decomposes it, expels its carbonic acid, and

---

QUESTIONS.—When is bread said to be slacked baked? Can the alcohol evolved in bread-baking be collected? What action does the yeast exert on the starch of the flour? What is the general effect of fermentation on the constituents of the bread? What effect has common salt on bread? What percentage of water does bread contain? What is stale bread? What occasions this change? What effect does toasting have upon bread? Is there any way of rendering bread light and porous without fermentation?

unites with the soda to form common salt. The result is the production of a light, spongy dough, as in ordinary fermentation, while the salt formed and remaining in the dough, renders the addition of this substance, in the first instance, unnecessary. The most serious objection to this plan is the difficulty of procuring pure hydrochloric acid, and of regulating the proportions of acid and soda. Tartaric acid may be substituted in the place of hydrochloric acid, and the so-called yeast powders are generally prepared by mixing bi-carbonate of soda and tartaric acid in proper proportions. The carbonate of ammonia is also not unfrequently used (§ 525).

731. **Sources of Alcohol.**—Alcohol is not a principle existing in nature, elaborated and stored up by the plants; but is always a product of the destructive decomposition of highly-organized matter. The principal sources from which crude alcohol is obtained, are the most valuable of our cereal grains, immense quantities of which are annually used for this purpose, and of course to the same extent the aggregate supply of food for man is diminished. The waste of raw material which accompanies the manufacture of alcohol from grain is also very great, since the nitrogenized elements of the grain do not enter into its composition, and are accordingly lost for any useful purpose; while the starchy and saccharine constituents are converted to the extent of half their weight into valueless carbonic acid and water. Woody fiber, it will be remembered, has identically the same composition as starch, and, like it, may be converted by the action of acids into grape sugar, which is capable of furnishing alcohol. This process, however, by reason of its expense, is not practically useful; but its consideration has much of interest, since the discovery of a cheap and ready method of converting woody-fiber, and bodies of like composition and character, into glucose, to be used in the manufacture of alcohol, would prove one of the most valuable discoveries in the annals of science.

732. **Products of the Action of Acids upon Alcohol. Ether.**—When equal weights of strong alcohol and oil of vitriol are heated to ebullition in a retort, a colorless, highly volatile liquid distils over, which is known as *ether*, or *sulphuric ether*. As soon as the contents of the retort blacken and froth, the process must be discontinued, or otherwise the distillate will be contaminated by other substances.

The formation of this liquid may be explained as follows: alcohol is assumed to be the hydrated oxyd of an organic radical ethyle, its composition being represented by the formula $C_4H_6O_2 = C_4H^5O + HO$. When sulphuric acid is added to alcohol and heated, it unites with the oxyd of ethyle to form a bi-sulphate ($C_4H_5O,2SO_3$), and from this compound at a higher temperature the oxyd of ethyle (ether) separates and distils over as a vapor. The alcohol, therefore, is converted into ether by the simple loss of an atom of water. The prefix sulphuric, as applied to ether, is merely intended to indicate its

QUESTIONS.—What are yeast powders? What is said of the sources from which alcohol is manufactured, and of its production? How is ether prepared? What is the theory of its production? Does sulphuric ether contain any sulphuric acid?

origin and distinguish it from other bodies of like character, since it contains no sulphuric acid in its composition.

Ether is a colorless, transparent, fragrant liquid, exceedingly thin and mobile. It boils at 96° F. (or when exposed to the sun in summer), and may be frozen by exposure to severe cold. In the open air it evaporates with great rapidity, and occasions thereby a degree of cold sufficient even to freeze water (§ 164). This property may be illustrated by allowing a few drops of it to evaporate upon the hand. It is highly combustible, both in the state of liquid and vapor, and on this account should never be brought near a flame. With atmospheric air, or oxygen, its vapor forms explosive mixtures. This may be experimentally shown by pouring a few drops into a tumbler, and after a little time applying a burning taper. Ether mixes with alcohol in all proportions, but is very sparingly soluble in water. It dissolves most oily and fatty substances with great readiness, but its solvent powers generally are far more limited than those of alcohol.

When the vapor of ether, mixed with atmospheric air, is inhaled, it produces at first a species of intoxication, which is speedily succeeded by a kind of stupor, during which the system is nearly insensible to pain. This important property is not, however, confined to ether alone, but is possessed by nearly all the gaseous hydrocarbons, and by some in a much greater degree. Ether, however, was the first substance employed as an anæsthetic agent, and under all circumstances must be regarded as the safest, no accidents from its moderate inhalation having ever been recorded.

733. **Varieties of Ether.**—By distilling alcohol with various acids, different combinations of the radical ethyle may be produced, which are generally spoken of as kinds of ethers. Thus, by distilling a mixture of alcohol, sulphuric and acetic acids, we obtain an exceedingly fragrant, volatile liquid, acetate of the oxyd of ethyle, or acetic ether. The fragrant odor of this body may be evolved by slightly heating in a test tube a mixture of the above-named substances. In like manner, with the aid of nitric acid we may obtain a *nitrous ether*, which is much used in medicine under the name of *sweet spirits of niter;* and with butyric acid, a *butyric ether*, which has the odor of rum, and is now prepared for the purpose of imparting to alcohol this flavor in the fabrication of liquors.

734. **Products of the Oxydation of Alcohol.**—When alcohol or ether are burned in free air, the products of combustion, as with all similar hydrocarbons, are carbonic acid and water. Under certain circumstances, however, these substances undergo a partial oxydation, in which the hydrogen alone is oxydated or separated, leaving the carbon unaffected. The result is the formation of a series of compounds which are supposed to contain a new organic radical called *acetyle*, C_4H_3, derived from ethyle, C_4H_5, by the removal of 2 equivalents of hydrogen by oxydation.

QUESTIONS.—What are the properties of ether? What is said of its solvent powers? What of its anæsthetic properties? Is this property confined to ether? How may different varieties of ether be prepared? Illustrate this by examples. What is the product of the ordinary combustion of alcohol? What is acetyle? How is it formed?

735. **Aldehyde.**—The first known product of this series is a hydrate of the oxyd of acetyle, C_4H_3O+HO, called *aldehyde* (from *al*, alcohol, *de*, from which, *hyd*, hydrogen, is taken). It is a limpid, colorless liquid, possessing a peculiarly suffocating odor, and may be prepared by distilling a mixture of alcohol, oil of vitriol, and the peroxyd of manganese. It may also be easily produced, and its characteristic odor illustrated, by plunging a coil of fine platinum wire heated to redness into a vessel containing a mixture of alcohol or ether vapor, and atmospheric air. (See Fig. 228, also § 469.) The aldehyde is formed in this experiment because the oxydation is not sufficient to occasion a complete combustion of the alcohol vapor. Aldehyde dissolves sulphur, phosphorus, and iodine, and is especially remarkable for its affinity for oxygen, in consequence of which it is capable of reducing many of the metallic salts. The addition of a little aldehyde in water to an ammoniacal solution of nitrate of silver, occasions the immediate precipitation of the silver as a brilliant white metal.

FIG. 228.

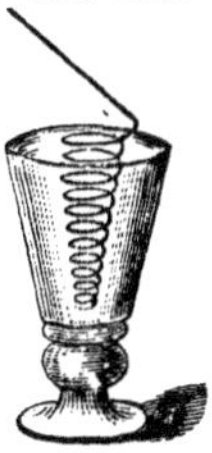

736. **Acetic Acid** is well known as the acid of vinegar, which latter substance is, in fact, a very dilute acetic acid, containing also much saccharine and mucilaginous matter. Acetic acid is regarded as a hydrated teroxyd of the same radical, *acetyle*, which enters into the composition of aldehyde—its composition being represented by the formula $C_4H_3O_3+HO$.

Alcohol, when pure, or merely mixed with water, undergoes no change when exposed to the air; but the presence or contact of various foreign substances, dispose it to absorb oxygen. Thus, if a few drops of strong spirits of wine be let fall upon a little platinum black, the oxygen condensed in the pores of the latter unites so rapidly with the alcohol, as to occasion its instant inflammation. Under the same circumstances, when the spirit is mixed with a little water, oxydation still takes place, but with less energy, and the alcohol is converted into acetic acid. In these transformations the platinum itself experiences no change. The oxydation of alcohol, through the agency of platinum black, may be experimentally exhibited, also, by placing a capsule containing platinum black upon a plate by the side of a small vessel of alcohol, and exposing the whole, covered with a bell-glass, to the sunshine. In a short time, the vapor of acetic acid will be observed to condense on the sides of the jar, and run down in drops; and by occasionally admitting fresh air, the whole of the alcohol may in a few hours be acidified.

The oxydation of alcohol, at the expense of the oxygen of the air, is also effected by the presence of almost any azotized matter (ferment) susceptible of putrefaction. Cider, wine, and beer naturally contain such substances, and

---

QUESTIONS.—What is aldehyde? What are its properties? How is it formed? What is the acid contained in vinegar? What is its chemical composition? Under what circumstances is alcohol oxydated? How may the transformation of alcohol into acetic acid be illustrated? What is the action of ferments on alcohol? Why do cider, beer, etc., turn sour by exposure to the air?

therefore readily undergo acetous fermentation when exposed to the air, at a moderate temperature, and become converted into vinegar (acetic acid). During this fermentation of alcoholic liquors, a mucilaginous substance, consisting chiefly of albuminous matter, is separated, which, from its influence in exciting or promoting acetous fermentation, is popularly termed the *mother of vinegar*. Acidification of this character occurs most readily immediately after a spirituous fermentation which has taken place at too high a temperature; hence brewers, during the summer months, experience much trouble in preventing their fermenting wort and mash from turning sour.

737. Vinegar is now manufactured, on a large scale, directly from alcohol, by diluting it with water, adding a little yeast, and exposing the mixture to the air. This last is best effected by causing the liquor to trickle slowly through a cask filled with shavings of beech-wood, and arranged as is represented in Fig. 229. The head of the cask, *d*, is closed with a shelf, *c*, perforated with many small holes, through which threads are passed to conduct the liquor downward, and distribute it evenly over the interior. The shavings, first soaked in vinegar, are placed loosely in the cask, a free circulation of air between them being provided for by means of holes, *a*, in the sides of the cask. In this way the alcoholic liquor is caused to present an immense extended surface to the action of the air, and oxydation takes place so rapidly, that very frequently, by the time the liquid has trickled to the bottom of the cask, it no longer contains any alcohol, but is entirely converted into vinegar. Usually, however, it is necessary to pass the liquid through the apparatus a number of times before this change can be completely effected. The presence of acetic acid itself assists the action of acidification, and it is for this reason that the shavings are soaked in vinegar before using. This process is known as the quick method of making vinegar.

FIG. 229.

The pyroligneous acid (or wood vinegar), obtained by the distillation of hard-wood in close vessels (§ 680), is an impure acetic acid, and as such is largely used in dyeing and calico-printing; the presence, however, of certain empyreumatic substances extracted from the wood, impart to it a disagreeable, smoky odor, and render it unfit for purposes of domestic economy.

The acid liquids obtained by the above-mentioned processes, are not pure acetic acid, but merely solutions of it in water. This may be concentrated, but

QUESTIONS.—What is the mother of vinegar? Under what circumstances does acidification occur most readily? Describe the quick process of making vinegar? What other product is a source of acetic acid? Is the acid liquid obtained by the oxydation of alcohol pure acetic acid?

if we attempt to obtain the acid free from any water by distillation, it is decomposed. Acetic acid in a separate state is prepared by neutralizing vinegar with soda or lime, evaporating to dryness, and distilling the solid residue in connection with sulphuric acid. The evolved vapors condensed, furnish a colorless, intensely-sour liquid, which possesses a pungent, fragrant odor, and blisters the skin. It mixes with water in all proportions, forming vinegars of different degrees of strength. Common table vinegar usually contains from 3 to 5 per cent. of acetic acid. The salts of vinegar, sold by druggists as a reviving scent in sickness and fainting, consist of sulphate of potash, impregnated with acetic acid. Acetic acid dissolves many organic bodies, such as gluten, gelatine, gum, resins, the white of eggs, etc.; hence, its use as vinegar, in moderation, promotes digestion. When vinegar is exposed to cold, the water contained in it is frozen before the acetic acid is; hence, weak vinegar is made stronger by partial freezing.

738. **Salts of Acetic Acid.**—Acetic acid unites with most bases to form an important class of salts called *acetates*, all of which are soluble in water. *Acetate of lead*, PbO, $C_4H_3O_3$, the *sugar of lead* of commerce, is a white salt formed by dissolving oxyd of lead (litharge) in acetic acid. It possesses a very sweet astringent taste, and is often employed in medicine, but when taken internally in any other than minute quantities is a poison. *Acetate of copper* constitutes verdigris (§ 613). Acetates of alumina and of iron are salts much used in dyeing and calico printing.

739. **Methylic Alcohol.**—In connection with the pyroligneous acid obtained by the distillation of wood in close vessels, there also passes over a volatile inflammable liquid, which is allied to alcohol in its composition and properties. This substance in its pure state is known as methylic alcohol, or wood-spirit, and is supposed to be the hydrated oxyd of a radical called *methyle*, the constitution of which is represented by the formula C_2H_3, and that of its alcohol by C_2H_3O+HO. Crude pyroligneous acid contains about 1-100th of its weight of this substance, which is separated from it by distillation. It occurs as an article of commerce, and is often substituted for alcohol in various processes in the arts. Its odor, however, is quite different from that of ordinary alcohol.

740. **Formic Acid.**—As alcohol by oxydation, under the influence of finely divided platinum, gives rise to acetic acid, so wood-spirit, under similar circumstances, produces an acid product which has been called *formic*, from the circumstance, that a similar acid may be extracted from ants by distilling them with water. As acetic acid is regarded as the hydrated teroxyd of the radical acetyle, so formic acid is considered as the hydrated teroxyd of a new radical *formyle*, which is derived from methyle as acetyle is from ethyle—the formula of formyle being C_2H, and that of formic acid,

---

QUESTIONS.—How is acetic acid prepared? What are the properties of acetic acid? What are salts of vinegar? What is said of vinegar? What are acetates? What is sugar of lead? What are other important acetates? What is said of methylic alcohol? is its chemical constitution? What is formic acid? What is its composition?

C_2H,O_3+HO. Formic acid also unites with bases to form salts, which closely resembles the acetates.

741. **Chloroform,** $C_2H\,Cl_3$.—This substance, which is regarded as a terchloride of the radical formyle, is best obtained by distilling alcohol, or wood-spirit, with a solution of chloride of lime (bleaching powder). It is an oily, colorless liquid, of an agreeable, ethereal odor, and of a sweetish taste. An alcoholic solution of chloroform, prepared by distilling chloride of lime with an excess of alcohol, is known in medicine by the incorrect name of *chloric ether*, and is the liquid which is now generally sold and used under the name of chloroform. The vapor of chloroform, when inhaled with atmospheric air, produces anæsthetic effects, like the vapor of ether. It is, however, much more potent and agreeable than ether, and has to a considerable extent replaced the latter agent in surgical practice. Chloroform, unless prepared from perfectly pure alcohol, is liable to contain certain foreign and volatile compounds, which exert a most deleterious and often fatal effect upon the system when inhaled. No person, therefore, should sell or use chloroform which is not known to have been properly prepared. Chloroform is with difficulty kindled, and burns with a greenish flame.*

742. **Amylic Alcohol.**—In the process of distilling whiskey from potatoes, there is generated, in connection with the crude spirit, a volatile, oily body, possessing a pungent, disagreeable odor. This substance, the complete separation of which from the crude spirit is a matter of difficulty, appears to be analogous, in its composition and chemical reactions, to alcohol, and is regarded as the hydrated oxyd of a radical, termed *amyle*,—the hydrated oxyd itself being called *amylic* alcohol, or more generally, *fusel oil* (from the German *fuseloel*), or oil of potato spirits.* Amylic alcohol, in a pure state, has the appearance of a thin, colorless oil; it is highly volatile; and the inhalation of its vapor, in even a minute quantity, is attended with very deleterious effects; the fatal accidents which have sometimes resulted from the use of chloroform being generally ascribed to its admixture with this compound. It exists in almost all ordinary alcohol in small quantity, and is the occasion of the persistent and somewhat faintly-disagreeable odor which alcohol leaves upon a surface after evaporation.

The extraordinary character of the compounds and derivatives of amylic alcohol (fusel oil), render it one of the most interesting products of organic

---

* A most efficient and economical apparatus for disinfecting apartments and purifying the air, may be arranged by burning chloric ether in a simple camphene lamp provided with one wick. In dissecting-rooms, in cellars where vegetables have decayed, or where drains allow the escape of offensive gases, and in outbuildings, no more effective and agreeable method of purifying the air can be resorted to.

* Amyle derives its etymology from the Latin *amylum*, starch, the substance being a product of the fermentation of starch.

QUESTIONS.—What is chloroform? How is it prepared? What are its properties? What is the so-called chloric ether? When is chloroform liable to be injurious? What is amylic alcohol? What other name is applied to it? What are its properties? What is a characteristic feature of this substance?

chemistry—most of the substances into which its constituents enter as components being characterized by very singular and remarkable odors. For example, when amylic alcohol is warmed, and dropped upon platinum black, it oxydizes and forms an acid, which bears the same relation to its alcohol that acetic acid does to ordinary alcohol. This compound possesses in an intense degree the odor of *valerian*, and is believed, furthermore, to be identical with the acid which imparts to the root of the plant valerian its odor and medicinal properties: it has hence been called *valerianic acid*, and has been advantageously employed in medicine in place of the natural extract.

By distilling amylic alcohol, under proper circumstances, with various acids, we obtain odoriferous compounds, which, during the last few years, have become familiarly known as "fruit extracts," or "essences," and as "liquor flavoring materials." Thus amylic alcohol, distilled with sulphuric acid and acetic acid (acetate of potash), yields an oily product which possesses most perfectly the odor of the "Jargonelle" pear; chromic acid, substituted in the place of acetic acid, gives oil of apples; while other acids yield products possessing the flavors of the banana, the orange, and many other fruits. In the same manner, the flavoring principles which characterize spirituous liquors may be obtained, and indeed are now manufactured and sold extensively under the names of "*oil of cognac*," "*oil of wine*," etc., for the fabrication of almost any kind of liquor or wine, from pure alcohol.* Although prepared from a poisonous basis (fusel oil), these extracts do not appear to possess any injurious qualities, when used in moderate quantities as flavoring agents; and the position has even been taken by some chemists that they are identical in composition with the perfumes which nature carefully elaborates in different fruits and plants. In addition to perfumes the most agreeable, however, odors of the most disgusting and nauseous character can also be produced by like means, as, for instance, the odor of the bed-bug, squash-bug, and of many disagreeable plants and weeds. The basic radical employed for this purpose is not, however, in all instances amyle, as the same properties are characteristic to some extent of a number of analogous radicals.

743. **Sulphur Alcohols, or Mercaptans.**—By various indirect processes, the oxygen of wine, methylic and amylic alcohol, may be replaced by sulphur, their other constituents remaining unaltered, and in this way a series of bodies may be produced, which from their resemblance in

* A few drops of oil of cognac, added to a glass of water colored with burnt sugar (caramel), will convert it, so far as appearance and odor is concerned, into a fair article of dark brandy. Manufacturers, in fabricating spirituous liquors from alcohol, by the aid of these flavoring extracts, find it necessary to use an article of spirits from which every trace of fusel oil has been previously extracted, as this substance, in a separate state, seems to destroy flavoring extracts which contain its elements as constituents. This separation of fusel oil from alcohol is now accomplished by distilling the crude spirit in connection with permanganate of potash.

QUESTIONS.—What is valerianic acid? What are other derivatives of this body? What are the so-called sulphur alcohols, or mercaptans?

composition to alcohol, have been called sulphur alcohols, or by reason of their great affinity for mercury, *mercaptans* (*mercurium captans*). Thus the composition of wine alcohol being $C_4H_6O_2$, its mercaptan would be $C_4H_6S_2$. These products in their properties closely resemble the oily compounds which impart to garlic, the onion, and other plants of like character their offensive odors, and in fact may be considered as artificial oils of garlic. The mercaptan produced from methylic alcohol is a colorless liquid, with a most offensive and concentrated odor of onions, which penetrates and obstinately adheres to every substance with which it is brought in contact.

744. If we replace the sulphur existing in these fetid compounds with arsenic, we produce new volatile substances which are not only insufferable in their smell, but rank among the most deadly poisons known to chemists.

Such a compound is *kakodyle* (from κακος, evil, and ὑλη, principle), formed by the union of arsenic with the radical methyle, and which, from the circumstance that it fulfils in composition the part of an element in a very remarkable manner, has been studied by chemists with great minuteness.* United with cyanogen, it forms cyanide of kakodyle, a compound possessed of most deadly qualities. "When exposed to the air it rises in the form of vapor, which by contact with moisture is instantly decomposed, its arsenic uniting with oxygen to form fumes of arsenious acid, while the cyanogen by combination with hydrogen forms prussic acid; and thus at the same instant the air is impregnated with vapors of the two most deadly poisons with which we are acquainted." The evaporation of a few grains of cyanide of kakodyle in the atmosphere of a room, is said to produce almost instantaneous unconsciousness. In addition to these substances, many other compounds of a somewhat similar character have been formed and described, and it has sometimes been proposed to employ them as ingredients in explosive war projectiles (asphyxiating bombs).

CHAPTER XX.

VEGETABLE ACIDS.

745. OVER two hundred distinct acid compounds, the products of the vegetable kingdom, have been isolated and described by chemists. They are all composed of carbon, hydrogen, and oxygen, with the latter element generally in large excess. They are not, however, usually found in plants in a

* A recent chemical authority has described the odor of this compound as far exceeding in offensiveness the fetor exhaled by any animal or vegetable.

QUESTIONS.—What are their properties? By replacing sulphur with arsenic, what compounds are formed? What is kakodyle? What are its properties? What is said of the number and distribution of the vegetable acids?

free state, but in combination with various bases derived from the soil, such as potash, soda, lime, etc. The salts thus formed are sometimes neutral, but more frequently acid in their character, and consequently impart to the portions of the plant containing them a distinctly acid taste and reaction. When the salt is sparingly soluble, it often accumulates in the cells of the plant in the form of minute crystals, which are readily discernible by the microscope. Fig. 230 represents crystals of this character found in the onion, and Fig. 231 crystals of oxalate of potash occurring in the rhubarb.

Fig. 230.

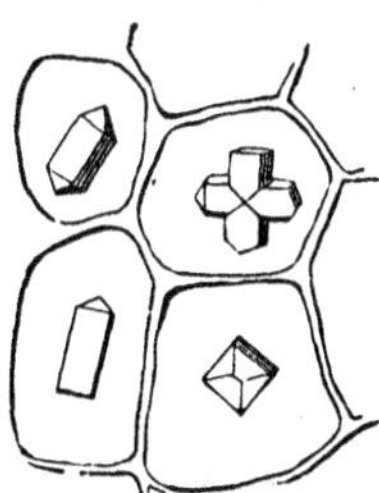

Fig. 231.

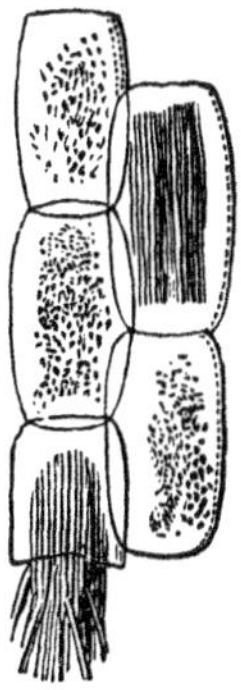

Some of these acids are very widely diffused in the vegetable kingdom, but the majority occur in only a few particular plants, and in minute quantities. The most important of them only require special consideration.

746. **Oxalic Acid, C_2O_3HO.**—This acid is found abundantly in many plants in combination with potash and lime, and is the principle of acidity in the leaves of the sorrel and the rhubarb (*pie-plant*). It is also a constituent in certain minerals. For practical purposes it is prepared artificially by digesting sugar with strong nitric acid; thus, when these two substances are gently heated in connection, in the proportion of 1 part of sugar to 8 of acid, violent action ensues, accompanied with a disengagement of copious fumes of nitrous acid; and the solution remaining after the cessation of the action, furnishes, by evaporation, crystallized oxalic acid. Starch, woody fiber, and many other organic substances, treated in the same manner, yield the same product.

In its pure state, oxalic acid is a crystalline solid, not unlike Epsom salts, for which it is not unfrequently mistaken. It possesses, however, an intensely sour taste (which Epsom salts does not), is freely soluble in water, and when taken internally, is highly poisonous, occasioning death in a few hours. The proper antidote for it is the administration of chalk or magnesia, suspended in water.

Oxalic acid is extensively employed in calico-printing, and to some extent by straw and Leghorn bonnet-makers, for the purpose of cleansing their wares. It is also used in chemical analysis as an exceedingly delicate test for the presence of lime, or any of its soluble compounds. The salts formed by oxalic acid are termed *oxalates*. Binoxalate of potash, which is often extracted from certain species of sorrel, is sold under the name of "salts of sorrel," or "essential salts of lemons," for the purpose of discharging iron-rust, or ink-stains from linen. Its use for this purpose depends upon the fact, that

---

QUESTIONS.—What is said of the occurrence of oxalic acid? How is it obtained for industrial purposes? What are its properties? What its uses? What are its salts called? What are salts of sorrel or of lemons? How are they operative in removing ink-stains?

oxyd of iron (the basic coloring matter of ink) is readily soluble in oxalic acid, and therefore leaves the fiber and forms an oxalate of iron. The corrosive powers of the acid are not sufficient to injure the fibers of the linen, if it be speedily removed by washing.

747. **Tartaric Acid**, $C_8H_4O_{10}, 2HO$, in combination with potash, exists in many fruits, and is especially the acid principle of grapes. When the expressed juice of the grape is fermented, as in the manufacture of wine, the tartaric acid, in combination with potash, forming an impure tartrate of potash, gradually separates from the liquor, and deposites itself as a crust upon the interior of the casks, and in this condition is known in commerce as *argals*, or *crude tartar*. The pure acid obtained from this source is a white, crystallized solid, freely soluble in water, and of an agreeable, acid taste.

Tartaric acid forms with potash two salts,—the neutral tartrate, containing 2 atoms of alkali to 1 of acid, $2KO, C_8H_4O_{10}$; and the acid, or bi-tartrate, in which an atom of potash is replaced by an atom of water, thus, $KO, HO, C_8H_4O_{10}$. This latter salt is the well-known "cream of tartar." By saturating a solution of cream of tartar with soda, a double tartrate of potash and soda is formed, which is extensively used in medicine as a mild purgative, under the name of "Rochelle salts," or "powders." Tartaric acid, mechanically mixed with bi-carbonate of soda, constitutes the so-called "soda powders," or the ingredients of the ordinary effervescing draughts. Tartaric acid is chiefly employed in dyeing.

748. **Citric Acid** is the principal acid which imparts sourness to the lemon, orange, and the cranberry; but also exists in many other fruits, as the currant, gooseberry, etc. It is readily obtained from the juice of the lime and lemon (citron), and is used in calico-printing, in medicine, and in domestic economy, as a flavoring material. Citric acid, by heating, passes into aconitic acid, an acid which occurs native in the plant called "monk's hood."

749. **Malic Acid** was first obtained from the juice of the apple (hence its name from the Latin *malum*, an apple). It is the most widely diffused of all the vegetable acids, and is the cause of acidity in most unripe fruits. For practical purposes it is usually obtained from the berries of the mountain-ash, though it exists abundantly in the stalks of the rhubarb, in the pear, the cherry, the raspberry, the strawberry, and many similar fruits.

750. **Tannic Acid**, or *Tannin*, is the general name given to various substances (probably of somewhat different composition), which are extensively diffused in plants, and which are characterised by a well-known puckering and astringent taste. They are regarded as acids, since they possess an acid taste, and are capable of uniting with bases to form salts. Tannic acid exists in almost all vegetables, in the bark and leaves of trees, and in the seeds of fruits. It is, however, most abundant in the bark of the oak and the

---

Questions.—What is said of tartaric acid? What are argals? What is cream of tartar? What are Rochelle powders? What are soda powders? What is said of citric acid? What of malic acid? What is tannin or tannic acid? In what substances is tannin most abundant?

hemlock, in the fruit and stems of the sumach, and especially in nut-galls, which are excrescences produced upon the branches and leaves of certain species of oak, by the puncture of insects. Green and black teas contain from 8 to 10 per cent. of tannin, which imparts to them their strong, astringent qualities. Tannic acid is freely soluble in water, and is readily obtained in solution, by digesting the portions of the plant containing it in water.

751. **Leather.**—The most remarkable feature of tannic acid, is its property of uniting and forming insoluble compounds with albumen, gluten, gelatin, and with the skin and tissues of animals in general. Such compounds will not putrefy, and are unchangeable in water. This principle is practically applied in the manufacture of leather, which is effected by steeping the skins of animals, which consist chiefly of gelatin, in aqueous infusions of barks containing a large percentage of tannic acid.* Some varieties of skins may be tanned in a few days, or even hours; but for the production of the best qualities of leather, they are allowed to remain in contact with the tan liquor from 9 to 15 months, and often for a period of years.

752. **Inks.**—When a solution of tannin is brought in contact with salts of the sesquioxyd of iron, it yields a deep bluish-black precipitate—the pertannate of iron—which is extensively employed for dyeing fabrics of a black or brown color, and in the manufacture of inks. Common writing-ink is formed by adding to a clear infusion of nut-galls a solution of protosulphate of iron (copperas). To prevent the precipitate from settling, and for thickening the fluid, a mucilage of gum-arabic is also added. Ink thus prepared consists at first principally of the tannate of the protoxyd of iron, and is too pale for use; by exposure to the air, however, it gradually absorbs oxygen, and is converted into the tannate of the sesquioxyd—the liquid, at the same time, deepening in color, and finally becoming black. Mouldiness in ink may be prevented by the addition of a small quantity of the oil of cloves, creosote, or corrosive sublimate: the latter, in small amount, is probably more efficient than all the others; but it should be remembered that it is a deadly poison. Faded writings can be restored in a measure by washing them with an infusion of galls.†

* Oak bark contains from 5 to 6 per cent. of tannin; and in this, as well as in all other astringent barks, the tannin is contained solely in the inner, white layers, next to the sap-wood, or alburnum. From 4 to 6 pounds of oak-bark are required for the production of 1 pound of leather. Leather tanned with oak-bark is considered superior to that made from any other tanning material, but the process is slower. Nut-galls contain more tannic acid than any other substance, the quantity varying from 30 to 40 per cent. Sumach is used in the manufacture of the lighter and finer kinds of leather. Sicilian sumach contains about 16 per cent. of tannin, and that grown in the United States from 5 to 10 per cent.

† The cause of the browning and fading of ordinary inks, results chiefly from a per-oxygenation of the iron, and its separation from the acid combined with it. No salt of iron, and no preparation of iron, equals the common sulphate (that is, commercial copperas) for ink-making, and the addition of any persalt, such as the nitrate or chloride of

QUESTIONS.—What is its most remarkable property? How is leather prepared? What is the reaction of tannin with sesquioxyd of iron? How is ink prepared? Why does ink grow dark by exposure to the air?

The permanent black color of the grain side of the leather used in the manufacture of boots and shoes is also a tannate of iron, produced by washing the leather when in a moist state with a solution of the acetate of the sesqui-oxyd of iron.

753. **Gallic Acid.**—This acid is found naturally associated with tannin in vegetable tissues, and is also formed from tannic acid by exposing a solution of the latter for some time to the action of the air. It produces, like tannin, a dark precipitate with the salts of the sesqui-oxyd of iron, but does not unite with gelatin to form insoluble compounds, and is consequently of no value for the manufacture of leather. When added to the salts of silver, gold, and platinum in solution, it occasions a precipitation of the metal in a state of minute subdivision. The most successful compounds for coloring the hair are founded upon this principle—the hair being wet in the first instance with a solution of gallic acid, and afterward with a solution of a salt of silver in ammonia. The reduced metal imparts to the hair a fine black or brown color, which is extremely permanent.

In addition to the substances mentioned which afford tannin, there are several others which afford it and constitute important articles of commerce. Among them may be mentioned the following: catechu, cutch, and terra japonica are the dried aqueous extracts of a species of acacia growing in India; kino is a product of like character; divi-divi is the pod of a leguminous shrub, a native of South America. These substances will be found mentioned in almost every commercial price current. The best gall nuts are exported from Asia Minor.

In addition to the acids which are secreted by living vegetable tissues, a great number have been also recognized by chemists which do not exist naturally in plants, but are the result of vegetable decompositions taking place either under natural or artificial conditions. The acids included in the substance called *humus*, and many of the products resulting from the action of mineral acids upon the constituents of coal-tar, are examples of this nature.

---

iron, or of logwood, impairs the durability of the ink. Experiments recently detailed to the Scottish Society of Arts, show that the quality and durability of ink is greatly increased, however, by the addition to it of a small quantity of sulphate of indigo, and the following receipt was given as affording an ink that was superior to all others for writing purposes: 12 ounces powdered nut galls, 8 ounces sulphate of indigo, 8 ounces of copperas, a few cloves, and 4 to 6 ounces of gum arabic per gallon of ink. Documents written with steel pens are less durable than those written with quill pens, as the contact of iron or steel with ink, injures it to a greater or less extent.

QUESTIONS.—How is a black color given to leather? What is said of gallic acid? What are its characteristic features? What products are commercially important on account of their tannin? What other acid compounds are considered in this connection?

CHAPTER XXI.

ORGANIC ALKALIES.

754. The terms *organic alkalies*, *vegetable alkaloids*, and *organic bases*, are applied to a peculiar class of organic substances which resemble in certain of their properties the alkalies of inorganic chemistry; that is to say, they neutralize acids, unite with them to form salts, and in most instances, when in solution, restore the blue color of reddened litmus. They all contain nitrogen as an essential constituent, and are exceedingly complex in their constitution. They are sparingly soluble in water, but dissolve freely in hot alcohol, from which they often crystallize on cooling in a very beautiful manner. The taste of these substances in solution is usually intensely bitter, and the majority of them are active and virulent poisons; when given, however, in small doses, they rank among the most powerful medicines.

Of the organic alkaloids, about one hundred are now known to exist in plants as natural products, always in combination with vegetable acids. They were formerly supposed to be exclusively the result of vegetable organization, but within a comparatively few years some seventy or eighty compounds of a similar character have been artificially prepared from organic products by chemists. These last are termed the *artificial organic alkaloids*, and do not occur in nature. The true vegetable bases have not yet been artificially imitated.

Most of the vegetable alkaloids are prepared by boiling the vegetable matter containing them in water acidulated with hydrochloric acid, when the alkaloid unites with the acid to form a soluble salt, and enters into solution. From this it is precipitated in a separate state by the addition to the liquid of a stronger base—*i. e.*, lime, potash, ammonia, etc. The plants which by treatment furnish alkaloids, are generally characterized by possessing poisonous or active medicinal qualities, which in turn are supposed to be due to the alkaloids contained in them. The following are some of the most important of the alkaloids extracted from vegetable products.

755. Morphia.—*Morphine.*—This alkaloid is the chief active principle of opium, which is the dried juice of certain species of the poppy. It exists in opium in combination with meconic acid, and the best qualities of opium contain about ten per cent. of it. It crystallizes in small, colorless prisms, is devoid of smell, and possesses a bitter, unpleasant taste. It is powerfully narcotic and poisonous, and is an invaluable remedy in medicine, in small doses, for soothing nervous irritation and allaying pain. A full dose of pure morphia

QUESTIONS.—What are the organic alkalies? By what other names are they designated? What are the general properties of these substances? What is their number? Are any of them prepared artificially? How are the vegetable alkaloids obtained? What are characteristics of the plants which furnish them? What is morphia? What is opium? What are the properties of morphia?

for a grown man is one eighth of a grain; and in the state of acetate or muriate of morphia (in which condition it is generally used in medicine) one fourth of a grain. It is a singular circumstance that this substance, which is so poisonous to man, has comparatively little effect upon animals, even when administered in large doses. The composition of morphia is represented by the formula, $C_{35}H_{20}NO_6 + 2HO$.

Opium also contains, in addition to morphia, eight other alkaloids, the principal of which are termed *narcotine*, *meconine*, and *thebaine*. They are all narcotics and poisons, and exist to some extent in laudanum, which is an alcoholic extract of the active principles of opium.

756. The dried juice of the common lettuce plant has considerable resemblance to opium, and contains an active principle (supposed to be an alkaloid), called *lactucin*. It is this substance which gives to lettuce, when freely eaten as a salad, its narcotic properties.

757. **Quinine** and **Cinchonine** are the alkaloids which impart to Peruvian bark its medicinal virtues. Quinine is a white, crystallized substance, of an intensely bitter taste. It is one of the most valuable and reliable of medicinal agents, and is generally administered in the form of a sulphate.

758. **Strychnia** and **Brucia** are extracted from a variety of plants belonging to the genus strychnos, and especially from the berries (*nux vomica*) of a small tree of this genus growing in India. These alkaloids are remarkable for being the most powerful of all vegetable poisons—a single grain of the former being a fatal dose for an adult man; while a sixth of a grain has proved fatal to a dog in thirty seconds. Its influence seems to be exerted principally upon the nerves and spinal marrow, producing violent spasms, which increase in frequency and severity until death. The celebrated *woorara* with which the natives of Guiana, S. A., poison the tips of their arrows, and the poison of the celebrated Upas tree of the island of Java, are varieties of strychnias.

Pure strychnia crystallizes in small, but exceedingly brilliant, colorless crystals, and is slightly soluble in water. It possesses the property of bitterness in a more marked degree, probably, than any other substance, and its taste can be distinctly recognized when dissolved in 600,000 times its weight of water. Vegetable matters containing this alkaloid are sometimes employed for imparting bitterness to beer, but their use should be considered criminal.

759. Among the other important alkaloids may be mentioned *Nicotine*, the poisonous principle of tobacco; *Aconitine*, or *Aconite*, extracted from the plant "monk's hood;" *Conicine*, prepared from hemlock; *Veratrine*, from the plant hellebore; *Hyoscyamine*, from henbane; and *Solanine*, from several species of the genus solanum, and from the white sprouts of the potato. All these are most virulent poisons, only inferior in their action to strychnia and brucia.

QUESTIONS.—What is its composition? What is laudanum? Is there an active principle in the lettuce plant? What are quinine and cinchonine? From what sources are strychnia and brucia obtained? For what are these alkaloids remarkable? What is said of the poisonous influence of strychnia? What are varieties of this poison? What are the other properties of strychnia? What are some of the other alkaloids remarkable for their poisonous qualities?

Among the alkaloids less injurious in their action on the animal economy, are *Emetine*, the medicinal agent of ipecac (ipecacuanha); *Piperine*, extracted from ordinary black pepper; and *Caffeine*, or *Theine*, the enlivening principle in coffee and tea.

The organic alkaloids are, almost without exception, precipitated from their solutions, by tannic acid, in the form of insoluble tannates, and consequently the most efficient remedies in cases of poisoning by them, are liquids containing tannic acid, such as decoction of oak-bark, tincture of gall-nuts, concentrated infusion of green tea, etc., etc.

The detection of their presence in the animal organism, in cases of death by poisoning, is extremely difficult, strychnia excepted, and the testimony of the most experienced chemists ought only to be relied on in such cases.*

760. **Vegetable Extracts.**—This name is applied to a very large class of substances extracted from plants, which do not possess sufficiently marked features, in a chemical point of view, to allow of their incorporation with any of the more well-defined groups of organic compounds. Some of them, however, possess active and medicinal properties, as, for example, the intensely bitter principle of wormwood, aloes, etc., the purgative principle of the root of the rhubarb, and the aromatic bitter of the hop, sweet-flag, etc. They have for the most part a bitter taste, and often occur crystallized; and are generally regarded as mixtures of various vegetable products.

CHAPTER XXII.

ORGANIC COLORING PRINCIPLES.

761. THE organic coloring matters, with the exception of the red dye obtained from cochineal, are all of vegetable origin. They do not, as a class, possess many chemical characters in common, and are considered under one general head, by reason of their common industrial applications. Many of the most valuable vegetable coloring agents do not exist naturally in plants, but are formed by subjecting certain vegetable products to specific chemical treatment. The most brilliant and splendid of the vegetable colors, as those of flowers, for example, are exceedingly evanescent, and are generally destroyed by any treatment employed to extract them; they also exist in the vegetable

* A few years since a man was convicted in Albany, N.Y., of murder, by the administration of tincture of aconite, upon what was supposed to be reliable chemical testimony, but which was afterward shown to be so utterly unreliable, that the means adopted for detecting the poison must have completely removed it, if present, from the matters tested.

QUESTIONS.—What are more medicinal than poisonous? What are antidotes for these poisons? What is said of their detection in the system? What are vegetable extracts? What examples of these substances? From what source do we derive organic coloring agents? What is said of them?

tissue, in very minute quantities. Coloring matters extracted from those parts of the plant which are removed from the immediate influence of the light, as the wood, bark, etc., are much more permanent, but less brilliant.

762. The art of dyeing consists in impregnating cloths and other textures with coloring substances, in such a manner that the acquired colors may remain permanent under the common exposure to which the articles may be liable. This is effected by producing a chemical union between the materials to be dyed and the coloring matter. Different fibrous materials present very different attractions for dye-stuffs, and absorb coloring matter in very different proportions: wool appears to have the greatest attraction; silk comes next to it; then cotton, and, lastly, flax and hemp. While the former, therefore, are dyed with very little difficulty, the latter can only be made to permanently combine with coloring substances, through the agency of indirect and complicated processes.

763. All coloring substances used in dyeing are divided into two classes, viz., *substantive and adjective colors.* A substantive color is one that imparts its tint directly to the substance to be dyed, without the intervention of a third substance. An adjective color, on the contrary, is one that requires the intervention of a third substance, that possesses a joint attraction for the coloring principle and for the substance to be dyed.

Most of the substances used in dyeing belong to the adjective colors; and if we except indigo, there is scarcely a dye-stuff in extensive use that imparts its own color directly; and by far the greater number of dyes have so weak an affinity for cotton fabrics, that alone they communicate no color sufficiently permanent to deserve the name of a dye.

The intermediate third substance which is used to effect a union between the dye and the cloth, is called a mordant, from the Latin word *mordeo,* to bite, from an idea the old dyers had that these substances bit or opened a passage into the fibers of the cloth, and allowed the color to penetrate. The action of a mordant may be illustrated by the method of procedure followed in dyeing cottons black, by an extract of logwood. An aqueous solution of logwood is very deeply colored, but imparts no permanent dye to cotton. If, however, the cotton be previously impregnated with a salt of oxyd of iron (as copperas), and then dipped into the extract of logwood, the coloring principle of the latter, by reason of its great affinity for oxyd of iron, unites with it, and the two are precipitated upon the fibers of the cloth in the form of a black precipitate or dye. A dye thus effected is usually a *fast color,* since it is formed in and incorporated with the whole structure of the fiber itself, and is not merely upon its surface; so that the color will only disappear when the texture and fiber of the cloth are destroyed. The use of mordants, further-

---

Questions.—In what does the art of dyeing consist? What fibrous substances have the greatest attraction for dyes? Into what two classes may dyeing principles be divided? What are substantive colors? What are adjective colors? To which class do the dyes in ordinary use generally belong? What is the derivation of the word mordant? Explain the use of mordants?

more, adds greatly to the resources of the dyer; because a single coloring substance will impart very different colors with different mordants: thus, an extract of logwood will dye with iron, black; with a solution of tin, violet; and with other mordants, all the shades of color included between a yellowish-white and a violet, a lavender and a purple, or a slate-brown and a black.

764. **Calico-Printing.**—The general process of calico-printing is as follows: The cloth is first prepared, by bleaching and other treatment, to receive the colors. The pattern is then stamped or printed upon it from plates or rollers, which have been previously covered with different mordants, in the same way that ink is applied to types. The cloth is then passed through a solution of dye, when those parts which have been printed with the mordant seize upon and retain the colors. The cloth is afterward washed, when all the color not combined with mordant disappears from its surface, and the pattern impressed is brought out with distinctness.

765. The most important coloring matters used in dyeing are as follows:—

Red and Violet Coloring Substances.—*Madder* is the ground-up root of the plant *rubia tinctorum.* Its most beautiful coloring constituent (*madder red, Turkey red*), called *garancine*, is not a natural product, but results from subjecting the root to the action of sulphuric acid. The action of the acid in this instance is often cited as an example of catalysis, as it does not enter into combination with the coloring principle of the root, but effects a change in it, apparently by its mere presence.

Madder is used to a greater extent in dyeing and printing cottons than any other substance, and with different mordants it furnishes very bright and durable reds, yellows, violets, and browns. The other important coloring substances of this class are "Brazil wood," "safflower" (the flowers of the red saffron), sandal-wood ("red-sanders"), and cochineal. The last is a dried insect, the *cocus cacti*, which lives upon several species of cactus, peculiar to warm latitudes, and especially to Central America. It yields the most brilliant scarlet and purple colors.

766. **Blue Dyes, Indigo.**—The most important of the blue dyes is *Indigo*, which is obtained from several species of American and Asiatic plants, particularly from those belonging to the genus *indigofera.* The juice of these plants is colorless, but when their leaves are digested in water, and allowed to ferment, a yellow coloring substance is dissolved out, which, by exposure to the air, gradually becomes blue, and is deposited from the solution in the form of a thick sediment. This washed and dried, constitutes the indigo of commerce.

Commercial indigo is far from pure, and in addition to the blue coloring matter, or pure indigo, it contains at least one half its weight of foreign substances. Pure indigo is quite insoluble in every liquid, with the exception of

---

QUESTIONS.—How do they increase the resources of the dyer? What is the process of calico-printing? What is madder? What is said of its coloring principles? What is cochineal? What colors does it furnish? What other dye-stuffs furnish red colors? What is the most important of the blue dyes? How is indigo prepared?

fuming sulphuric acid (Nordhausen), with which it forms a compound quite soluble in water, called *sulphindigotic acid.* When indigo is brought in contact with water and deoxydizing agents, it becomes converted into a soluble and colorless substance, known as *indigo white*, which, by exposure to the air, again absorbs oxygen, and resumes its blue color. This circumstance is taken advantage of in dyeing; thus, the indigo is mixed, in a state of fine powder, with hydrate of lime and copperas, and the whole digested in water. Under these circumstances, the hydrated protoxyd of iron, resulting from the action of the lime, abstracts oxygen from the indigo, and reduces it to a state of a yellow liquid (white indigo). Cloths steeped in this liquid, and exposed to the air, readily acquire a deep and permanent blue tint, by the formation of the blue, insoluble indigo in the substance of the fibers, and it is in this way that the fine indigo-blue colors are produced. What is called *Saxon blue*, a brighter color than ordinary indigo, is imparted by boiling the fabrics in sulphindigotic acid. Among other prominent blue dyes, may be mentioned *litmus*, which is obtained from several species of lichens, by treatment with ammonia—the plants themselves being destitute of color. Archil and cudbear are substances allied to litmus.

767. **Yellow Coloring Substances.**—The most valuable dye-stuffs of this class are *fustic*, the rasped wood of a West Indian tree; quercitron, obtained from the bark of the American black oak; the berries of the buckthorn; annotto, prepared from the pulp of certain South American seeds; and tumeric, the root of an East Indian plant.

768. **Chlorophyle** is the name given to the green coloring matter of the leaves and stems of plants. It exists in them in very small quantity, and is extracted with difficulty in a state of purity. It appears to be united in the vegetable tissue with a substance resembling wax, and is insoluble in water, but dissolves in alcohol and ether; hence all tinctures in pharmacy, prepared from the fresh stalks and leaves of plants, possess a green color. Chlorophyle appears only in those parts of plants which are exposed to the action of light; hence this agent is supposed to exercise an important and direct influence on its formation. Plants grown in the dark are nearly destitute of color, but when removed into the sunlight become rapidly green. The red and yellow colors which leaves assume in autumn, are supposed to be due to the decomposition and oxydation of the chlorophyle, and the formation of an acid compound; but the information we possess on this subject is very limited.

Most of the greens used in dyeing are of a mineral origin, *i. e.*, the salts of chromium and of copper.

No genuine black substantive color has ever been obtained from plants.

QUESTIONS.—What is said of the solubility of indigo? What is indigo white? How is it employed in dyeing? How is "Saxon blue" imparted? What other blue dyes are used? Enumerate some of the principal yellow dye-stuffs? What is chlorophyle? What are its solvents? What agent influences its formation? What is the character of the greens used in dyeing?

CHAPTER XXIII.

OILS, FATS, AND RESINS.

769. **Connection between Oils and Fats.**—The oils and the fats, whether of animal or vegetable origin, are regarded as belonging to the same general class of organic substances; and with the exception of the volatile oils, they are all closely allied to each other in their chemical properties, and are composed of the same elements, viz., carbon, hydrogen, and oxygen, united in various proportions. As a class, they are all, however, characteristically poor in oxygen, but rich in hydrogen, and some few of the volatile oils contain no oxygen. The distinction between a fat and an oil is founded merely upon the circumstance, that the former is solid at ordinary temperatures, while the latter continues more or less liquid; an oil, therefore, may be called a liquid fat, or a fat a solid oil.

The fats and the oils are all highly combustible, and burn with a brilliant flame; they are insoluble in water; but dissolve with more or less readiness in alcohol or ether, and when brought in contact with paper leave a greasy mark, and render the paper translucent. The oily substances secreted by plants, are principally accumulated in the seeds and coverings of the fruit, although no portion of the plant is entirely destitute of them. The proportion existing in some seeds is very considerable. Thus flax-seed contains about 20 per cent. of oil, Indian corn 9 per cent., rape-seed 30 to 40 per cent., while the seed or bean which furnishes castor oil contains as much as 60 per cent.

770. **Division of the Oils.**—All oily substances are divided into two classes, viz., the fixed and volatile oils; the former when exposed to the air do not diminish in bulk, while the latter under the same circumstances, readily evaporate.

771. **Volatile Oils.**—The volatile oils do not possess the greasy feel of the fat oils, and are almost always characterized by a strong aromatic odor, and a pungent burning taste. Many of them, also, are highly poisonous. With alcohol they form solutions called "*essences*," and from this circumstance the oils themselves are very frequently termed "*essential.*" They also dissolve in ether and acetic acid, and mix in every proportion with the fixed or fat oils. They do not, however, like the fat oils, form soaps, but when exposed to the air they are frequently changed by the absorption of oxygen, and converted into resins.

These oils (with the exception of a few which have been formed ar-

---

QUESTIONS.—What is said of the class of oils and fats? What of their composition? What constitutes the difference between an oil and a fat? What are their properties? What is said of the distribution of the vegetable oils? Into what two classes are oily substances divided? What are the characteristics of the volatile oils? What are essences? From what sources are the volatile oils derived?

tificially) are almost exclusively the products of the vegetable kingdom, and are *generally* obtained by distilling the plant with water; in some instances, however, they are extracted from the cellular tissue containing them by pressure, as in fresh orange or lemon peel. The boiling points of almost all these oils are above that of water, but their vapors are carried over mechanically in distillation by steam at 212° F., and condense with it in the receiver. In this way are obtained the oils of roses, orange flowers, lemons, lavender, winter-green, peppermint, and many others which in smell and taste more or less resemble the fresh plants from which they are derived. The greater portion of the oil floats upon the surface of the water which distils over with it, but the latter usually retains a small quantity of the oil in solution, and thus acquires its peculiar taste and odor. Waters thus impregnated are termed "medicated," or "perfumed waters;" rose-water, lavender-water, peppermint-water, etc., being examples of this character.

The various perfumes and odors which plants emit, are believed to be mainly due to the presence of some one or more of the volatile oils in their structure, which gradually evaporate, and diffuse themselves in the atmosphere. The quantity of oil, however, yielded by some plants which possess a marked odor, is exceedingly small—a thousand fresh roses, for example, affording by distillation less than two grains of oil (attar of roses). In some flowers, as the jasmin, the violet, and the tuber rose, the oil which imparts fragrance is, moreover, so evanescent and delicate, that it is destroyed by any process of distillation, and in such cases the perfume is obtained by arranging the flowers in layers between cotton imbued with some fixed oil; which latter gradually absorbs the volatile oil or perfume of the flower, and in turn imparts it to alcohol—thus forming an odoriferous essence. Fatty bodies perfumed in this way have received the name of *pomatums.*

The volatile oils do not appear to be uniformly diffused throughout the whole plant. In the mint and thyme they reside principally in the leaves and stems; in the sandal and cedar trees they are in the wood; in the rose, violet, etc., in the leaves of the flower; in the vanilla, anise, and carraway, they are in the seed; and in the ginger and sassafras in the root. Different parts of the same plant not unfrequently furnish different oils; thus, the flowers of the orange-tree furnish one kind, the leaves another, and the rind of the fruit a third.

772. **Composition of the Volatile Oils.**—The volatile oils differ materially from each other both in their composition and chemical reactions, and are conveniently divided into three classes, viz., those composed of carbon and hydrogen only; those composed of carbon, hydrogen, and oxygen; and those which contain in addition sulphur and nitrogen. Most of them contain at least two proximate principles, one of which, termed *stearop-*

Questions.—How are they obtained? What are medicated waters? What is supposed to be the origin of the odors of plants? What of the quantity of volatile oils yielded by plants? What are pomatums? Are the volatile oils uniformly diffused throughout plants? Illustrate this. What is said of the composition of the volatile oils?

tene, is less fusible than the other, and may be separated by cold in the form of a camphor-like substance. The more liquid constituent, termed *elaioptene*, on the contrary, may be often obtained in a separate condition by distillation at a low temperature.

773. **Oils composed of Carbon and Hydrogen.**—This class embraces a large number of the odoriferous essences of plants, and furnishes some of the most remarkable examples of isomeric bodies known in chemistry. Thus, the oils of turpentine, lemons, oranges, juniper, copaiba, citron, black pepper, and many others which possess entirely different properties, contain exactly the same constituents, united in exactly the same proportions—100 parts of each by weight containing 88·24 of carbon and 11·76 of oxygen. These proportions are expressed by the formula C^5H_4, or by some multiple of it, as $2(C_5H_4)$. In addition to their identity in chemical composition, these substances also agree as regards their density and boiling points. The fact, however, that the internal arrangement of their molecules or particles is different, is strikingly shown by their diverse influence on a ray of polarized light, some of them causing it to diverge to the right, some to the left, while others transmit it unaltered, or directly.

774. **Oil or Spirits of Turpentine.**—This substance, which may be regarded as the type or representative of the volatile oils containing only carbon and hydrogen, is obtained by distilling with water the semi-fluid sap or pitch called in commerce *crude turpentine*, which exudes from incisions made in the wood of various species of pine. The product left after distillation is a resinous solid, which is popularly termed *rosin*.

Oil of turpentine is a thin, colorless liquid, which is highly inflammable, and possesses a well-known and powerful odor. It has a specific gravity of 86, and boils at 312° F. It is nearly insoluble in water, but dissolves freely in alcohol or ether.

Camphene, which is extensively used in lamps as a substitute for oil, is spirits of turpentine purified by repeated distillations. *Burning fluid* is a solution of rectified turpentine or camphene in alcohol—the tendency of the turpentine to smoke being diminished by the addition of alcohol.

Camphene and burning fluid, although highly inflammable, are not in themselves explosive; a mixture, however, of the vapor of these liquids with atmospheric air is highly explosive, and igniting at a distance at the approach of the slightest spark or flame, is apt to communicate fire to the liquids themselves. Burning fluid being much more volatile than camphene, is much more dangerous, and its use as an illuminating material should be discountenanced and forbidden. The explosive character of the mixture of its vapor and air may be illustrated without danger by allowing a small quantity of the fluid

---

QUESTIONS.—What is the first class remarkable for? What are examples of this fact? What substance is regarded as the type of this class? How is turpentine obtained? What is the residue left after distillation? What are the properties of oil of turpentine? What is camphene? What is burning fluid? Are these liquids explosive? Why then is their employment so dangerous? What is the comparative volatility of the two? What experiment illustrates the explosive character of the mixed vapor of burning fluid and air?

to evaporate in a tin can or vial, and then applying a lighted taper to the mouth. If, however, the mouth of the can be tightly corked, under these circumstances, and flame be applied through a minute orifice in the side, an explosion of great violence is occasioned. As a matter of ordinary precaution in using these liquids, no attempt should ever be made to fill or replenish lamps that are lighted, neither should any vessel containing camphene or burning fluid be ever opened in the vicinity of a flame.

When a current of dry hydrochloric acid gas is passed through oil of turpentine cooled by a freezing mixture, a white solid is formed which resembles common camphor in odor and appearance, and is termed *artificial camphor.* It is from this circumstance, probably, that the term camphene, as applied to spirits of turpentine, first originated.

Oil of turpentine is extensively used as a solvent for resins in the manufacture of varnish, in the preparation of paints, and to some extent in medicine. Many substances, also, like India rubber, etc., which are insoluble in alcohol, readily dissolve in it.

The other more important oils of this class have been mentioned above as isomeric with oil of turpentine.

775. **Essential Oils containing Oxygen.**—The principal oils of this nature are the oil of bitter almonds, of cinnamon, of roses, lavender, bergamot, and peppermint. In this class the proportions of the several constituents are rarely the same in two different oils.

Common camphor, prepared by distilling the wood of the camphor-tree (found principally in Borneo and Japan), is regarded as a solid oil, or volatile fat of this class. It partakes of the general properties of the volatile oils, may be distilled without decomposition, and evaporates in the air at ordinary temperatures. It is nearly insoluble in water, but dissolves freely in alcohol, forming what are termed *camphorated spirits.* On adding water to this, nearly all the camphor is thrown down in a minutely divided state. Camphor taken internally in other than very small doses, acts as a poison, a hundred grains being sufficient to cause death. "When small pieces of perfectly clean camphor are allowed to fall upon the surface of pure water, they rotate and move about with great rapidity, sometimes for several hours together; but if while the camphor is rotating, the surface of the water be touched with any greasy substance (a glass rod dipped in turpentine answers best), all the floating particles quickly dart back, and are instantly deprived of all motion." This phenomenon appears to be due to the continued escape of vapor from the surface of the camphor.

776. **Essential Oils containing Sulphur.**—The oils obtained by distillation from black mustard-seed, from assafœtida, onions, horse-

---

QUESTIONS.—What precautions should always be observed in the use of these liquids? What is the reaction between turpentine and hydrochloric acid gas? What are the uses of oil of turpentine? What are other important oils of this class? What are the principal essential oils of the second class? What is common camphor? What are its properties? What are examples of oils of the third class?

radish, garlic, and hops, belong to this class. Many of them are characterized by nauseous and offensive odors, which at the same time are remarkably persistent. The bad smell imparted to the breath by eating onions or garlic, for example, is occasioned by the continued presence of a minute quantity of the volatile oils of these vegetables in the air exhaled from the lungs.

The volatile oils which are produced by the destructive distillation of vegetable and animal substances, are as a class called *empyreumatic.*

The essential oils are mainly used in the preparation of essences, perfumes, and cordials. The latter are generally made of brandy, flavored with various aromatic oils, and afterward sweetened. In the preparation of perfumery, a single oil or essence is rarely used by itself, but the best result is obtained by a skillful mixture of the odoriferous principles of many plants. *Eau de Cologne*, called the perfection of perfumery, owes its excellence to the application of this principle.*

777. **Fats.—Fixed Oils.**—The fixed oils are mostly destitute of either taste or smell; but the presence of certain volatile acids imparts odors to some of them: thus, butter contains *butyric acid;* goat's fat, an odorous acid called *hyrcic acid;* while the nauseous smell of whale oil is due to the presence of an acid called *phocenic acid.* They do not evaporate in the air, are decomposed by the action of heat, and are unctuous and greasy in their feeling. They dissolve readily in ether and in the essential oils, but are not soluble to any great extent in alcohol, and are entirely insoluble in water. All the oils have an attraction for oxygen, and when exposed to light and air, absorb it rapidly, and give out carbonic acid and hydrogen; this action may be sufficiently energetic to produce ignition (spontaneous combustion), especially when the oil is distributed over porous substances, tow, cotton, etc.

---

* "Odors resemble very much the notes of a musical instrument. Some blend easily and naturally with each other, and produce a harmonious impression, as it were, on the sense of smell. Heliotrope, vanilla, orange-blossoms, and the almond blend together in this way, and produce different degrees of nearly similar effect. The same is the case with citron, lemon, vervain, and orange-peel, only these produce a stronger impression, or belong, so to speak, to a higher octave of smells. And again, patchouly, sandal-wood, and vitivert form a third class. It requires, of course, a nice or well-trained sense of smell to perceive this harmony of odors, and to detect the presence of a discordant note. But it is by the skillful admixture, in kind and quantity, of odors producing a similar impression, that the most delicate and unchangeable fragrances are manufactured. When perfumes which strike the same key of the olfactory nerve, are mixed together for handkerchief use, no idea of a different scent is awakened as the odor dies away; but when they are not mixed upon this principle, perfumes are often spoken of as becoming *sickly* and *faint*, after they have been a short time in use. A change of odor of this kind is never perceived in genuine *eau de Cologne.* Oil of lemons, juniper, and rosemary are among those which are mixed and blended together in this perfume. None of them can, however, be separately distinguished by the ordinary sense of smell; but if a few drops of ammonia be added to their solution, the lemon smell usually becomes very distinct."

QUESTIONS.—What are their peculiarities? What are empyreumatic oils? For what purposes are the essential oils chiefly used? What are cordials? In what does the perfection of perfumery consist? What are the properties of the fats and fixed oils? To what are their odors owing? What is said of their attraction for oxygen?

778. The fixed oils, according to the changes which they undergo through the absorption of oxygen, are divided into two classes, viz., the *drying* or *siccative* oils, and the *unctuous* or *greasy* oils; or those which become hard and resinous by exposure to the air, and those which remain soft and sticky under the same circumstances.

779. **Drying Oils.**—**Linseed Oil**, or the oil obtained by expression from the seeds of the flax plant, is a representative of this class; and is the oil generally used for mixing with paints, and in the manufacture of varnishes. Its drying properties, which especially recommend it to the painter's use, are greatly increased by boiling it for some time, with the addition of a little *litharge* (protoxyd of lead). As thus prepared, it is known as "*boiled oil*," or *linseed-oil varnish*. *Oiled silk* is silk to which successive coats of purified linseed oil have been applied. *Glazier's putty* is prepared by kneading together boiled linseed oil and pulverized chalk (whiting).

The other principal drying oils are those extracted from rape-seed, poppy-seed, the seed of the castor-oil plant, and from walnuts.

Printer's Ink is prepared by igniting linseed oil in suitable vessels, and allowing it to burn until it becomes thoroughly charred, and acquires a viscid consistency; it is then mixed with a certain proportion of the finest varieties of lamp-black.

780. **Unctuous Oils.**—This class includes the oils expressed from the fruit of the olive and the palm, and most of the oils and fats of animal origin. These oils, by exposure to the air, are liable to become sour and rancid, but they do not solidify.

781. **Composition of the Fats and Fixed Oils.**—Most fats and fixed oils, vegetable and animal, are mixtures of two, and generally three, distinct compounds, each of which, taken singly, has all the properties of a fat or an oil. The first of these substances, called *stearine* (from στέαρ, tallow), is solid at common temperatures; the second, *oleine* (from ἔλαιον, oil), is liquid at common temperatures; the third, called *margarine* (from μάργαρον, a pearl), on account of its pearly appearance, is also a solid at common temperatures. All the fats and fixed oils, therefore, may be regarded as mixtures of the fluid *oleine* with the solids *stearine* or *margarine*. If the solid be in larger proportion than the fluid, then the compound at ordinary temperatures is a solid, and resembles tallow or lard; if, on the contrary, the fluid constituent prevails, the compound has the characters of an oil. For example, when olive oil is subjected to a cold of 40° F., it deposits a solid fat, *margarine*, which may be separated by filtration and pressure; the largest portion of the oil, however, consisting of *oleine*, retains its fluidity at a much lower temperature. Again, by subjecting mutton tallow to pressure, a per-

QUESTIONS.—Into what two classes are the fixed oils divided? What are drying oils? What are unctuous oils? What is a type of the drying oils? What is linseed oil? For what is it principally used? What is boiled oil? What is oiled silk? What is putty? What are the other principal drying oils? What is printers' ink? What are included in the class of unctuous oils? What is the composition of the fixed oils? When will a fatty body have the characters of a solid, and when of a liquid?

manently fluid oil, *oleine*, is extracted, while the solid which remains has its melting point raised, and is much harder than the original tallow; it consists of stearine and margarine, the latter of which melts at a much lower temperature than the former.

Stearine, margarine, and oleine are, however, susceptible of further analysis. They are, in fact, true salts, composed of an organic or fat acid, united to a base; the acid being peculiar to each fatty principle, while the base with which it is naturally united, is almost always the same. The name given to this base is glycerine, and it is regarded as the hydrated oxyd of a radical, *glyceryle.* In stearine the existing acid is called *stearic acid;* in margarine, *margaric acid;* and in oleine, *oleic acid.* Stearine is accordingly the stearate of the oxyd of glyceryle, and margarine and oleine the margarates and oleates of the oxyd of glyceryle. Olive oil also must be described as a mixture of much oleate of the oxyd of glyceryle, and a little margarate of the same base.

It may seem singular to the student that bodies of an acid character should exist in fats and oils. Such, however, is the case, and they exhibit acid properties in marked degree, such as reddening litmus paper, neutralizing alkalies, and uniting with bases to form salts.

782. Soaps.—When fixed fatty or oily bodies are brought in contact with alkaline solutions at high temperatures, they undergo a change called *saponification;* that is to say, the strong alkaline bases (potash or soda) displace the weak base glycerine, and unite with the acids existing in the fats or oils to form a homogeneous mass, called soap. Soaps, therefore, are true salts, combinations of stearic, margaric, or oleic acid, with an alkaline base.

Soaps are of two kinds, *hard* and *soft.* The former are made with soda alone, or a mixture of potash and soda, while the latter are made exclusively with potash. Soaps made with potash are soft, mainly by reason of the deliquescent character of the potash, which is unable to harden in the presence of any considerable quantity of water. A soda soap, on the contrary, may be made to absorb more than its own weight of water without becoming fluid. Besides, in a potash soap, the glycerine, which before saponification was combined with the fat acids, remains mechanically mixed with the soap, and promotes its fluidity. In the manufacture of soda soaps, the soap is obtained in a nearly pure state by the addition of common salt to the watery solution in which the soap is suspended. Soap not being soluble in salt water, immediately separates from the water and the glycerine contained in it, and floats upon the surface, and in this condition may be removed, while the spent lye, glycerine, and salt are allowed to run to waste. When this treatment is applied to a potash soap, another change is occasioned which is purely chemical. The salts which the fatty acids form with potash are decomposed by chloride of sodium, and a mutual interchange of acids takes

QUESTIONS.—What is the constitution of stearine, margarine, and oleine? What is glycerine? What are stearic, margaric, and oleic acids? What is understood by saponification? What is a soap? When are soaps hard or soft? Why are potash soaps soft? How are soda soaps made hard? What is the chemical composition of hard and soft soaps?

place; and hence, when a potash soap is mixed with a solution of common salt, both the soap and the chloride of sodium are decomposed, and a soda soap, and chloride of potassium are formed.

Hard soaps are generally made of tallow, and are mainly mixtures of stearate and margarate of soda; soft soaps are, on the other hand, usually made of oils, soft fats, and are mainly oleates of potash, with glycerine mechanically mixed with them. *Castile soap* is manufactured of olive oil and soda, its mottled appearance being produced by the addition of oxyd of iron. Resins form with the alkalies, salts, which possess characters allied to those of the soaps, and in the manufacture of common soaps, a quantity of resin (rosin) is mixed, on the ground of economy, with the fats. Such soaps have a yellowish appearance, and are known as yellow soaps.

Soaps, by reason of their strong attraction for water, always retain a considerable quantity of it in their composition; the proportion in the best hard soaps varying from 25 to 30 per cent. It is possible, however, to prepare a solid soap containing more than its own weight of water. Such soaps look well when fresh, but contract greatly on drying. Dealers generally store their soap in cellars and damp places, since it is for their interest to sell as large a proportion of combined water as possible.

Soap is freely soluble in pure water, but in salt water, and all other saline solutions, it is insoluble; soap made from the oil extracted from the cocoa-nut, is, however, an exception to this rule, as it dissolves freely in strong brine, and is hence much used as marine soap. When a solution of a soap having an alkaline base is mixed with a salt of any other base, double decomposition ensues, and an insoluble compound of the fatty acids with the earthy or metallic bases is precipitated. The salts of lime and magnesia contained in natural waters act in this manner, and their presence in a water renders it *hard* and unfit for washing.* The slimy scum which is formed by the addition of soap to such water, is a compound of the fatty acids with lime or magnesia. The strong acids decompose both soap and fats, uniting to their bases, and setting free the fatty acids.

Ammonia acts far more feebly upon fatty bodies than either potash or soda,

783. **Cleansing Properties of Soaps.**—The detergent, or cleansing action of soap depends entirely upon its alkaline constituents. The impurities upon the skin, or on articles of clothing, always contain a certain proportion of oily matter, which exuding from the pores of the sys-

* The hardness of a water may be easily tested by adding to it a few drops of a solution of soap in alcohol (tincture of soap). If the water remains clear, it is perfectly soft; if it becomes cloudy, it may be regarded as hard—the degree of hardness being proportioned to the degree of cloudiness occasioned.

QUESTIONS.—What is said of the use of resin in the manufacture of soaps? What percentage of water is contained in soap? Why will not soaps wash in salt water? When a solution of an alkaline soap is brought in contact with an earthy or metallic base, what happens? Why will not soaps wash in hard water? What effect have acids upon soaps and fats? What is the action of ammonia? To what are the cleansing properties of soaps due?

tem, and existing in the perspiration, acts as a cementing agent with whatever dust or dirt is brought in contact with it. Water alone, by reason of its total want of affinity for all fatty or oily substances, is unable to dissolve these impurities, and effect their removal. An alkali, on the contrary, readily unites with the greasy and organic matter, and renders it soluble.

When a soap is dissolved in water, a portion of its alkali is set free (by the substitution of water as a base), and uniting with the impurities intended to be removed, partially saponifies them, and renders them soluble or miscible with water. The fatty acids also, by their lubricity, cause the dissolved matter to wash away more easily. An alkali used alone would act more powerfully than any soap as a detergent, but it would tend to destroy the texture of the organic substance to which it was applied, and also to remove the colors of dyed fabrics. When used in the form of soap, its solvent powers are partially neutralized. In washing the surface of the body with soap, its alkaline constituent not only effects the removal of the dirt, but also dissolves off the cuticle, or outer layer of the skin itself, which being mainly composed of albumen, is soluble in alkaline solutions; and thus every washing of the skin leaves a new and sensitive surface.

What are called *washing fluids* are merely solutions of the caustic alkalies. They facilitate washing simply by providing an excess of alkali. When the water employed in washing is somewhat hard, their use in moderate quantity may be recommended, as they precipitate the earthy salts present in the water, and render it soft. An excess of free alkali, however, in washing always tends to injure fibers and occasion them to shrink. Camphene (rectified spirits of turpentine) is also employed to some extent in washing; it acts as a solvent for grease, and its use is in no way injurious to fabrics.

784. **Stearic Acid** is a milk-white solid, inodorous, tasteless, and highly crystalline. Mixed with some margaric acid, it is extensively used for the manufacture of candles, which are sold under the name of *stearine*, or *adamantine* candles. It is obtained for this purpose mainly from tallow and lard, by heating them by steam in vats with a mixture of lime and water. Under these circumstances an insoluble lime soap is formed, while the glycerine remains dissolved in the water. This soap is then heated separately with dilute sulphuric acid, which unites with the lime to form an insoluble sulphate, and leaves the fat acids in a separate state floating upon the surface of the liquid. These, when cold, are submitted to pressure, by which the oleic acid is removed, and the stearic and margaric acids left in a nearly pure condition. Stearic acid melts at a temperature of 158° F. *Margaric acid* closely resembles stearic acid, but is more fusible, melting at a temperature of about 140° F. *Lard oil*, extracted from lard by pressure, is nearly pure *oleine*.

Questions.—Why will not pure water answer as a detergent? How does a soap act in removing dirt? Why do we not use alkalies alone as detergents? What are washing fluids? What is their use? What is the appearance of stearic acid? What are stearine candles? How is stearic acid prepared? What is said of margaric acid? What is an example of nearly pure oleine?

We apply the term *lard* to those animal fats which at common temperatures have a soft and unctuous consistency, and *tallow* to those which remain hard; the only difference between the two is in the proportion of the constituent, oleine, which is greater in lard than in tallow. The fats of carnivorous animals and of birds are soft (lard), while that of ruminating animals is hard (tallow). Fish, or train oil, is obtained from the blubber of whales, seals, and various fishes. Spermaceti is a peculiar fat found in cavities of the head of the sperm whale. It differs from other animal fats, inasmuch as it does not contain glycerine, but another basic substance termed *ethal*, while the fat acid combined with it is called ethalic acid.

Olive oil, or the sweet oil of commerce, is obtained by pressure from the fruit of the olive-tree. It is composed chiefly of oleine and a little margarine. *Palm oil*, which within a comparatively few years has become an important article of commerce, is obtained principally on the West Coast of Africa, in immense quantities, from the fruit of a species of palm-tree. It has, when fresh, a deep orange-red tint, and an agreeable odor, and at ordinary temperatures has the consistency of butter. It consists of a fluid fat, oleine, and a crystallizable solid, resembling margarine, which has been called *palmatine*, and which consists of palmatic acid and glycerine.

Human fat contains palmatine, margarine, and some oleine.*

784. **Glycerine** is a sweet, syrupy liquid, not volatile, and readily soluble in water and alcohol. Until within the last few years its properties have been overlooked, and it was not regarded as applicable to any useful purpose. In its solvent power, however, with respect to the metalloids, the salts, and the neutral organic bodies, it equals, if not surpasses, that of alcohol or water. Exposed to the air, it does not become rancid, or readily dry up. It also possesses remarkable antiseptic properties, and preserves animal tissues immersed in it in all their natural colors. It has recently been extensively applied in medicine for the dressing of wounds, burns, and sores, as a solvent for various medicinal principles in the place of alcohol or oils, and as a remedy for insect bites. It may be obtained in a nearly pure state by saponifying tallow with lime, and by various other processes.

785. When glycerine is strongly heated it is decomposed, and evolves a volatile, extremely pungent substance termed *acroleine*, which causes lachrymation. The formation of this body occasions the disagreeable smell noticed

* Bodies buried in churchyards, or submerged for a long time in water, are sometimes entirely converted into a peculiar substance resembling fat, termed *adipocere*. In the removal of the extensive cemetery, les Innocens, in Paris, in 1767, more than 1500 bodies, which had been interred in one pit, were found in this condition, and were to some extent disposed of to soap-boilers, and manufactured into soap.

QUESTIONS.—How does lard differ from tallow? From what sources are lard and tallow obtained? From what sources is train oil obtained? What is spermaceti? What is olive oil? What is palm oil? What are its constituents? What does human fat consist of? What are the properties of glycerine? What is said of its solvent powers? What is acroleine?

during the smoldering of a candle-wick, and it may be also perceived during the imperfect combustion of all kinds of fats.

786. **Wax.**—The term wax is applied by chemists to substances derived from various sources, which resemble in composition and properties the wax forming the solid portion of honeycomb. It has long been a matter of dispute among naturalists, whether the bee merely collects the wax formed by plants, or secretes (manufactures) it from honey (sugar) in the tissues of its body. The latter view of the case is now generally adopted. The constituents of wax are the same as those of the fats and oils, viz., carbon, hydrogen, and oxygen—the formula for bees-wax being $C_{34}H_{34}O_2$.

Bees-wax, in its natural state, is yellow, but is bleached white (white wax) by exposure in thin ribbons to the action of light, air, and moisture. It fuses at a temperature of 150°, and is soluble in ether and spirits of turpentine. When heated with boiling alcohol, it separates into different proximate principles, *myricine* and *cerine*, the last of which separates from the alcohol on cooling in delicate needle-like crystals. It is doubtful whether these bodies are susceptible of saponification. Wax digested with oils, forms a kind of ointment termed *cerates*. Wax also occurs in all plants, especially in the glossy coating or varnish observed upon the surface of leaves and the skins of fruit (as in the skin of the apple). From some species of plants it is obtained in sufficient quantities to constitute an article of commerce; as the *bayberry tallow*, or *myrica wax*, which is obtained by steeping the leaves and fruit of a species of myrtle in hot water. The great demand for wax is for the manufacture of candles, which are first molded by the hand and then shaped by rolling upon a hard surface. Wax burns with a beautiful clear light, and is the most expensive material employed for illumination.

787. **Resins.**—Resinous substances are found in greater or less abundance in almost all plants, and are regarded as the products of the oxydation of the essential oils. Many of them exude naturally from fissures or incisions in the bark or wood. They are all insoluble in water, but dissolve readily in alcohol, ether, and the essential oils. When pure and free from essential oils, they have no odor except when rubbed or heated. They are also good insulators of electricity, and become electric by friction. In color they are pale brown or red.

788. **Colophony.**—Common pine resin (*rosin*), also termed *colophony*, which is the residue left after the distillation of crude turpentine, is a good example of this class of resins. It contains two distinct bodies having acid properties, called *pinic* and *silvic* acids, which may be separated from each other by treatment with alcohol. These acids unite with bases to form salts, and their combinations with the alkalies are true soaps (rosin soaps). Rosin

---

QUESTIONS.—What is wax? What is the origin of bees-wax? Into what two principles may wax be divided? How is white wax formed? Under what circumstances is wax found in vegetables? What is bayberry tallow? What is said of the occurrence of resins? What are their general properties? What is a characteristic example of this class of bodies? How is rosin obtained? What is its chemical name? What is its composition? What is pine oil?

yields by distillation a great variety of products, the most important of which is a fixed oil, which is extensively used for lubrication and somewhat for illuminating purposes, under the name of *sylvic*, or *pine* oil. Rosin is extremely brittle, and may be easily reduced to a fine powder, in which condition it is used to increase friction, as it renders the surfaces to which it is applied rough and adhesive; its application to the bows of violins, and to the cords of clock weights to prevent their slipping, are familiar illustrative examples. Rosin ignited for a time and then extinguished, is converted into a soft, black, pitchy substance, generally known as ship's pitch, or shoemaker's wax.

789. **Lac**.—This important resinous substance, which is exported from the East Indies to the extent of half a million of pounds annually, is produced by the puncture of the bark of certain species of trees by an insect, and by its elaboration of the exuding juice into cells for its eggs. It occurs in commerce under three forms. Thus the broken off twigs of the trees incrusted with lac constitute *stick lac*, removed from the twigs it is *seed lac*, and when refined by melting and straining it is *shellac*. Stick lac, owing to the presence of the dead insect in its structure, yields by proper treatment a dye which is nearly, or quite as bright as that obtained from cochineal. Lac is also extensively employed in the preparation of varnishes, in the manufacture of hats (for stiffening the hat body), and as the principal ingredient in sealing-wax.* What is called *gold varnish* is a solution of shellac in alcohol, colored yellow by gamboge and tumeric.

Mastic, "*Dragon's blood*," so called from its deep red color, and *Sandarac*, are also resins largely employed for the manufacture of varnishes. *Copal* is exceedingly hard, and of a light yellow color; it differs from the other resins in being almost insoluble in alcohol and the essential oils. Copal varnish is made by first fusing the resin, and then adding spirits of turpentine and linseed oil. *Gum guiacum*, much used in medicine, is the product of the lignum-vitæ tree of the West Indies.

190. **Amber**.—The source of amber was for a long time uncertain, by some it was supposed to be a carbonaceous mineral, but it is now universally considered to be a fossil vegetable resin, the product of a species of the pine family now extinct. Whenever found in its natural location in the earth, it is associated with carbonized wood or coal. It is chiefly found on the shores of the Baltic Sea, and is apparently washed out of the sand by the waves. The largest block known is in the Royal Museum of Berlin,

* Common red sealing-wax is usually made of 4 parts of lac, 1 to 1½ of Venice turpentine, and 3 parts of vermilion, the whole being fused together by a moderate heat. By substituting different coloring principles, different colored varieties of sealing-wax are prepared.

QUESTIONS.—What is shoemaker's wax? What is lac? What are its varieties? What are its uses? What other resins are largely employed for varnishes? What is said of copal? What is amber? Where is it principally found?

and weighs 13 lbs. Amber often contains insects so perfectly and delicately preserved, that they could not have become incorporated in the mass, except it was once in the condition of a volatile oil or a semi-fluid resin. It is the hardest of all the resins, has a yellowish color, and is slightly acted upon by alcohol or the essential oils. Being commonly translucent, and susceptible of a fine polish, it is often made into ornaments, such as necklaces, the mouth-pieces of pipes, etc. The beautiful black varnish used by coach-makers is a very carefully-prepared compound of amber, asphaltum, linseed oil, and turpentine. Amber is a compound of several resinous principles, and a peculiar acid called *succinic acid.*

791. **Balsams.**—Many resinous substances, as they exude from trees or shrubs, are mixed with an essential oil, which either evaporates on coming in contact with the air, or is converted into resin by the absorption of oxygen. Such mixtures of resins and essential oils are called *balsams.* The crude turpentine or pitch which exudes from the pine is an example of a true balsam, since by distillation it is separated into a volatile oil—turpentine and hard resin. Among the other important commercial balsams are "*Canada balsam,*" the product of the silver fir, "*Venice turpentine,*" the product of a species of larch, *Copaiba balsam, balsam Tolu, Peru,* and *gum benzoin.* The three former are merely natural varnishes, *i. e.,* resins dissolved in volatile oils; the latter contain in addition an acid principle. This acid in gum benzoin is called benzoic acid, and is chemically interesting by reason of the number and marked character of the salts which it forms with bases. Benzoic acid may also be obtained artificially as a product of the oxydation of the oil of bitter almonds. The gum itself is very fragrant, and is the chief ingredient in the incense burnt in Catholic churches.

792. **Gum Resins.**—This term is applied to a class of vegetable products which contain in addition to a resin and an essential oil, a portion of gum and various other extractive matters. When they first escape from incisions in the stems or branches of trees and shrubs, they are fluid and of a light color, but gradually harden, and become of a deeper hue. Most of them also possess a strong odor, and a warm, acrid taste. Owing to their mixed composition, they are not perfectly soluble in either water or absolute alcohol, but are completely dissolved by proof spirit. This class of substances includes many valuable medicinal principles, such as myrrh, asafœtida, aloes, gamboge (the well-known coloring agent), scammony, and others. Opium is also included in this class.

793. **Varnish** is a solution of a resinous substance which is applied to the surface of bodies for the purpose of investing them with a hard, transparent, lustrous coating. When the solvent for the resin is alcohol, the product is termed spirit varnish, and when an essential or drying oil, *oil var-*

Questions.—What are its properties and uses? What is said of its chemical composition? What are balsams? What are examples? What is said of gum benzoin? What are gum resins? What are their properties? What are some of the principal bodies of this class? What is varnish?

nish. French polish is an alcoholic solution of shellac, with a little oil added.

794. **The Elastic Gums.**—Two varieties of this class only are known in commerce—caoutchouc or India-rubber, and gutta percha. There are, however, several other vegetable products of a like character which have not been made practically available.

Caoutchouc is obtained from the milky juice afforded by several species of tropical plants, in which it exists in the form of small globules suspended in an aqueous liquid, precisely in the same manner as the little globules of oily matter float about in milk. When the juice is exposed to the air, the caoutchouc gradually separates, and hardens into a white elastic mass, insoluble in water or alcohol. The usual black color of India-rubber is a discoloration occasioned by the smoke of the fires over which the fresh product is dried. The addition of a little ammonia to the milky juice temporarily prevents the separation of the caoutchouc, and under these circumstances the caoutchouc may be exported in tightly-corked bottles in its natural condition. A short exposure to the air, however, soon occasions its separation as a milk-white solid.

The physical properties of caoutchouc are well known. It is soluble in pure ether, naphtha, benzole, oil of turpentine, and the bi-sulphide of carbon. At a temperature a little above the boiling point of water it melts, but does not regain its solid, elastic state on cooling. Caoutchouc contains no oxygen, and is composed of carbon and hydrogen united probably in equal proportions.

When caoutchouc is heated in connection with sulphur, it incorporates a quantity of the sulphur into its structure, and undergoes a remarkable change, becoming what is called *vulcanized rubber.* In this condition it is less liable to be hardened by cold or softened by heat, and is also rendered more elastic and insoluble in ether and the essential oils. It is from this material that almost all India-rubber goods are now fabricated. Vulcanized rubber, by mixture with a proportion of bituminous or pitchy matter, and some earthy material like magnesia, may be converted into a hard, black, lustrous substance, which works like ivory, and is extensively used for the manufacture of combs, pencil-cases, knife-handles, etc.

FIG. 232.

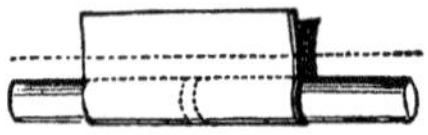

As caoutchouc is unaffected by most chemical agents, and is at the same time supple and flexible, it admits of many useful applications in practical chemistry. Short flexible tubes for the connection of apparatus are easily formed by wrapping a piece of sheet rubber over a glass tube or rod (see Fig. 232), and cutting off the superfluous portions with a pair of scissors (see Fig. 233). On pressing together and gently warming

QUESTIONS.—What substances are included in the class of elastic gums? From what source is caoutchouc obtained? What is its natural condition and color? What is said of its solubility? What is its chemical composition? What is vulcanized rubber? What effect has the addition of sulphur upon the qualities of rubber?

Fig. 233.

the fresh-cut edges, they cohere, and form a tube, which firmly tied at both ends, binds two separate glass tubes air-tight with each other. (See Fig. 234.)

Most of the caoutchouc at present used is obtained from the country bordering on the banks of the Amazon, South America.

795. **Gutta Percha.**—This substance, like caoutchouc, is obtained from the milky juice which exudes from several species of trees peculiar to Southern Asia. At ordinary temperatures it is slightly elastic and as tough and hard as wood; but when immersed in warm water, it softens and becomes highly plastic and ductile, regaining its original hardness on cooling. This property allows it to be molded with great facility into many articles of utility and ornament. Gutta percha possesses a dirty white color, and a peculiar leathery smell; it is highly inflammable, and is insoluble in water or alcohol, but dissolves in ether, the essential oils, chloroform, and bi-sulphuret of carbon.

Fig. 234.

CHAPTER XXIV.

THE NUTRITION AND GROWTH OF PLANTS.

796. **Elements of Vegetable Organization.**—The elements which constitute the organic structure of plants are, as has been already stated, carbon, hydrogen, oxygen, and nitrogen—the three former being largely in excess.

In addition to these, all plants contain various inorganic, or rather mineral substances, the presence of which in their structure is essential to a healthy growth and organization. The number and the nature of these mineral substances are ascertained by analysis of the ashes (the incombustible part) which plants yield by combustion. They are mainly *potassa, soda, lime, magnesia*, and *sesquioxyd of iron*, combined with *carbonic acid, sulphuric acid, silicic acid, phosphoric acid*, and various chlorides. The ashes of all cultivated plants contain these mineral substances; but the proportions vary with the nature of the plant. Thus silica abounds in the stalks of grains and grasses, phosphoric acid in the seeds of grain-bearing plants, potash in leaves and many edible roots, and lime in leguminous plants, peas, beans, etc.

Questions.—What is gutta percha? What are its properties? What are the elements that make up the organic structure of plants? What other substances are regarded as essential constituents? How may we ascertain the nature of the mineral substances which enter into the composition of plants? Do all plants contain the same mineral constituents? Illustrate this.

The mineral constituents of plants do not necessarily exist in the living tissues in the same form as in the ashes afforded by the combustion of these tissues. Thus, sulphur and phosphorus appear to exist uncombined in albuminous matter, while the earthy bases are very generally united in the structure of plants with vegetable acids. In the process of combustion, however, the latter become converted into carbonates, while the sulphur and the phosphorus unite with oxygen to form acids, which in turn generally unite with one of the bases present to form salts characteristic of these elements.

797. **Sources of Nutriment to Plants.**—Plants obtain their nutriment partially by their leaves and partly by their roots. The former are furnished with a great number of microscopic *pores*, or *stomata*,* while in the latter the nutritious matter penetrates the cell-walls of the rootlets by the force of endosmosis. It must be, therefore, evident that the plant can only absorb its nutriment in a liquid or aeriform condition.

798. The *hydrogen* and *oxygen* which plants contain are derived principally from water which is absorbed as a liquid by the roots from the earth, or as vapor, from the air, by the leaves. The substances which make up the great bulk of the structure of all plants, viz., cellulose, lignine, starch, sugar, and gum, contain oxygen and hydrogen in exactly the same proportions as they exist in water, and they may in fact be regarded as merely compounds of carbon (their other constituent element) with water. The presence of water in a liquid condition in the plant is, moreover, indispensable to its development, since all the solid ingredients of plants are assimilated from the sap, which is rendered liquid by water. Plants, however, absorb through their roots much more water than is applied to the enlargement of their structure, and in such cases a constant evaporation takes place from their leaves.

799. The *carbon* existing in plants is entirely derived from carbonic acid, carbon itself being insoluble in water. Plants absorb carbonic acid principally from the air through their leaves. Although but 2 measures of this gas are contained in 5,000 of air, its aggregate supply, by reason of the great extent of the atmosphere, is very large, and has been estimated to exceed seven tons for each acre of the earth's surface. The immensely-extended surface presented by the leaves of plants enables them to withdraw carbonic acid from the atmosphere in a very rapid manner.

* In the leaf these pores are found mainly upon the under side. In the white lily, where they are unusually large, and are easily seen by a simple microscope of moderate power, there are about 60,000 to the square inch on the epidermis of the lower surface of the leaf, and only about 3,000 in the same space upon the upper surface. More commonly, there are few or none upon the upper side, direct sunshine being unfavorable to their operation. Their immense numbers make up for their minuteness. They are said to vary from less than 1,000 to 170,000 to the square inch of surface.—Gray.

Questions.—Do they exist in the tissues in the same form as in the ashes of plants? Through what organs do plants obtain their nutriment? In what conditions is nutriment only absorbed by plants? From what source do plants derive oxygen and hydrogen? What is said of the existence of water in plants?

Carbonic acid is also supplied to plants from the soil through their roots. Humus, in the course of its decomposition, continually evolves carbonic acid; and the air in all soils rich in decaying vegetable matter always contains a much larger proportion of carbonic acid than an equal bulk of the general atmosphere. Carbonic acid does not, however, enter into and circulate in the structure of plants as a gas, but always in a state of solution. In the leaf the moisture with which the tissues are saturated becomes the medium of its absorption; in the case of the root, it is taken up naturally in solution in water. Some chemists maintain that the soluble forms of humus (crenic and apocrenic acids) are directly absorbed by roots, and thus become sources of nutriment to the growing plant. This theory, from the fact that it has been strenuously opposed by Liebig and other authorities, has not been generally received, but the most recent investigations appear to substantiate its correctness.

The carbonic acid absorbed by the plant, either by its leaves or roots, is decomposed; its carbon constituent being retained and assimilated, while the oxygen originally combined with it is restored to the atmosphere. This decomposition takes place mainly in the leaves of plants, and is effected solely under the influence of light. It goes on most actively when the plant is exposed to the direct action of the rays of the sun, but is entirely suspended during the night. It is also checked in a very marked degree during the daytime, when the light of the sun is intercepted by thick clouds.

Plants, therefore, in the daytime continually absorb carbonic acid and exhale oxygen.

In the night this process is to a degree reversed; carbonic acid is absorbed as before, but the influence of light being withdrawn, it is again restored to the air unchanged. Oxygen, also, as the result of certain processes allied to oxydation, is at the same time abstracted to a very small extent from the atmosphere. The action of oxygen under such circumstances is illustrated by the fact that the leaves of certain plants which are bitter in the evening are sour in the morning, inasmuch as the products formed during the day become acid by oxydation at night; when, however, the assimilation of carbon is recommenced under the influence of light, the excess of oxygen is neutralized, and the original bitter properties are restored. Furthermore, if during the night a plant be covered by a bell-glass, the atmosphere contained in it will be found to contain a larger amount of carbonic acid than before. This is occasioned by the oxygen of the air surrounding the plant effecting an oxydation on its surface, and thus producing a certain quantity of carbonic acid; the amount, however, is very unequal in different plants, and is most abundantly produced by such as contain a large proportion of easily oxydizable volatile oil in their glandular vessels. Flowers and fruits also form an

QUESTIONS.—What is the source of carbon to plants in the soil? In what condition does carbonic acid exist in plants? What becomes of the carbonic acid absorbed by plants? Under what circumstances does its decomposition take place? State the action of plants by day. What takes place at night? How may the decomposition of carbonic acid be illustrated?

exception to the usual action of vegetation, as they absorb oxygen from the atmosphere, and evolve carbonic acid.*

The decomposition of carbonic acid by the green portions of plants may be easily demonstrated by placing fresh leaves in a bell glass partially filled with water, and partially with carbonic acid gas; on exposing the glass to the sunshine, the carbonic acid disappears, and after some time is replaced by a rather smaller quantity of oxygen, which may be tested in the usual manner.

The carbonic acid withdrawn from the air by the action of vegetation is constantly reproduced and restored to the atmosphere by the respiration of animals, and by the processes of decay and combustion; and these two classes of phenomena so completely compensate and balance each other, that the proportional quantity of oxygen and carbonic acid present in the atmosphere remains ever essentially unchanged. (§ 330.)

800 It is the generally received opinion that plants derive their *nitrogen* entirely from the soil, by means of their roots, in the form of ammonia, although certain eminent French chemists maintain that this element is in part supplied directly from the atmosphere. The sources of supply of ammonia to soils are numerous; it is absorbed and condensed from the atmosphere by dew, rain, and snow, and also by the clay and humus of the soil itself. It is an abundant product of the decomposition of all nitrogenized animal and vegetable substances, and is undoubtedly produced to some extent by the direct contact of humus with the nitrogen of the air.

In what manner the assimilation of ammonia takes place in the vegetable kingdom is not certainly known. Its decomposition, however, furnishes plants with an additional source of hydrogen. The quantity of nitrogen contained in plants is comparatively small, and it is found chiefly in the sap and in the seeds. In 2,500 lbs. of hay there are 984 lbs. of carbon and only 32 lbs. of nitrogen.

801. Plants derive their *mineral*, or *earthy constituents* from the soil, and the solution of these substances in water, which is necessary for their absorption by root-fibers, is greatly facilitated by the action of carbonic acid. (§ 432).

802. Soils owe their origin to the disintegration or gradual crumbling down of rocks, by the action of water, air, frost, and various other agencies. Through the action also of air, moisture, and carbonic acid, the stony parti-

* There is a common belief that plants in flower at night deteriorate the air, and that, therefore, their presence in sleeping apartments is objectionable. The ill effects noticed, if actually occurring, are probably due, not to the formation of carbonic acid, but to the volatilization of certain volatile oils, many of which, in even very small quantities, act powerfully upon the animal system.

QUESTIONS.—How is the carbonic acid withdrawn from the air by plants restored? From what source do plants obtain their nitrogen? Do plants absorb nitrogen directly from the atmosphere? From what source do plants derive their mineral constituents? What is the origin of soils? Through what agency are certain of the mineral constituents of a soil rendered soluble?

cles which make up a soil are chemically decomposed, and certain of their mineral constituents, potash, soda, etc., are rendered soluble and capable of assimilation by plants. The most abundant constituent of soils is silica (sand), which frequently forms nine tenths of their entire weight. Good arable land, however, always contains a large proportion of alumina (clay), and in soils underlayed by limestone or calcareous rocks, the proportion of carbonate of lime present is often very considerable.

The relative proportions of sand, clay, and lime in soils give to them certain peculiar physical characters. A soil in which sand predominates is light and porous; an excess of clay, on the other hand, renders it heavy and retentive of moisture. The best soils are those in which the earthy constituents are so proportioned that the light, porous qualities of one are balanced by the close, retentive properties of the other.

The quantity of organic matter (humus) derived from the decomposition of animal or vegetable substances present in a soil, essentially modifies its character. The *average* amount of organic matter contained in soils is about 5 per cent. Fertile alluvial soils, or those deposited from water, are generally characterized by the presence of a much larger proportion, and in some peaty soils, the amount may exceed 70 or 80 per cent.

803. Although plants obtain a large proportion of their nutriment from the air, yet as they abstract from the soil considerable quantities of earthy matter, which is only replaced naturally by the slow disintegration of mineral substances, it is evident that the long-continued cultivation of the same plant upon the same soil may so far exhaust its soluble mineral constituents as to render it unfruitful. This is especially the case where large crops are raised year after year, and entirely removed from the soil to furnish food for men and animals. As different plants, however, require for their nourishment different mineral substances, or different quantities of the same substance, a soil which has become unfitted for the growth of one plant, may still contain the elements necessary for the support of another; and hence a succession of crops of different vegetables may be raised upon the same soil, when two successive crops of the same vegetable could scarcely be obtained. This system of cultivating different plants in succession, upon the same soil, is termed the *rotation of crops*, and the period of time over which the rotation is allowed to extend is usually several years. During the interval which, under these circumstances, elapses between two successive crops of the same nature, the soil has time to renew itself; or in other words, it regains through the gradual decomposition of its insoluble, stony compounds, the constituents originally abstracted from it. In England, wheat is ordinarily grown upon the same soil only once in four or five years, the intermediate crops being

---

QUESTIONS.—What is the most abundant constituent of soils? What influence has an excess of sand upon a soil? What clay? What is the composition of the best soils? What is the average quantity of organic matter in soils? How does the growth of plants tend to impoverish a soil? What is understood by the rotation of crops? How does the system of rotation tend to benefit a soil?

turnips, barley, oats, and potatoes, crops which require but a small quantity of the mineral constituents which are essential to the growth of wheat.

The resuscitation of an exhausted soil is also often effected by allowing it to *fallow*, or remain without a crop, exposing it at the same time (by plowing) to the action of air and moisture.

804. **Manures.**—The method, however, of obtaining from the soil the largest produce, consists in presenting to the plants cultivated upon it all the materials requisite for their nutrition in sufficient quantity, and in the condition which will most readily admit of their absorption. This is accomplished through the agency of manures.

The most valuable and energetic of all manures are the excrements of men and animals, inasmuch as they are capable of yielding to the soil, through their decomposition, a large quantity of ammonia and carbonic acid, and the principal mineral substances which enter into the composition of plants. By acting as ferments, they also assist in rendering useful, materials which without them would be far less beneficial. The flesh and blood of dead animals, fat and oily matters, hair, wool, skin, horns, hoofs, and bones, are also highly efficacious as manures. *Guano*, which is the decomposing excrement of sea-birds, owes its value principally to the ammonia and phosphate of lime which it is capable of yielding to plants. These two ingredients, in the best varieties of guano, constitute about one third of its entire weight. Animal substances which decompose most readily, such as excrement, blood, flesh, etc., yield ammonia and carbonic acid most rapidly, and constitute the most powerful manures; those, on the contrary, which decompose more slowly, are less powerful, but more lasting in their effects.

Animal manures exposed to air are liable to deterioration by the volatilization and escape of their ammonia. They may also, when incorporated with the soil, prove injurious by evolving a greater quantity of ammonia and carbonic acid than plants require or can absorb. Agriculturalists express this when they speak of a manure as being too strong. These evils may be in a great measure prevented by incorporating with the strong manure a considerable quantity of vegetable refuse, straw, weeds, leaves, peat, etc., which substances, being less prone to decomposition, check the otherwise too rapid putrefaction. The animal products at the same time react upon the vegetable substances, and gradually bring them into such a state as renders them also most valuable additions to the soil. Common farm-yard manure is an example of a mixture of this character. The loss of ammonia may also be effectually prevented by adding to manures a small quantity of a weak solution of any acid, or gypsum (sulphate of lime), or copperas (sulphate of iron).

Questions.—What is fallowing? In what manner can the largest produce be obtained from the soil? What are the most valuable of manures? Why are animal manures especially valuable? What is guano? To what does it mainly owe its value as a manure? What is said of the comparative effect of different animal manures? Under what circumstances will animal manures deteriorate? When are they said to be too "strong?" How may these evils be obviated?

805. *Vegetable manures*, under which head are included vegetable refuse of all kinds, straw, leaves, sea-weed, and green crops which are merely sown to be plowed in, yield by their decomposition, when mixed with the soil, carbonic acid and small quantities of ammonia and the mineral constituents of plants. They also render a soil porous and retentive of moisture and ammonia. They are most advantageously used when employed in combination with some kind of animal manure.

806. *Mineral manures* are generally used for specific purposes. Of these the most important is lime. This substance acts mechanically by giving a proper consistency to soils, and chemically, by facilitating the decomposition and promoting the solubility of the more insoluble mineral and vegetable compounds. Quicklime is especially useful in soils rich in humus—peaty or mossy soils. Soils of this kind generally contain an excess of acid, which greatly interferes with their fertility; this acid is neutralized by the addition of lime. Quicklime, however, should never be mixed with animal manures, as it tends to promote the escape of ammonia. Gypsum, or marl which contains lime in combination, may be used in such cases with beneficial results. *Wood ashes* act upon soils and manures in the same manner as lime, they are, however, more valuable than lime, as they contain alkaline salts and phosphoric acid. *Hard coal ashes* have but little value as manures; they do not contain any appreciable quantity of alkaline salts or phosphoric acid, and consist mainly of silica, alumina, oxyd of iron, and a small percentage of sulphate of lime. *Phosphate of lime* is an exceedingly valuable manure, and as it is found in almost all plants, it may be applied with advantage to almost all cultivated soils. It exists abundantly in bones and in guano, and in smaller quantity in all organic manures and in the ashes of plants. Phosphate of lime is the special mineral constituent of wheat, and its presence in a soluble condition in a soil, is necessary for the successful cultivation of this cereal. *Gypsum* or sulphate of lime is a valuable addition to soils which do not contain it. It is partially useful as supplying lime and sulphuric acid, and partially as an agent for fixing ammonia. It is especially adapted for clover, bean, and pea crops.

A thorough tillage, or a complete pulverization and separation of the particles of a soil, will go far toward compensating for a lack of manures. With every increase in the comminution of the particles of a soil, an increased power is given to the soil for the absorption, retention, and condensation of moisture, ammonia, and carbonic acid, an opportunity for the free permeation of atmospheric air, a facility to the rootlets of plants for extension, and a consequent increased facility for receiving and appropriating nourishment. This fact is strikingly illustrated by a comparison of the sterile soils of New

QUESTIONS.—What is the action of vegetable manures? How may they be most advantageously used? What is said of lime as a manure? Upon what soils is the use of lime especially beneficial? When should lime not be used? What is said of the fertilizing action of wood ashes? What of hard coal ashes? What of phosphate of lime? What of gypsum? What is said of the importance of thorough tillage and pulverization of a soil? How is this illustrated?

England and the fertile ones of the West. Both have been formed from the disintegration of the same varieties of rocks, and both contain the same mineral constituents in nearly the same proportion. In the former, however, the mineral particles are extremely coarse, but in the latter they are nearly in the state of an impalpable powder. The fertile soils of the West also contain a large percentage of humus in an advanced stage of decomposition, while very often the humus in the soils of New England is in a state allied to charcoal, and completely insoluble.

CHAPTER XXV.

ANIMAL ORGANIZATION AND PRODUCTS.

807. **Animal Organization.**—Inasmuch as all animals derive their sustenance, either directly or indirectly, from the vegetable kingdom, the elements which enter into their composition are essentially the same as those contained in plants. Most animal substances are, however, more complex in their nature than substances of vegetable origin, and as a necessary consequence, they are less permanent, and the products of their decomposition are more numerous. Water and fat are almost the only substances which contain but two or three elements that exist in the animal organism—almost all the others being also rich in nitrogen, sulphur, and phosphorus.

808. **Proximate Animal Constituents.**—The chief proximate constituents that are found in the animal system are albumen, fibrine, caseine, gelatine, fat, water, and phosphate of lime. The proportions of solids and fluids in the animal body are very unequal. A man of 154 lbs. weight contains 116 lbs. of water, and only 38 lbs. of dry matter. By slow desiccation this water may be got rid of, when the body will assume the condition presented by the mummies of Egypt and Peru. The fluids of the body, as they exist in the living tissues, are not simply water, but watery solutions of various organic and inorganic substances.

Of the proximate animal constituents named above, albumen, fibrine, and caseine appear to have essentially the same composition and properties as the substances of the same name originating in vegetable tissues. The two first are diffused throughout the whole body; the third is found only as a special secretion.

809. **Albumen.**—The best example of animal albumen is to be found in the white of an egg. This, when evaporated to dryness, yields about one

---

QUESTIONS.—What are the elements of animal substances? In what respects do animal substances differ from vegetable? What are the chief proximate constituents of the animal system What is the relation between the solids and the fluids in the animal body? What is said of the composition and distribution of animal albumen, fibrine, and caseine? What is the best example of animal albumen?

eighth of solid albumen, the rest being water. The ashes of albumen thus obtained contain common salt, carbonate, phosphate and sulphate of soda, and phosphate of lime, which saline substances constitute about 5 per cent. of the weight of the white of the egg, or $1\frac{1}{2}$ per cent. of the weight of the dried albumen. The *yolk of eggs* consist essentially of albumen, holding in suspension drops of yellow oil. This oil forms about two thirds of the weight of the yolk in a dried state, and may be extracted from the coagulated yolk by pressure, or by digestion in alcohol.

When albumen is agitated with water, little solid bodies are formed, which under the microscope resemble the cells which make up the cellular tissue of animals, and are perhaps the nearest approach to an organic structure that man has yet been able to produce artificially.

810. **Fibrine** is found in the animal body in two distinct states, viz., in a solid condition in muscular flesh, and as a fluid in the blood. A piece of lean beef washed in cold water until it is perfectly white, affords us an example of fibrine in the first condition, associated with membraneous matter, nerves, fat, etc. It may be extracted from the blood in a purer condition, by strongly agitating that fluid, in its recent and warm state, with a bundle of twigs. The fibrine adheres to these latter in the form of long, elastic strings, and is removed and cleansed by washing with cold water. In this condition it contains only a little fat, which may be extracted by ether.

FIG. 237.

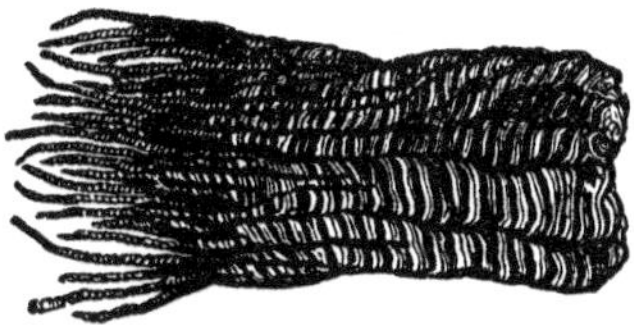

The lean part of the muscles of all animals consists chiefly of fibrine, and it is, therefore, the principal constituent of animal flesh. Fig. 235 represents the structure of muscle as seen under the microscope, the cross wrinkles showing the way in which the fibers contract in the living animal. Fibrine derives its name from its peculiar fibrous appearance, but under the microscope it appears to be composed of small globules arranged in strings. When pure, it is quite tasteless, and insoluble both in hot and cold water, but by long-continued boiling it is partially dissolved. By drying it shrinks pro digiously in volume, loses about 80 per cent. of water, and becomes transparent and horny, and in this condition may be preserved for an indefinite period. Fibrine, when in solution, assumes the solid form spontaneously, as as soon as it is withdrawn from the influence of life. It is this which causes blood to coagulate almost as soon as it is drawn from the veins—the coagulation being a net-work of fine fibers of fibrine inclosing the liquid se-

QUESTIONS.—Of what does the white of an egg consist? What is the composition of the yolk? What phenomenon of albumen is mentioned? In what conditions is fibrine found in the animal economy? How may it be prepared in a state approaching to purity? Of what part of the animal system does it form the principal part? What is the origin of its name? What are its properties? What is said of fibrine in solution? What causes the coagulation of the blood?

rum and coloring principle of the blood. Owing to this circumstance, little or nothing is known of fibrine in the soluble state, but it is believed that the chemical composition of soluble and insoluble fibrine is somewhat different. Its composition is represented by the formula $C_{400}H_{310}N_{50}O_{120}PS$.

811. **Caseine** in the animal system occurs only in milk. Its composition and properties have been already described. (§ 706.)

812. **Gelatine.**—Various parts of the animal body, particularly the skin, the tendons, cartilage, and the soft portions of the bones, dissolve completely by long boiling in water, and produce a liquid which solidifies on cooling to a jelly. The substance so produced is termed *gelatine.* Chemists do not regard it as existing naturally in the system, inasmuch as it is never found in the fluids of the body, as might be expected from its ready solubility in warm water; but it is supposed to be produced by a specific chemical change of some of the albuminous principles by the action of the hot water and the oxygen derived from the air. The gelatine extracted from cartilage appears to differ somewhat from that extracted from animal membranes proper, and has received the distinctive name of *chondrine.* The term *cartilage* is applied to a dry, elastic tissue, very widely distributed in the animal economy, which sometimes serves to connect the ends of bones which move upon each other, and sometimes constitutes prolongations of the bones themselves, as for example, in the ribs, thus increasing their elasticity.

Gelatine is an important constituent of the animal body, and is obtained from almost all solid parts of it, but more especially from the tendons, ligaments, the inner skin, and from bones and horns. It is very rich in nitrogen, and contains some sulphur, but it is not allied to the proteine group of substances. Its formula is $C_{13}H_{10}N_2O_5S$. Gelatine is exclusively an animal product, and is never found in plants, pectine being the vegetable jelly principle.

Common glue is dried gelatine, and is prepared by boiling refuse skin and bones, and evaporating the solution. The liquor yields on cooling a thick jelly, which is cut by wires into thin layers, and dried by exposure to the air. *Isinglass,* which is the purest variety of gelatine, is the dried swimming, or air-bladder of several varieties of fish, especially of the sturgeon. Gelatine is also extracted from the tender and ligamentous part of calves' feet, for the purpose of forming the well known "calves' foot jelly."

A dilute solution of gelatine prepared from clippings of hides constitutes the *size* which is usually applied to paper to fill up its pores, and thus prevent the spreading of ink. The difference between writing and printing paper consists simply in the fact, that the former is sized, while the latter is not.*

* A cheaper kind of sizing for paper is also prepared by boiling resin with a strong solution of potash. This is first added to the paper pulp, and when it has become thoroughly incorporated, a solution of alum is poured in. The alumina displaces the potash in combination with the resin, and forms a more insoluble compound in the fibers of the paper.

QUESTIONS.—What is said of caseine? How is gelatine prepared? What is gelatine? What is said of the distribution and composition of chondrine? What is cartilage? What is glue? What is isinglass? What is size? For what purpose is size applied to paper?

Gelatine is largely employed as an article of food, in soups, jellies, etc., but it possesses very little nutritive value. In an indirect way, under the conditions of a restricted diet usually met with in a sick room, its administration in the form of jellies, etc., appears to be beneficial, as it seems to protect some of the constituents of the body from waste.

Gelatine united with tannic and gallic acids produces insoluble compounds, and the application of this principle to the manufacture of leather has been already noticed. Skins may, however, be converted into leather by other methods; as by impregnating them, after they have been freed from fatty matters by digestion in alkalies, with a solution of common salt and alum, and then working them with various oils. Glove leather is prepared in this manner; the still softer chamois, or wash-leather is obtained by working the skins for a long time with the brains of certain animals or the yolks of eggs—the effect in both instances being due to the action of certain peculiar oily or fatty substances.

813. **Glycocoll.**—By boiling gelatine with dilute sulphuric acid, and afterward separating the acid by chalk, a very remarkable change is effected—the gelatine being converted into a sweet, crystallizable substance, which is termed *glycocoll*, or sugar of gelatine.

814. **Brain and Nerves.**—The substance of the brain, nerves, and spinal marrow differs from that of all the other animal textures. It appears to be albumen in a peculiar state, associated with certain remarkable fatty substances, and in the brain especially a large amount of unoxydized phosphorus is believed to be present. Only about one fifth part of the nervous tissue, however, is solid matter. The phosphorus contained in the brain is said to amount to 3 or 4 per cent. of its entire weight.

815. **The Skin** of animals consists of two layers, the skin proper, called also the *cutis*, and the *derma*, which envelopes the muscles and the bones; and the outer layer, the *epidermis*, or *cuticle*, which originates from the former, and consists mainly of albuminous cells, which losing their liquid contents by evaporation, gradually become flattened scales at the surface. These undergo constant exuviation, and are constantly replaced from beneath, the superficial ones becoming dry and horny (scarf skin), and serving as a protection to the sensitive or true skin underneath. The lowest portion of the cuticle, resting on the cutis, is called the *rete mucosum*, and contains the pigment which in the dark races imparts color to the skin. This pigment seems to be produced by the agency of sun-light and continued high temperature, and contains a large percentage of carbon.

The cuticle, or outer skin of most animals is perforated by numerous small orifices, through some of which hairs pass, while others give passage to the fluids of perspiration, or allow certain oily fluids to exude. "In the human

---

QUESTIONS.—What is the nutritive value of gelatine? How is leather formed other than by tanning? What is glycocoll? What is said of the composition of the brain and nerves? Of what does the skin consist? How is the epidermis formed? What is the rete mucosum? What is said of the pores of the skin?

system the pores are more numerous in some parts of the body than in others, but the outer skin of a full-grown man is sprinkled over with about seven millions of them, while the united length of certain spiral vessels connected with them is reckoned at 28 miles." Through the pores of the skin, also, air enters and escapes continually in a healthy state of the body, as it does from the air-vessels of the lungs.

Fig. 236.

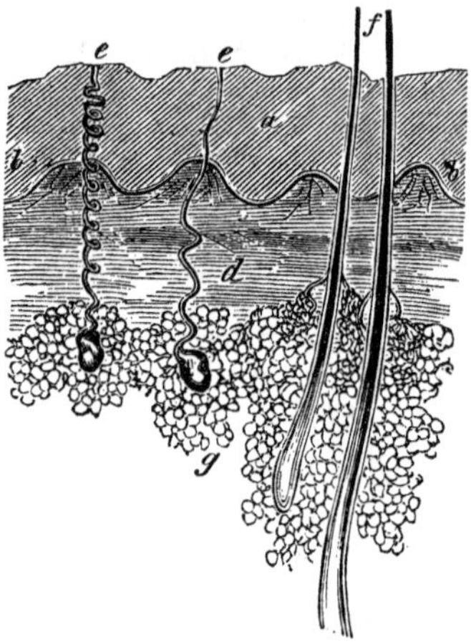

Fig. 236 represents a vertical section of the skin greatly magnified, *a* being the cuticle, or outer skin, *b d* the true skin, *e* the sweat glands and their ducts, the outlets at the surface being pores, *f* hairs, *g* cellular tissue.

816. **Horny Matter.**—Hair, wool, bristles, feathers, nails, claws, and hoofs of animals are regarded as having the same general chemical composition as that of the epidermis, of which they may be considered as appendages. They are insoluble in water, but soften in boiling water, and entirely dissolve by continued digestion in caustic alkalies. They contain several oily or fatty substances, generally colored, from which they derive their peculiar hue.

Each hair originates in a little flask-shaped follicle (*f*, Fig. 236), which is formed by a depression of the cutis, and lined by a continuation of the cuticle. The hair grows by constant prolongation from this follicle, and its color is due to a peculiar colored oil, which in black hair contains a considerable quantity of iron. The surface of the hair is scaly, and not smooth, as it appears to the naked eye; and in the case of wool, which is a modification of hair, the edges of the fiber, seen under a microscope, have the appearance of a fine saw, with the teeth sloping in a direction from the roots to the points. Were a number of thimbles with uneven edges inserted into each other, a cylinder would result not dissimilar in outline from a filament of merino wool, the appearance of which, under the microscope, is represented by Fig. 237. This peculiar structure of wool gives it the property of *felting*, so that when a mass of wool is alternately compressed and relaxed, the little imbrications or scales of the fibers lay hold of and match into each other, and thus compact the whole into a solid tissue, or *felt*. Some varieties of hair, included under the term *fur*, have also sufficiently roughened surfaces to enable them

---

Questions.—What is said of the composition of hair, horns, etc.? What of the origin of hair? To what does hair owe its color? What is the external structure of hair? Why does wool felt?

to felt. Fig. 238 exhibits the appearance of the hair of the seal (*a*) and of a species of caterpiller (*b*), when viewed under the microscope.

FIG. 237.

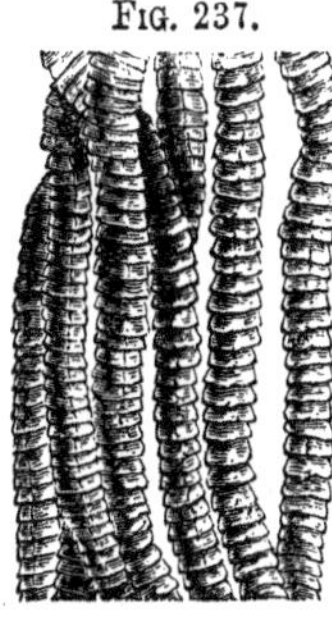

FIG. 238.

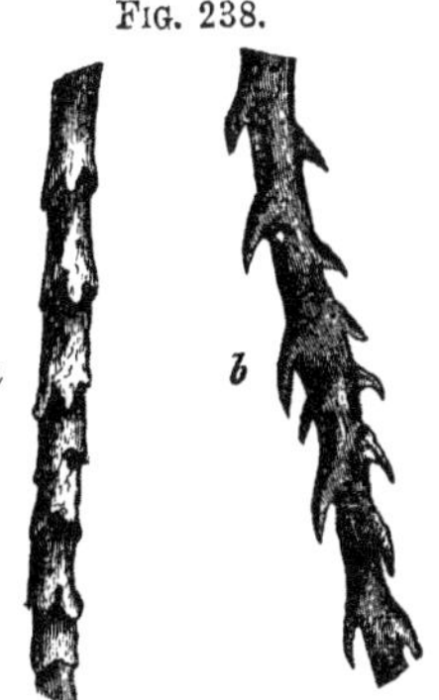

817. **Bones.**—The bones of animals are composed of organic matter, which is essentially the same as cartilage, and of earthy matter, consisting chiefly of phosphate and carbonate of lime—the latter constituting in mammalia about two thirds of the weight of the bone.* The organic and earthy bases contained in bones may be easily separated from each other. Thus when a bone is digested for some days in a dilute solution of hydrochloric acid, the earthy salts dissolve out, leaving the cartilage soft and flexible, but retaining exactly the shape of the bone. To accomplish this perfectly, it is necessary to renew the liquor several times, and finally to wash the cartilage with fresh water until no trace of acid remains. The cartilage may also be removed by heating the bone for some time in an open fire with free access of air—the organic matter in this way being burned away, while the bone-earth remains.

The bones of mammalia and of birds agree very closely in chemical composition, but the bones of fishes vary considerably as regards the relative proportions of contained earthy and organic matter. In what are called the *cartilaginous fishes*, sharks, etc., the bones are almost entirely destitute of calcareous salts, and in the bones of all fishes the proportion of cartilaginous matter is always greater than in those of other vertebrated animals; hence the flexibility of the bodies of fishes. The composition of fish-scales resembles that of bone, since they contain from 40 to 50 per cent. of phosphate of lime, from 3 to 10 per cent. of carbonate of lime, and from 40 to 55 per cent. of organic matter.

---

* The following is an average composition of the bones of a healthy adult man:—

Cartilage	32·17
Blood vessels	1·13
Phosphate of lime	51·04
Carbonate of lime	11·30
Fluoride of calcium	2·00
Phosphate of magnesia	1·16
Soda, chloride of sodium	1·20
	100·00

QUESTIONS.—What is the composition of bones? How may the constituents of bones be separated? How do the bones of mammalia, birds, and fishes correspond?

818. **The Teeth** have essentially the same composition as the bones, except that they contain less cartilage. The white external part of the tooth beyond the gum, called the enamel, is almost wholly composed of phosphate of lime, carbonate of lime, and a small quantity of fluoride of calcium, and contains only a trace of animal matter.

819. **Shells** are composed of a mixture of carbonate and phosphate of lime. The shells of crustacea, lobsters, crabs, etc., usually contain from 50 to 60 per cent. of carbonate of lime, from 4 to 5 per cent. of phosphate, and the balance animal matter. The shells of mollusca, oysters, clams, etc., on the contrary, are nearly pure carbonate of lime, and contain scarcely any phosphates or organic matter.

820. **Milk.**—This peculiar liquid is secreted by the female of the class mammalia for the support of its young, and seems to contain the same constituents, although in somewhat different proportions, in all the different species of animals producing it. Milk is wonderfully adapted for the office it is naturally intended to discharge, viz., that of providing materials for the rapid growth and development of the young mammalian animal; inasmuch as it contains caseine, a nitrogenous matter nearly identical in composition with muscular flesh, fat, sugar, and various salts, the most important constituent of the latter being phosphate of lime. This last is held in complete solution in the slightly alkaline liquid, and sustains an important relation to the formation and growth of bone. The following analysis exhibits the composition of 1,000 parts of cow's milk in a fresh state:

Water	873·00
Caseine	48·20
Fat (butter)	30·00
Milk sugar	43·90
Phosphate of lime, magnesia, and iron	2·30
Chlorides of potassium and sodium, with a little free soda in combination with caseine	2·10
	1,000

Woman's milk contains more sugar, but less caseine and butter than the milk of the cow. The latter is not so well adapted to the functional wants of the child, but may be improved by diluting it with water and sweetening it with sugar, the effect of which is to reduce the percentage of the nitrogenized element, the caseine, and render it more suitable for digestion and assimilation. "Milk, moreover, is not suitable as the sole nourishment of adult life, since it does not contain in sufficient quantity those phosphorized compounds which are necessary to repair the waste of the tissues, which at this period are more active than in infancy."

When milk is viewed under a microscope of moderate power, it is seen to consist of a perfectly transparent liquid, in which are suspended numerous

---

QUESTIONS.—What is the composition of the teeth? Of shells? What is milk? What is its natural office? What is its general composition? In what respect does woman's milk differ from cow's? What is the appearance of milk under the microscope?

globules of fat, as is represented in Fig. 239. These globules are the butter, and mainly give to milk its opaque, white appearance. When milk is allowed to stand, the globules, by virtue of their low specific gravity, rise to the surface, and form a layer of cream, and by strong agitation or churning, they may be further made to coalesce into a mass, and form "butter." It is also believed that each fat globule is inclosed in a little sack of caseine, which is ruptured by the agitation. During the operation of churning, oxygen is absorbed from the air, the temperature rises, and the milk, if not already acid, turns sour.

FIG. 239.

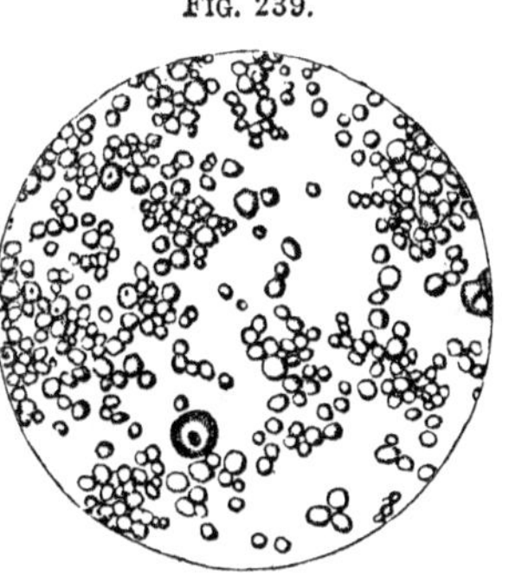

Butter consists of a mixture of margarine, oleine, and a peculiar volatile, odoriferous principle termed *butyrine*, which contains butyric acid. In order that butter should keep well, it is necessary that the buttermilk should be thoroughly freed from it, since the caseine and albumen contained in this readily undergo decomposition, and produce an acid fermentation which separates the butyric acid and other volatile acids, and imparts to the butter a disagreeable, rancid taste. This same object, *i. e.*, the preservation of butter, can be also attained by melting the butter, when the watery part subsides and carries with it the azotized matter. The flavor of the butter is, however, somewhat impaired by this process.

821. Milk, when in a fresh state, is always feebly alkaline; but it soon sours in the air, particularly in warm weather—lactic acid being developed. The presence of this acid causes the caseine to coagulate, or become insoluble, when it separates in clots, carrying the fatty globules with it. Milk in this condition is said to be *turned*. This change may be prevented, without injuring the quality of the milk, by the addition of a minute quantity of carbonate of soda.

822. **Cheese** is a mixture, in various proportions, of coagulated caseine and butter. The caseous matter is separated in the form of cheese, by leaving the milk for some little time at a temperature of 120° F. in contact with a piece of the lining membrane of the stomach of a calf, which is called rennet. This by its presence is believed to cause a sort of acid fermentation, which causes the milk to separate into a solid white opaque *curd*, and a thin, translucent *whey*, the former consisting chiefly of caseine and butter, and the latter of water, holding in solution most of the saline constituents of the milk, together with the milk sugar. The coagulum thus obtained is sepa-

QUESTIONS.—In what condition does butter exist in milk? How is butter collected in a separate state? What is its composition? Why is butter containing butter-milk liable to deteriorate? How does the melting of butter tend to preserve it? What is the chemical condition of milk? What causes it to coagulate? What is cheese? How is it manufactured?

rated from the whey by straining; then drained, mixed with a portion of salt, and sometimes other condiments, and subjected to pressure. The product is cheese, which, when kept for several months in a cool situation, undergoes a kind of putrefaction, and obtains thereby a peculiar taste and odor. The goodness of cheese depends upon the proportion of cream left in the milk, and upon the method of its manufacture.

823. **Blood.**—"The blood is the general circulating fluid of the animal body, the source of all nutriment and growth, and the general material from which all the secretions, however much they may differ in properties, are derived. It also serves the scarcely less important office of removing and carrying off from the body principles which are hurtful, or no longer required."

In all vertebrated animals, viz., man, mammals, birds, reptiles, and fishes, the blood has a bright red color; while in the invertebrata, as insects, the crustacea, mollusca, and zoophytes, it is very often colorless, but sometimes tinged with red, yellow, green, or other hues.

824. **Composition of the Blood.**—The blood, as seen under the microscope, circulating in the vessels, appears to consist of a colorless liquid, holding in suspension little globules, called *corpuscles*, or cells. Some of these, in man, are white, but most are red, and give to the blood its color. The red corpuscules vary in size and shape in different animals, and the microscopist, taking advantage of this circumstance, is enabled, even after the lapse of years, to distinguish in the dried stain, human from animal blood, and also to pronounce with certainty whether a particular spot is occasioned by blood or some other liquid. In man they appear as circular flattened disc, having an average diameter of 1-3200th of an inch, and a thickness of 1-124,000th. In reptiles they are elliptical and larger than in man. Fig. 240 represents their appearance in human blood magnified 500 diameters, and Fig. 241 their appearance in the blood of a frog, magnified 250 diameters. When dried, they form, in man, on an average, about 13 per cent. of the whole weight of fresh blood.

FIG. 240.

FIG. 241.

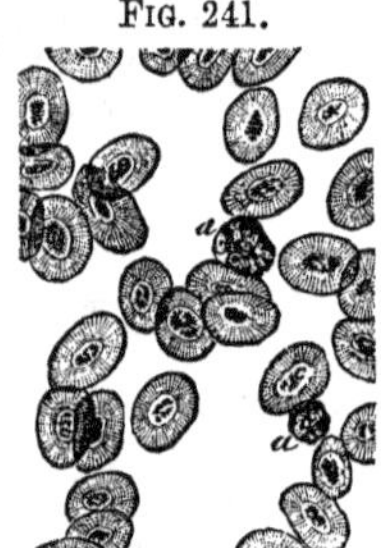

In man and all warm-blooded animals, the color of the blood in the arteries is of a bright scarlet, while in the veins it is dark red. These changes of color are primarily due to the action of atmospheric oxygen upon the blood, while passing through the lungs.

QUESTIONS.—What is the blood? What is said of its color? What is its appearance under the microscope? What is said of the blood corpuscles? What is said of the color of the blood in the veins and arteries?

The fluid of the corpuscles contains the coloring matter of the blood, which is called *hæmatine*, particles of fat, a colorless substance called *globuline*, which resembles caseine in its properties and composition, and various saline matters. *Hæmatine* is remarkable for containing, as an essential ingredient, *oxyd of iron*, which may be easily extracted and tested by igniting a little dried clot of blood in a crucible, and digesting the residue with hydrochloric acid; the solution thus obtained, gives Prussian blue, with ferrocyanide of potassium.

The colorless corpuscles of the blood are supposed to contain principally fat.

The colorless liquid surrounding the blood corpuscles is water, holding in suspension or solution a great number of different substances, viz., albumen, fibrine, fat, and a great number of salts, such as the phosphates of soda, lime, and magnesia, the carbonates and sulphates of potash and soda, and the chlorides of potassium and sodium. It also contains several gases, oxygen, carbonic acid, and nitrogen, arising from the action of air in the lungs. A healthy, full-grown, average sized man, contains about 20 lbs. of blood; 1,000 parts of which consist of 700 to 790 parts water, 60 to 70 albumen, 2 or 3 fibrine, 1·4 to 3 of fat, and 10 of mineral salts.

The heat of the blood depends in a great degree upon the activity of the process of respiration. In man, when in a state of health, its temperature remains, under almost all circumstances, in the extreme cold of the polar regions and under the tropics, at about 98° F. In birds, the temperature is sometimes as high as 108° F. In fishes, it is about that of the water in which they live. Animals whose temperature is but little higher than the medium in which they live are called cold-blooded, while those whose temperature is warmer than the air which surrounds them, are called warm-blooded.

In its ordinary state, the blood has a decidedly alkaline reaction, a saline taste, and a peculiar odor. When taken from the living animal, it soon undergoes spontaneous coagulation, and separates into two portions; one, a pale, yellowish, slimy fluid, called the *serum*, the other a gelatinous, red mass, called the *clot*, or *coagulum*. The former contains nearly all the albumen and saline constituents of the blood, while the latter, as before stated, is produced by the coagulation of the fibrine, which, "although constituting, when dry, a very small proportion of the whole, yet in the bulky and swollen condition in which it separates, is voluminous enough to entangle in its net-work of fibers the whole of the coloring matter, and cause its mechanical separation." The cause of the coagulation is not fully determined; the addition of certain saline substances, such as a saturated solution of chlorids of sodium, either retards or prevents it; while alum, and the oxyds of zinc and

QUESTIONS.—What is the composition of the blood corpuscles? What of the fluid surrounding the corpuscles? What quantity of blood is contained in a healthy man? What are the general constituents of the blood? What is said of the heat of the blood? What are warm and cold-blooded animals? What change does the blood undergo when drawn from the veins? What is the serum? What the clot?

copper, promote it. The blood of persons also who have died a sudden, violent death by some kinds of poison, or from mental emotion, is usually found in a fluid state.*

825. **Nutrition.**—The constant waste of the animal body consequent on the discharge of the various functions necessary to the support of life, requires that an equally constant supply of new material should be afforded, from which the repairs and renewals of the system may be effected. This end is accomplished through the agency of food, which in all animals consists of *protein* in its various forms (albumen, fibrine, caseine, etc.), starch, sugar, gum, and fat, to which, in the case of flesh-eating animals, gelatine must be added. Food, or nourishment from without, can, however, be only made available for the wants of the system by being first converted into blood, and this is effected through the agency of various processes, which are collectively termed *digestion.*

826. **Digestion.**—The various acts of the function of digestion are as follows:—From the mouth, where the food is chewed by the teeth and moistened by the saliva, it passes into the stomach.

The *saliva* is secreted by glands which open into the interior of the mouth, and consists chiefly of water, holding in solution about 1 per cent. of saline matter. The quantity of saliva produced in a full-grown, healthy man, in the course of 24 hours, varies from 8 to 21 ounces. Its chief office seems to be to dilute the food and assist mastication and deglutition; but it is also supposed to act chemically, through the agency of a peculiar organic substance contained in it, termed *ptyaline*, which, like diastase, is capable of converting the starch and gum of the food into sugar. Its action, however, in this respect, is probably very limited.

The food, having reached the stomach, is subjected to the action of a pe-

* "No other component part of the organism," says Liebig, "can be compared to the blood, in respect to the feeble resistance which it offers to external influences. It is not an organ which is formed, but an organ in the act of formation; indeed it is the sum of all the organs which are being formed. The chemical force and the vital principle hold each other in such perfect equilibrium, that every disturbance, however trifling, or from whatever cause it may proceed, effects a change in the blood. In fact, it possesses so little permanence, that it can not be removed from the body without immediately suffering a change, and can not come in contact with any organ in the body without yielding to its attraction. The slightest action of a chemical agent upon the blood exercises an injurious influence; even the momentary contact of the air in the lungs, although effected through the medium of cells and membranes, alters the color and other qualities of the blood. Every chemical action propagates itself through the mass of the blood: for example, the active chemical condition of the constituents of a body undergoing decomposition, fermentation, putrefaction, or decay, disturbs the equilibrium of the chemical force, and the vital principle in the circulating fluid. Numerous modifications in the composition and condition of the compounds produced from the elements of the blood result from the conflict of the vital force with the chemical affinity, in their incessant endeavor to overcome each other."

QUESTIONS.—What is the office of food? Of what does the food of animals consist? What change is necessary to render food efficacious? What is the first process of digestion? What is the saliva? What is its constitution? What its action?

culiar fluid, called the *gastric juice*, which flows out of minute openings in the inner surface—or mucous membrane, as it is called—of the stomach. This fluid possesses the power of dissolving, at the temperature of the body, the nitrogenized alimentary principles, such as albumen, fibrine, etc., but exerts no solvent action upon starchy or fatty substances. These last, however, through the joint action of the saliva and the uniform warmth and motion of the muscular walls of the stomach, are all brought into a semi-fluid state. In what manner the gastric juice is enabled to effect the reduction of nitrogenized food to a nearly fluid condition, is not known. It is said to contain free hydrochloric acid, and an organic principle called *pepsin*, and to the joint influence of these two the solvent power of the gastric juice has been attributed.*

The amount of gastric juice secreted by the stomach of a well-fed, grown man, has been estimated at from 60 to 80 ounces in every 24 hours.

Digestion generally commences immediately after the introduction of food into the stomach, and is usually finished in about four hours—the food being converted into a grayish, gruel-like, slightly acid pulp, called *chyme*. This chyme passes from the stomach into the upper part of the small intestines, called the *duodenum*, where it is moistened by two saliva-like liquids, the *bile* and the *pancreatic juice*, which are secreted by peculiar organs termed respectively the *gall-bladder* and the *pancreas*. The action of the bile on the food is not well known, but the pancreatic juice acts instantaneously on the non-nitrogenous alimentary substances, converting starch, etc., into sugar, and the fatty matters into an emulsion which renders them fit for absorption. After undergoing the action of these liquids, the nutritious matter presents a uniform milky appearance, and is termed *chyle*. In this condition it is nearly all absorbed by a system of vessels called the *lacteals*, which terminate in a common reservoir—the *thoracic duct*—which in man is about the size of a large goose-quill. The thoracic duct terminates in a large vein near the left shoulder, and into this the chyle is discharged and passes forward to the lungs, where it assumes a red color and becomes blood.†

* Some years since a French Canadian by the name of St. Martin, was severely injured in the side by the explosion of a gun, but the wound finally healed, leaving a permanent orifice in the walls of the stomach through which food could be introduced, and all the phenomena of the digestion observed. From the stomach of this person, also, gastric juice has been taken out by means of a little cup, and chemically examined. Professor F. S. Smith, of the Pennsylvania Medical College, who examined the gastric juice thus obtained in 1857, states that it contains hardly any hydrochloric acid, but much lactic acid; and to this latter agent he ascribes the constant acid reaction of the stomach. It has also been shown by observations made through this subject, that the food introduced into the stomach is caused to revolve continually around its interior, the revolutions requiring a period of from one to three minutes.

† It is not to be understood that all food lingers in the stomach for the space of several

QUESTIONS.—What change does the food undergo in the stomach? What is the gastric juice? What quantity of gastric juice is secreted by the stomach? What period is usually required for digestion? What is chyme? What takes place when the food leaves the stomach? What is the function of the bile and pancreatic juices? What is chyle? What becomes of the chyle?

That part of the food (chyle) which is insoluble, or unfit for assimilation, is left unabsorbed by the lacteals, and passes off through the intestines in the form of excrementitious matter. "How effectual the digestive process is in exhausting what we eat of its nutritive matter may be judged of from the fact, that a healthy, grown man, fed with ordinary diet, rejects of undigested and of wasted or used up matter, both taken together, only from four to six ounces. And this rejected matter consists of—

Water	3 to 4½	oz.
Organic matter	0¾ to 1⅛	"
Mineral matter, chiefly phosphates	0¼ to 0⅜	"
Total	4 to 6	oz.

Or he discharges one to one and a half ounces of dry solid matter daily."—JOHNSON.*

***. **Respiration.**—All animals as well as vegetables require, for the proper performance of their various functions and their continued existence in a living state, a free supply of atmospheric air as well as a supply of food. It is also necessary that this air should have free access to the interior of their structure, and the act or process by which this is accomplished is termed *Respiration.*

The organs by which the act of respiration is performed differ essentially in different species of animals. In the lowest types of the animal kingdom, as the polypes, respiration is accomplished exclusively through the skin. Insects also draw in air into their system, or in other words, breath, by means of organs called *tracheæ*, or wind-pipes—tubes which penetrate in various directions through their bodies, and terminate externally in little orifices called *stomata.* If we smear the body of an insect, as a wasp, with thick oil, we close up the stomata, and the insect speedily dies of suffocation. All vertebrate animals are endowed with localized organs of respiration, which are termed *lungs*, or *gills.* In man and the higher animals, the "lungs consist of two rounded, oblong, somewhat flattened masses, of very cellular substance, situated in the cavity of the chest, and communicating with the at-

hours. Soups and nutritious fluids which require no "breaking down" in the stomach, pass from the stomach into the intestines in a very short period. Neither is nutriment taken up wholly through the lacteals of the intestines, but a certain portion, in a fluid state, by the action of endosmotic force, passes through the walls of the stomach, and is mingled with the general blood.

* The cause of the peculiar odor of fœcal matter is in great measure unknown, although scientific ardor has induced some chemists to undertake most repulsive investigations with a view of obtaining information on the subject. By treating fresh night soil with alcohol two principles have been extracted, viz., a crystalline, slightly alkaline substance, named *excretine*, and an acid called excretolic acid; but little, however, is known concerning them. It has also been ascertained that when albuminous compounds are heated with hydrate of potash, and the residue distilled with sulphuric acid, an odor characteristic in an intense degree of fœcal matter is produced.

QUESTIONS.—What becomes of the unassimilated matter? What is respiration? How is respiration effected in different animals? What is the constitution of the lungs in man and the higher animals?

mosphere through the wind-pipe, or tracheæ. The general form of the human lungs is represented in Fig. 242. The air or wind-pipe, *a b*, as it descends from the throat, branches off into large (bronchial) tubes, *c c*, and these again into smaller and still smaller, and finally into hair-like, or capillary vessels. These capillary tubes, in turn, communicate with little air-cells contained in an elastic membrane, so minute that the number existing in the lungs of a full-grown man is estimated at 600 millions, and between, or imbedded in these cells, blood-vessels equally minute are distributed in every direction. The appearance of the air-cells and blood-vessels of the lungs, as seen under the microscope, is represented in Fig. 243.

FIG. 242.

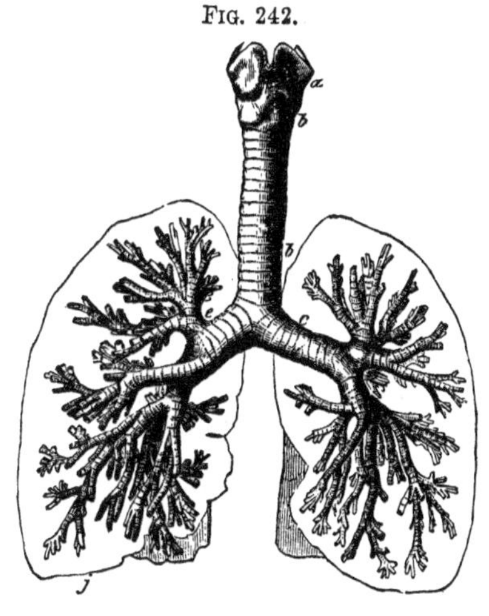

FIG. 243.

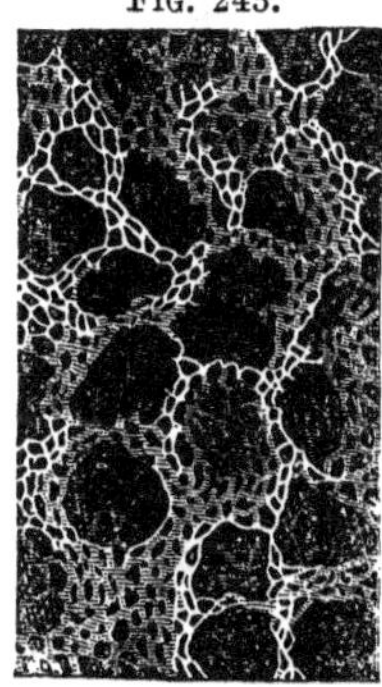

The motion of the lungs in respiration is analogous to the motion of the leather of a pair of bellows. When we *inhale*, the cavity of the chest or thorax is expanded by muscular action, and a *vacuum is formed around the lungs*, in consequence of which the external air instantly rushes in and penetrates to the remotest parts of the cellular substance. When we *exhale*, the thorax contracts, and the air contained in the lungs is expelled, the muscles of the wind-pipe at the same time contracting in order to assist the process. In ordinary respiration, a man makes 17 or 18 respirations per minute, during each of which he draws in about 20 cubic inches of air, or between 3 and 4 thousand gallons per day. In man also the skin is to some extent a respiratory organ, through which air enters and escapes, as it does from the air-vessels of the lungs, though less rapidly.*

* When a portion of the skin has been burned, it is no longer capable of exercising the function of respiration, and the lungs are therefore obliged to perform extra duty, and suffer in consequence. Hence diseases of the lungs are a frequent result of extensive burns.

QUESTIONS.—What is the mechanical action of breathing? What amount of air enters the lungs by respiration? Is the skin a respiratory organ?

The composition of the air which escapes from the lungs is not the same as that which enters, and is found to contain a greatly increased quantity of carbonic acid and vapor of water, and a diminished percentage of oxygen; the quantity of nitrogen, however, remains nearly unaltered.

The amount of pure carbon which is thrown off from the lungs of a full-grown man, in the form of carbonic acid, in a space of 24 hours, varies from 5 to 15 ounces; while the quantity which escapes from the skin also during the same period, by respiration, is estimated at from 50 to 60 grains. The amount of water exhaled from the lungs and skin in 24 hours probably averages about 3 or 4 pounds.

The lungs extract or absorb from the air which enters them from one seventh to one fifth of its oxygen, and the absolute weight of the oxygen thus introduced into the system in a day, is estimated to be equal to about one fourth of the weight of the whole food, solid and liquid, which an animal consumes. The absorption of oxygen takes place in the minute air-cells of the lungs, through the thin membraneous walls of which it passes by the action of endosmosis into the adjacent blood-vessels, and combines with the blood contained in them, imparting to it the bright scarlet color which is characteristic of arterial blood.

827. **Uses of Respiration.**—From what has been already said, it must appear evident that the principal object of respiration is to introduce oxygen into the blood, which contains the nutritive portion of the food taken into the stomach. The purpose which oxygen subserves in the blood is three-fold:—

I. *It assists in building up the substance of the body.* The composition of gluten, albumen, and the other nitrogenized vegetable principles, is, as has been before stated, very nearly the same as that of the corresponding principles in animal tissues; yet chemical investigations have shown that the former require to be combined with a certain proportion of oxygen before they can become incorporated in the substance of the body. This oxygen is supplied through the lungs, but the quantity thus used for restorative purposes is small.

II. *It assists in removing waste and effete matters from the system.* The expenditure of every kind of force in the animal system is accompanied by, or requires an expenditure or change in animal matter. The particles of matter which have once undergone such change, or have once discharged their functions, become inoperative, or waste, and their removal from the system is necessary to a continuance of healthy action. Now the agent which mainly effects the change in the first instance, and removal of the waste pro-

---

QUESTIONS.—What is the composition of the air which escapes from the lungs? What amount of carbon passes from the system by respiration? What amount of water is exhaled from the lungs and skin? What proportion of oxygen is absorbed by the lungs from the air? In what part of the lungs does the absorption of oxygen take place? What is the use of respiration? What purpose, subserved by oxygen in the blood, is first mentioned? What is the second end attained to? How does oxygen remove waste matters from the system?

ducts in the second, is the oxygen absorbed by the blood in the lungs. Thus muscle, by the addition of oxygen, becomes decomposed, and passes in a state of solution into the veins, from whence it is secreted by various organs, and finally thrown out from the system.

Urine.—The channel through which most of the products of the decomposition of the *azotized bodies* and many of the waste mineral salts pass out of the body, is the urine. This liquid, which is secreted by the kidneys from the blood, also serves to remove any superfluous water from the system. Its principal constituents are two complex organic substances termed *urea* and *uric acid*, which are composed of carbon, hydrogen, nitrogen, and oxygen, and readily furnish by their decomposition various salts of ammonia. In addition to these products, urine contains phosphates of lime, magnesia, and soda, sulphates of potash and soda, chloride of sodium, lactic acid, and certain imperfectly known organic principles, including a coloring and an odoriferous substance. All these substances exist in the urine dissolved in water, which constitutes more than nine tenths by weight of the whole secretion.

III. *The absorption of oxygen produces animal heat.* This is accomplished by the oxydation or combustion of the constituents of the non-nitrogenized food existing in the blood. The reasons which lead us to this inference may be briefly stated as follows:—

If a fat animal be deprived of nourishment for some days, it will rapidly diminish in weight. This result is the necessary consequence of the fact, that the animal is continually throwing off carbonic acid and water from the lungs and skin, and urea and mineral constituents through the excretory organs, and receiving no food to replace them.

If we examine the condition of an animal after this period of starvation, we find the loss of weight and substance is most remarkable in the fat of the body, which has diminished in far greater proportion than any of its other constituent substances. Careful examination also shows that this fat has not passed off as liquid or solid excrement, but has been converted in the blood, by oxydation, into carbonic acid and water, and in this condition has been breathed away through the lungs and skin. If, however, instead of starving the animal, we give it abundance of fat in its food, then the fat of its own body will suffer no diminution, but the oxygen taken into the blood will transform the fat of the food into carbonic acid and water, and these will be breathed out of the lungs as before. The same end will also be attained if instead of fat we give food, like starch and sugar, which is analogous to fat in its composition.—JOHNSON.

Now when carbon and hydrogen compounds, *i. e.*, fat, starch, sugar, etc., are oxydized or burned in the open air, carbonic acid and vapor of water are produced, and *heat is evolved.* The same action must necessarily be attended with the same results in the body, and we have, therefore, an explanation of

QUESTIONS.—What is urine? What is the composition of this secretion? What third purpose is subserved by oxygen? How does the absorption of oxygen occasion animal heat? What reasons lead to this inference?

the phenomenon of *animal heat.* Furthermore, all experiments show that the amount of heat generated by burning (oxydating) a certain quantity of fat, etc., is the same, whether the combustion takes place in a furnace or in the animal system.

The oxydation of fat and the other constituents of the blood is supposed to take place mainly in the minute vessels or passages, termed capillary vessels, which unite the ultimate subdivisions of the veins and arteries, and are distributed over every part of the body where nervous influence is perceptible. In these, the arterial blood, coming from the lungs and possessing a scarlet color, gives up its oxygen to the substances with which it is brought in contact, and receives in return the products of oxydation, carbonic acid and water. It also changes in color from a bright to a dark red, and returning through the veins to the lungs, through the action of the heart, passes into the minute blood-vessels of the lungs, which are surrounded by the air-cells. Here the carbonic acid and excess of water pass out through the walls of the membraneous tissue inclosing them through endosmotic action, and by the act of exhalation are forced into the air; while at the same time oxygen from without is by similar means carried inward, and the blood, restored to its arterial condition, returns upon its circuit to effect the same changes and undergo the same transformation.

Animals whose respiratory organs are small and imperfect, and which, therefore, consume but a comparatively small amount of oxygen, possess a bodily temperature but little elevated above that of the medium in which they live; animals, on the contrary, whose lungs are large in proportion to their bodies, and respire frequently, possess the highest bodily temperature. In man, the mean temperature of the body is about 98° F. The temperature of a healthy child, who consumes proportionally more oxygen and respires more frequently than an adult person, is somewhat higher, 102° F. In birds the temperature is from 104° to 108° F. The temperature of the same animal also at different times, varies with the activity of the respiration. When the blood circulates slowly, and the temperature is low, the quantity of oxygen consumed is comparatively small; when, on the contrary, the circulation by vigorous exercise or labor is accelerated, a large quantity of oxygen disappears, and the animal heat rises.

828. **Nature and Functions of Food.**—A careful consideration of all the facts connected with the subjects of nutrition and respiration, has led to the division of all animal nutriments into two great classes, viz., those which are devoted to the repair and nutriment of the body, and those whose duty it is to furnish animal heat by combustion in the blood. The former have been termed by Liebig the *plastic elements of nutrition,* and the latter the *elements of respiration.*

QUESTIONS.—Explain the manner in which the oxydation of matter takes place through the circulation. What relation exists between the frequency of respiration and the animal temperature? How does vigorous exercise increase the temperature of the system? Into what two classes are all animal nutriments divided? What are called plastic elements of nutrition?

The substances included in the first class are exclusively the protein compounds, viz., *vegetable fibrine* (*gluten*), *vegetable albumen*, *vegetable caseine*, *animal flesh and blood*. These only have the power of reproducing muscular and nervous material, and these only can afford nourishment and support in the strict sense of the term. In a state of great purity, these bodies, however, are not alone sufficient for the due maintenance of the vital powers. The experiment has been frequently tried on animals, and always with a negative result. Certain of the non-azotized substances, and certain saline compounds which are always present in natural food, are also required.

The elements of respiration are *fat*, *starch*, *gum*, *sugar*, *alcohol*, etc. Gelatine also probably belongs to this class, inasmuch as it has never been found in the blood, and is supposed to be converted in the process of digestion into sugar and ammonia compounds. These substances alone, are still less capable of supporting life than the simple protein principles.

829. The quantity of food required by an animal for purposes of nutrition or respiration varies greatly under different circumstances. When the waste of muscular or nervous material is great, a large supply of nitrogenized food, or that rich in the elements of nutrition will be required. When the body is exposed to severe cold or to violent exercise, the loss must be met by a proportionate increase in food rich in the elements of respiration. In the food most abundantly provided by nature for animals, the cereal grains, vegetables, and ordinary meat, both forms of nutriment abound. In tropical countries, where the loss of animal heat is small, and where muscular power and motion are less required and employed, the waste of the body is greatly diminished, and a comparatively small quantity of food, both for *fuel* and nourishment, is required. The inhabitants of such countries, therefore, live mainly on rice and fruits—substances which contain a large amount of oxygen, and are therefore less adapted to furnish animal heat by oxydation in the blood. The desire for animal food, under such circumstances, is very slight, and is sometimes altogether absent. In cold countries, on the contrary, a greater quantity of the elements of respiration is needed to generate the proper amount of heat, and at the same time, as the air is much colder and therefore more condensed, a larger quantity of oxygen is taken into the lungs at each inspiration. The inhabitants of such countries, therefore, consume enormous quantities of food of a fatty nature—substances rich in hydrogen and eminently combustible, and which, weight for weight, generate a larger amount of heat, when oxydated or burned in the blood, than any other products that can be taken as food. Navigators exposed to the intense cold of the Arctic regions, share to a certain extent with the Esquimaux, the same liking for blubber and train oil, which in milder latitudes they regard with aversion.

830. The fat and oils found in animal tissues appear to be stores of respi-

QUESTIONS.—Can these substances in a state of purity alone suffice for food? What are the elements of respiration? What is said of the quantity of food required by animals for the purposes of nutrition or respiration? What effect has climate on the wants of the system? What is said of the accumulation of fat and oils in the animal system?

ratory food, laid up by nature against time of need. They accumulate most in the system when fat itself, or the compounds containing its elements, are supplied in excess as food, and when the animal, through lack of active exertion, absorbs but little oxygen, and consequently experiences but little waste.*

When the supply of food is wholly withheld from the animal, the fat, as the most combustible substance, and the one most capable of supplying carbon and hydrogen to meet the wants of respiration, rapidly disappears. When this has all been consumed, the muscles are next attacked, and last of all the substance of the brain and nerves; then insanity intervenes, and the animal dies, like a lamp or candle that has been burnt out.

831. The main difference between beef and bread, which two substances may be regarded as the representatives, or types of animal and vegetable food, is, *first*, that the flesh does not contain starch, which is so large an ingredient in vegetable products; and *second*, that the proportion of fibrine in ordinary flesh is about three times greater than its corresponding element, gluten (vegetable fibrine), is in bread. It therefore follows, that a pound of beef-steak is as nutritive as three pounds of wheaten bread, in so far as the nutritive value of food depends upon this one ingredient. In meat, also, fat to a certain extent represents and replaces the starch of vegetable food.

The relative nutritive value of the different meats is as follows: beef is the most nutritious, then chicken, pork, mutton, and veal. Of vegetable productions, the cereals *generally* rank first as respects nutritive value; after them come the seeds of leguminous plants, peas, beans, etc.; then the cabbage, onion, turnips, carrots, potatoes, rice, and watery fruits. "The dried potato is less nutritive, weight for weight, in the sense of supporting the strength and enabling a man to undergo fatigue, than any other extensively used food of which the composition is known, with the exception of the rice and of the plantain." Fish in general contains more fibrine and less fat than flesh-meat, and is highly nutritious.

Salted meat is less nutritious than fresh meat. The application of salt to meat causes the fibers to contract, and the juices to flow out from its pores. Hence fresh flesh over which salt has been strewed is found, after the lapse of a little time, to be swimming in its own brine, although not a drop of water has been added. The juice thus extracted contains a large proportion of the nutritive constituents of the meat, *i. e.*, albuminous compounds, with the alkaline and earthy phosphates. Hence the continued and exclusive use of salt provisions occasion a disease called the scurvy, in which the blood becomes impaired mainly through a lack of the soluble mineral salts which are removed from the meat by the brine.

* This principle is applied in the fattening of animals, by compelling them to remain inactive by confinement in stalls or pens, and at the same time supplying them plentifully with rich, oily food.

QUESTIONS.—What takes place when the animal is deprived of food? What is the difference between beef and bread as respects nutritive qualities? What is the relative values of different meats and vegetables? What is said of salted meat?

The preservation of fresh meat by salting is due to a separation of its water, to an exclusion of air through a contraction of the fibers of the meat, and upon the formation of a compound of the flesh and the salt, which does not readily undergo decay.

832. **Relation between Animals and Plants.**—All the various forms of matter which are essential to the existence of living organisms are in a constant state of circulation. Thus, the essential constituents in the formation of vegetable products are carbonic acid, ammonia, and water. Plants absorb these from the soil or from the atmosphere, and, under the influence of sun-light and the vital principle, rearrange and organize them into vegetable tissue, starch, sugar, fat, and the protein compounds. These substances constitute the food of animals, and after employment in their systems, and after passing through various decompositions, they are again restored to the earth and the atmosphere in the form of carbonic acid, water, and ammonia: and are once more rendered capable of assimilation by plants. Thus an uninterrupted and perpetual chain of vital phenomena is established from inanimate matter to the living plant, and from the living plant to the living, sentient animal, and the products of one order of beings become the sustenance of the other.

833. **Conclusion.**—"What has been called organic chemistry is nothing but a name, and a wrong one. There is really no such science; it is only the chemistry of inorganic forms, of substances that have been living but are now dead—of the mere refuse and remains of organization. The composition of those favored materials from which the vegetable world weaves its tissues—water, carbonic acid, and ammonia—is known. The composition of the proximate principles which are extractable by easy processes from dead plants and animals, is also known. But the composition of the truly living tissues neither is, nor can be understood. They die the moment chemistry puts her finger on them. She can trace the constructive elements into the structure of the living animal or plant, and out of it, but not in it. What may be their mode of arrangement, or of their possible ingredients in matter which is genuinely alive, scientific investigation fails to reveal. The living frame of the meanest animal or plant is sacred and enchanted ground, where the chemist can only take the shoes off his feet and confess the sanctity and inviolability of life."

Questions.—How does salt preserve meat? What is said of the relation of animals and plants? What does organic chemistry really consider? Do we actually know the composition of a living tissue?

APPENDIX.

Apparatus.—The apparatus essential for illustrating and facilitating the study of chemistry, need not be of necessity expensive or complex. With the somewhat popular idea, however, that a course of experimental chemistry can be successfully conducted with an apparatus improvised from a few bottles, tobacco-pipes, and glass tubing, the author has no sympathy. Chemical experiments are most easily and successfully performed with apparatus especially constructed for the purpose, and what is saved in expense by using imperfect and unsuitable materials, will be more than lost in time and vexation of spirit. It is no doubt true that many eminent chemists have instituted important investigations, and performed brilliant experiments, with exceedingly simple or imperfect apparatus; but it is also equally true, that the tact and ability required to overcome the inherent difficulties of such an undertaking, have been deemed sufficiently singular to occasion especial comment. In short, it is only the operator rendered skillful by long experience and practice who is able to work successfully in chemistry with poor materials, and not the tyro.

We believe, therefore, the most practical advice that can be given to teachers and students who are lacking in experience, is to procure the very best apparatus their resources will admit of, as being in the end the cheapest and most serviceable.

In purchasing apparatus it will be found advisable, also, to first send to some one or more of the prominent dealers in Boston, New York, or Philadelphia, for an illustrated and priced catalogue of their stock. In this way the purchaser will be enabled to make his selections most judiciously and economically.

The following articles will be found most serviceable and indispensable for a short course of chemical experimentation:—A copper flask, with adjustable tube and collar, for generating oxygen gas; a retort stand with movable rings of various sizes; a glass (4 oz.) spirit-lamp; 2 dozen test tubes and stands; 2 wide-mouth, stoppered glass jars, or receivers; 2 tall and plain cylindrical air-jars (see Fig. 87); 4 to 6 flat-bottom, thin glass flasks, suitable for generating hydrogen, hydrosulphuric and carbonic acid gases (see Figs. 101, 126, 130); 1 one quarter pint stoppered retort and receiver; 1 one half pint do., plain; a gas-bag, provided with stop-cock and bubble-pipe; a set of small porcelain basins; glass tubing and small glass rods for stirrers, etc.;

2 small glass funnels; a deflagrating ladle or spoon; a small wedge-wood mortar and pestle; a blow-pipe; platinum foil and wire; filtering-paper; test papers; set of cork-borers; a steel spatula; a strip of sheet caoutchouc; a round and a three-cornered file for filing corks, cutting glass tube, etc.; a nest of earthen crucibles; and two small porcelain crucibles.

A pair of gasometers, oxygen and hydrogen, arranged in such a way as to admit of being used conjointly as a compound blow-pipe (see Fig. 102) are almost indispensable. They are now made of small size, and at a very moderate expense, and constitute an exceedingly durable, serviceable, and ornamental article of laboratory furniture. A Berzelius spirit-lamp (see Fig. 173) will obviate, to a great degree, the necessity of ever using a furnace. The operator can easily arrange a pneumatic trough after any of the models given on page 197, to suit his own convenience.

In addition to the articles thus specified, there are many others, such as a small galvanic battery, an apparatus for decomposing water, specific gravity bottles, thermometers, scales and weights, etc., etc., the necessity for which will depend in a great measure upon the extent and fullness of the course of experimentation prescribed or adopted.

In regard to chemical reagents, the following is a list of the more important: the acids, sulphuric, hydrochloric, nitric, acetic and oxalic; potassium, sodium, ammonia (*aqua*), carbonate of ammonia, sal-ammoniac, phosphorus, caustic potash, carbonate of soda, black oxyd of manganese, chlorate of potash, alum, sulphur, bone-black, iodine, bleaching powder, acetate (sugar) of lead, iodide of potassium, sulphate of copper (blue vitriol), sulphate of iron (green vitriol), borax, bi-chromate of potash, ferrocyanide of potassium (yellow prussiate of potash), fluor spar, arsenious acid, metallic antimony, fine iron-wire, sheet zinc, tin foil, copper turnings, chloride of barium, chloride of strontium, lime water, metallic mercury, chloride of mercury (corrosive sublimate), saltpeter, nitrate of silver, alcohol, ether, and bees-wax.

Of the above-mentioned reagents, it is recommended to have the following (in solution) arranged upon a convenient stand, or tray, in clear glass bottles, fitted with ground glass stoppers, and of the capacity of about a half pint: sulphuric acid dilute; do. strong (oil of vitriol); nitric dilute; do. concentrated; hydrochloric acid; acetic acid; oxalic acid; aqua ammonia; carbonate of ammonia; chloride of ammonium (sal ammoniac); chloride of barium; lime water; caustic potash; caustic soda; carbonate of soda; sulphate of copper; ferrocyanide of potassium; chlorine water; chloride of mercury (corrosive sublimate); bi-chromate of potash; sulphindigotic acid; acetate of lead; perchloride of iron; alcohol and ether. Also the following in solution in 1 or 2 ounce bottles: iodide of potassium; nitrate of silver; chloride of platinum. Reagent bottles suitable for this purpose, with printed labels and formula, may be obtained of all dealers in chemical apparatus.

Most of the reagents needed for ordinary chemical experiments are exceedingly cheap, and may be procured of any druggist.

The teacher would, however, do well to bear in mind, that if his resources in apparatus and chemical reagents are limited, he can supply himself, almost without cost and with but little trouble, with abundant materials for rendering his instructions both interesting and practical. Thus, he has in the common varieties of coal, gas-carbon, plumbago (black-lead), coal-tar and coal-oils, all readily accessible—the best materials for illustrating the study of carbon; and in wood-ashes, common potash, carbonate of soda, lime, marble, spar, oyster-shells, gypsum, chalk, Epsom salts, common salt, and alum, the best illustrations of the alkalies, the alkaline earths, and their compounds. In like manner, specimens of most of the ores, the common metals, and their oxyds, the products of the smelting furnace, the glass-house, and the pottery, with a great variety of organic compounds, may be easily collected; and it is by such simple and common objects that the applications of chemistry to the wants and employment of every-day life are made most familiar.

The operator will also find it an advantage, in preparing and arranging apparatus, to have some work on chemical manipulations for consultation; such as Morffit's, Noad's, or Williams' Chemical Manipulations, or Bowman's Practical Chemistry.

The present work constitutes the third of a Series of Educational Text-books on Scientific Subjects, arranged upon the same general plan by the same author—the two others being "Wells' Natural Philosophy," and "Wells' Science of Common Things."

It has been the aim of the author to render these works, in the highest sense of the term, practical, and at the same time interesting to the student. Advantage has also been taken of the very latest results of scientific discovery and research.

INDEX.

D

E

F

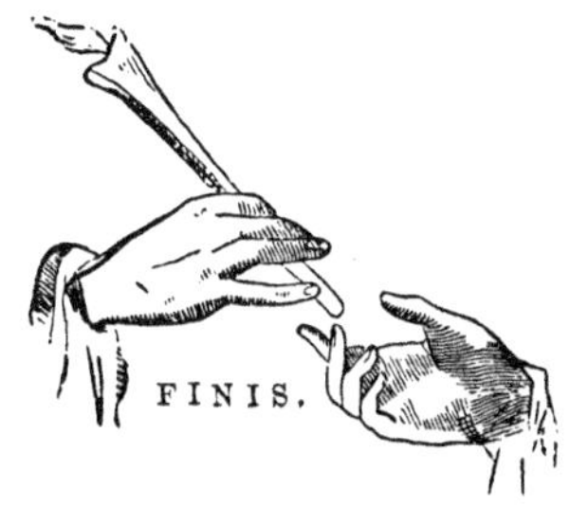
FINIS.

www.ingramcontent.com/pod-product-compliance
Lightning Source LLC
LaVergne TN
LVHW021315110826
845150LV00003B/581

* 9 7 8 1 4 2 5 5 5 7 8 9 8 *